Servicing RCA/GE Televisions

Servicing RCA/GE Televisions

By

Bob Rose

PROMPT® PUBLICATIONS

PROMPT© Publications is an imprint of Howard W. Sams & Company, A Bell Atlantic Company, 2647 Waterfront Parkway, E. Dr., Indianapolis, IN 46214-2041.

International Standard Book Number: 0-7906-1171-6
Library of Congress Catalog Card Number: 98-67777

Acquisitions Editor: Loretta Yates
Editor: Pat Brady
Assistant Editor: J.B. Hall
Typesetting: Pat Brady
Proofreader: Ann Jackson
Cover Design: Christy Pierce
Graphics Conversion: Jason Higgley, Christina Smith
Illustrations and Other Materials: Courtesy of the Author

PRINTED IN THE UNITED STATES OF AMERICA

9 8 7 6 5 4 3 2

Dedication

Servicing RCA/GE Televisions is dedicated to my best friend and life companion,

Vicki.

Acknowledgments

The author gratefully acknowledges the following, without whom this project could not have been completed:

Ms. Loretta Yates of Howard W. Sams for her editorial expertise,

Thomson Consumer Electronics for permission to use excerpts and illustrations from factory service literature and technical training material for the CTC167, CTC169, CTC175/176/177, CTC185, CTC187 and CTC195/197 chassis,

Mr. Jeff Murray of Sencore for permission to refer to several of their "tech tips" and articles in *Sencore News*, and

Mr. Conrad Persson, editor of *Electronic Servicing & Technology* for permission to use his magazine in several bibliographical references.

An Expression of Gratitude

A project like this is never the product of just one person, rather a collaboration among many. I want, therefore, to add to the acknowledgments "a special thank you" to the many who have supported and sustained me in this effort:

Loretta Yates of Howard W. Sams for her encouragement and direction,

Billy White of *The Service Center* for his oft-repeated feelings for RCA/GE products,

Donald Bush of *Bush's TV* for his encouragement,

Emory Florence of *Florence Electronics* for his assurance that such a book is really needed,

Robert Graves for carrying a little more than his usual "load" at work when I became bogged down in my writing,

and my wife, who understood and supported me unconditionally.

Contents

CHAPTER 3
CTC175, CTC176, CTC177, and CTC187

INTRODUCTION

We are in the problem-solving business. It is our business, our livelihood—and if we fail to solve the problems which are brought to us, our livelihood suffers. The method we use to solve problems is called "troubleshooting." Therefore, our problem-solving success depends on how adept we are at troubleshooting. A friend of mine used to talk about troubleshooting as "the art of drawing increasingly smaller circles around the problem we are trying to solve until we have drawn a small circle around the exact cause of the difficulty." I think that is about as good a description of the process as I have ever heard.

The ability to troubleshoot depends on at least four factors:

First, you simply have to have a disposition to work. I have known more than one technician who would not get to work before ten or eleven o'clock in the morning and would quit about three o'clock in the afternoon. I have many times been in the shops of these peopled and often couldn't wait to leave. Junk was piled so high I felt as if I were going through a maze just to get from the front door to the work area of the shop. The workbench was cluttered with jobs started and never finished. Parts were stuffed here and there without a thought to orderly storage. Tools were scattered everywhere. Test equipment was covered with dust. Literature, such as it was, was piled here and there like waste paper that never made it to the waste bin. If I had been a potential customer, I would have taken my merchandise somewhere else for repair because I would have been thinking, "If he can't take care of his own stuff, can I trust him to take care of mine?"

I am not saying organization guarantees success because that certainly is not true. Laziness can be the companion of the organized as well as the disorganized. I *am* saying if you want be or become an efficient problem solver, you must have the desire to apply yourself. One sure sign of applying yourself is how well you keep the place and the things that contribute to your livelihood.

Second, you need some basic tools. Any vocation or profession has a set of tools its practitioner has to master. Ours is no different. What kinds of tools am I talking about? I believe the following list is the minimum required: an isolation transformer and variac, a digital multimeter (DMM), a dual-trace oscilloscope, an adequate soldering station, a signal generator, and a lighted magnifier. You will also need a modest assortment of hand tools like screwdrivers, diagonal cutters, wire strippers, needlenose pliers, etc. Our profession also requires a few supplies like solder, a small quantity of chemicals, and an assortment of odds and ends (swabs, lint-free wipes, tape, heat-shrink tubing, etc.).

I consider a few other items, though not necessities, certainly helpful to have around. For instance, I prefer an analog meter (the old VOM) for checking solid-state devices like transistors and diodes—so I am more likely to reach for my trusty Simpson 260 for that job than the Fluke. Somewhere along the way, a CRT tester should be mentioned. I'm not sure how "essential" it is—I operated my shop for a while without one, but I operate a whole lot more efficiently with it than I did without it. It also takes the guesswork out of certain repair jobs, which causes me to feel more comfortable about condemning a picture tube than I used to feel. If you are considering buying one, I suggest you get a CRT tester that will let you reliably rejuvenate weak and/or marginal picture tubes. Your purchase will pay for itself over the course of its active life.

Of course, it's hard to do much of anything without a computer. As you will find out, if you don't already know, a computer has become almost a necessity for servicing consumer electronics. Access to the software and alignment procedures of today's televisions, camcorders, and VCRs is gained more-and-more by a computer, a software program, and some type of interface device. Before the end of the century, shops that don't have at least one computer will more than likely be out of business, or at least on the way out. As a matter of fact, the time is coming when a computer will be at the heart of your bench, which means a veritable host of other pieces of equipment will be tied to it.

If you haven't purchased a computer, I suggest you get the best you can afford. If you don't, you'll have to spend money almost yearly having it upgraded. I speak from experience. As a matter of fact, I got so tired of having the main shop computer upgraded that I purchased a new one that had as many features as I could afford, and I'm glad I did.

You may opt to keep all of your records on computer—tracking, billing, accounts receivable, accounts payable, parts, etc. I don't. I use my computers strictly for service. My wife keeps the books, and I store the parts. The reason is that there is no substitute for "hands on." My method keeps me up-to-date. Your experience may not parallel mine. You may find it easier to computerize the whole thing, especially if you are just starting out.

I don't need to belabor the point about tools. You get the point. The list I have given you is basic, and will vary depending on the products you service. If, for instance, you service VCRs as well as televisions, you will need to add to the basic list. I repeat—I am *not* giving you a hard-and-fast list of tools and equipment you need. I could spend the rest of my time talking about that. Obviously, you have to have a certain assortment of tools and equipment, and a few others added to the list of necessities makes the repair go easier. Ultimately, you will have to decide for yourself what is best for you in your situation. Look on my comments as a stimulus to get you to thinking.

Third, you must have literature. Working on modern electronic equipment without good literature is like taking a trip to a place you have never been without the benefit of a road map. You might get to your destination without a map, but chances are you won't. Even if you do arrive, you will probably have taken far more time just to get there than you can afford.

The subject of this book is "Servicing RCA/GE Televisions." There are two sources of literature on which I have learned to depend. The first is from Thomson Consumer Electronics; the second is the Sams PHOTOFACT®. Thomson publishes a wide variety of service literature and service aids, the most helpful being technical training manuals and factory service literature, which includes schematics and voltage and waveform charts.

Until just recently, circuit descriptions, troubleshooting guides, and part numbers were stored on microfiche, which required a microfiche reader to access. Thomson changed the format by putting everything in one paper manual, which made their literature really quite useful. And they changed things yet again—right now (1999), everything comes on CD ROM.

While on the subject of "electronics literature," I will call your attention to two software packages you might want to investigate. The first is a program called FixFinder™. As the name implies, it helps you locate "fixes" for a variety of problems you will encounter as you service Thomson products, and contains information about the new, "middle-aged," and "elderly" televisions in the RCA/GE line. The second software package is called PartsFinderII™. It delivers replacement parts information in a format which is quick, convenient, and easy to use. The program also includes service data publications by model, parts pricing, and technical bulletins. FixFinder sells for $89.95; PartsFinderII for $149.95 (1998 prices).

Of course, you can opt for a subscription to Thomson's literature. The subscription will be delivered on two CD ROMs. The first contains the software packages I just mentioned plus additional information. The second CD contains the information needed to repair a Thomson product. It includes schematics, circuit board views, alignment procedures, parts lists, technical bulletins, etc. In other words, it has all the information you will need to work on a particular product. The CDs will be mailed directly to the subscriber as new models or chassis are introduced.

You will, however, need a computer which meets certain <u>minimum</u> standards: 486DX2 processor, 16 MB RAM, 4X CD-ROM, a printer, Windows® 95, Microsoft® Explorer 4.0, and Adobe Acrobat®. IE 4.0 and Adobe Acrobat will be supplied on one of the CDs you receive. Mine came on the CD that had information about the CTC195 chassis.

The literature is also surprisingly affordable. TV literature for 1998, for example, lists for $80.00. In other words, you get a lot for your $80.00. You cannot get the 1998 literature in any format except CD-ROM, but you can get prior years in paper and/or microfiche.

I include in the appendix a list of the training manuals sold by RCA. If you want to order one or more of the manuals or just want to know what Thomson offers, write to them at this address:

TCE Publications

10003 Bunsen Way

Louisville, Kentucky 40299

The telephone number is: 502-491-8110. If you prefer, use the fax number: 502-499-2194. At the present time, you cannot order products on-line, but you can see what products are available on their website at *www.rca-electronics.com*.

Sams is another source for good literature at a reasonable price. I think their subscription fee is a little more than $56.00 a month now. For the $56.00+ dollars, you get eight fairly complete television schematics and a VCR schematic, both of which are loaded with technical information, and a little booklet called 'POM Subscriber Bonus," designed to keep you up on what is happening in the industry. The subscription fee also fetches for you an index which you use to locate the particular set of "FACTS" you need.

Sams, through its Prompt® Publications imprint, publishes several books a year and has for sale a fairly extensive lists of titles. I just bought one on switching power supplies and have found it well worth the price. Their books are designed to teach you basic technology, keep you up on the latest developments in the field, and help you hone your troubleshooting skills.

If you want the book list, have questions about their monthly subscription program, or just want general information, write to them at this address:

> Howard W. Sams and Company
>
> Attention: Customer Service
>
> 2647 Waterfront Parkway, East Drive
>
> Indianapolis, Indiana 46214-2041

Trade publications are another excellent source of service information. One of the best is *Electronic Servicing and Technology*. It is designed for people like you and me who make their living servicing consumer electronics. It publishes helpful articles, keeps you abreast of books published in the field, previews parts suppliers, publishes a monthly schematic, and does a number of quite useful things. If you are interested in getting a subscription started, contact:

> Electronic Servicing and Technology
>
> 76 North Broadway
>
> Hicksville, New York 11801

The telephone number is: 516-681-2922. Their fax number is: 516-681-2926. If you want to query the editor about something in particular or about things in general, use this address:

> Conrad Persson, Editor
>
> Electronic Servicing and Technology
>
> Post Office Box 12486
>
> Overland Park, Kansas 66212

The telephone number for the editorial office is: 913-492-4857.

Several magazines contain information about what's new in consumer electronics. I particularly like *Popular Science* because I often find information in it that doesn't show up in other sources until the product is on the market. One of the best and most informative pieces I have read on so-called "high-definition television"—that is, "Advanced Television System Committee" (ATSC)—appeared in the

October 1998 issue, but not in the usual trade publications. There are other magazines out there too. It often pays to stop by the local book store and browse through the offerings.

I would be remiss if I didn't mention the internet. I will mention over the course of this book several web sites that offer superb information. One of the leading sites is *www.repairworld.com.* It offers a variety of services, including a chat room where technicians can gather on-line, a bulletin board where you can post your problem and ask for help, and "hot tips of the week." It isn't free, but it won't cost you "an arm and a leg." If you get help with one repair, you will have paid the user fee for a year! Other web sites offer useful information about today's integrated circuits. You might check out, for instance, *www.questlink.com.* You will be surprised what you can find. (By the way, this is another benefit of having a computer in your shop.)

There is another bit of literature which is very easy to overlook. I'm talking about the catalogs you get more often than you like. Would you believe I have learned to fix more than one problem simply by reading the catalogs the mailman brings? Moreover, I can keep up with which jobber offers which part at a good price, and I learn about parts which I did not know existed. For instance, I learned by reading an MCM catalogue that I can buy RF connectors for those tuner-on-board chassis. I no longer have to worry with the old one or try to engineer one to fit the tuner wrap. I can buy an exact replacement. RCA did not teach me this—a parts catalog did! So, don't overlook the lowly catalog as a source of useful information.

I should also call your attention to professional organizations/societies. They exist for the benefit of the industry, of which we are a part. NESDA is one of the leaders in the industry, and publishes *ProService Magazine,* which I have found extremely helpful. It is has become for me "the newspaper" of the consumer electronics industry because it highlights trends which I might otherwise miss. It is also chock full of helpful information. If you want to know more, write to:

NESDA

2708 West Berry Street

Fort Worth, Texas 76109

The phone number is: 817-921-9061. The fax number is: 817-921-3741.

Fourth, you need to have a certain amount of technical knowledge to which you are constantly adding. Knowledge alone is not enough, but you won't solve many problems without it. In other words, knowing how a circuit works won't guarantee you the ability to fix it—and that's a fact—but you have to know how it works to fix it—and that's a fact too!

There are many ways to deepen your knowledge about and understanding of the products you service. Reading this book and others like it is one way. Watching training videos is another. For example, I subscribe to a "video magazine" Philips Consumer Electronics releases four times a year. I also get training videos from almost every manufacturer for whom I do warranty work. Nowadays, I get quite a bit of stuff on CD-ROM, which is at least as useful as the video tapes.

My favorite form of training is what I call "the live training event." I get to go somewhere and meet with other people! Last year (1998) I attended training events sponsored by JVC, Zenith, and Philips. I've always favored these sessions above all others, and I am saddened that manufacturers are phasing them out because they are "cost-prohibitive." The live events give us an opportunity to meet with the factory representative, who is usually quite knowledgeable, pick up useful information from him/her, and meet with fellow techs, where we share information just as valuable as that from the factory rep.

Sometimes manufacturers conduct seminars for a fee. I know how difficult it is coming up with the money, because I've had to scrape it together myself. But I consider the time and money I have to invest well worth the effort. Maybe one picture will be worth a thousand words. I attended a one-day session in Atlanta three years ago. The motel room, air fare and ground transportation added up to about $300.00, not counting the time missed from the shop. Was it worth it? I picked up several ideas. Just one in particular has helped me repair many dozens of VCRs, and it is a tech help which I am still using. Money spent getting training translates into money earned with each repair job it helps you complete. Training also helps your self-image.

The better you do, the more competent you feel. It's a cycle that is a pleasure to get caught up in!

I have spent quite a bit of time talking about conventional and electronic literature and training events of various sorts because I believe the technician who does not keep up will be the technician who will soon be out of a job. Would you go to a physician whose knowledge has not advanced, who still works with medical data ten years old? That's a foolish question, isn't it? Considering the advances in electronics and the fact that consumer products make radical changes about every six months, the question becomes pertinent, doesn't it? I spend about an hour a day reading, watching videos, or previewing CD-ROMs as a way of keeping up. I'm also proud to say I don't miss a training event unless I'm too sick to go. And I don't regret a minute of the time because I expect to be around and active well into the next millennium.

I will include under this heading the need to find a way to store information. The computer is really a storage device that gives you an opportunity to store your own service tips and well as the tips most manufacturers offer. RCA, for instance, has a parts program and a Fixfinder program that is worth the money. Zenith offers what it calls Z-Tips™, and other manufacturers offer similar software packages. You can even get ECG and NTE indexes on the 3.5-inch floppy disks.

Call me old-fashioned if you like, but I prefer to store my personal information on 4x6" note cards. They are easy to use and take up very little space. I have thought about putting the information on computer, but decided to keep the system as it is. If I were starting out in the business today, I think I would opt for computer. But each to his/her own.

There is one other subject I want to cover in the introduction, and it is "the newest of the new." Thompson will be adding a new feature in the 1998 CTC195E, F, K, and L projection televisions and the CTC197E, F, K, and L direct-view models. The new feature is called a "warranty clock." The "date first used" will be established from a GEMSTAR or XDS broadcast signal. The month, day, and year will be recorded, and a counter will keep track of the number of hours the product is powered on. The data will be stored in the EEPROM and cannot be reset. The information will be display in two ways: (1) on the menu screed accessed via the remote control which the consumer can use; or (2) on the screen of the technician's PC when connected to the chassis via the Chipper Check™.

If you operate an authorized service center, you will be required to post the warranty clock twelve-character code on all warranty claims. Chipper Check software will be necessary to access the claim code data.

Well, enough of this. Let's get on with *Servicing RCA/GE Televisions.*

CHAPTER 1
THE CTC167 CHASSIS

I will begin the discussion of RCA/GE televisions with a look at the CTC167 chassis. As usual with Thomson, there are several variations of the basic chassis, each variation incorporating certain features. The Thomson Consumer Electronics Service Data Index lists six variations (CTC167C, CN, CS, E, M, and R) which include about 59 models! Factory service literature includes the basic CTC167 manual plus five supplements (S1 through S5) and the microfiche which contains circuit descriptions, troubleshooting guides, schematics and parts numbers. The fiche is therefore a necessary accessory. Of course, Sams PHOTOFACT® will have pretty much the same information. The Sams I will be using is PHOTOFACT 2971(April 1992), the literature for CTC167C, CN, CS, and R, which encompasses ten models. You might like to know that Thomson has just made available a newly published field service guide for the CTC166/67 which contains a complete set of schematics, key voltages and waveforms, field service adjustments, service bulletins, and parts list. Use the address I gave you in the introduction. Ask for FSG 12.

Component Number System

Before getting into the particulars of the CTC167, I will give you a useful tool for locating parts on the PCB. RCA enhances serviceability by "roadmapping" the circuit board on the top and bottom and assigning each component a number which relates it to a general circuit area. For example, a part with a number in the 4400 range will belong to the horizontal output circuit. The component numbering system is:

0000	The Tuner	2700	Luminance Processor	4400	Horizontal Output
1000	The TV Processor	2900	RGB Bias and Drive	4500	Vertical Deflection
1100	Remote Receiver	3100	System Control	4600	Standby Rectifier
1200	Audio IF	3300	Analog Interface	4700	Secondary Power Supplies
1400	Audio/Video/ Hi-Fi Outputs	3400	Front Panel Assembly	4800	Pincushion Correction
1500	S-Video Inputs	3600	Tuner Control	4900	X-Ray Protection
1700	Stereo Audio Demodulator	4000	AC Input	5000	Kine Driver
1900	Audio Output	4100	B+ Regulator	7500	FM Radio
2300	Video IF	4200	Degauss		
2400	Sync Separator	4300	Horizontal Oscillator		

The Power Supply Circuits

Remember, these TVs have hot/cold grounds and the use of an isolation transformer for service is a must. It protects you, the customer's merchandise, and your test equipment. The AC input, horizontal output, and B+ regulator are referenced to hot ground. A good common test point for checking voltages in the hot ground area is the negative lead of C4007 or the junction of the anodes of CR4001 and CR4002.

A discussion of the power supply necessarily includes the AC input, standby power source, the B+ regulator, and the scan-derived power supplies. I do not intend to give you a detailed description of how these circuits work. You can dig that out from the sources I have mentioned. I will, however, go into enough depth to give you a feel for how the circuits work. I proceed on the assumption that knowing how a circuit works does not necessarily imply the ability to repair it! But you still need to know how it works if you are going to fix it. To that end, I recommend a very nice article by Homer Davidson in the September 1997 issue of *Electronic Servicing and Technology*, entitled "Servicing RCA's CTC166 Power Supply Circuits." You probably already know that the CTC166 and CTC167 chassis are remarkably similar.

The AC Input

AC is applied to a full-wave bridge rectifier (CR4001-4004) via F4001 and R4001 to develop the raw B+ supply and the 33-volt supply (via CR4104). Both of these voltages are utilized by the B+ regulator to develop regulated B+ for the horizontal output circuit. K4112, RT4201, and Q3302 perform the degauss operation. See *Figure 1-1* for details.

The Standby Supply

Figure 1-2 is Sam's representation of *Figure 1-3*, which is from RCA literature. AC is applied to the primary of T4601. CR4601-4604 rectify the stepped-down voltage to produce 22 volts. The 22 volts is used to provide start-up voltage for Q4301 (horizontal driver) and to produce 12 volts via Q4161. The 12 volts is fed to Q4160 which produces the 5-volt standby source. The 12-volt supply is also used to generate start-up current for the 6.8-volt source.

The B+ Regulator

Figures 1-4 (RCA version) and *1-4A* (Sams version) are the schematic for the B+ regulator. Its "heart" is SCR4101. The 129 volts for the HOT is developed by controlling the ON time of the SCR with a control circuit consisting of Q4101, Q4102, Q4103, Q4104, and associated components. When the TV is off, the regulator acts like a free-running oscillator. C4108 charges from the 33-volt source through constant-current source Q4103, producing a sawtooth wave at its collector. As C4108 charges, the voltage at the collector of Q4103 decreases until Q4101 (the oscillator) turns on. Its turn-on time is determined by the resistors in its base circuit. When Q4101 turns on, Q4108's collector goes low and turns Q4102 on. When Q4102 turns on, it discharges C4108 through T4101, the SCR driver transformer. Q4101 then turns off. The voltage at its collector rises, turning Q4102 off, which interrupts the

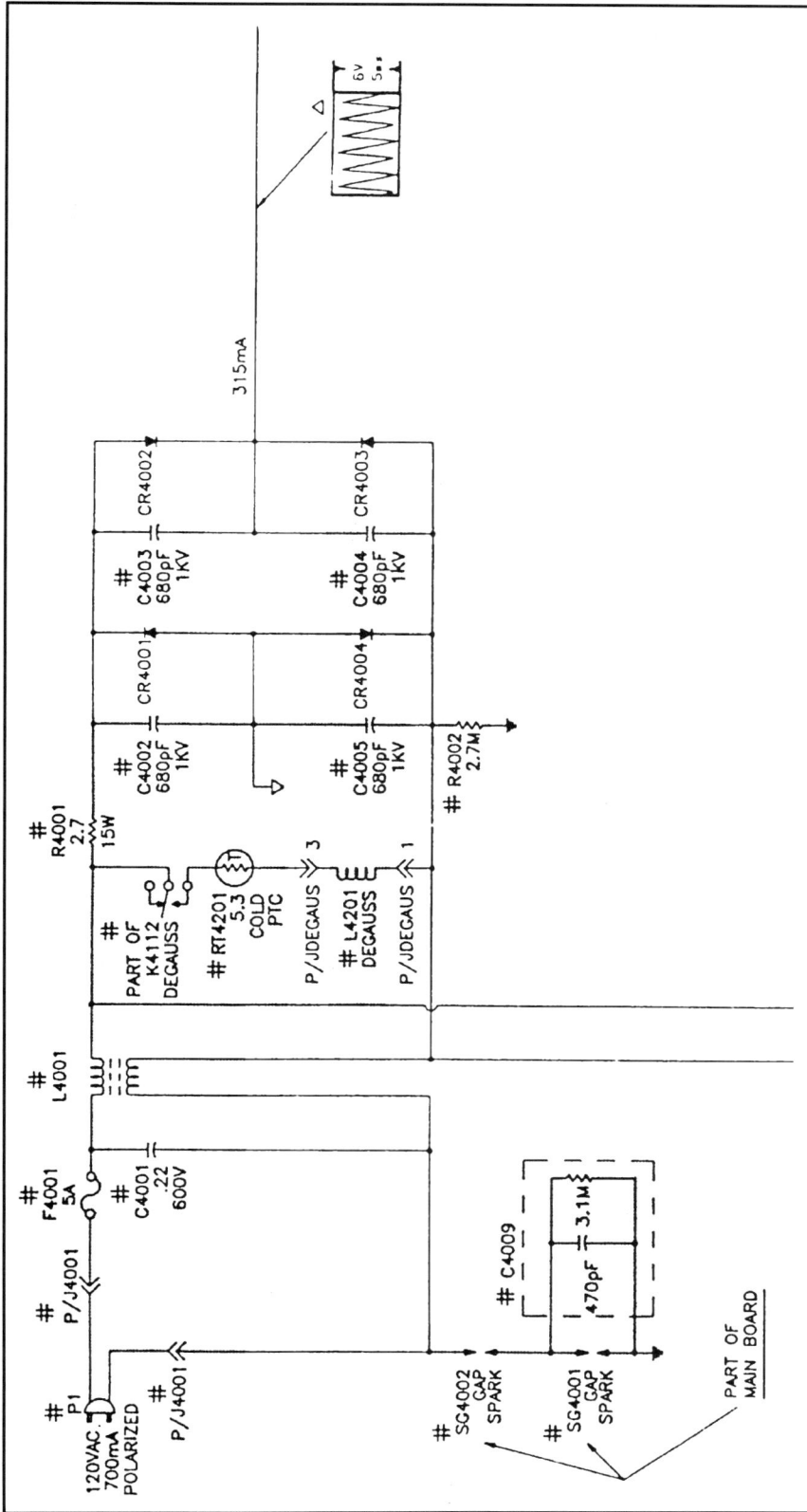

Figure 1-1. AC input.

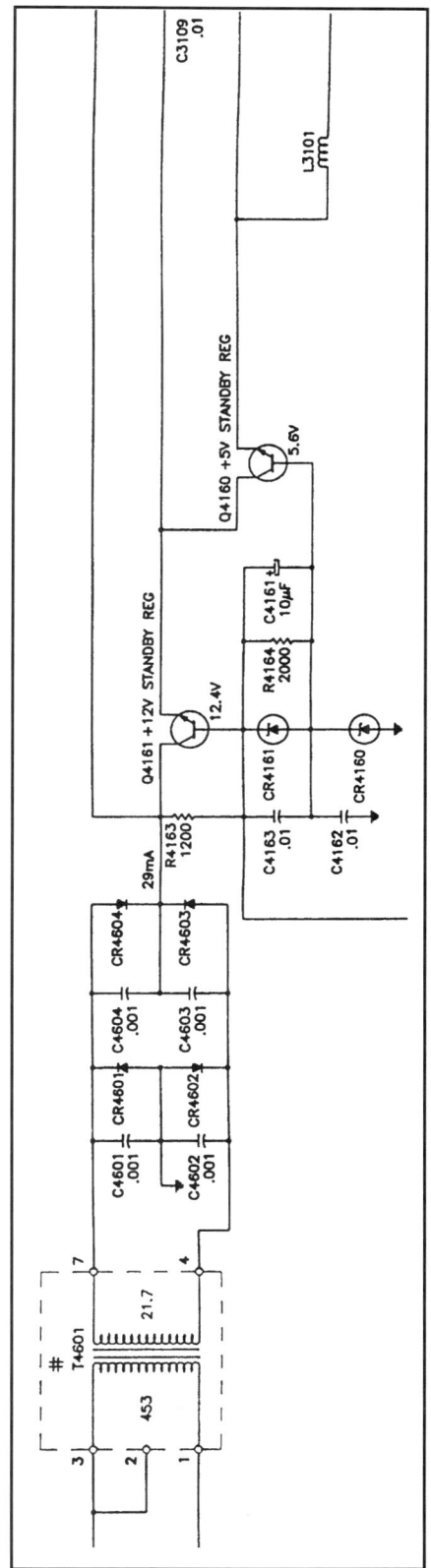

Figure 1-2. Standby power supply.

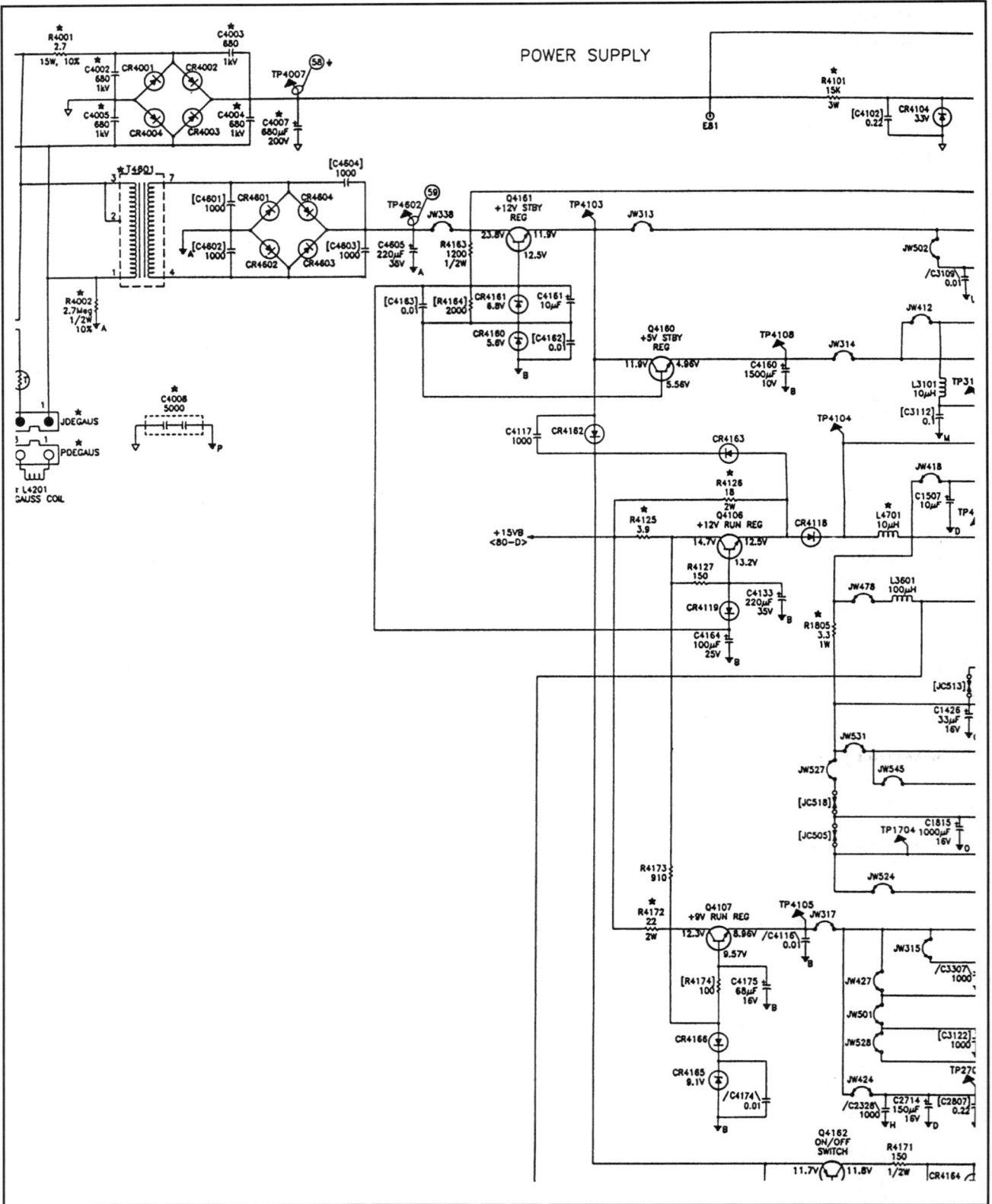

Figure 1-3. Standby power supply.

discharge path for C4108, causing it to charge again. The charge-discharge cycle creates a pulsing current through T4101, which is coupled to the gate of the SCR to turn it on.

Remember, the TV is in standby at this point. The load on the B+ regulator is so low that SCR4101 does not turn on. The small amount of current needed from this source in standby is provided by the raw B+ source through R4113.

When the TV turns on, the regulator is synchronized to the horizontal deflection via feedback from pin 1, T4401. A negative-going retrace pulse from T4401 turns on Q4202 through C4109 and R4104, which discharges C4108. Note that the gate pulse produced by discharging C4108 during retrace does not cause the SCR to conduct, since the same pulse used to discharge C4108 is used to turn it off. C4108 will charge and discharge, gating the SCR before retrace occurs. Once the SCR is gated (turns on), it latches until retrace occurs. The retrace pulse then turns off the SCR and discharges C4108.

Regulation is maintained by controlling the timing of the SCR gate pulse that occurs during trace. The earlier the pulse occurs, the longer the SCR stays on, transferring energy from raw B+ to regulated B+. A part of the regulated B+ is divided by a precision resistor network and compared to a fixed reference (CR4120) by Q4104, the error amplifier. An increase in regulated B+ causes Q4104 to conduct more, which increases the base voltage on Q4103. The decreased base voltage reduces the charge current for C4108, which delays the turn-on of Q4101 and therefore delays the gating pulse for the SCR. The reverse happens if regulated B+ decreases.

Well, enough of that. Thank goodness the regulator works and doesn't give much trouble! I will have a word about troubleshooting it in just a few pages. If you want a more detailed discussion, please consult the microfiche (CTC167-S5, Feb94, 1C1-1C2) that accompanies the factory service literature.

Secondary Supplies

There are five secondary or scan-derived voltage sources. *Figure1-5* will give you the particulars. I will more-or-less list the secondary supplies:

(1) A 200-volt source generated by the pulse at pin 3 of the flyback and rectified by CR4702. It supplies power to the video output transistors.

(2) A 44-volt source at pin 9, rectified by CR4703. This source provides power for vertical deflection (4501 and Q4503) and the 33-volt tuning source for the tuner.

(3) A 26-volt source via pin 5 of the flyback, which is the power source for the audio output circuits and the vertical deflection IC.

(4) A 15-volt source at pin 10 of the flyback. The 15 volts is used to develop the 12-volt (via Q4106) and 9-volt (via Q4107) run sources. The 12 volts powers the audio processing, aux video input, and the tuner circuits. The 9 volts powers the IF and luma/chroma processing circuits.

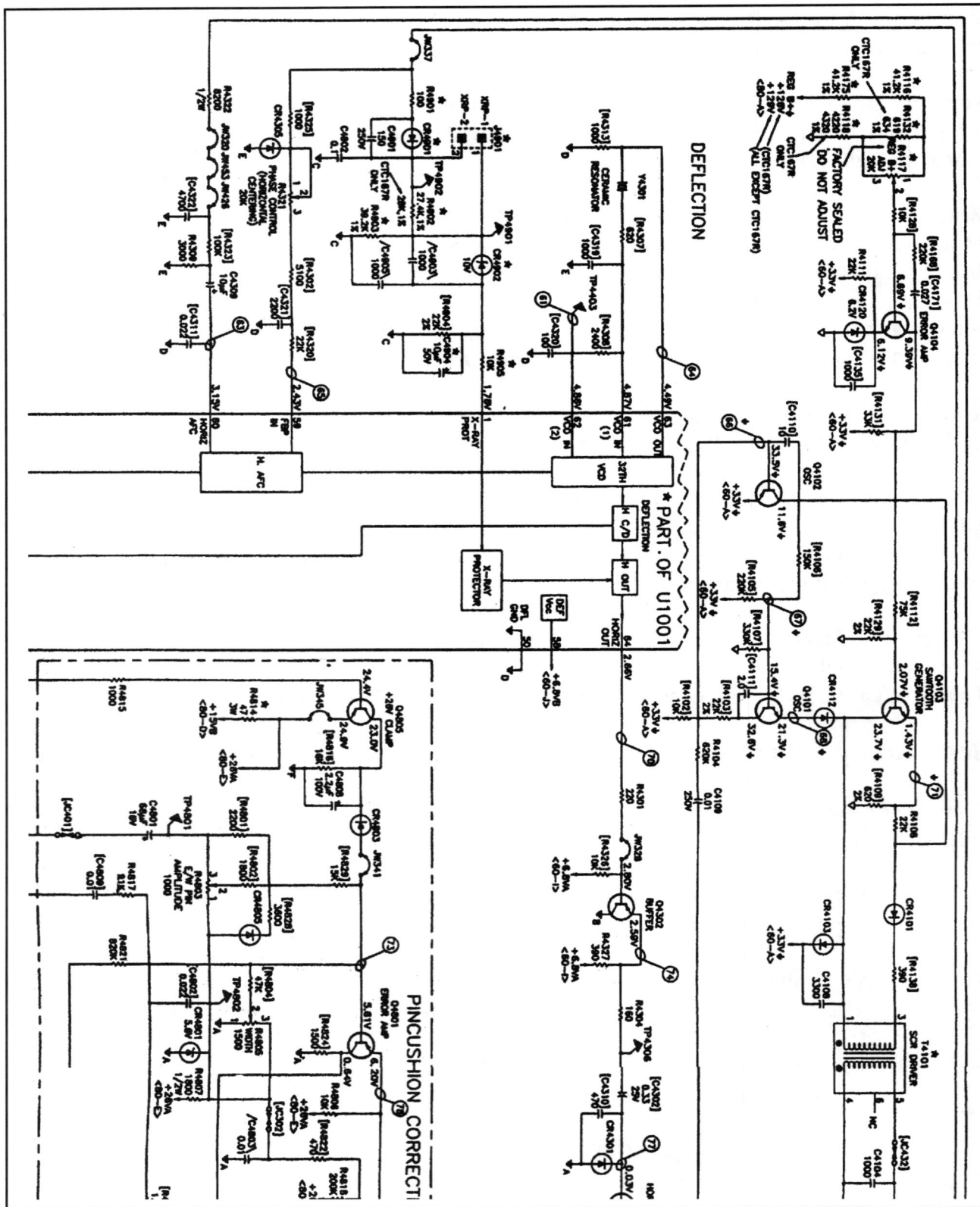

Figure 1-4. B+ regulator.

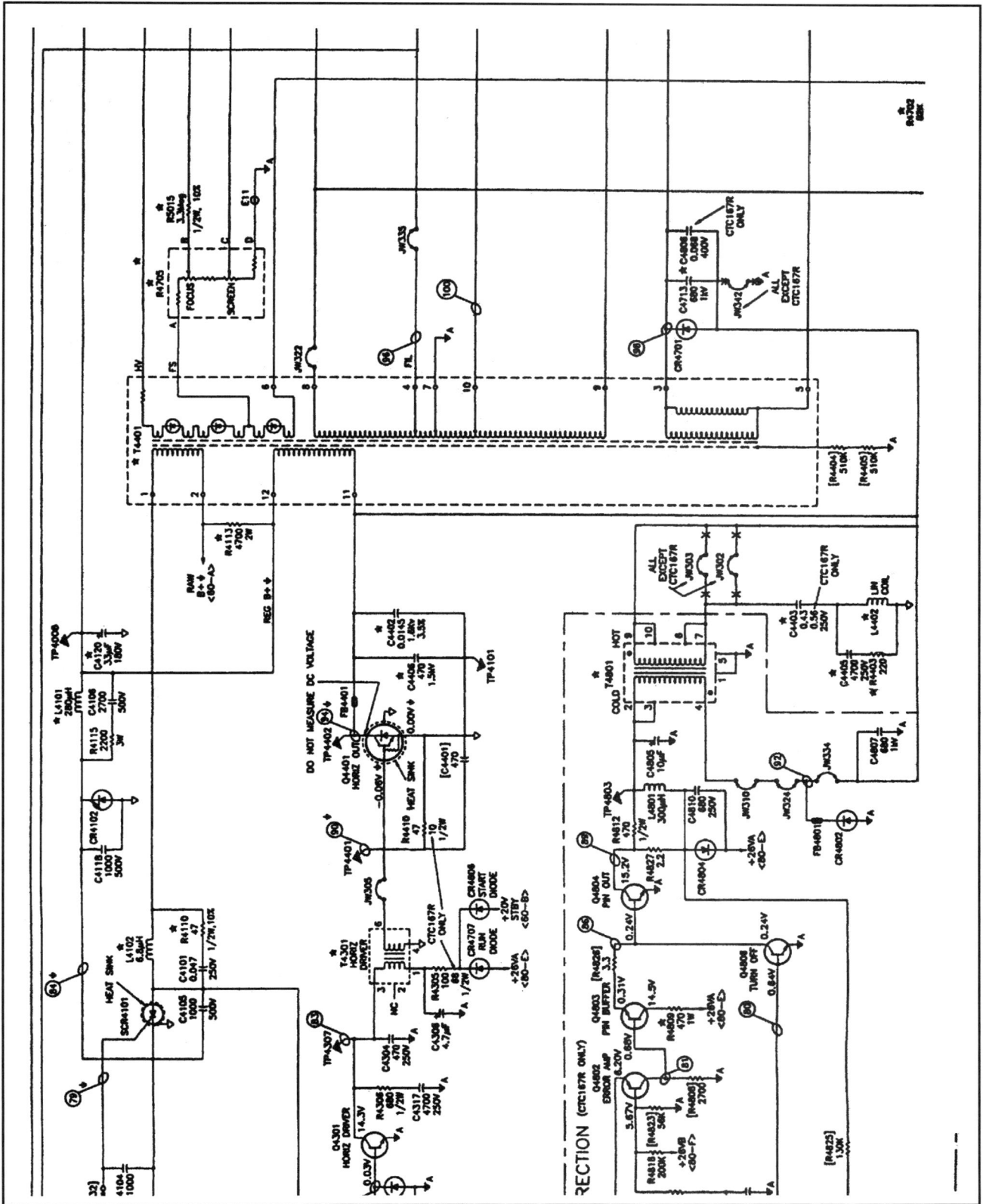

Figure 1-4. B+ regulator (continued).

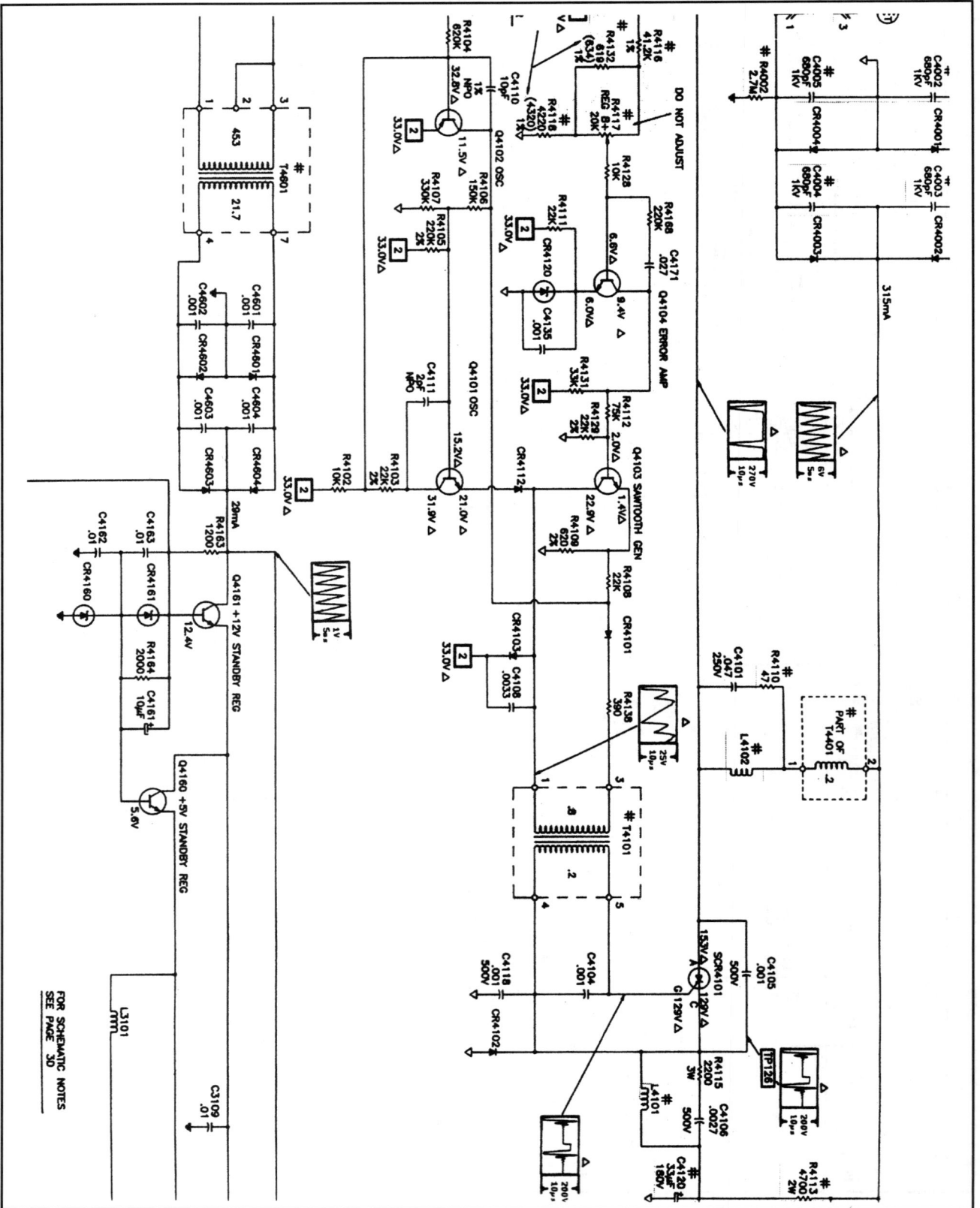

Figure 1-4A. B+ regulator.

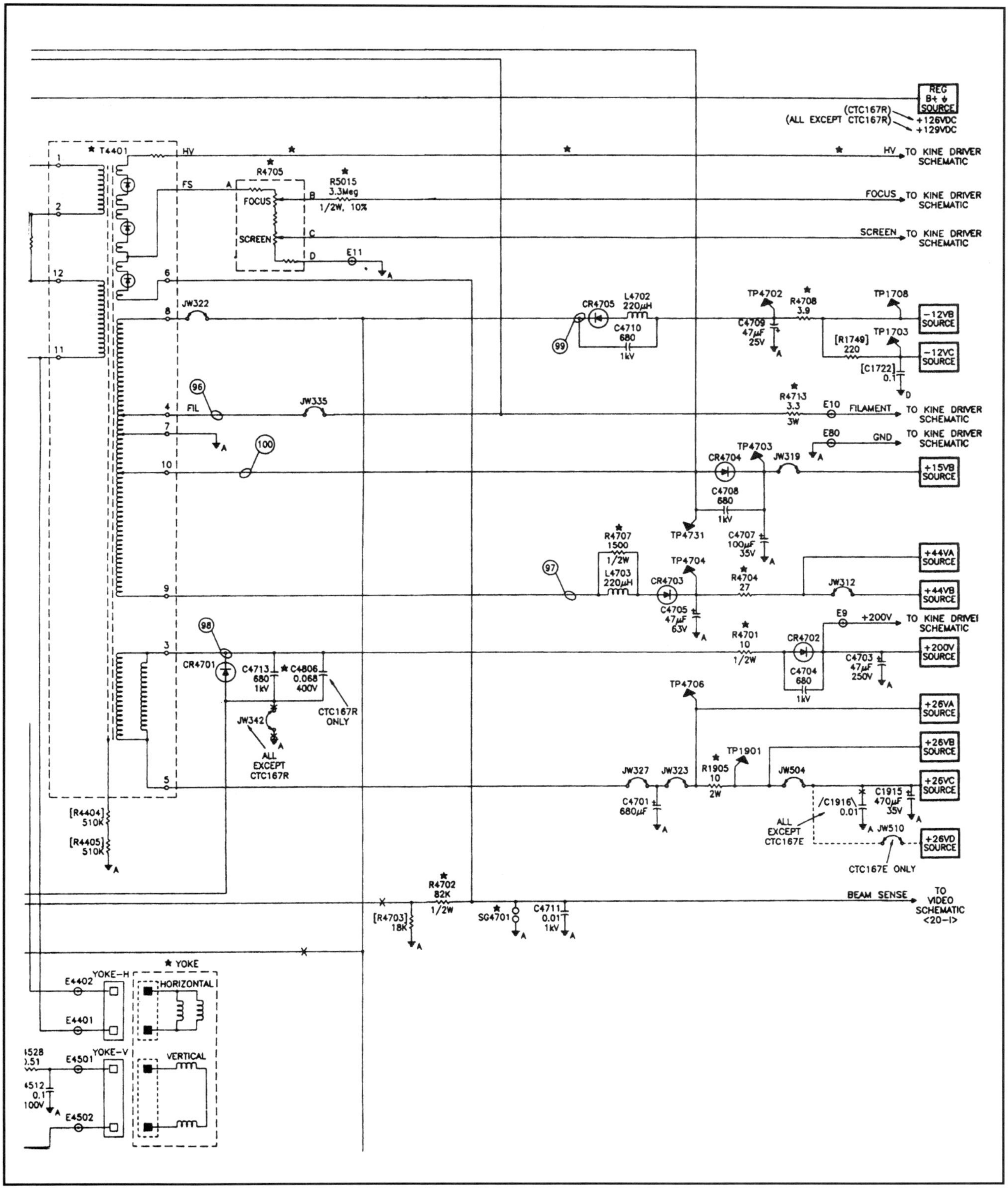

Figure 1-5. Scan-derived sources.

(5) A negative 12-volt supply (at pin 8 of the flyback). The -12 volts supplies power to the audio processing circuits and tuner.

Troubleshooting the Power Supplies

Now the big question: How do you troubleshoot the power supply? I am going to use a chart based on a procedure recommended by RCA and modified by me. Call it a "flowchart" if you like. I'm not sure I like the terminology, even though I use it. Sometimes flowcharts are helpful, and sometimes they are not. I suggest you have a copy of a schematic as you work through the troubleshooting chart. It might not be necessary, but it certainly makes things easier. If you are not familiar with these chassis, you will save lots of time just by having something that will help you to locate parts and components on the circuit board. Some of the parts are exceedingly small and necessitate the use of an illuminated magnifying glass to find and service.

Dead Set

Suppose you have a CTC167 in your shop which is dead. Let's attempt to find the problem by asking and answering a series of questions.

(1) Do you have 20 volts at the collector of Q4161? If the answer is no, check the fuse, T4601, and CR4601-4604.

(2) Do you have about 12 volts at the emitter of Q4661? If the answer is no, check the transistor for an open and the components in its base circuit, especially the zener diodes, CR4161 and CR4160.

(3) Do you have 5 volts at the emitter of Q4160? If not, check the transistor.

(4) Do you have a raw B+ of about 150 volts? If not, check R4001 and diodes CR4001-CR4004.

(5) Do you have regulated B+?

    (a) If the answer is no, check R4113.

    (b) If the answer is yes, check for a sawtooth waveform at the collector of Q4103. The waveform should have a frequency of about 16 kHz.

        (1) If the waveform is absent, proceed to regulator troubleshooting chart.

        (2) If the answer is yes, check the SCR, CR 4102, L4101 and L4102. Do not overlook the possibility that something is loading the B+ supply down, like a shorted horizontal output transistor.

If these voltages appear good, then we have to move to the B+ regulator. There are several places to start. Wherever you start, be sure to use a systematic approach. If you don't like the one I am about to suggest, find one you do like. Whatever procedure you adopt, make it systematic. Please note that the following voltages are taken while the set is in the standby mode. Remember to use the correct ground. The reference I will be using is *Figure 1-4*.

(1) Check CR4104 for 33 volts. If this voltage is absent, check CR4104, CR4103, and R4101.

(2) Is there a sawtooth wave form at the collector of Q4103 at a frequency of 16 kHz?

    (a) If the answer is yes, the problem is in the secondary of T4101.

    (b) If the answer is no, is the emitter of Q4103 at 0 volts?

(3) If the emitter of Q4103 is not at 0 volts, is the collector of Q4104 at about 6.1 volts?

    (a) If the answer is yes, check C4108, R4109, T4101, R4112, and R4129.

    (b) If the answer is no, check Q4104, R4131, and CR4120.

(4) If the emitter of Q4103 if at 0 volts, is its collector greater than about 15 volts?

    (a) If not, check Q4103, Q4104, R4109, R4110, R4112, R4129, and the regulated B+ feedback.

    (b) If it is, check the base of Q4101 for a voltage between 0 and 15 volts.

(5) If the base of Q4104 is between 0 and 15 volts, check C4101, CR4112, C4108, CR4103, Q4102, T4101, R4102 and 4103, and CR4101.

(6) If the base is NOT 0 to 15 volts, check R4106, Q4102, R4107, and R4103.

Fortunately, the regulator circuit does not give a lot of trouble. I say "fortunately" because it can be exceedingly difficult to troubleshoot. But it can be done if you use some kind of system. For example, I just worked on one that would go off a second or so after it came on. I checked the regulated B+ and found it to be about 161 volts, which of course is far too high. Remember, when the regulated B+ rises, the feedback voltage to the base of Q4104 rises, which ultimately results in lowering the B+. I used a variac to set the regulated B+ to about 120 volts. While monitoring the voltage on Q4104, I alternately raised and lowered the AC. The voltage on Q4104's base did not track. The problem seemed to be in the feedback loop that is from the 120-volt source through the precision resistor divider network to the base of Q4104. I found that R4116, a precision 41.2k resistor, had increased in value to 220k. A new resistor restored the B+ regulator circuit.

Case Histories

Now let's look at a few case histories.

The first case concerns a TV that pulsed on and off and would not come on and stay on except at high AC input. The problem was a defective C4605, the 220 µF filter capacitor in the standby power supply.

The second case concerns a set that at turn-on pulsed on and off for a few seconds and then completely shut down. There was no 33 volts to the B+ regulator circuit because CR4104 was leaky. A new zener diode put the set in good working order. However, a word of caution may be in order. The set that pulses on and off and then turns off may also be lacking horizontal drive. I worked on one just this week that exhibited the same symptoms. I quickly checked for B+ on the collector of the HOT and found evidence that it was there. The same check also told me there was no horizontal drive. I traced the problem to a defective horizontal driver transistor (Q4301). I won't say Q4301 fails often, but it fails often enough that I keep a few in my parts bin at all times. As a matter of fact, I have seen lightning blow these transistors apart and do no other damage to the TV.

Then there was the set that was dead because the 20-volt line was low. Q4301, the horizontal driver, was leaky. We haven't covered it yet, but the 20-volt standby circuit also supplies start-up voltage for horizontal deflection. A leaky Q4301 will most certainly pull the 20-volt supply low!

I have had the following happen several times. Let's suppose you change the horizontal output transistor because it shorted. You turn the set on and notice that the picture has a kind of "bend" in the raster. You feel the horizontal output transistor's heat sink and realize it's getting hotter by the second. Then the television goes off because the new transistor fails. What has happened? The SCR in the B+ regulator shorted or developed leakage, permitting higher than normal B+ to the output transistor. You would think the x-ray protection circuit would trip and turn the set off before the new transistor failed, but it won't. The moral is: Check the SCR when you replace a shorted horizontal output transistor. By the way, it's also a good idea to check R4410 in the base of the transistor because it may either have opened or increased in value when the output transistor failed.

I have also seen some of these sets that had the raster pulled in from each side, a symptom of low B+ or pincushion circuit problems. You can tell which is which by measuring the regulated B+. If it is low, service the regulator. If it checks okay, go to the pincushion circuit. If it is a pincushion problem, you might want to start at Q4804. I serviced one where Q4804 was literally burned to a crisp. I installed a new transistor only to see it go up in smoke after the TV had played for a while. I called RCA technical support because I couldn't find anything wrong. The tech to whom I talked asked me to check for a horizontal pulse on both sides of C4809. C4108, a surface-mount component, was completely missing! It had evidently cracked and fallen off the board!

I have serviced several of these chassis that will come on without picture or sound. If I turned the G2 voltage up, I could see a dim raster without the hint of video in it. A quick check revealed no 12-volt or 9-volt run voltage. I traced the problem to a leaky CR4118. I won't call it a common problem, but I have seen quite a few sets that suffered from this malady.

System Control

The second subsystem I want to look at is system control. I will reference my remarks to *Figure 1-6* which I took from Sam's PHOTOFACT. RCA's literature is too cumbersome to reproduces here because it divides the system control processor (U3101) into segments and groups a particular segment with the circuit it controls. For instance, eleven pins of U3101 are displayed with the part of the schematic dealing with the power supply. Such a procedure is often helpful, but I still like to see everything at a glance because it helps me to understand what to expect and saves time flipping through pages in a manual.

Troubleshooting Procedures

Before getting to system control, I want to say a few words about troubleshooting microprocessors. First, you must have a reasonably good scope. I know there are shops that try to get by without one, but the complexity of today's televisions mandates owning and using one. Second, you have to know what you are doing. Guesswork will get you just so far and no farther. Ever buy an expensive chip, install it, and still have the same problem? We've all "been there and done that" and will probably do it again. Even though nothing guarantees it won't happen, you can "hedge your bets" by using a systematic approach in your troubleshooting. You may not like my way of troubleshooting system control microprocessors. If you don't, find your own. If you do, feel free to use it.

I am not aware of much literature to which I can refer you, but I have discovered a few helpful pieces here and there. The article that started me thinking about a systematic approach to system control problems is Sencore's *Tech Tip* #109. The writer presented an approach to troubleshooting based on the use of Sencore equipment, but it is helpful regardless of the brand of equipment you use. There is a sort of related article in *Sencore News*, December, 1991 (issue #155), which deals with troubleshooting VCR sensor and system control problems. The points the article makes are also applicable to TV system control problems. The folks at Sencore have always been helpful when I have called, and I have every reason to expect that they would furnish you with Tech Tip #109 and possibly the article in Sencore News just for the asking. They will also try to sell you something, but who can blame them? They are also in business. If you want a more detailed explanation of microprocessor troubleshooting techniques, see my article in the August 1998 issue of *Electronic Servicing and Technology*. And I suggest you read "New Generation System Control Circuits," by Steven Babbert in the September 1998 issue of the same periodical.

Now to the matter of troubleshooting a suspected microprocessor problem. You will need to ask and answer five questions.

(1) Is there B+ on pin 42 (V_{DD})?

You should find about 5 volts, which is developed by Q4160 from the 12-volt source (Q4161). RCA says the voltage is 4.96 volts. If the voltage varies more than about ten percent, the microprocessor just might not work, or certain of its functions might not work, or it might work in very strange ways.

We often think of a microprocessor not working because the B+ is too low. Don't forget that a higher than normal voltage can also cause problems. I remember a particular Philips chassis that had a very dark-colored OSD instead of the bright blue to which we are accustomed. The picture was also washed out, as if the contrast were set too high. The 5-volt regulator had failed and was generating 6 volts instead of 5.

A CTC177 I just worked on illustrates the typical low B+ problem. The customer complained that his TV would not turn on, and it wouldn't. B+ checked good the first time I tested it. The oscillator for the microprocessor was outputting 3.5 volts peak-to-peak instead of the 5 volts it should have. Moreover, there was no information on the data line to the EEPROM. I was satisfied the microprocessor was defective. But for some reason, I remeasured the 5-volt supply and discovered it had dropped to less than 3 volts. I disconnected the microprocessor from the 5-volt line and discovered the voltage was still less than 3 volts. The 5-volt regulator had, in fact, failed.

In other words, look for the stated voltage (about 4.96 volts) and be absolutely certain it is as ripple free as possible. A leaky filter capacitor in the standby 5-volt line can lead you on the merriest wild-goose chase you have ever been on.

(2) What is the voltage on pin 21 (Vss)?

"What?" you say, "But that is ground." You are correct. It *is* ground, and the voltage must be 0, which indicates a good ground. I have had the rare exception of finding a voltage on the ground pin, indicating a loose ground connection. The TV would sometimes work and sometimes wouldn't, and it was more likely to fail after it had played a while. The problem was a cold solder connection on the ground pin of the microprocessor!

(3) Is the oscillator working?

Check pins 31 and 32 for an 8 MHz signal of about 5 volts peak-to-peak. Also check for the correct DC levels. Incorrect DC levels can indicate the oscillator is not working. (I know—RCA says, "Do not measure DC voltages.") If you don't find the 8 MHz signal or if it is low, you will have either a defective IC or crystal (Y3101) and/or in very rare instances defective ceramic caps. The IC is more likely to fail than the crystal, though that isn't "carved in stone."

(4) Do you have the correct voltage on the reset pin (33)?

The voltage for the CTC167 should be about 3.7 volts, and is developed from the 12-volt standby source (source #5, according to Sams) via CR3101 and a resistor network. The reset voltage permits the chip to initialize at the beginning of its program. If the voltage is absent or grossly incorrect, the TV will present you with a "dead set" symptom. I have never encountered a reset problem in a CTC167, but I have in other RCA chassis.

(5) Do you have correct information on the data IN and data OUT lines?

The data input lines are pins 10-15. Correct DC voltages indicate nothing is loading down the data input lines. A loaded input line can cause the micro to appear dead and/or to have some really funky symptoms. The voltage on pins 10 and 11 needs to be about 2.5 volts; pins 13, 14, and 15 need about 4.9 volts. Pin 37 is the IR input. Pins 2 through 5, 24 and 25 are data out lines. These pins control the various functions of the set—color, brightness, audio, etc. You don't need much time to check data input and data output lines for correct information. The time spent may save you time and money in the long run. I certainly do suggest you check them for correct information before you condemn the chip.

An Overview of U3101

All system functions are controlled by U3101.

Pins 1-6 control volume, sharpness, color, tint, brightness, and contrast. These pins output a 31.25 kHz pulse width modulated signal, which is filtered to obtain the DC voltages for audio and video control. The DC voltages will vary between 0.0 to a maximum of 6.0 volts, depending on which function you check.

Pin 7 is the tuning indicator used to detect the presence of an FM signal. The pin is used only in the CR and M versions of the CTC167.

Pin 8 is usually not connected in the TVs we service. It is a bidirectional line used in commercial applications of the CTC167M for data transfer to and from an external device.

Pin 9 is for AFT. It detects AFT crossover during the tuning process. The voltage can vary between 0 and 5 volts. The crossover point is 2.5 volts. A mistuned AFT circuit can cause a variety of problems, like a TV that will not autoprogram. To align the AFT, plug the TV in, disconnect the antenna, and turn it on. After the set has come on, select a station, and connect the antenna. While you monitor the voltage on pin 9, adjust L2306 for a reading of about 2.54 volts. The fiche will give you a different procedure for aligning the AFT circuit (frame1-B1), but I think you will find my procedure just as effective. At least, it has never failed me.

Pins 10-15 are the data input lines. There are two keyboard drive lines and four keyboard sense lines.

Pin 16 is the fault detect input; in other words, the x-ray protection input. It monitors the 9-volt source.

Pin 17 controls the front-panel power LED. When the TV is off, pin 17 floats to prevent the LED switch (Q3101) from turning on. An ON command pulls the pin low, turning on Q3101 which turns on the LED.

Pin 18 turns the TV on and off. An ON command pulls the pin low, turning on Q4162 which supplies 6.8 volts to the deflection circuits inside U1001.

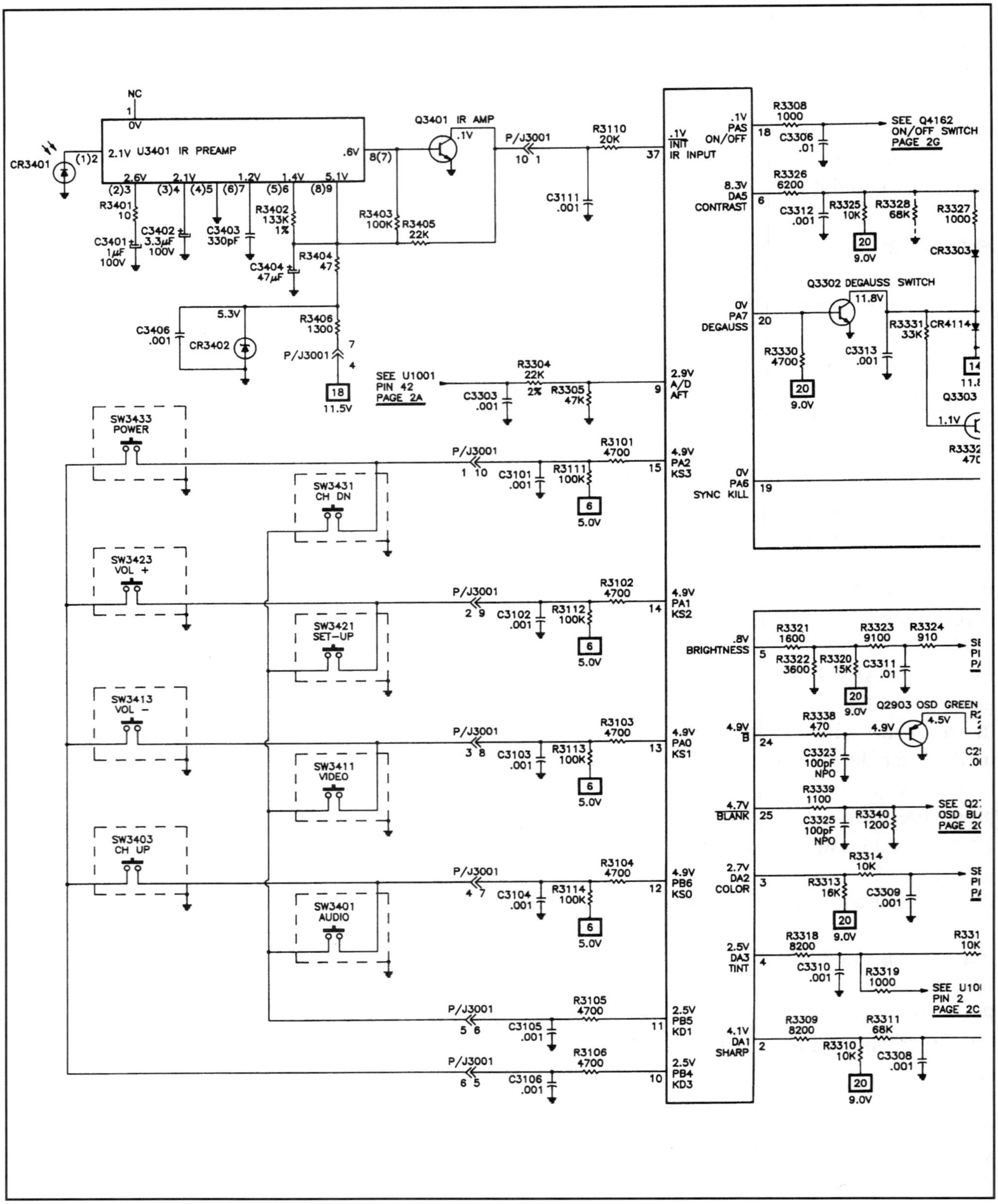

Figure 1-6. System control.

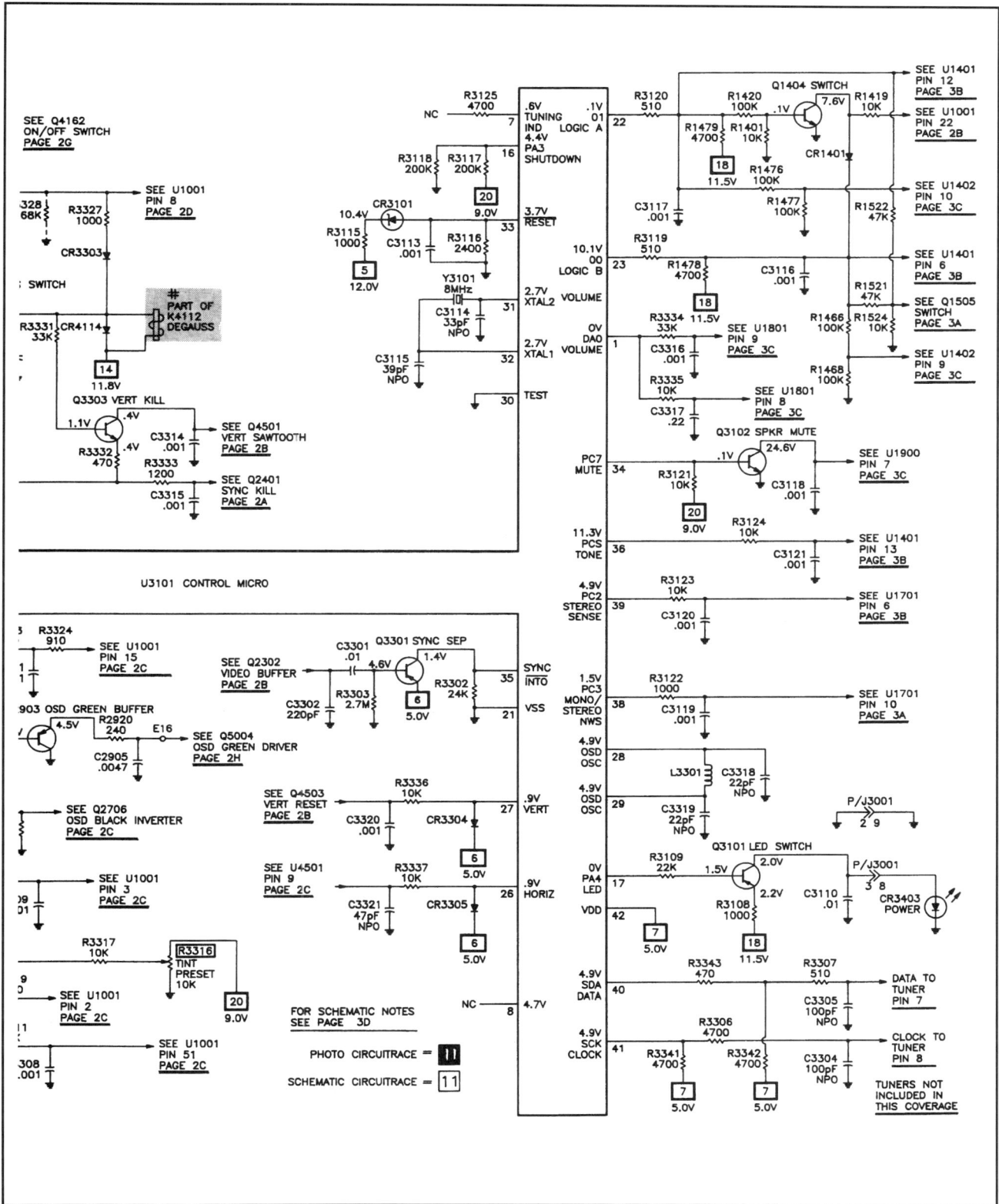

Figure 1-6. System control (continued).

Pin 19 is sync kill. The pin is held low during normal operation. When the tuner searches, as in autoprogramming or channel change, pin 19 goes high and causes a reduction in vertical scan (via Q3303).

Pin 20 operates the degauss function. During normal operation, it is held low, keeping degauss transistor Q3302 from turning on. It also keeps the vertical kill switch Q3303 from turning on.

Pin 21 is V$_{DD}$; that is, ground.

Pins 22 and 23 are used to select auxiliary audio and video. The audio is switched via U1402, whereas video switching is shared by several circuits.

Pin 24 is on screen display (OSD).

Pin 25 is OSD black. Along with pin 24, it supplies the information necessary to generate on-screen information.

Pin 26 is horizontal sync input for OSD.

Pin 27 is vertical sync input for OSD.

Pins 28 and 29 are the OSD oscillator.

Pin 30, labeled "test," is not used.

Pins 31 and 32 provide drive for the 8 MHz crystal.

Pin 33 is reset. The 12-volt standby source is fed to pin 33 via zener diode CR 3101 and keeps it low until the 5-volt standby turns on the microprocessor. Then the zener conducts, providing the reset pulse to initialize the chip.

Pin 34 is speaker mute. It turns off the internal speakers in sets that are equipped with Hi-Fi outputs (C, CN, CR, CS, M, and R series). A high on pin 32 turns on Q3102, which disables the outputs of U1900, the audio output chip.

Pin 35 is the sync input used during channel tuning to detect valid sync. Composite video from Q2302 is routed here via Q3301 (sync separator). This transistor also supplies the inputs for pins 26 and 27.

Pin 36 is used to control a pair of low-pass filters at the input to the volume control chip (IC 1801), permitting the viewer to have control over the tone of the audio.

Pin 37 is the IR input.

Pin 38 selects between mono and stereo audio. It is also pulled low during channel change to avoid false indications of stereo reception.

Pin 39 is the stereo sense input. It informs the microprocessor which audio signal the TV is receiving.

Pin 40 sends data to the tuner and is logic high when inactive.

Pin 41 is the clock output. It is also logic high when inactive

Pin 42 is V$_{DD}$, the B+ connection. It should always read about +5 volts.

An Interesting Case History

Almost all of my encounters with system control problems have been pretty much routine—lightning damaged TV, for example, in which Q4160 shorts collector-to-emitter and takes out the microprocessor. The obvious symptom is a dead set. A few quick checks let the technician know the microprocessor has been damaged and needs to be replaced.

A few, however, have been more than routine. I think, for instance, of the CTC167 that would come on but with no raster and no audio. When I turned the G2 voltage up, I noticed the thin white horizontal line that indicated no vertical defection. I used the remote control to see if the channel change function worked, and I found no indication of channel change. The line will flicker under normal circumstances when the channel is changed. And there was no audio whatsoever. The front controls, with the exception of power on/off, had no effect on the TV. What was going on? Every symptom pointed to system control problems. I had good ground connections and a solid, ripple-free 5 volts. The oscillator was working. Reset was okay. But other voltages were off, like 0 volts on pins 26 and 27. I replaced the micro, and set fired up and worked fine. I have encountered the problem several times since then, and have discovered it may be the only evidence of lightning damage.

Replacing Defective Microprocessors

Before I conclude this section, I should mention the procedure for installing and initializing new microprocessors. The parts list will give you numbers for two different microprocessors. The part number for the CTC167M is 206049; the number for the CTC167C, CN, CS, E, and R is 206038. Be sure to order the correct part for the TV you are working on. When it arrives, you will find a yellow sheet enclosed in the package. The yellow sheet will be titled "Field Replacement Instructions, U3101 Microprocessor," and it will tell you that you MUST configure the new IC for the chassis in which it is being installed. You get just one chance to do it right.

Install the IC, turn the TV on with the power button. DO NOT turn the TV off with the remote control at this point. The screen will display "Configure Code: FF." The "FF" is the configure code stored in hexadecimal format. Use the enclosed data to configure the micro for the chassis you are working on. The instructions will tell which code to use. Use the number keys on the remote to enter the code. You will see the "FF" change. When the correct code has been entered, press "power off" on the remote, and the configure code will be written to memory and permanently stored. *Figures 1-7* and *1-8* are a reproduction of the instructions that accompany the replacement microprocessor.

Special Instructions

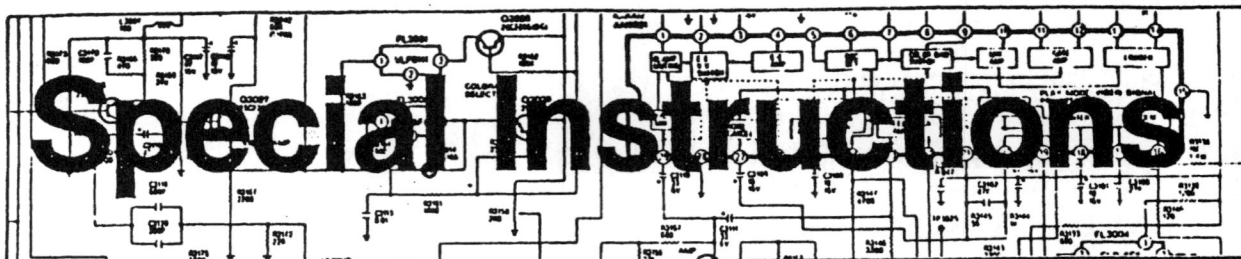

SPS 4261-1
Stock No. 206038, 206049,

Description: Field Replacement Instructions, U3101 Microprocessor

Note: The configuration code must be stored into the memory of the microprocessor to inform the microprocessor of the features in each specific chassis version.

Warning: Pressing one or more number keys and then the TV or Power key may cause an incorrect configuration code to be stored, resulting in possible loss of features in the television.

1. Install the replacement microprocessor and make all necessary connections to operate the set. Do not press any remote control number key until you are ready to set the correct configuration.

2. Using only the buttons on the front of the set, turn the set on, set the volume, and select a channel.

3. The set should display "CONFIG CODE: FF" across the lower part of the picture. You may now set the configuration code for the chassis version according to Table 1. Pressing the number keys indicated will change the configuration code from "FF" to the code indicated in the table. Do not turn the set off, or press the TV button on the remote transmitter until the correct configuration code is displayed.

Warning: Once stored, the configuration code cannot be changed without replacing the microprocessor. You do not get a second chance. However, the following procedure for setting the code will give you ample opportunity to review and verify that the code is correct before you store it.

4. If you realize that you have pressed an incorrect number key, simply press that number key again to toggle it to the previous state. If the displayed configuration code is incorrect and you are not sure how you got there, refer to Table 2 to determine which keys have been pressed. The configuration code consists of two characters which each represent a particular combination of number keys. The left character represent a combination of keys 7, 6, 5, and 4. The right represents a combination of keys 3, 2, 1, and 0. Pressing a number key either adds it to the combination (if it had been absent) or removes it from the combination (if it had been present).

Table 2 details the number combinations represented by each of the 16 possible displayed characters. For example: a "5" displayed as the left character indicates a combination with keys 7 and 5 present, a "4" displayed as the right character indicates a combination with keys 3, 1, and 0 present, but with key 2 absent. Putting this all together therefore, shows that a configuration code of "54" indicates a combination of 7-5-3-1-0.

RCA **ge** Thomson Consumer Electronics
Accessories & Components Business / Deptford, NJ 08096-2088

Figure 1-7. U3101 replacement.

If the displayed configuration code is incorrect, simply use Table 2 to determine what combination of numbers is represented by the desired code, and what combination is represented by the displayed code. If the displayed code lacks a number found in the desired code, or if the displayed code has a number not found in the d ired code, simply press the corresponding key to "toggle" the number in or out of the combination. For example: if the desired code was "0D" (keys 7-6-5-4-1) but the displayed code was "19" (keys 7-6-5-2-1), you would press the number keys 4 and 2, because those are the keys where the two codes differ.

5. After confirming that the correct final configuration code is displayed, press the TV button on the remote hand unit to store the displayed configuration code into memory. The procedure is now completed.

Table 1 -- Desired Final Configurations (by chassis)

CHASSIS	INITIAL CONFIGURATION	PRESS KEYS TO OBTAIN	FINAL CONFIGURATION
CTC166A, CTC167A	FF	0,1,2,3,4,5,6,7	00
CTC166B	FF	2,3,5,6,7	13
CTC166C, CTC167C	FF	2,3,6,7	33
CTC166CR, CTC167CR	FF	2,3	F3
CTC166CS, CTC167CS	FF	2,3,7	73
CTC167CN	FF	2,3,7	73
CTC166H, CTC166J	FF	1,2,4,5,6	89
CTC166JC, CTC166JD	FF	1,4,5,6,7	0D
CTC167E	FF	3,4,5,6,7	07
CTC167M	FF	1,2,6	B9
CTC167R	FF	2,3,7	73

Table 2 -- Explanation of Key Combinations

Character displayed	Keys represented by left character				Keys represented by right character			
F				(none)				(none)
E			4					0
D		5					1	
C		5	4				1	0
B	6				2			
A	6		4		2			0
9	6	5			2		1	
8	6	5	4		2		1	0
7	7				3			
6	7		4		3			0
5	7	5			3		1	
4	7	5	4		3		1	0
3	7	6			3	2		
2	7	6	4		3	2		0
1	7	6	5		3	2	1	
0	7	6	5	4	3	2	1	0

Figure 1-8. U3101 configuration codes.

Horizontal Deflection

The CTC167's horizontal deflection circuit is rather straightforward. *Figure1-9* gives you the particulars. Composite video enters U1001 at pin 57 and is processed by a sync separator. Horizontal sync is routed to the horizontal AFC block inside the IC. The AFC block controls the phase of the horizontal VCO with respect to the incoming video signal. The AFC block receives two feedback signals, one at pin 60 which is used for horizontal AFC, and another at pin 59 which is used for APC (automatic phase control). The latter ensures the picture is centered within the raster. Both are used to affect the phase of the 32xH VCO which is counted down to produce the horizontal drive signal at pin 64. The drive signal is buffered by Q4202 and fed to the base of the horizontal driver transistor Q4301. The output of Q4301 is transformer-coupled to the horizontal output transistor (Q4401), which drives the yoke and flyback.

How does the horizontal deflection start up when the TV receives an ON command?

When AC is applied, raw B+ becomes available to the B+ regulator and the horizontal output transistor. The 20-volt standby source provides B+ to the horizontal driver transistor via CR4606. Powering up the TV then means supplying a voltage to U1001. When the microprocessor receives an ON command, it pulls pin 18 low, turning on Q4162, and the 6.8-volt source to U1001 becomes active, permitting the horizontal oscillator to begin operating. Once the horizontal circuits become active, the scan-derived sources power the remainder of the chassis.

X-Ray Protection Circuit

I suppose this is as good a place as any to comment on the x-ray protection circuit. *Figure 1-10* is a partial schematic for the circuit. The filament pulse from pin 4 flyback (T4401) is rectified by CR4901 and applied to R4902 and R4903, a precision resistor network, the product of which is applied to pin 1 of U1001. If the resulting voltage exceeds the breakdown value of CR4902, the x-ray protection block inside U1001 becomes active and shuts off the horizontal oscillator, which powers down the TV. Note that the normal voltage at pin 1 is about 1.75 volts.

Problems in this circuit can be exceedingly difficult to ferret out. One way to detect x-ray protection problems is to attach a DMM with a sample-hold function to pin 1. Let the TV play while you go about your business. If the TV shuts down a minute or an hour later, the voltage will be recorded and held by your meter. You can then determine if the shutdown is x-ray protection related. Sometimes the TV will turn on and go immediately off. Is the problem here—or elsewhere? A DMM with a good sample-hold function will read and store the voltage in a matter of milliseconds. My meter can record a reading in 1/1000 of a second, but it needs a duration of a tenth of a second for an accurate reading. A good sample-and-hold meter may translate into fewer gray hairs and a quicker repair.

The system control micro (U3101) detects x-ray shutdown by monitoring the 9-volt source at pin 16. If the 9-volt source drops, the micro will turn the set off by interrupting the 6.8-volt source to U1001. When horizontal deflection shuts down, the voltage at pin 1 of U1001 drops, which causes the x-ray protection block inside U1001 to reset. In about two seconds, the micro turns the 6.8 volts back on. If the fault has been removed, the x-ray protection circuit remains inactive. If the fault is still there, x-ray

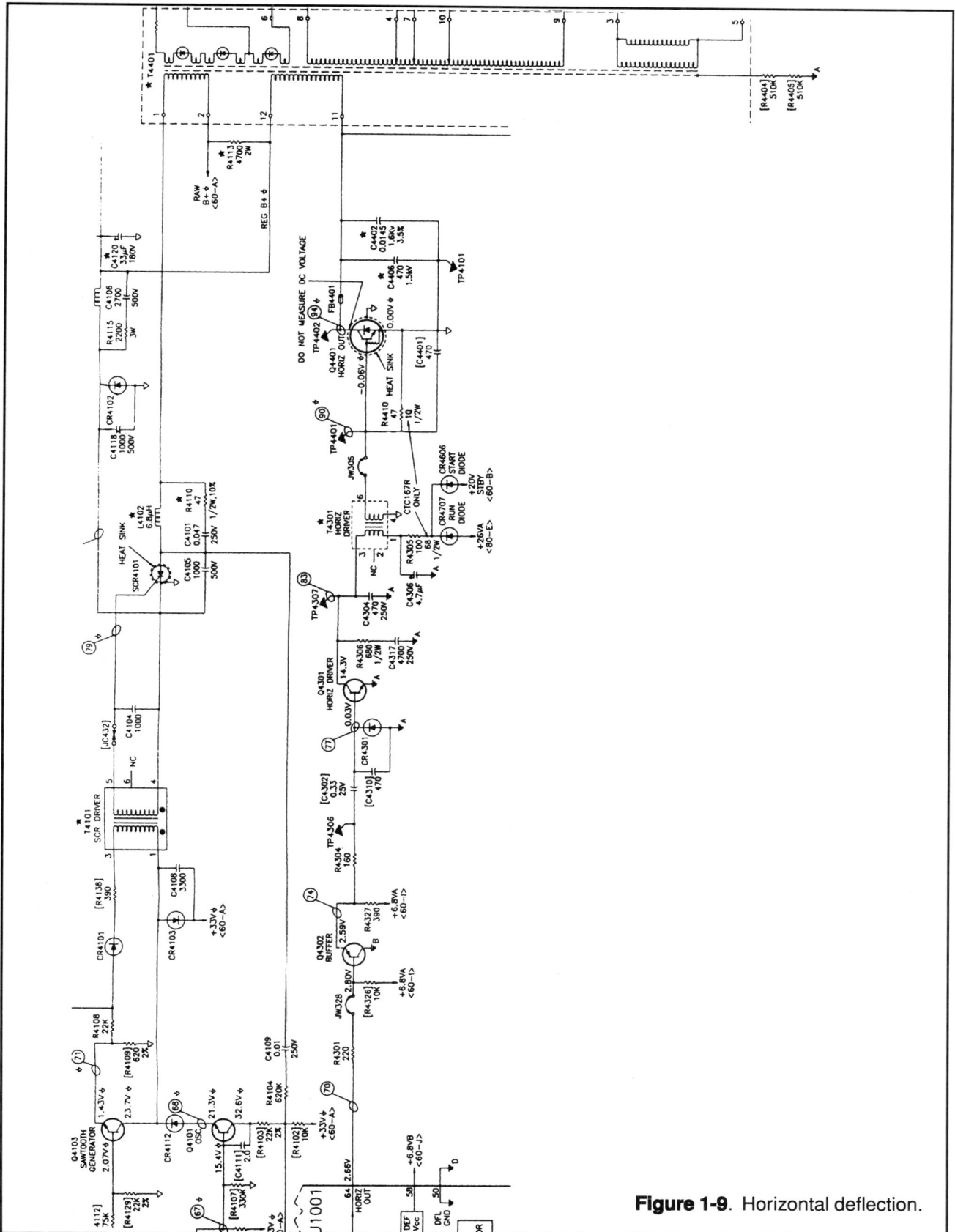

Figure 1-9. Horizontal deflection.

protection will respond again. The TV will begin to cycle on and off until the fault is removed or the TV stays off. If it stays off after a cycle of on-offs, the TV will require a power-on command to restart.

Again, a sample-hold instrument attached to pin 18 of the micro will give you good diagnostic information! At the very least, it will let you know if the on-off cycling is the result of system control shutdown.

Troubleshooting Horizontal Deflection Problems

Is there an easy way to troubleshoot horizontal deflection problems? I don't think so, but some ways are better than others. RCA suggests a procedure you might find helpful. It is found on frame 1-E3 of the microfiche.

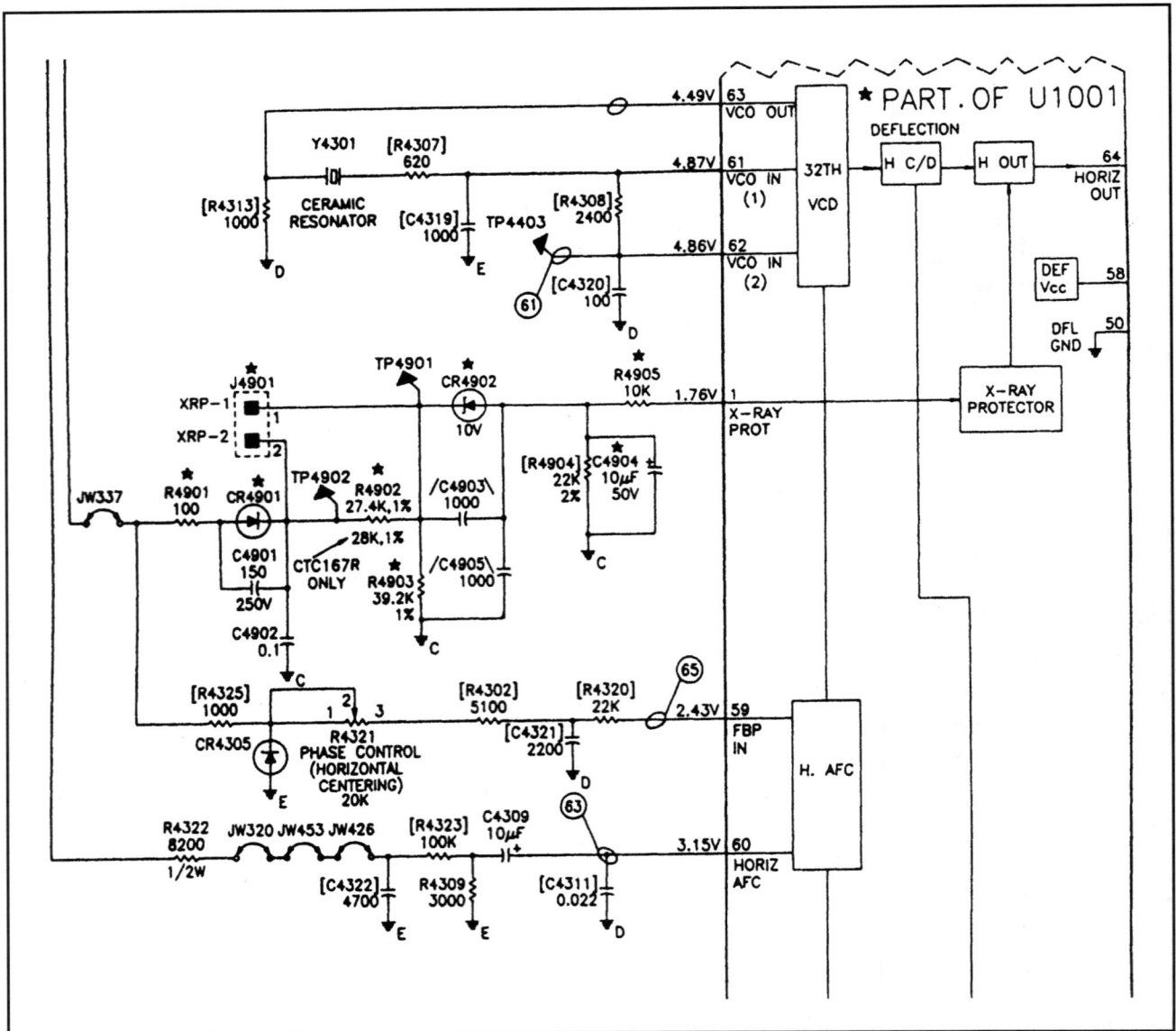

Figure 1-10. X-ray protection.

Assume you have encountered a dead set and that you have ruled out problems in the power supply and the B+ regulator. The recommended procedure is to disconnect the collector of Q4401 (horizontal output transistor) from the B+ supply by lifting one end of FB4401. Plug the cord into an AC outlet, and then apply +20 volts from an external DC source to the anode of CR4606. How do you locate CR4606? Remember our parts identification numeration? The 4600 number lets you know the part is a standby rectifier. The next step is to short the base of Q4162 to cold ground through a 1000-ohm resistor. Q4162 controls the 6.8-volt supply to U1001. By following this procedure, you are attempting to turn U1001 with the horizontal output transistor disabled.

You are now ready to begin troubleshooting.

(1) Check for the presence of 6.8 volts at pin 58 of U1001. If it is not there, suspect CR4164, Q4162, R4171, or the 12-volt standby supply.

(2) If the 6.8 volts is present, check for a 32H output at pin 63 of U1001. If it is not there, suspect Y4301, R4307 and R4308, R4313, C4320, C4319, and U1001.

All things are possible, I suppose, which means some of these parts could be bad. However, I have never known anything but U1001 to fail, and it is a highly reliable part. If you do change the chip, first install a 64-pin socket! A socket is cheap and easy to use. If the old chip turns out to be good, you can unplug the new one instead of unsoldering it.

(3) If the 32H signal is present, check for horizontal drive at pin 64. If it is not there, check resistors R4326 and R4301. If they are good, replace U1001.

(4) If there is drive at pin 64, check the collector of Q4301. You should see a signal of 20 to 30 volts peak-to-peak. If it is not there, look at Q4301, Q4302, R4327, R4304, C4302, and CR4301. If it is present, move on to the Q4401, T4401, and/or the scan-derived supply circuits.

Now, that's a systematic, thorough procedure. Is there a quicker way? Yes, I think so. A middle point is the horizontal driver, Q4301. Remember, the TV will pulse on and off about three times in an effort to start. Have someone depress the power ON control while you scope the collector of the driver transistor. If you are by yourself, use a remote control to enter the ON command. Is there a signal at the collector? If not, is there a signal at its base? If not, then try pin 64 of U1001. You can make these checks quickly and easily and can usually put your finger on the problem without fanfare. If you trace the problem to the IC, you may then elect to use the procedure RCA suggests. The one major benefit of using their procedure is that it ensures a stable B+ supply to the chip while you do your troubleshooting. The stable B+ supply eliminates, among other things, a triggering of the x-ray protection circuit. So, make sure the B+ supply is stable before you begin thinking about a defective U1001.

I operate on the assumption that anything you learn can be helpful; therefore, I refer you to two interesting articles by Glen Kropuenske, applications engineer for Sencore. One article is entitled, "Understanding And Troubleshooting TV Shutdown Circuits," in *Sencore News*, #167 Nov./Dec.1994. Kropuenske's article is "generic," but you can pick up useful information that will help you deal with the CTC167 and other TV shutdown circuits and problems. He has another article in issue #175 Sept./Oct. 1996. This one is called "The TV Shutdown Dilemma—Fear Or Opportunity?"

I am not going to cite case histories here because I have already covered a few of the major horizontal deflection problems. Let me emphasize that the horizontal driver transistor is a common failure item. Underscore the fact that you need to check SCR4101 whenever you replace a shorted Q4401. Chances are a failed output transistor will also have damaged the SCR. Be sure to check the 4.7-ohm resistor (R4410) in the base-emitter circuit of the HOT. It will sometimes open. I also recommend checking the solder connections around the pins of the horizontal driver transformer (T4301). Loose connections here often lead to the failure of the horizontal output transistor. It is better to be safe than risk a recall.

In case you need them, I am going to give you a few part numbers. The part number for the output transistor is 190483; the part number for SCR4101 is 197591; the part number for the horizontal driver transistor is 190482.

The Vertical Deflection Circuit

Figure 1-11 depicts the vertical deflection circuit. Refer to it as we talk about that circuit.

Q4501 is the sawtooth generator. When it turns off, C4506 begins to charge, producing a negative-slope sawtooth wave at the collector of the sawtooth generator. The sawtooth waveform generated by the charge on C4503 is coupled to Q4502, the error amp. Q4502 inverts, amplifies, and sends the signal to pin 6 of U4501, the vertical output IC, initiating vertical scan. Retrace is controlled by the vertical signal, which exits U1001 at pin 55. It turns off Q4503, which allows Q4501 to turn on, discharging C4506.

At the beginning of scan (the top of the picture), current flows from pin 4 of U4501 into the vertical yoke. Current flow decreases to zero as the electron beam approaches the center of the screen. Current then becomes increasingly negative (that is, flowing from the yoke into pin 4) until the scan reaches the bottom of the screen. When the beam reaches the bottom of the screen, U1001 sends a pulse to the vertical reset transistor, initiating vertical retrace by turning off Q4503.

In order to provide fast retrace, the CTC167 employs a boost circuit to increase the supply voltage to U4501. The boost circuit consists of C4502 and CR4504. C4502 charges from the 26-volt supply, and discharges during retrace through CR4504. Its discharge effectively doubles the B+ supply to pin 8 of U4501, ensuring rapid vertical retrace.

RCA adds some interesting circuits to its vertical deflection. For example, the CTC167 disables vertical deflection during start-up and degaussing. When it receives an ON command, the microprocessor pulls pin 20 high. The high turns on the degauss switch Q3302, which initiates the degauss procedure and turns off the vertical kill transistor Q3303. With Q3303 turned off, C4506 cannot charge. If C4506 cannot charge, there will be no vertical deflection. The literature says pin 20 of the microprocessor will remain high for about 0.8 seconds, at which time it will go low, turning off the degauss transistor which permits the vertical kill switch to turn on. *Figure 1-12* will give you the details.

The vertical kill switch has two other functions:

The first is to produce a service line for setup. To produce the service line, disconnect AC power and allow time for the 5-volt standby voltage to drop to zero. Then press and hold the front-panel "Setup" while you connect the TV to AC. Release the "Setup" button and the service line will appear when the OSD disappears. Be sure to set the brightness control to near minimum before to calling up the service line to prevent damage to the CRT while you perform the setup.

The second use of vertical kill is to stabilize OSD during channel change and autoprogramming. The micro pulls pin 19 to about 1.8V while the tuner searches. The 1.8V is not high enough to turn Q3303 off, but it is high enough to raise the voltage on its collector, which slightly reduces the amplitude of the vertical sawtooth. The low-amplitude vertical sawtooth reduces vertical scan but keeps OSD in view!

Troubleshooting Vertical Deflection Problems

First a word about service helps in the literature with which I am familiar. Glen Kropuenske has written "Understanding TV Vertical Circuits And How To Troubleshoot Them Quickly And Efficiently," which appears in *Sencore News*, issue #178 May/June 1997. His illustrations are mostly generic, but the information is applicable to all modern televisions. He takes a little time to explain how the new single-package vertical output ICs work. I suggest you also utilize the new ECG cross reference book for the information it has, and access web sites like *www. repairworld.com.* Quite a lot of manufacturers and jobbers have web sites which you can access and from which you can download reams of useful stuff. I just discovered a new source for obtaining pinouts and descriptions for ICs at *www.questlink.com.* Even though it is free, you will have to register to get to their files. The point is, knowing what's inside the package and what its voltage-signal requirements are makes finding the trouble it causes easier to pinpoint and fix.

The CTC167 does not lend itself to vertical deflection problems. Remember, I am speaking from my experience and from what I have read. But when one occurs, it can be a real bear to nail down. I suggest you begin by taking voltage measurements and checking waveforms around U4501. If the DC voltages are correct and if there is a good drive signal at pin 6, replace the chip. If DC voltages are correct and there is no vertical drive, troubleshoot the circuits from pin 55 of U1001 forward. Voltage and waveform analysis will help you find the problem quickly.

I said I have had very few problems out of the vertical deflection circuit, but four components in the circuit have given me trouble. The first is the vertical output chip itself. The second is CR4504. If it opens or becomes leaky, the voltages at pins 8 and 9 will be off. It is an easy problem to spot. The third is C4502, a 220 µF 35-volt capacitor, which can dry out and cause a peculiar set of symptoms. The TV will have perfect audio but no picture. If you turn up the G2 control, you will see that you have raster but no video, repeat raster but no video! In this event, check the DC voltage on pin 13 of U1001. If it is higher than 6.3 volts, change C4502 and check CR4504! The negative lead of the capacitor is connected to the line that supplies vertical-horizontal blanking to U1001. When it dries out, the capacitor permits a higher than normal DC voltage to appear on the V/H blanking signal, which biases pin 13 of U1001 OFF. The result is raster if G2 voltage is increased, but no picture.

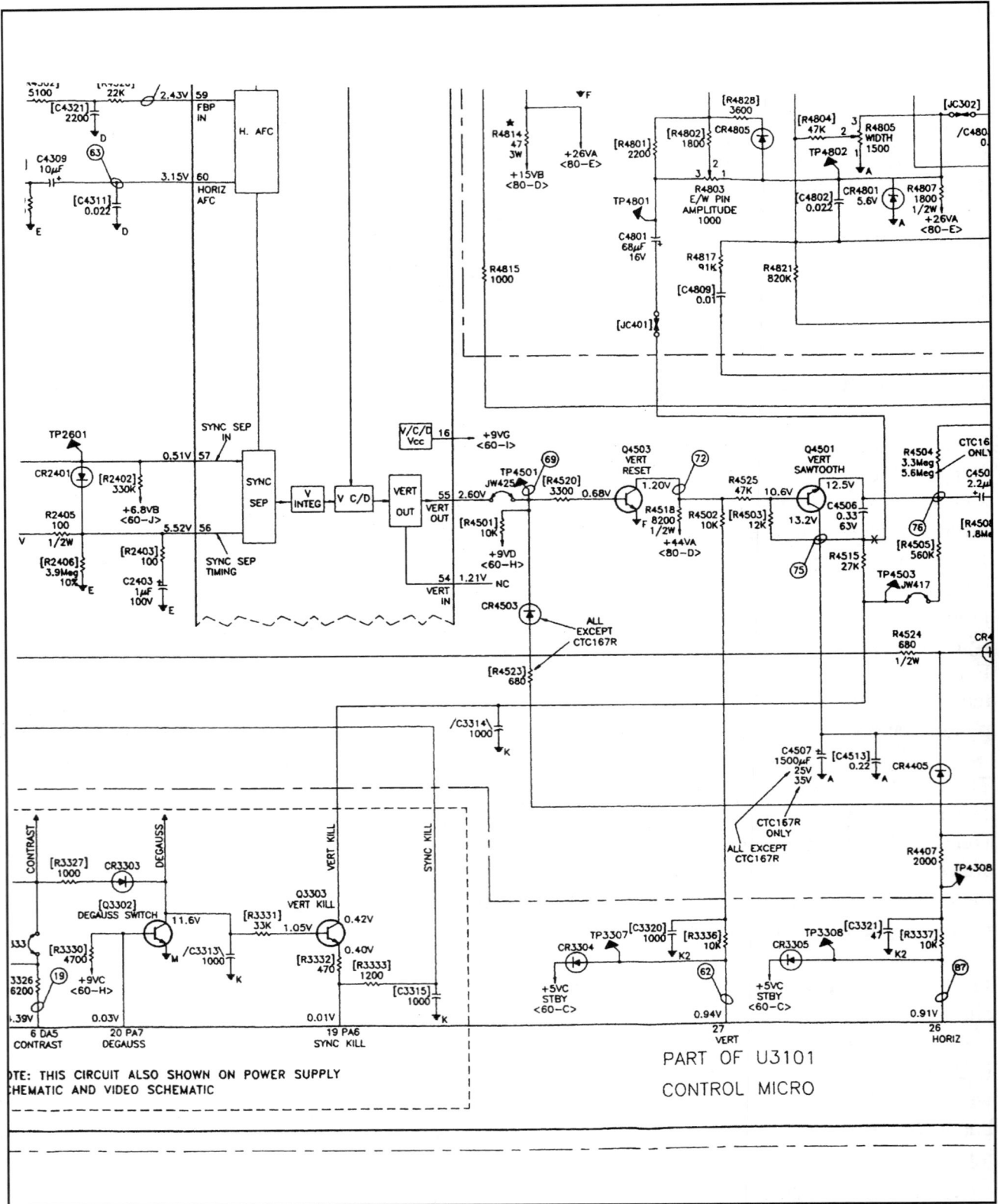

Figure 1-11. Vertical deflection.

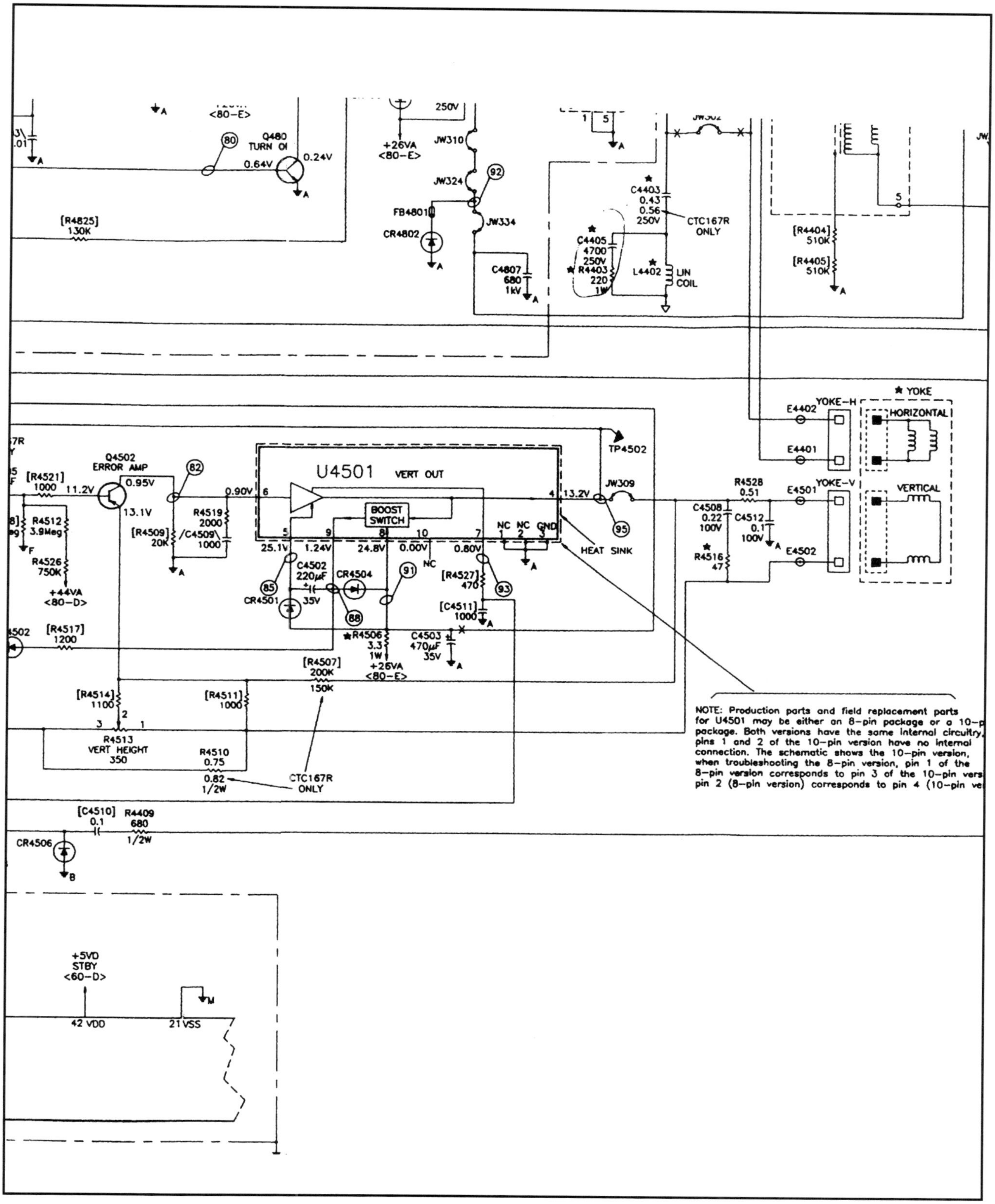

Figure 1-11. Vertical deflection (continued).

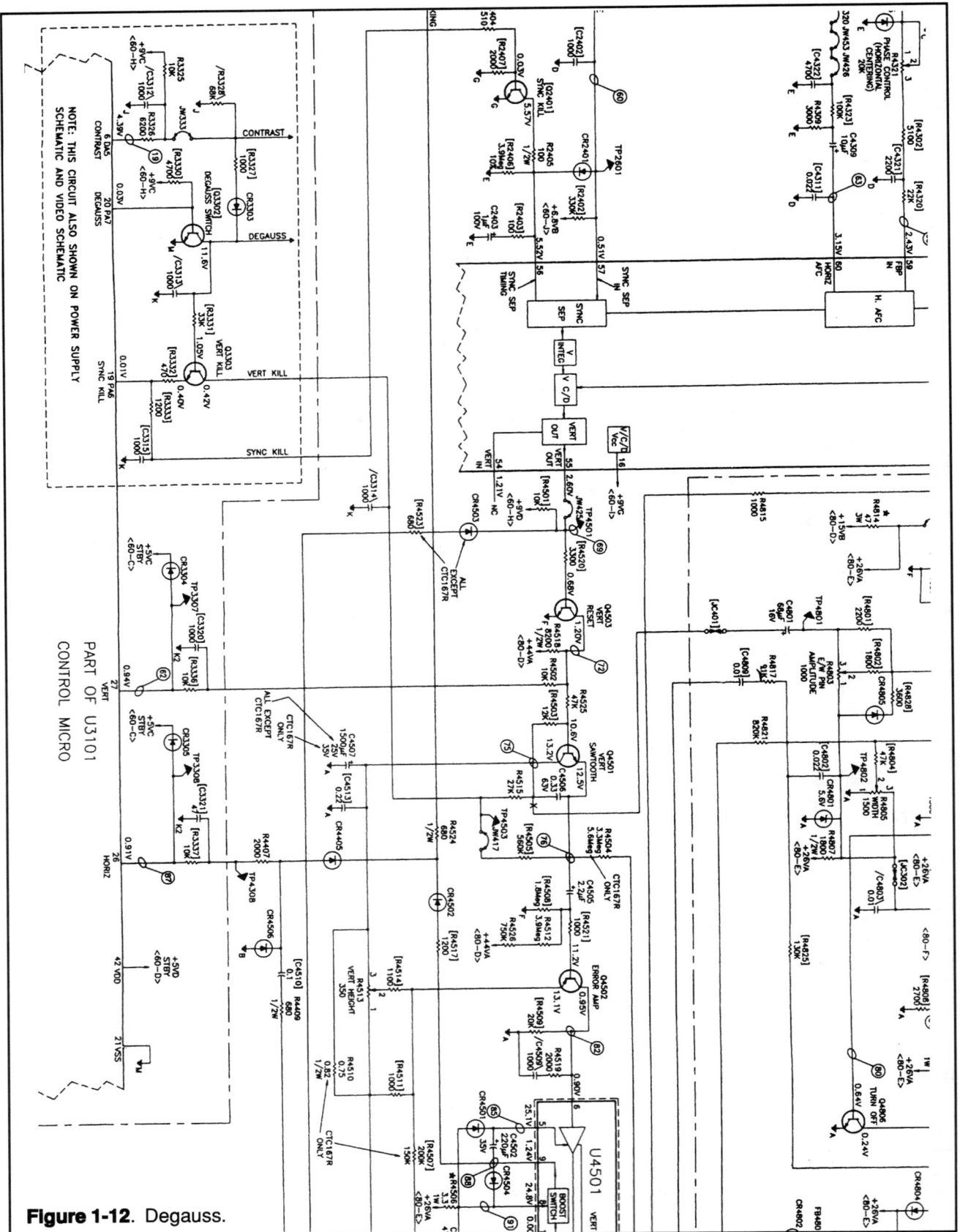

Figure 1-12. Degauss.

The fourth problem—and I have already alluded to it—is a defective microprocessor. If you have not read the book from the front to this point, then read what I said about system control problems.

The Video Processing Circuits

I will discuss the video processing circuits in the CTC167 by dividing the circuit into small segments.

The Tuner

With one exception, the CTC167 chassis uses the same tuner, MTP-M-2016. The exception is the CTC167M, the commercial version, which uses the MTP-M-2006. It provides additional levels of isolation that commercial applications often require. The following table gives you pinout information for the MTP-M-2016.

Pin 1 is the connection for RF AGC, which exits U1001 at pin 46.

Pin 2 is the +12-volt input.

Pin 3 is the IF output, which is routed to pin 20 of U1001 via the IF amp (Q2301) and the SAW filter (SF2301).

Pin 4 four is the tuning voltage source. It is developed from the 44-volt supply via CR3602, a 33-volt zener.

Pin 5 is connected to the -12-volt source.

Pin 6 is connected to the 5-volt source.

Pin 7 is the data line from pin 40 of the system microprocessor.

Pin 8 is the clock line from pin 41 of the system microprocessor.

There are at least four ground connections.

A word of advice. Before you change a suspected defective tuner, take a minute and do some voltage checks. With the exception of lightning, which does mortal damage to these tuners, I have found them to be reliable. I guess I ought to say "I have found them reliable for the most part." They do have a habit of losing one or more tuning bands, and it is usually channels 2-6.

Video IF

Figure 1-13 is a partial schematic for the video IF processing circuit.

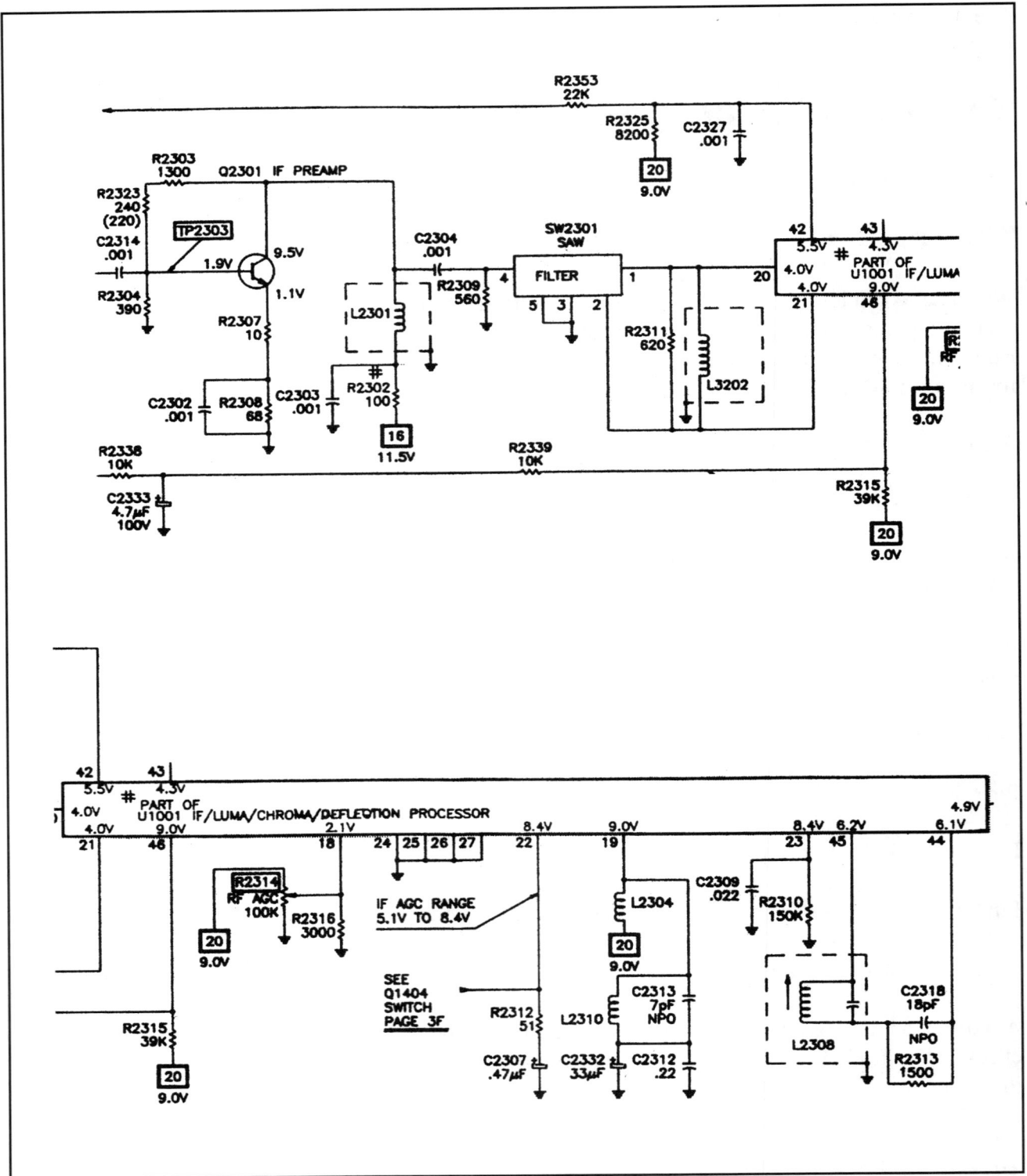

Figure 1-13. Video IF.

After processing by Q2301 and SF2301, the video signal enters a differential amplifier inside U1001 at pins 20 and 21. U1001 takes this signal and generates: (1) a baseband video signal, available at pin 47; (2) an AFT signal at pin 42; and (3) an AGC signal at pin 46. The AFT signal is sent to the microprocessor to allow for fine tuning of the incoming RF signal. The AGC signal is routed to the tuner to set the gain of the RF stages inside the tuner.

Baseband video information is buffered by Q2305 at pin 47 of U1001, where the 4.5 MHz audio is removed by the trap CF2301. The signal is then routed to Q2302, a video buffer, where it is taken off at the emitter and sent either to a video switch or back into pin 53 of U1001.

Video Switches

Depending on which CTC167 chassis you are working with, you may encounter a series of video switches between Q2302 and U1001. These switches don't give much trouble except in the case of lightning damage. They are easy to troubleshoot—it's usually a matter of checking B+, switching information from the microprocessor, and checking for the presence (or absence) of video information at entry and exit pins.

Luminance Processing

Wide-band luminance enters U1001 at pin 53 of U1001 (*Figure 1-14*). It is processed by an amplifier whose gain is controlled by a voltage from pin 6 of the system microprocessor (contrast control). High-frequency information is input to pin 52 and fed to an amplifier whose gain is controlled by the sharpness control. These signals are combined in the contrast block inside U1001 and then fed to the DC clamp circuit, where the DC level of the luma signal is controlled by the brightness control voltage from pin 25 of the microprocessor. The processed luminance signal exits U1001 at pin 13 and is then combined with the OSD Black signal from pin 25 of the microprocessor. The combined signal is buffered by Q2901 and fed to the bias/drive network.

The luminance processing network is prone to problems. For example, there was the CTC167 that had perfect audio but a bright screen with retrace lines. I checked the collectors of the video output transistors and saw no evidence of video information. The signal was present at pin 13 of U1001. Something had therefore happened between these two points. The culprit was the luminance buffer (Q2901). You might like to know this transistor is a rather high-failure part. Incidentally, an ECG159 makes a perfect substitute for it.

U1001 does not often cause problems, but it can. I recently repaired a CTC167 that had the set of symptoms I just described. This time, however, there was no luminance signal at pin 13 of U1001. Since the DC voltages were correct, indicating the luma amp was not biased OFF, I had to change the chip. The part number, if you need it, is 193082. When I change one of these chips, I always use a 64-pin socket—a practice I heartily recommend.

I have had a variation of the no-luminance problem, and it cropped up about five months ago. The customer said the TV would play fine for about an hour, and then the picture would get very bright and

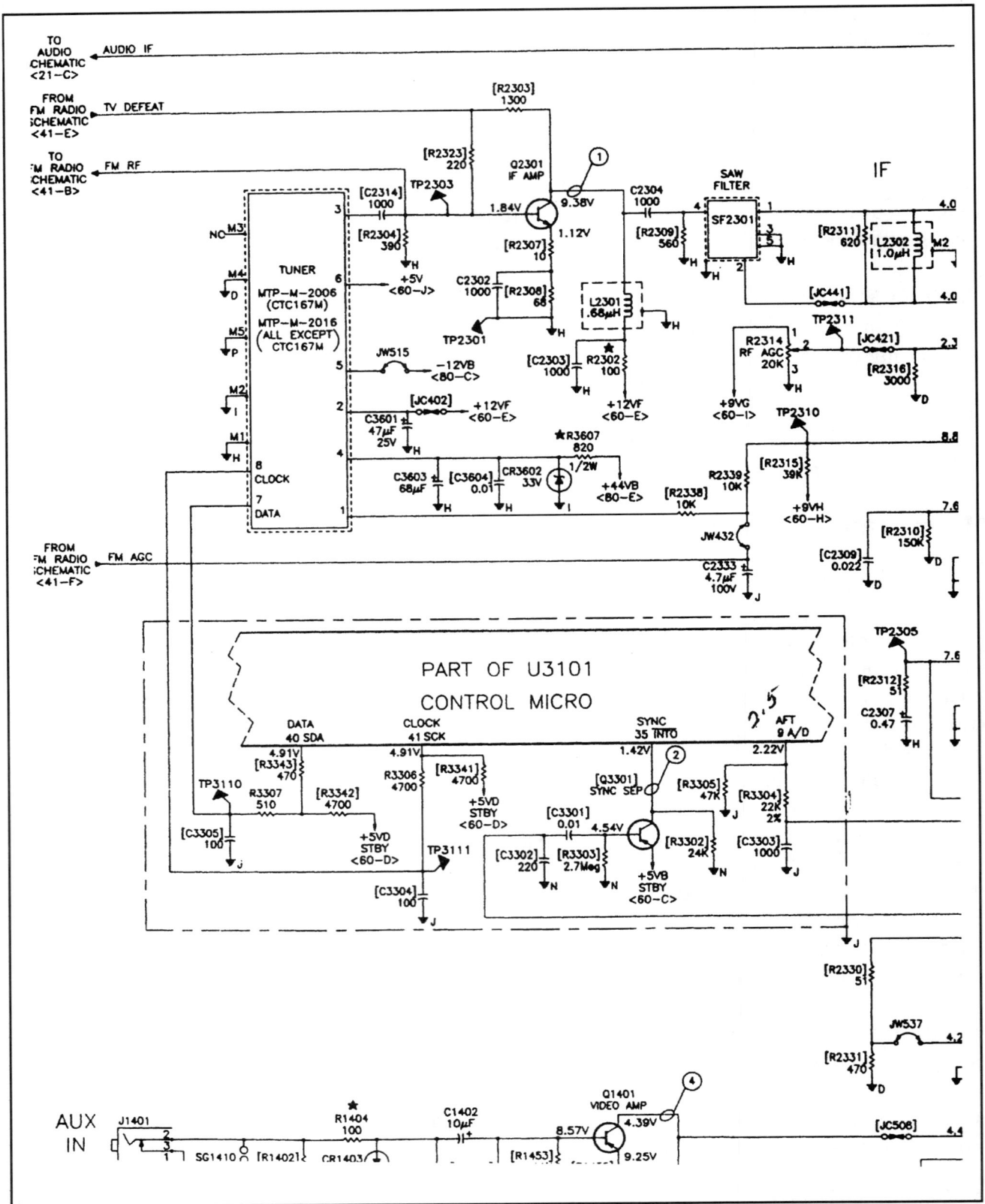

Figure 1-14. Luminance circuits.

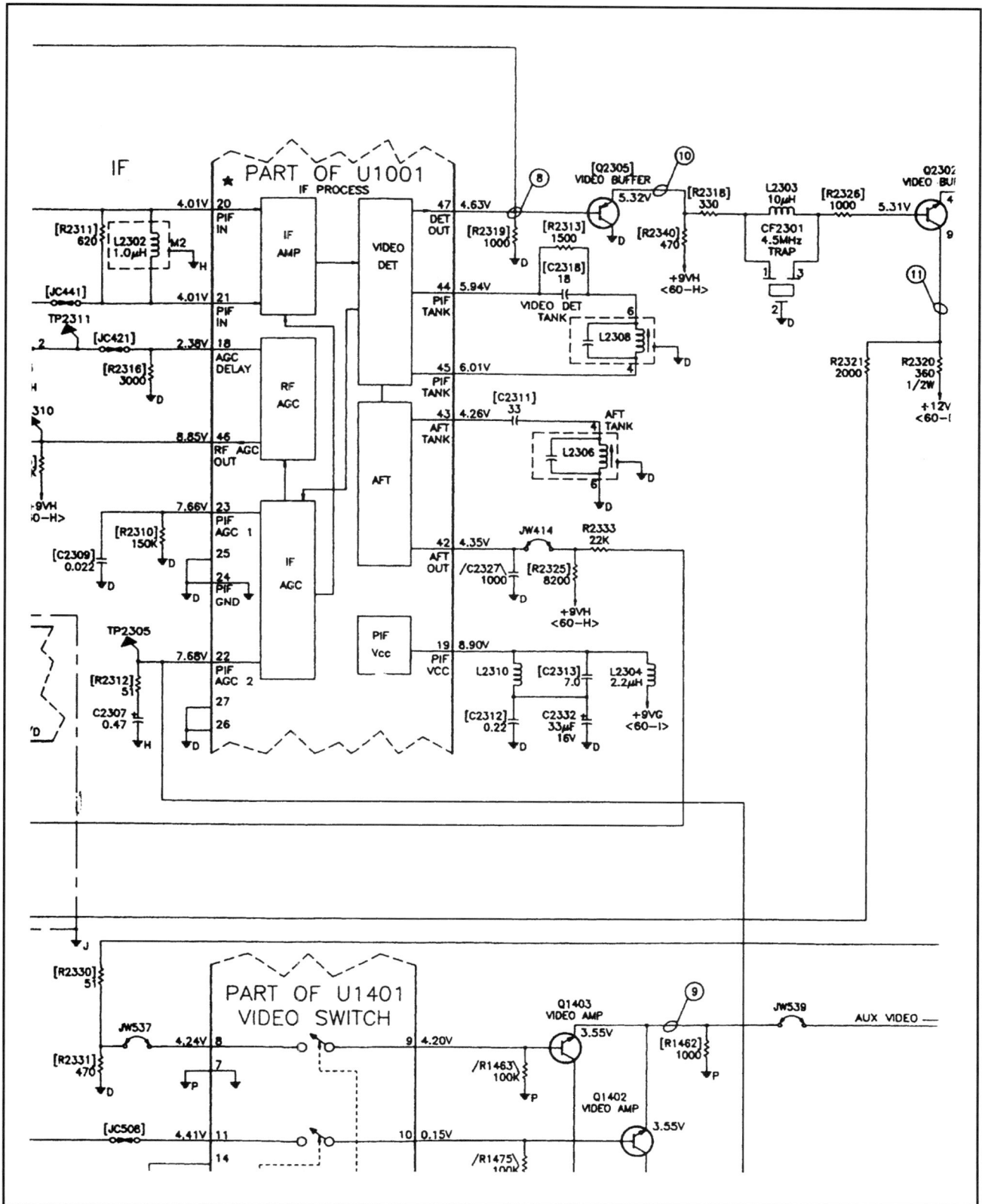

Figure 1-14. Luminance circuits (continued next page).

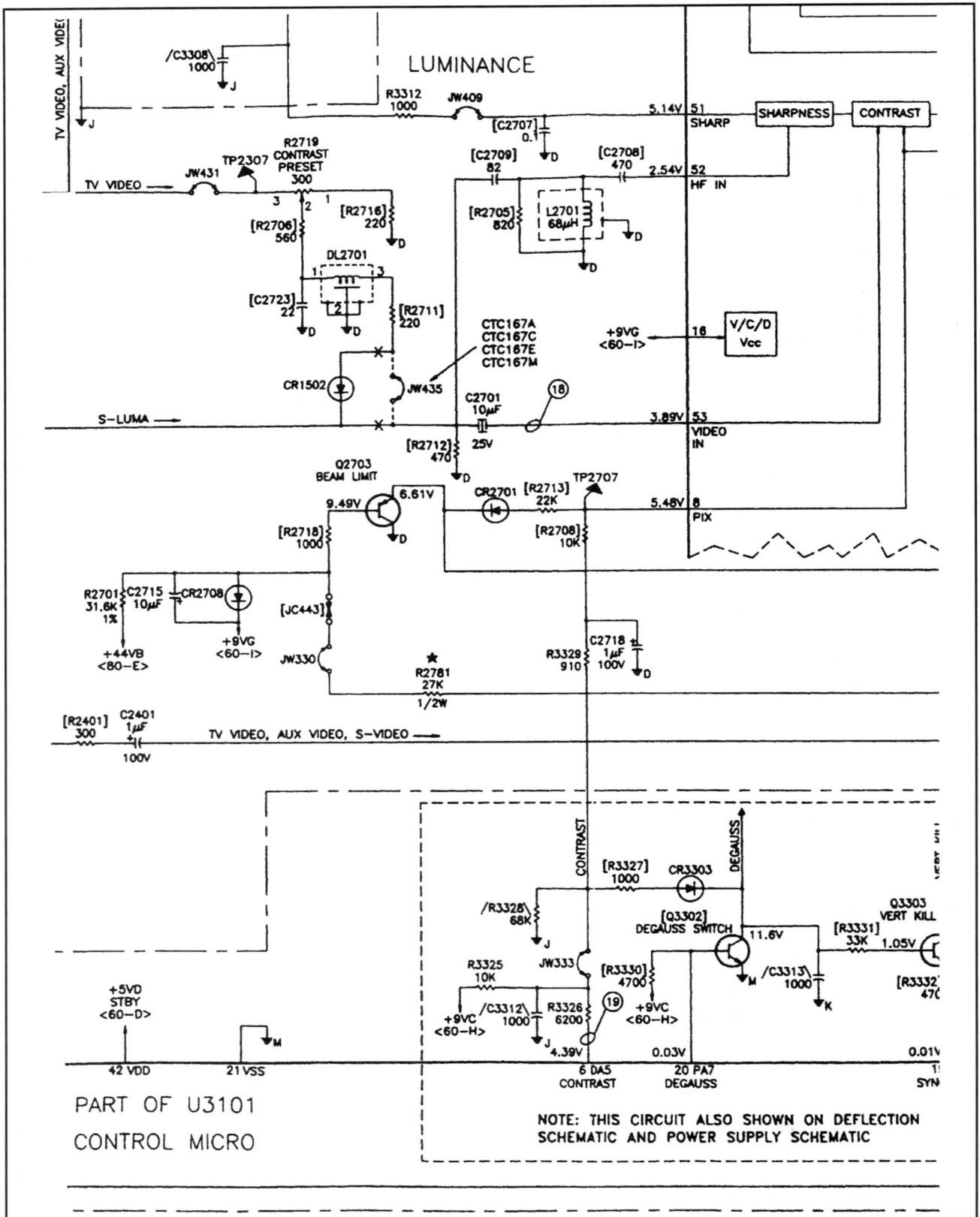

Figure 1-14. Luminance circuits (continued).

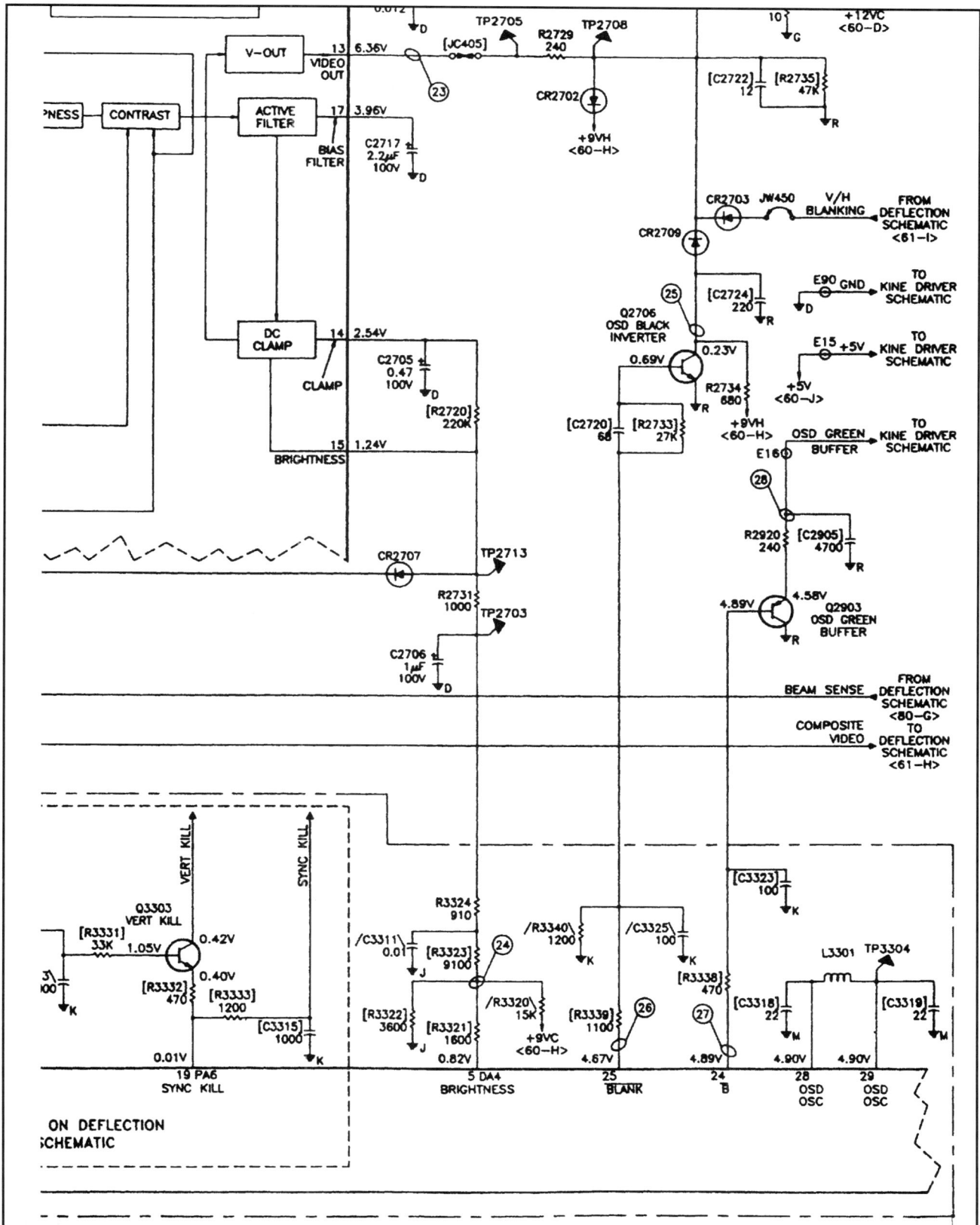

Figure 1-14. Luminance circuits (continued).

have lines in it. Sure enough, after about two hours of play, the TV acted up. "Easy fix," I commented. But there was video information at the at the emitter of Q2901, and its voltage readings were correct. Voltages readings on and around the video output transistors were so-so. What was going on? In a case like this, check the voltages on the pins of the CRT. Pin 5, which is the connection for the control grid (G1), is normally about 25 volts. When the TV acted up, the voltage shot up to 200+ volts! Where was that voltage coming from? It had to be coming from inside the tube!

One or more of the elements of the electron gun had to be shorting together. Sure enough, the problem was the CRT. Fortunately, I had a good used picture tube in stock, which saved the job.

If it is possible, use a signal generator as a signal source when you troubleshoot a video problem like the "no luminance" one I just discussed. An off-the-air signal is okay, but not ideal. You need a signal that has a constant peak-to-peak level, and you need one you can easily recognize. I use the "ten-bar pattern" signal from my Sencore generator because its peak-to-peak value does not change, and it presents a stairstep pattern which I can easily recognize even if it is distorted. Makes for easy troubleshooting.

Chrominance Processing

My references will be to *Figure 1-14*.

Chroma information enters U1001 at pin 49. It is fed to an amplifier whose gain is controlled by a voltage from pin 6 of the microprocessor. From there the signal goes to a second amplifier, whose gain is controlled by the color control voltage from pin 4 of the microprocessor. Demodulated color information appears at pins 9, 10, and 11 of U1001. The voltage at pin 2 controls the tint or hue of the signal. You can check for the presence of the 3.58 MHz signal at pin 4.

Kine Drivers

Refer to *Figure 1-15*.

Demodulated color information is fed to the bases of the red, green, and blue video output transistors (Q5001, Q5002, Q5003). Luminance information is combined with the color information to produce the red, green, and blue signals that drive the picture tube guns. OSD information is added only to the green output signal via Q5004.

The Audio Processing Circuits

The audio processing circuit will be the last system I will examine. What I have to say about it will apply to almost all models under the CTC167 umbrella. The exception will be the E series.

Let's begin by examining the stereo decoder, U1701. See *Figures 1-16* (RCA) and *1-16A* (Sams).

Wideband audio leaves U1001 at pin 34, goes to Q1702 which serves as a buffer, exits through R1726 (input level control) and enters the stereo decoder at pin 2. The signal exits at pin 3, where it is filtered

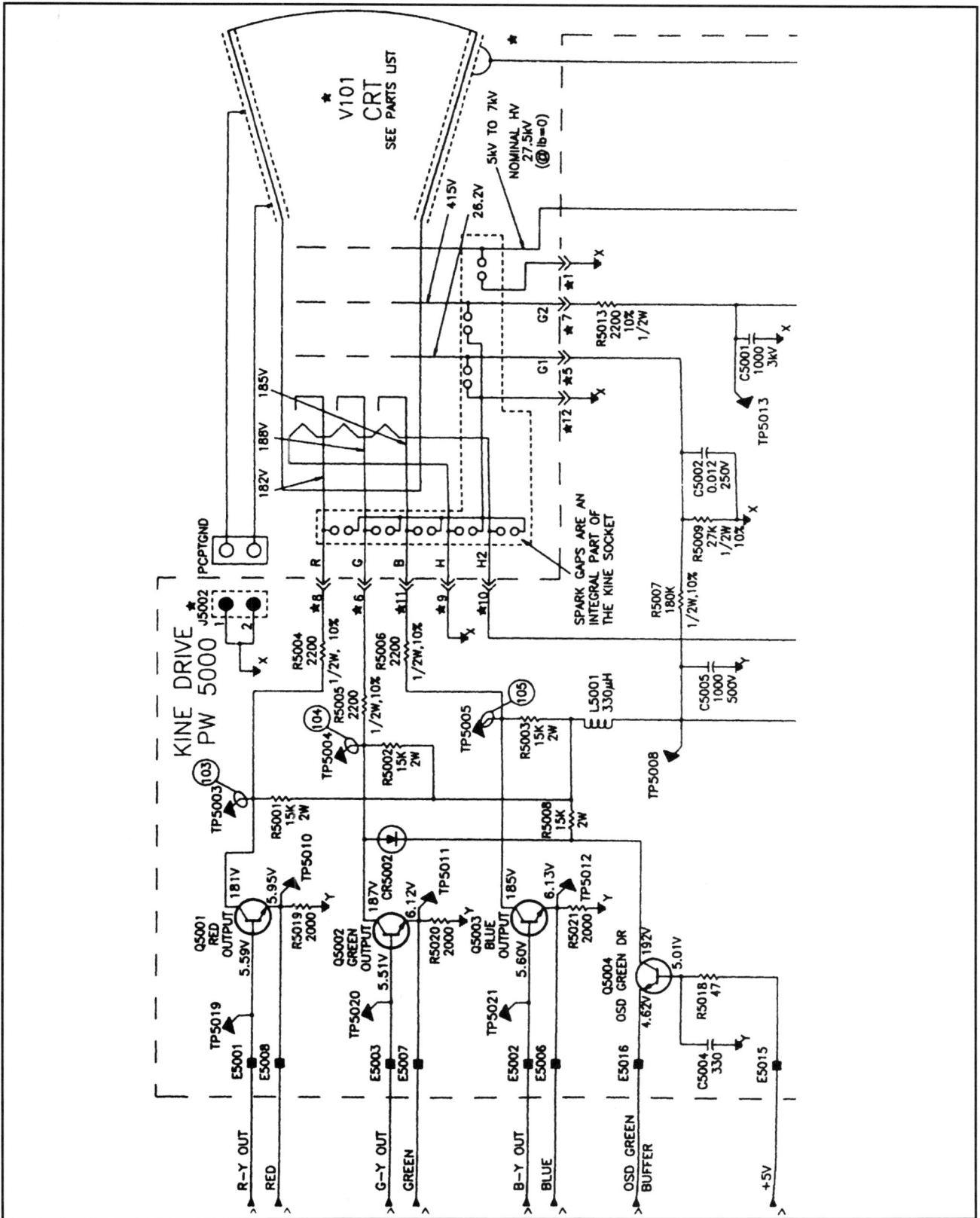

Figure 1-15. Kine driver.

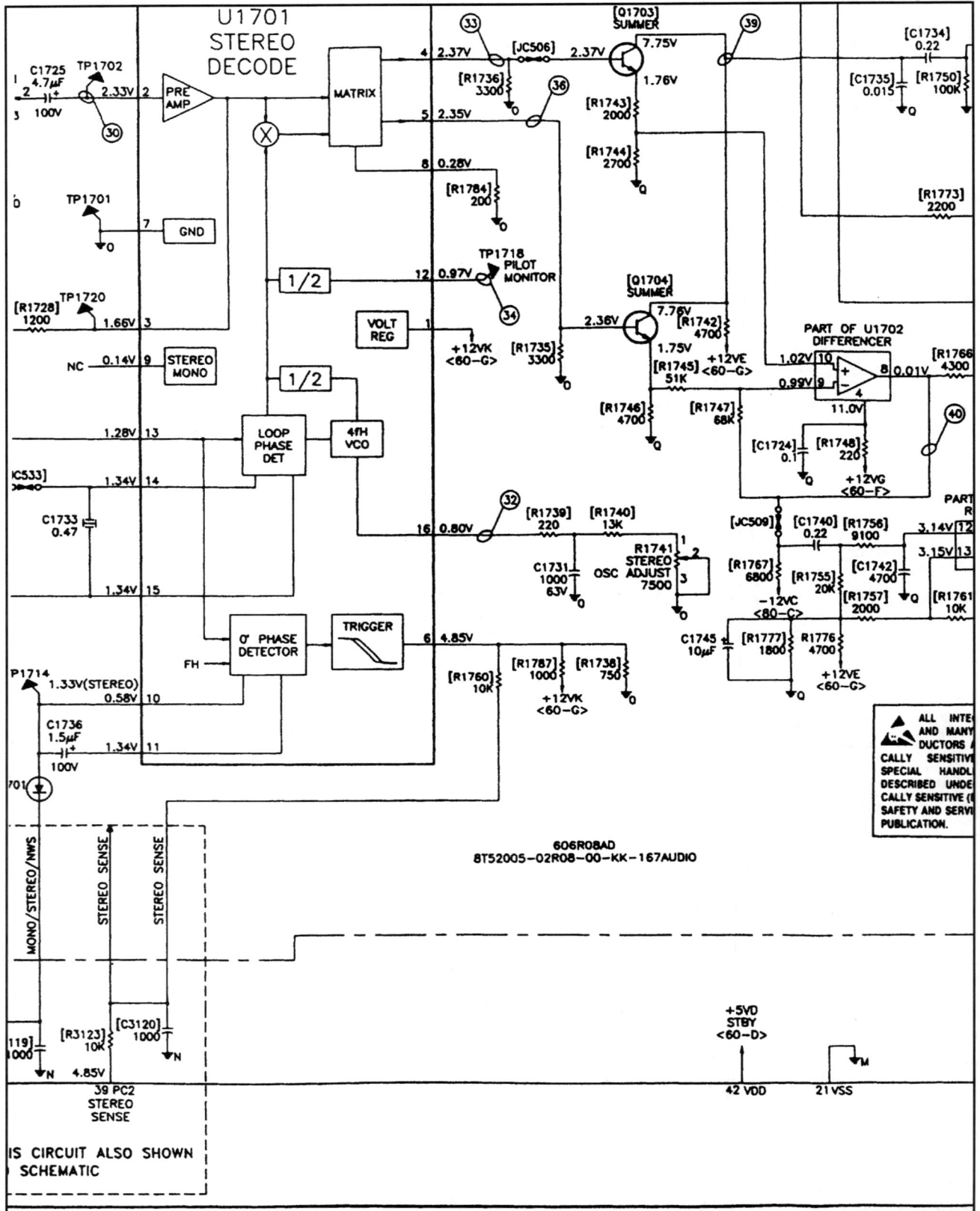

Figure 1-16. Audio decoder.

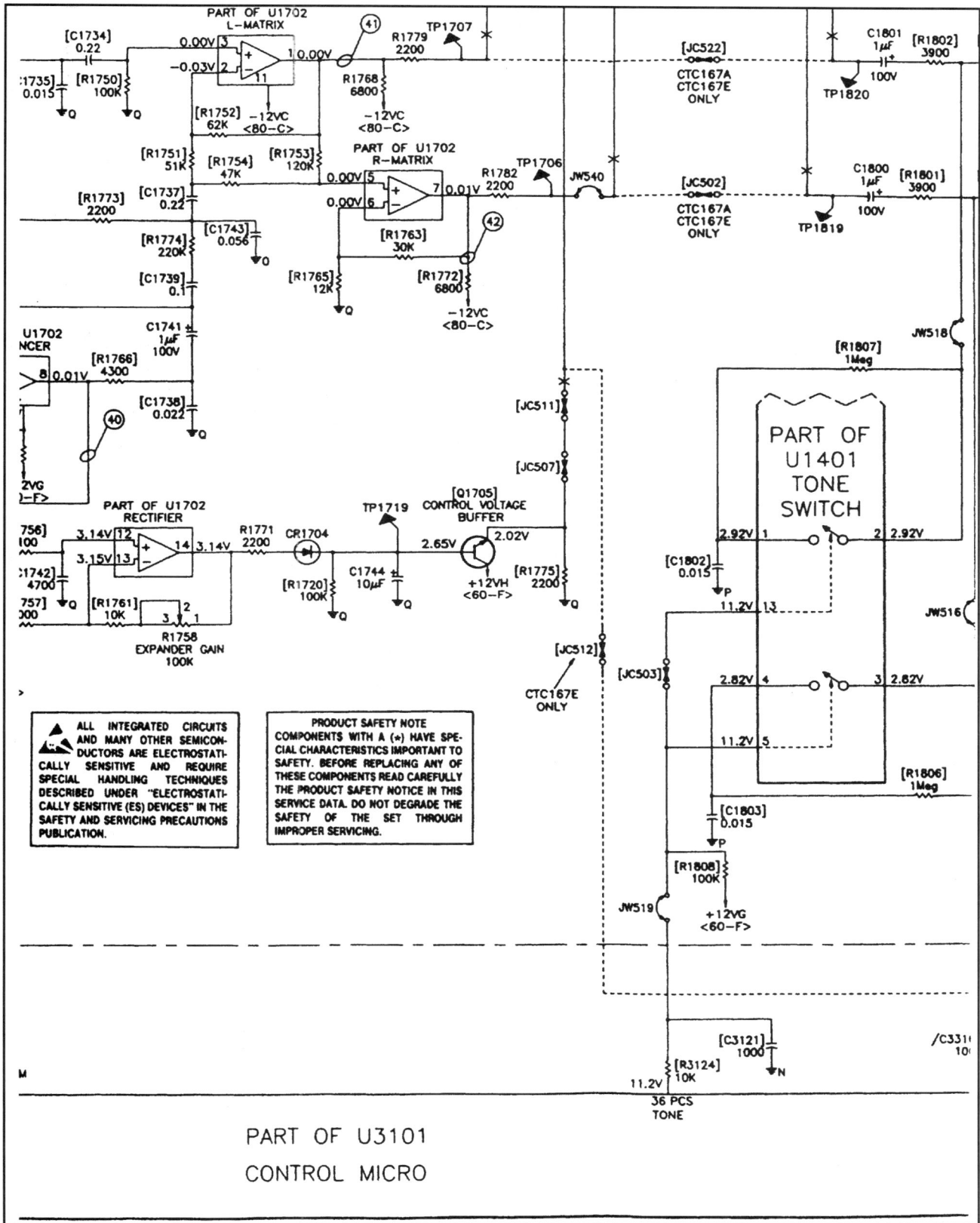

Figure 1-16. Audio decoder (continued).

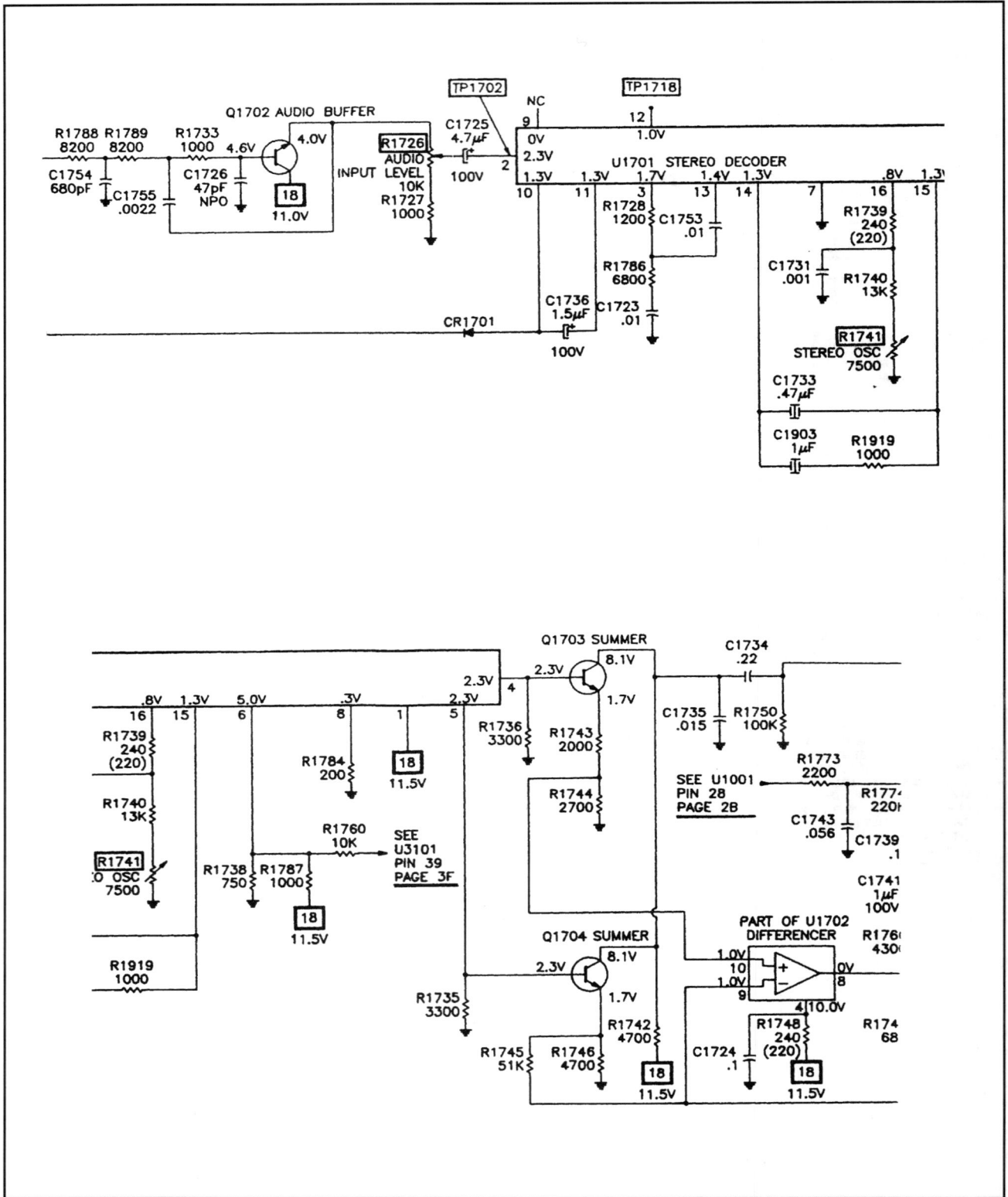

Figure 1-16A. Audio stereo decoder.

and reenters the chip at pin 13 for pilot detection. If it detects a stereo pilot signal, the chip pulls pin 6 low which signals the micro at pin 39 that stereo is present. If the pilot signal is not detected, U1701 holds pin 6 high. The customer can select mono even if stereo is present via a command from the remote control or front-panel controls. The mono command will cause the microprocessor to pull pin 10 of U1701 high, as if a stereo signal were not detected.

I won't go into the details of how the stereo decoder IC works. You can find that information easily enough. Let me simply say that the processed and encoded L and R signals go to the tone switch IC (U1401), which low-pass filters the signals when pin 36 of U3101(the micro) is high. The signals then proceed to the volume control IC (U1801). A signal from the microprocessor controls the peak-to-peak level of the signals before they reach the audio output amp. *Figure 1-17* is the schematic for the tone and volume control ICs. *Figure 1-18* gives you the details for the audio output amp. Note that the amp can be disabled by a signal from pin 34 of the microprocessor.

There is one other component in the audio processing stage I will mention—the audio switch U1402 (*Figure1-19*), which is used to select either auxiliary audio or the internal audio signal. The microprocessor controls input selection via logic A and logic B lines.

Troubleshooting

Troubleshooting is relatively straightforward. I suggest using a signal generator that produces a variable 1000 Hz sine wave because it's an easy signal to spot with a scope. If you are in doubt about what you are seeing, you can vary its amplitude and even its frequency. The audio problems I have repaired have been, for the most part, the result lightning and have involved the volume control IC and the audio output IC.

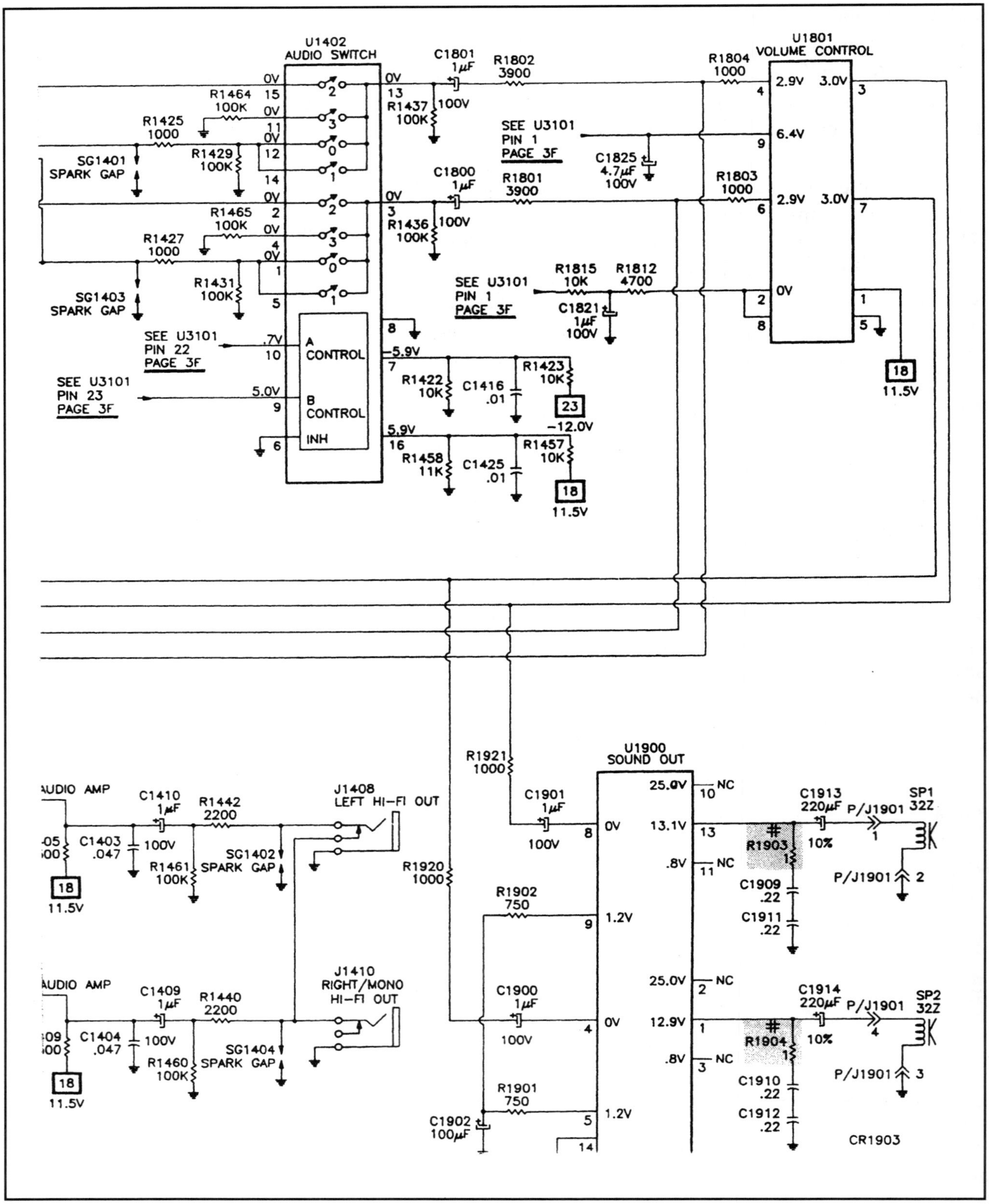

Figure 1-17. Tone and volume control.

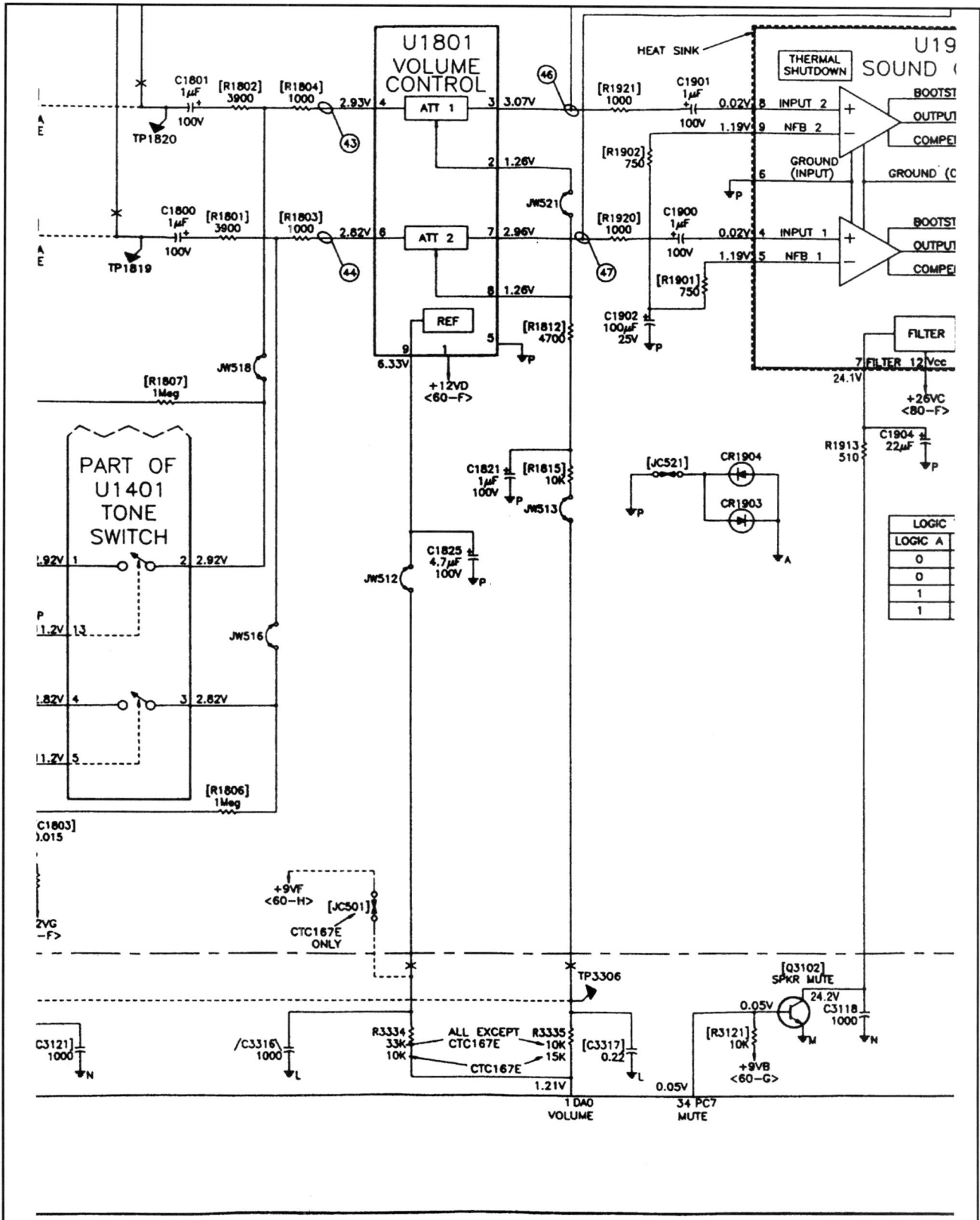

Figure 1-18. Audio output (continued next page).

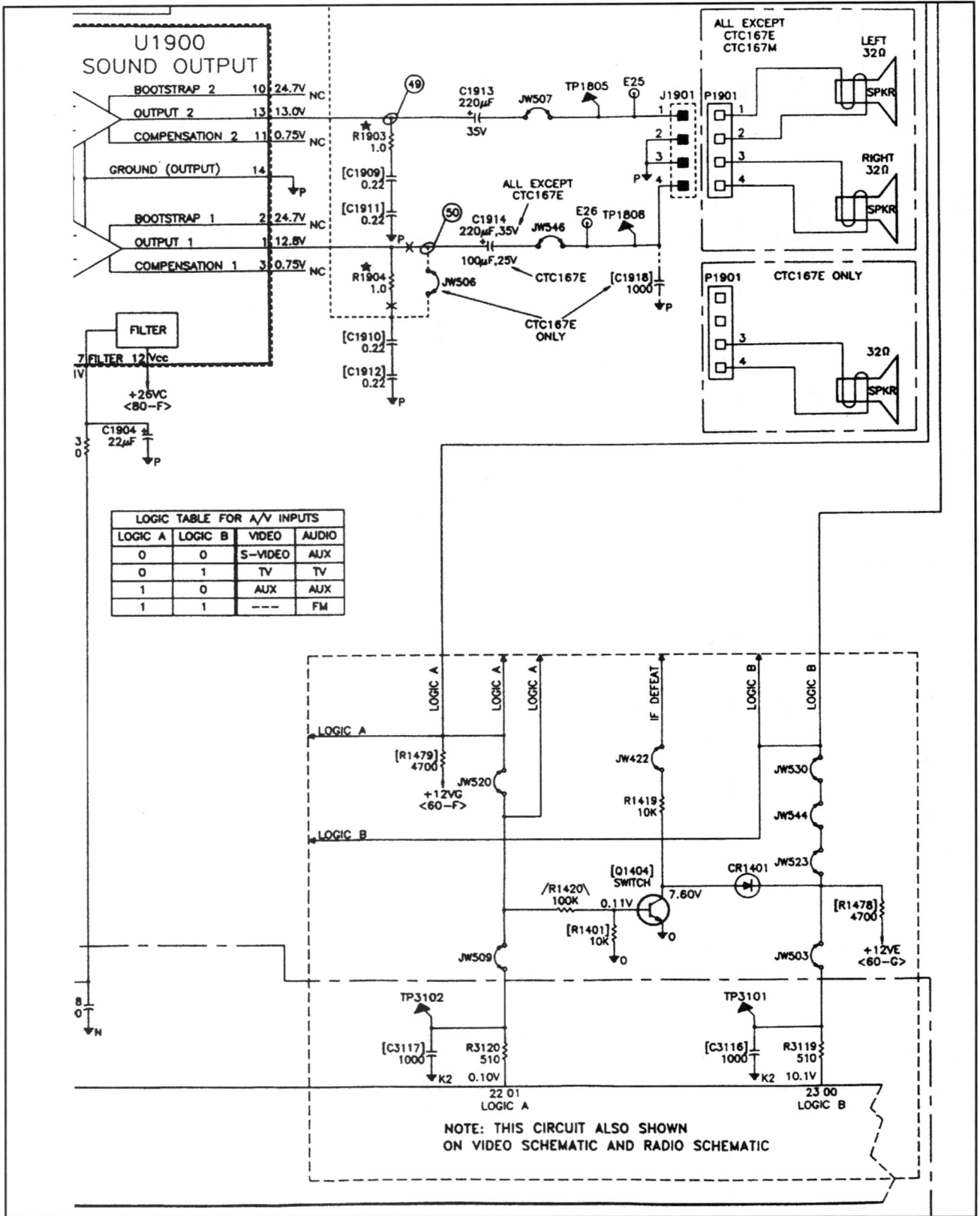

Figure 1-18. Audio output (continued).

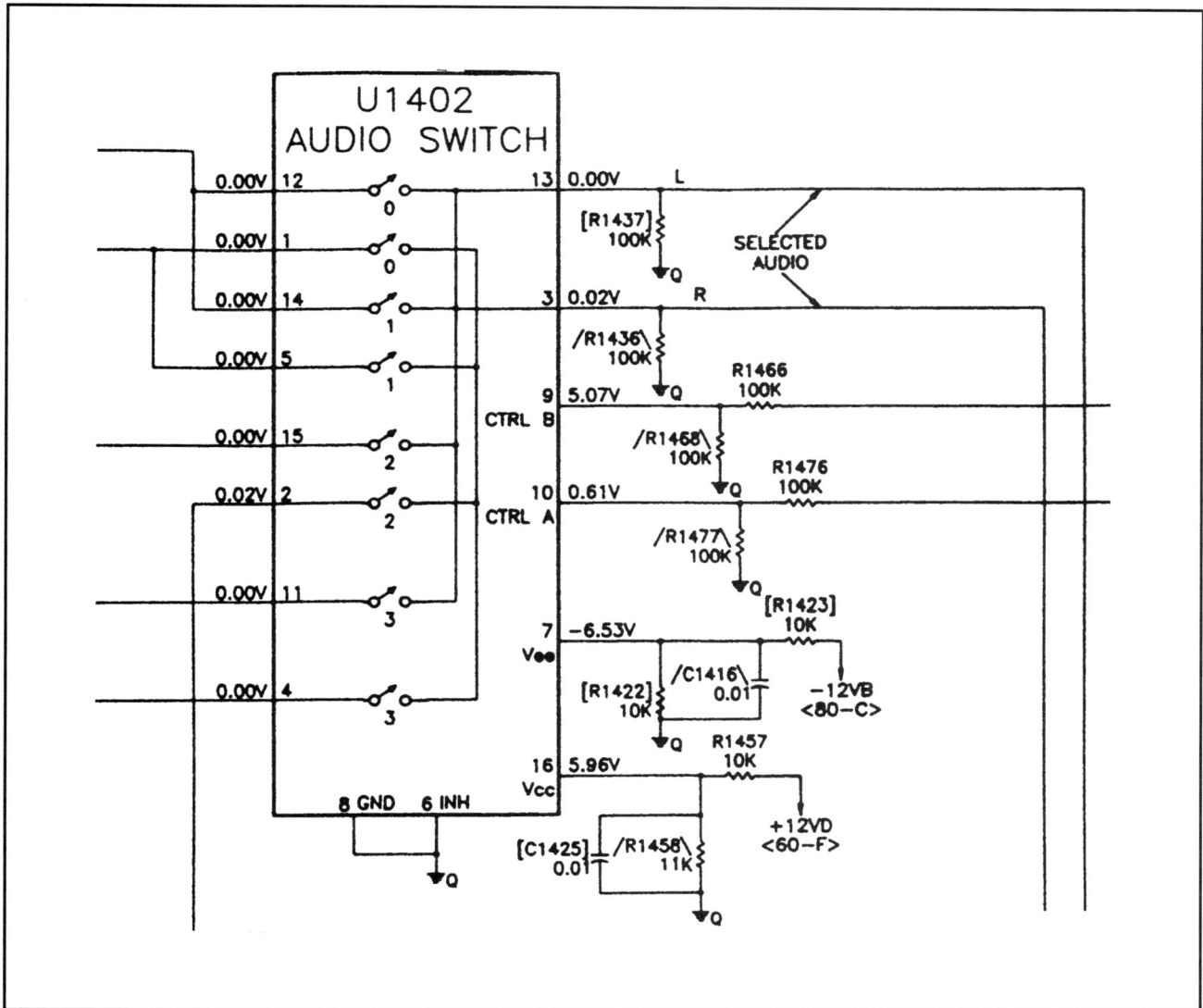

Figure 1-19. Audio switch.

CHAPTER 2

The CTC169 Chassis

Though I can't document it, I believe the CTC169 chassis has been one of the most widely used, if not *the* most widely used, chassis ever manufactured. I checked Thomson Consumer Electronics Service Data Index (1976-1997) to see how many listings it had for this venerable chassis and got tired of counting at about 250! In other words, it has been used in over 250 models. The number includes both direct view and projection televisions. RCA kept it in production until just recently when it was "retired" and replaced by the CTC195/197, but there are literally multitudes of them still out there in consumer hands—and they still give trouble.

Since there are so many variation of the CTC169, my discussion will necessarily be generic. Some of my comments will not apply to all CTC169 chassis, and there will be some features which I won't discuss. I hope you understand that it is due in part to limitations of time and space. I simply cannot discuss everything.

The literature is also the most extensive—and the most complicated—of all the makes and models of televisions I have in my library. I stacked the manuals one on top of the other just to see how high the stack would be, and it measured an even eight-and-one-half inches! Those are just the manuals. You will find additional information on the fiche that accompany the paper manuals. As if that were not enough, there are the technical training manuals and troubleshooting guides to be added to the list.

I will be more concerned with direct TV models than projection models, though information about the one will generally apply to the other. RCA divides the various models into "families," of which there are four. If you have access to the microfiche, you will find the family members listed in frames 1-A5 through 1-A8.

Before we get into the discussion of the CTC169, I will recommend two books you will find handy as you work with these televisions. The first is *The CTC 168/169 Technical Training Manual*; the second, *The CTC 168/169 Troubleshooting Guide*. Both are available from Thomson Consumer Electronics at the address I gave in the introduction. You can also order them through your local RCA distributor. I have no hesitation recommending these booklets because I consider them indispensable for anyone who wants to tackle these chassis. You can also order a new field service guide (FSG 13) which has schematics, voltages, waveforms, parts lists, etc. FSG 13 covers the direct-view models only. It appears to be an excellent buy.

There is a third booklet published by Philips Consumer Electronics that will help you understand the switch-mode power supply. It is entitled *Technical Training Manual: Switching Mode Power Supplies*. Philips reference number is ST1349. It deals with several power supplies as well as the CTC169 and is well worth reading. However, it basically rehashes the material presented in the second Thomson publication I mentioned. But there are times when a "rehash" lets us see things in a different light and facilitates understanding.

Component Numbering System

Before getting into the details of servicing the CTC169, I will give you the component numbering system but will not go into the kind of detail I did for the CTC167.

These numbers, however, will let you locate the area in which a part is found. I suggest you use a Sams PHOTOFACT® to help you locate specific parts. If you have ever worked on a CTC169, you know how filled with parts the main circuit board and its various daughter boards are. There have been times when I have spent more time looking for a part than I did repairing the problem.

1000 Series	Audio and AV In/Out
2000 Series	Signal Processing Circuit
3000 Series	System Control
4000 Series	Deflection and Power Supply Circuits
5000 Series	Kine Socket and Its Components
6000 Series	Digital Comb Filter
8000 Series	Picture-in-Picture Circuit

The Power Supply

The AC input, B+ regulator, and horizontal output circuits are referenced to hot ground. Never attempt service of a CTC169 without using an isolation transformer. Because it utilizes a hot ground/cold ground chassis you must remember to use the correct ground when you take voltage measurements. I use the negative lead of the main filter capacitor as the reference when I troubleshoot the B+ regulator and the horizontal output circuit simply because it is easy to locate. If you get on the wrong ground, you will not get accurate measurements.

The power supply is a switch-mode power supply, and it can be difficult to service under the best of circumstances. I believe the more you learn about switching power supplies the better you are equipped to service them. So, I will give you a modest bibliography to help in that direction. I have two articles in *Electronic Servicing And Technology*. The first is "Servicing Switching Power Supplies: A Five Step Approach," in the October 1997 issue. The second is "Switching Power Supplies: Representative Examples," in the November 1997 issue. Jurgen Ewert also has an article in the October 1995 issue, which he titles "Switched Mode Power Supplies." You will find a nice little article in Sencore's *Tech Tips* # 158, "Troubleshooting Switching Power Supplies."

There are also two articles in *Sencore News*, one by Glen Kropuenske, "Understanding and Trouble-shooting Switched Mode Power Supplies" in issue #153 (April/May 1991), and the other by Rick Meyer, "Switched Mode Power Supplies—How They Work And How To Troubleshoot Them," in issue # 151 (November/December 1990).

I checked the books published by Prompt and found one that you might find interesting, especially if you are into building. The book is written by David Lines and is entitled *Power Supplies: Projects for the Hobbyist and Technician*. When the last Sams PHOTOFACT subscription arrived, I found in it an excerpt from a new book just published by Prompt. The excerpt was so good, I ordered the book. It is called *Power Supply Troubleshooting and Repair*, written by Lannie Logan. Although the book deals with Philips' products, it is helpful for teaching how power supplies work.

I will use as a reference the second edition of *RCA/GE Color Television Service Data* for the CTC169 (direct view), dated 1990. The choice of a PHOTOFACT is purely arbitrary because there are so many models for the CTC169. I chose PHOTOFACT number 3330 for model F350STFM1/JX1 because I had just finished working on one and had the literature handy.

The AC Input

The 150-volt raw B+ is developed from 120-volt AC which is routed to the bridge rectifier circuit (CR4001-CR4004) by F4001 (a 5-amp fuse) and R4001 (a 15-watt resistor, the value of which will vary from 1.8 to 2.7 ohms). The rectified AC is filtered by C4007 (820 µF, 200 volt) and then applied to the collector of the chopper transistor (Q4101) through the primary of the chopper transformer (T4102). Look at *Figure 2-1* for the specifics.

An Overview of the Power Supply

The power supply begins to run as soon as the unit is plugged into an AC outlet. Start-up voltage for U4101, the regulator IC, is routed to pin 16 through R4003, a 33k resistor. When it starts, the IC provides a series of pulses to the base of Q4101 which turns it on, energizing the primary of T4102. The energy is coupled to the various secondary windings during turn-on and turn-off of Q4101. The power supply is now up and running and capable of providing all the voltages necessary to operate the TV. For example, run voltage for the regulator IC comes from pins 12 and 13 of the chopper transformer via CR4101 and R4149. Standby 5 volts for the system microprocessor and the remote receiver comes from the 12-volt standby source through the 5-volt regulator Q4601. And approximately 140 volts is applied to the collector of the horizontal output transistor.

The method of regulation depends on the state of the system. Power demand in standby is minimal; therefore, the pulses to turn on the chopper transistor come in short bursts at about a 20 kHz rate. CR4106 provides DC feedback through R4113 and a resistor network to an error amplifier inside U4101 at pin 6. The feedback voltage is compared to a fixed reference. The difference between the feedback voltage and the reference voltage enables the logic to determine if the output voltage is high or low.

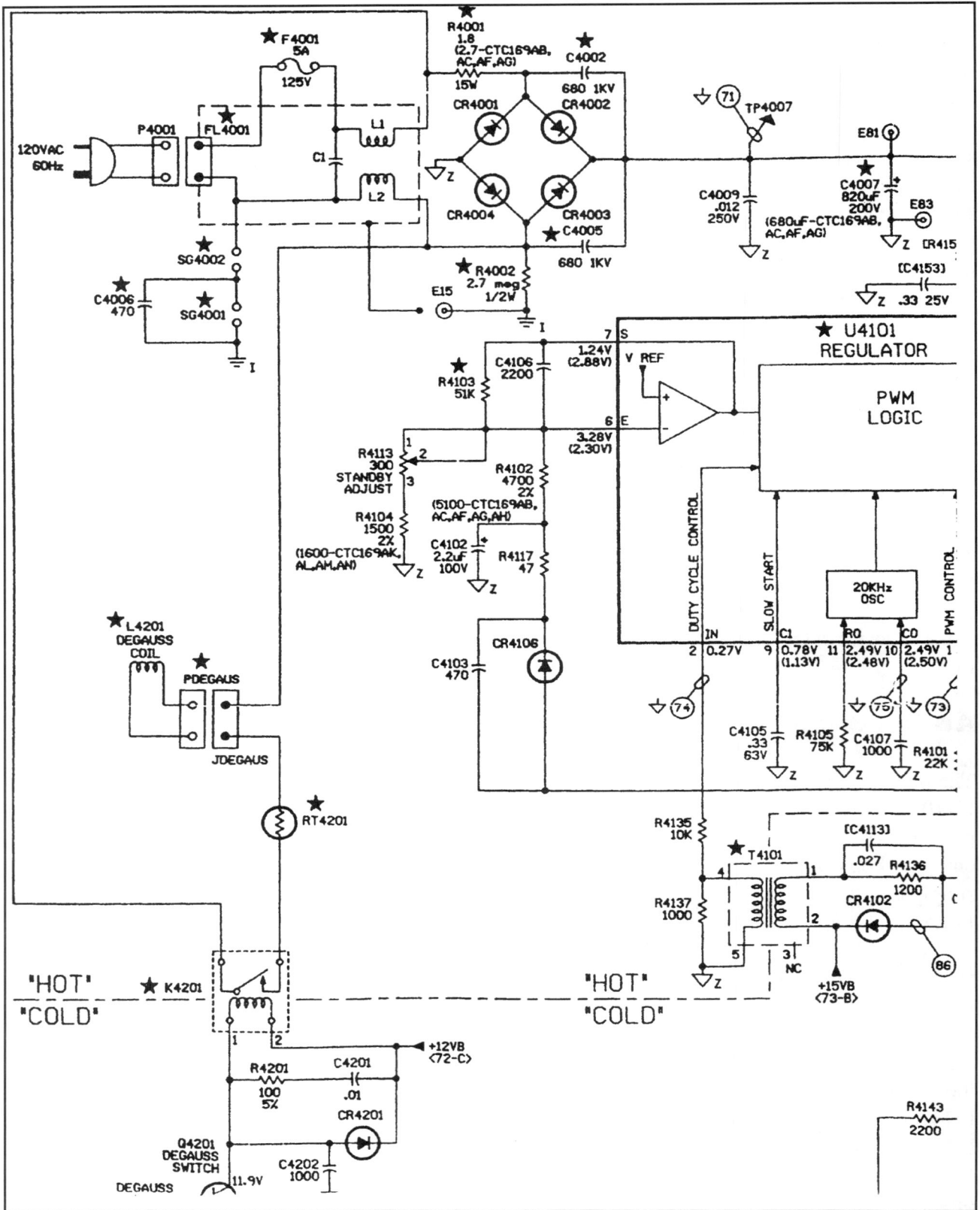

Figure 2-1. AC input.

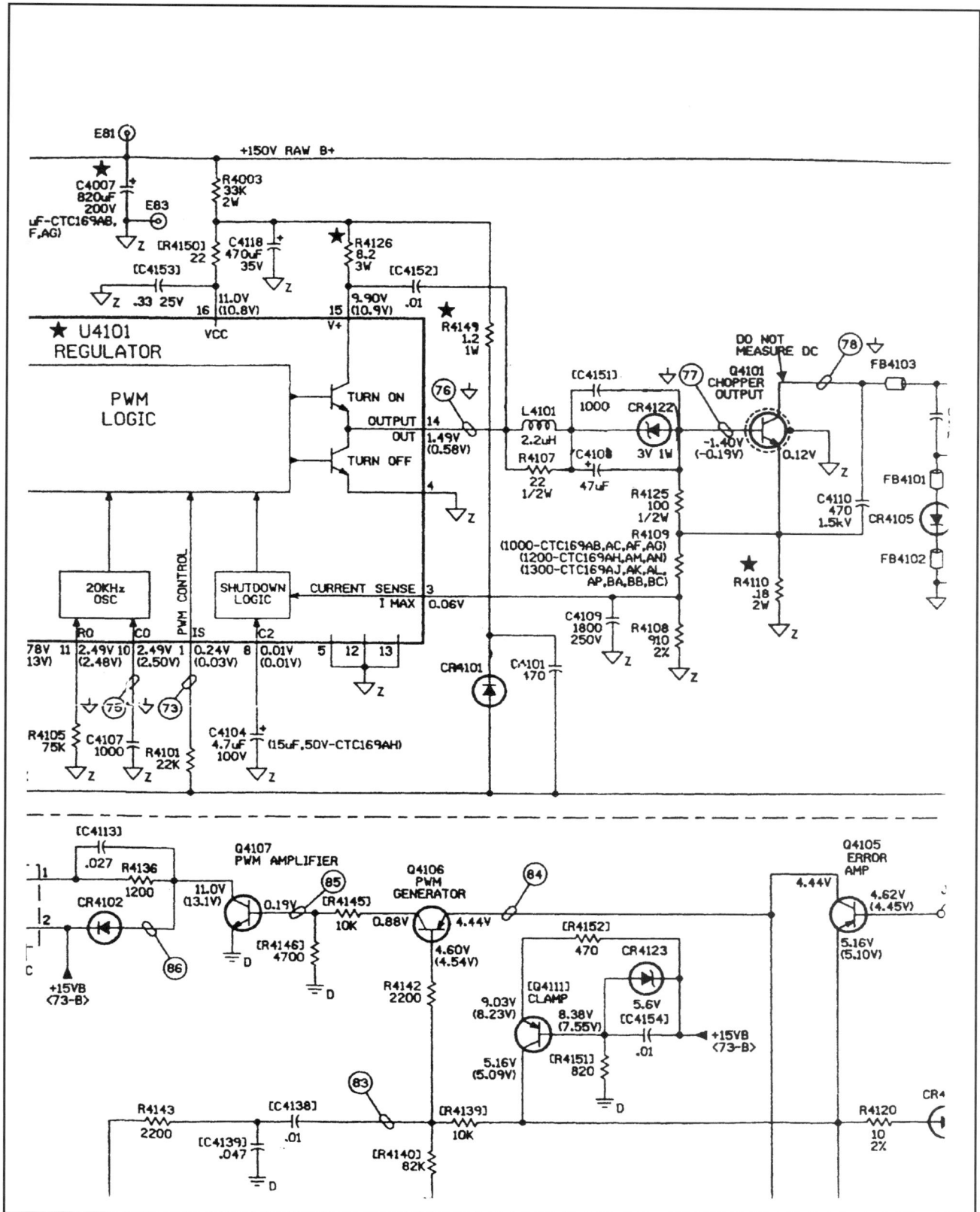

Figure 2-1. AC input (continued).

Power demand increases significantly when the TV receives and on command and begins to produce picture and sound. The internal reference voltage for U4101 switches from 2.25 volts to 2.5 volts. The full-power feedback signal is now provided by pulse-width modulation rather than a simple resistor network, as it did in standby mode.

Full power regulation relies on two signals. The first signal uses a resistor network (R4114, R4115, R4116, and R4118) to drop the regulated B+ and apply it to the base of Q4105, the error amplifier. The collector of Q4105 feeds the emitter of Q4106, the pulse width generator. The second signal is a horizontal pulse from the IFT which is applied to the base of Q4106 through R4144, JW306, R4143, C4138, and R4142. The output of Q4106, a combination of the two signals, drives Q4107 (pulse-width amp) which in turns drives T4101, the purpose of which is to isolate the two grounds. The signal from this transformer goes to pin 2 of U4101, where it is used for duty cycle control and regulation.

Two final comments about the regulator circuit. Q4109 is labeled "switch." It really doesn't have anything to do with the "on/off" control. The engineers at RCA put it there to ensure no pulses get to Q4106 when the TV is off. Q4110, the beam sense transistor, senses the state of the high voltage via the high-voltage return line. If beam current becomes excessive, Q4110 will conduct, turning off Q4108, which reduces the reference voltage supplied to the error amp transistor.

Shutdown Circuits

This is a good place to discuss the shutdown circuits. You will be pleased to know there are <u>five</u> of them. Two are inside the power supply, two in system control, and one in the horizontal circuit (x-ray protection).

Overvoltage and overcurrent are in the power supply. Overvoltage shuts down the power supply if the standby voltage rises to an unsafe level. Remember, we are dealing with dual regulation! Overcurrent shuts it down if the chassis places excessive current demands on the power supply. I will discuss these two problems in a few pages.

Reset and fault detect are in system control.

The microprocessor is reset at pin 1 when AC power is applied. Reset is accomplished by the 15-volt supply via R3114, R3115, and zener diode CR3101. Remember, the 5-volt standby is derived from this supply. The value of the zener diode (5.6 volts) allows the 5-volt standby source to be established before pin 1 becomes high. During AC dropout, pin 1 will also go low before the 5-volt standby source begins to drop. Now look at *Figure 2-2*. CR3102 and R3116 are connected to RESET and the oscillator. Their purpose is to stop the oscillator when the reset line voltage drops. This action conserves the charge on C4612. The discharge time of C4612 is the memory retention time, which is specified as a minimum of 10 seconds. The memory retention is typically several minutes.

If the reset circuitry detects a drop in the 5-volt source, it will pull pin 1 of U3101 low, placing the TV in the low-power mode. If reset shutdown activates, the TV might momentarily shut off and then turn itself back on. When it comes back on, it will go to the channel, volume level, and picture level it was

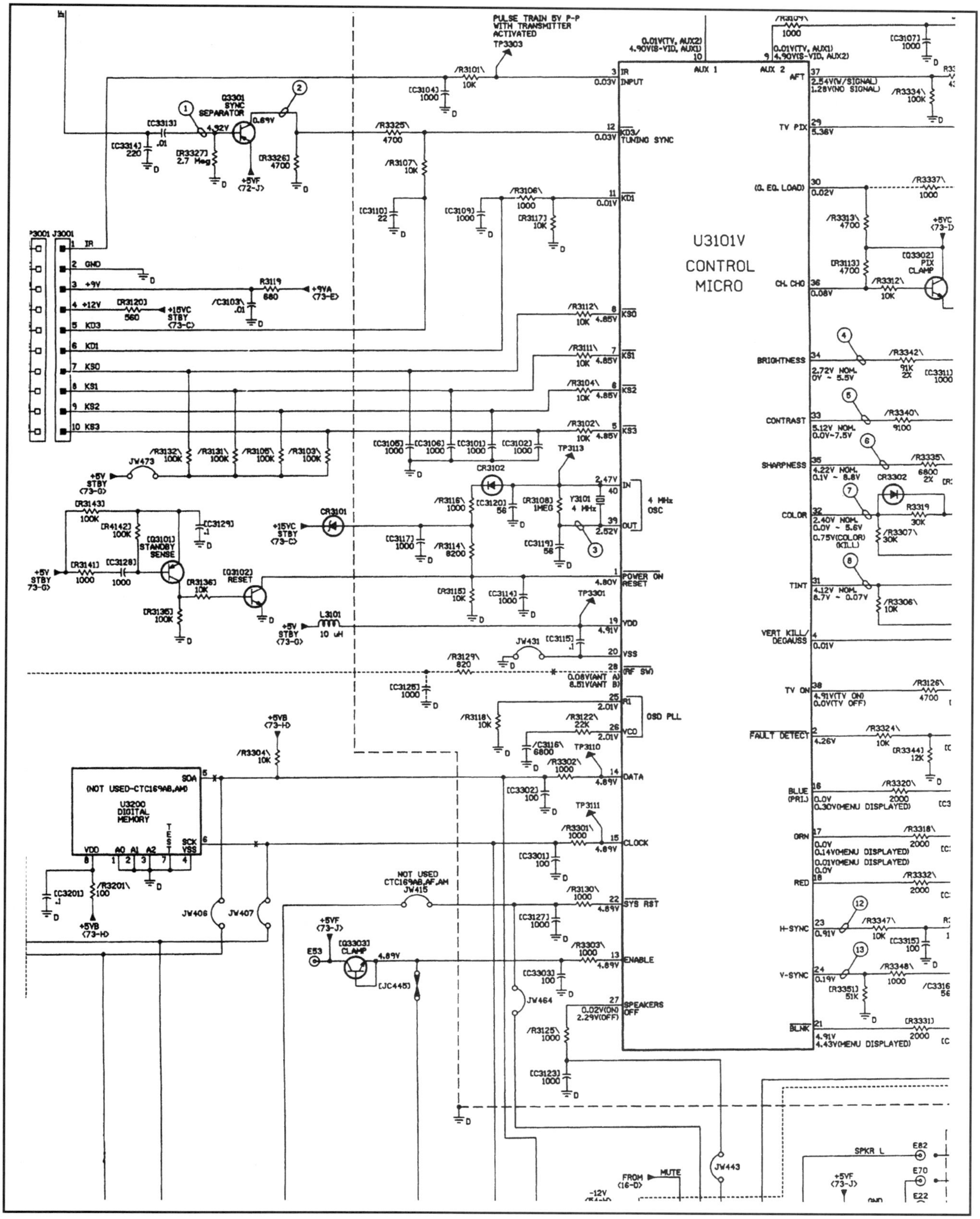

Figure 2-2. Shutdown circuits.

set to before the set went off. What might cause this kind of shutdown? Usually, CRT arcs or noise on the AC power line, like what might happen during a lightning storm. If reset is pulled low and remains low long enough for the microprocessor to lose data in memory, the user will have to turn the TV back on and reset the controls. If I understand it, reset shutdown can occur when the TV is off but still plugged into an AC outlet. It can occur if there is a spike or a significant drop in the AC line voltage.

Fault detect is active when the TV is on and operating. If it is triggered, the unit will power off and in about two seconds attempt to restart. The cycle will repeat itself until the fault is removed or the TV is turned off. Fault detect monitors the 9 volt run supply via pin 2 of the microprocessor. If the voltage rises above 2.5 volts, the microprocessor will assume the TV is operating properly. If the voltage drops below this level, the microprocessor will assume the TV has a problem and respond by powering down the set. What can cause the 9 volts to fail? A failure in horizontal drive which results in no 26 volts or a load on the 26 volt line, like a shorted audio output IC are two examples.

If you suspect fault detect shutdown, monitor pin 2 of the microprocessor while you turn the TV on. If the voltage doesn't rise to about 2.5 volts, you will know the TV is probably in fault detect shutdown. The very first thing you should do is disconnect the audio output IC. If the TV then comes on, you will know where your problem is!

The fifth shutdown circuit is x-ray protection. Refer to *Figure 2-3*. This circuit takes a pulse from pin 5 of the IFT (T4401), rectifies it by CR4901, applies it to the cathode of zener diode CR4902 and a precision resistor network (R4903 and R4904) at pin 22 of U1001. If the voltage exceeds 10 volts, the zener will conduct and the resulting voltage will activate the x-ray protection inside U1001, which will respond by turning off horizontal drive.

If you are presented with an intermittent shutdown problem, you might try the following. First, position the TV where you can monitor it. Then turn it on and select an active channel and mute the sound. "Mute" will appear on the screen. If the on-screen mute indicator disappears and the audio remains muted after the TV shuts down, the microprocessor has experienced a momentary shutdown, either through the fault or reset circuitry, but has not been off long enough to lose its memory. If the TV shuts off and can be turned on again by pressing the power button, the microprocessor has again experienced shutdown either by fault or reset, but has been off long enough to lose its memory.

If the TV shuts off and you must remove AC power before you can turn it back on, the power supply has shut the set down, and you will know to look at the power supply shutdown circuits as the likely source of the problem. I learned about this technique by calling RCA's tech help line. It can be very helpful because it permits you to isolate the type of shutdown you are dealing with. Since my conversation with tech line, I have learned it is discussed in some CTC169 technical manuals.

Troubleshooting the Power Supply

The power supply will likely be the most troublesome circuit to service. The question then is, "How can I effectively troubleshoot it?"

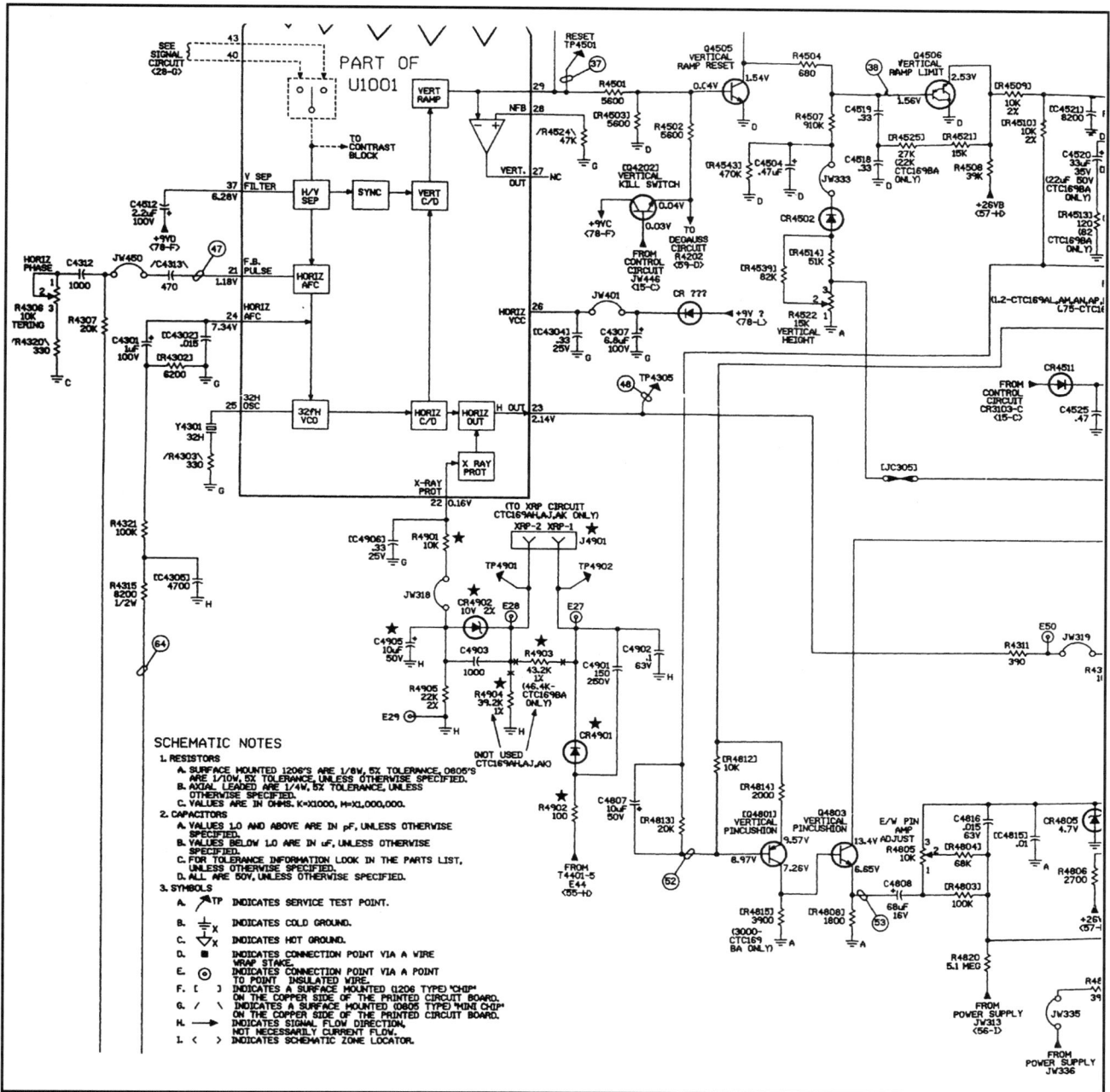

Figure 2-3. X-ray protection.

The most common power supply problem will be the dead set. Where do you go from there? Let's get the answer to two questions before we begin to probe the circuitry: (1) Is the power supply completely dead, or; (2) does it work fine in standby but fail to switch to full power when it receives an ON command? You can tell by observing the LED on the front panel. Does it fail to come on, or does it come on and go off one or more times? You might even hear the high-voltage pulse on and then off.

Take the back off if you haven't already done so, and prepare to make a quick check or two. I like to check at least two voltages to confirm power supply operation. So, let's check for 15 volts standby at the cathode of CR4118 and 140 volts at the cathode of CR4116 or the collector of the horizontal output transistor. The presence or absence of two or more voltages tells you a great deal about the state of the TV. If you confirm the presence of two voltages, the problem will more than likely lie outside the power supply. If you check just one voltage and find it missing, you might suspect the power supply when in fact the problem is just one missing voltage! Make it, then, a rule of thumb to check for the presence of at least two voltages before you assume the power supply is not up and running.

For the sake of discussion, I'm going to assume the power supply is completely dead. My reference will be *Figure 2-4*, which is a partial schematic for the power supply.

The first check is for raw B+.

If it is not present, check the usual—the AC fuse (F4001), R4001, and the diodes in the bridge rectifier. If these are good, check the chopper transistor. If it is bad, do yourself a favor and check R4110 in its emitter circuit. R4110 is a 0.18 ohm, 2-watt resistor which may either have opened or been stressed to the point where it has increased in value. I recently serviced a 52-inch projection set that would come on and play for maybe a minute or fifteen minutes before it shut down. The only problem I found was an increase in the value of R4110. It hadn't increased much, but the increase caused the overcurrent sensing circuit to shut the TV down. So, don't overlook it. If the chopper transistor is bad, chances are U4101 is also bad. My luck usually runs about 50/50, which means I replace both about half the time.

Here's an interesting scenario. You replace Q4101 and U4101 and apply AC. The power supply starts, but you hear a squeal. However, the TV set comes up to full power. Audio is okay; picture is fine. You check the chopper transistor heat sink and discover it is getting hotter by the second! Turn the AC off to let the heat sink cool. Apply AC and check the base of Q4101 (either in standby or full power) for a slight negative voltage. It should be on the order of -1.4 volts. If the voltage is positive, check CR4122 and C4108. I have not yet found a defective diode, but I have had to replace several of the electrolytic capacitors (usually about 47 µF).

Suppose you have raw B+, but the power supply is not producing any secondary voltages. The problem could be a defective regulator IC or one or more of the components tied to it or overcurrent or overvoltage shutdown. A few voltage and resistance measurements will usually tell you which of these circuits to troubleshoot.

First, check for start-up voltage at pin 16 of U4101. If the voltage is missing, check R4003 and R4150 for an open condition and C4118 and C4153 for a short. I have never discovered a shorted C4153, but I have seen problems with C4118. If the voltage is more than 0 and less than 2.5 volts, suspect a shorted IC or a component tied to it. With AC removed, check the resistance between pin 16 and COLD ground. The normal reading will be in excess of 40,000 ohms. Don't overlook CR4101, which fails more often than any component. Of course, there is also the possibility that the IC is shorted.

You can check the IC in-circuit by removing AC power and using an external DC source to supply about 13 volts to pin 16 while you monitor pin 14 for an 8-volt peak-to-peak waveform. If the pulses are there, check for base drive at the chopper transistor. Drive from the chip but not at the base obviously means an open circuit path. If output pulses at pin 14 are missing, check for the 20 kHz clock signal at pins 10 and 11. If the clock signal is missing, check C4107 and R4105. If these are good, change U4101. If you are like me, you will "cut to the chase" and change the IC because I have never found C4107 or R4105 bad, but that doesn't mean they don't go bad!

A word of caution. Be careful as you poke around U4101. If you accidentally short pins 15 and 16 together you will destroy both the chip and chopper transistor. Also remember to use hot ground as the reference for voltage checks.

If the voltage at pin 16 fluctuates between 7 and 11 volts, think in terms of overcurrent shutdown, a condition in which some component is placing heavy current demands on the power supply. There are some quick checks to help you to locate the source of the current drain. Using COLD ground as a reference, check the resistance of the following with a good DMM:

(1) the cathode of CR4116 (140-volt supply) for a reading of 50k or better;

(2) the cathode of CR4118 (+15-volt supply) for 500 ohms or greater;

(3) the cathode of CR4117 (+7-volt supply) for 4k or greater;

(4) the cathode of CR4119 (+17-volt supply) for 4k or greater;

(5) the cathode of CR4120 (-17-volt supply) for 4k or greater.

A low resistance reading at any of these diodes will point you to the problem area. For example, I have found more shorted horizontal output transistors than any thing else, so much so that it is the first reading I take. I have taken this information from *The CTC168/69 Troubleshooting Guide* (page 23), and I have found it to be generally accurate.

If the voltage is greater than 15 volts, suspect overvoltage shutdown. Overvoltage in this context means the cold side B+ regulator is regulating at too high a voltage. It can be caused by an open in the standby adjust circuit input to the error amplifier or by a spike on the AC line due to insufficient filtering. It can also be the result of a leaky C4118. Check the following to troubleshoot the problem: CR4106, C4102, R4117, R4102, R4113, and R4114. If the input voltage is still too high, check R4149 for an open or an increase in value. Remember, if the problem is overvoltage shutdown, the TV will go into shutdown and stay off until AC is removed and reapplied.

Sometimes the problem is really hard to locate. I remember one CTC169 that would play beautifully for, say, and hour and then shut down. It could be restarted by issuing an ON command and would play for a while before it shut down again. Every system I checked seemed to be without fault. Out of exasperation I checked the x-ray protection input to U1001 (pin 22). Sure enough, the voltage would ramp high just prior to shutdown. But the regulated B+ was not ramping high, which seemed to rule out the power supply. I talked to the techs at RCA, who suggested I change C4401 (.015 µF, 1.6 kV) in the collector circuit of the horizontal output transistor. It, the tech said, had a history of changing value, permitting the high voltage to go too high. I changed C4401, and the television operated like a champ from that time on. I have seen the problem several times since then.

Figure 2-4. Power supply.

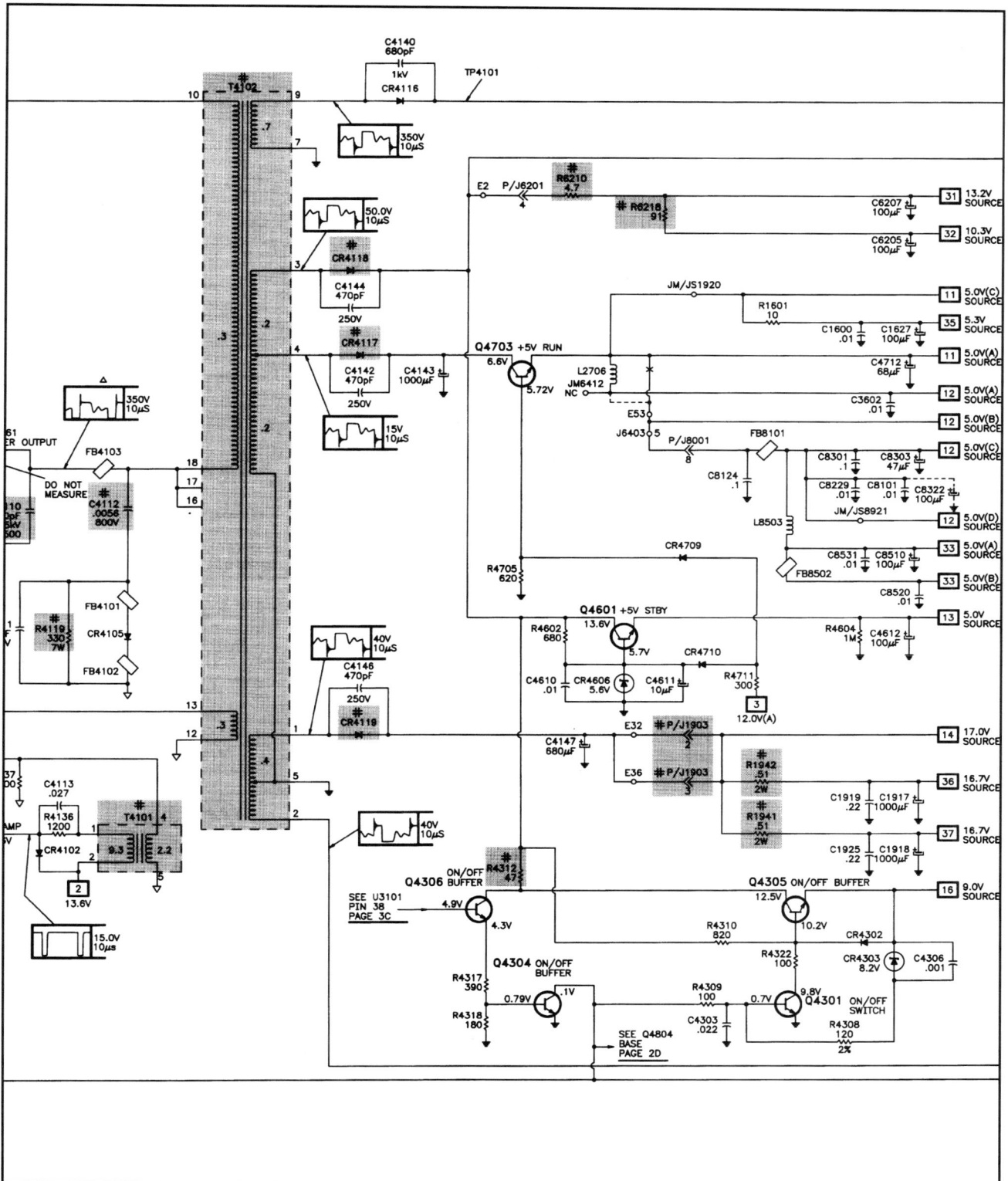

Figure 2-4. Power supply (continued).

You might be interested in the voltage reading for some of the other pins of U4101. For example, the peak-to-peak voltage at pin 3 should be less than 0.6 volts during normal operation. If it is between 0.6 and 0.9 volts, the regulator will cycle on and off. If it is greater than 0.9 volts, the IC will latch off until AC power is removed. The voltage at pin 8 should be nearly 0. If it greater than 2.5 volts, the supply will be in hard overcurrent shutdown.

I'm going complicate matters a little by noting that the no-start problem might lie outside the power supply. For example, is the horizontal on-off system working? You can quickly confirm it by grounding the collector of Q4304 and applying AC. If the set comes on, you have a system control problem. If it doesn't, you have a horizontal deflection problem. I know my experience will not have universal application, but I have seen lots of these sets and have seen just one—repeat, *one*—that had horizontal drive problems. RCA outlines a rather complicated process to check the horizontal circuits, which I will discuss later on.

Now, let me give you some part numbers that should be helpful. I urge you to use OEM parts whenever you can. Some generic parts will work; some won't. I have learned "the hard way" to go with an original part in these circuits.

Q4101	→	200165	U4101	→	200419
CR4101	→	164588	R4110	→	200183
C4108	→	193043			

This is by no means a complete parts list. It does, however, contain the parts that in my experience fail more often than others. I hope you find it helpful.

Troubleshooting the B+ Regulator

B+ regulation problems means the circuit is either producing too much voltage or too little. If your experience parallels mine, you will have one of those TVs that comes on but has a significantly reduced raster. In other words, the B+ regulator will not be producing enough voltage for the horizontal circuits to work effectively. *Figure 2-5* is a diagram of the regulation circuit, which also has a sample of the DC voltages and wave forms you generally should find.

If you suspect a problem with B+ regulation, first confirm proper operation of the standby circuit. It's a quick, easy step, which you should not overlook. You may "think" you have a regulation problem when in fact you do not!

Next, confirm the presence of the pulse-width modulation signal at pin 2 of U4101. If the signal is present, the regulator circuit is probably operating correctly and the problem is in some other area of the circuit. If the signal is missing, check for the 26-volt peak-to-peak horizontal ramp at the base of Q4106. If the ramp is missing, suspect a problem in the horizontal deflection circuit. If the ramp is present, check the DC voltage at the emitter of Q4106. If this voltage is low, check Q4109 or a malfunction in the on/off circuitry. Don't overlook the possibility of the reference voltage being defective. If the

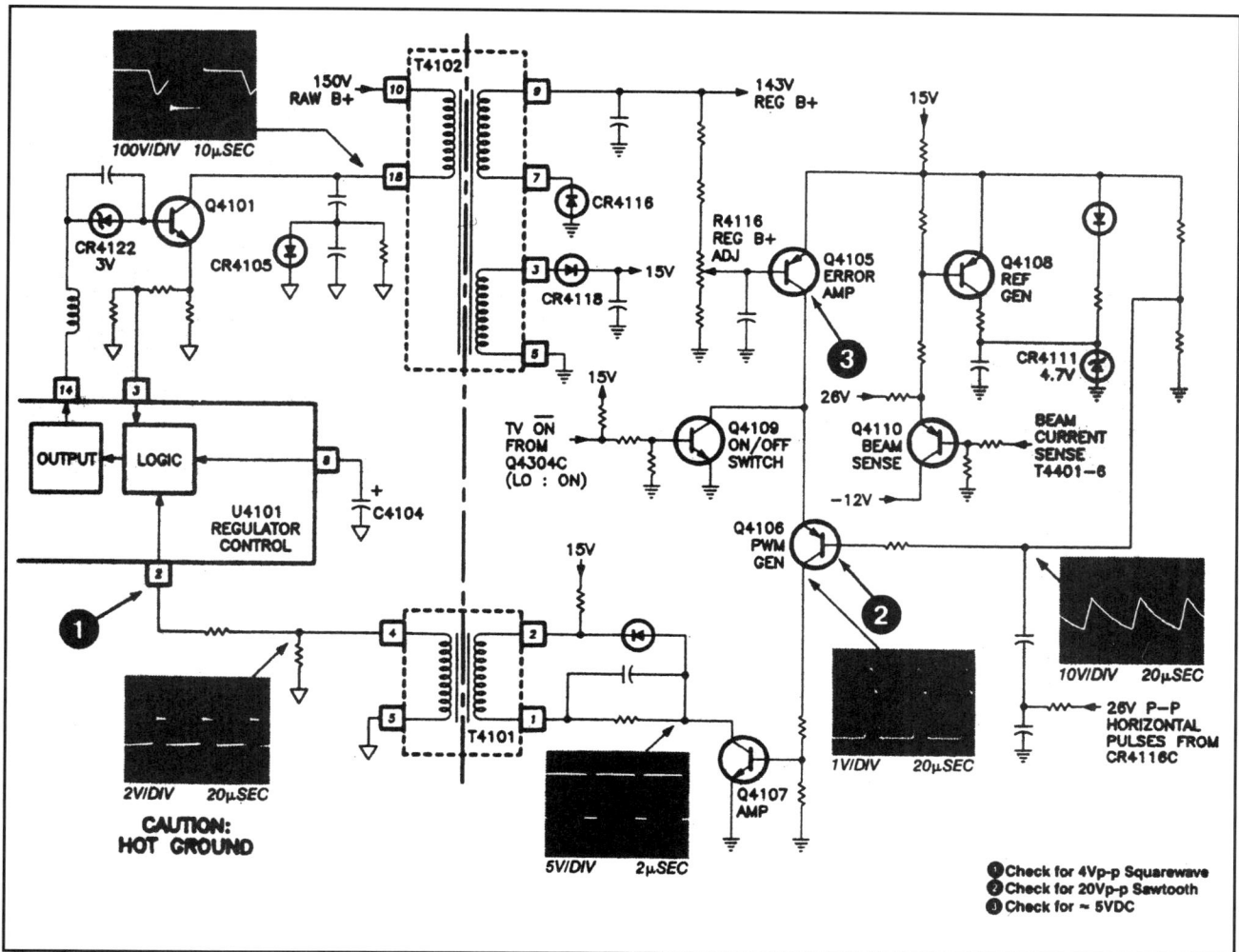

Figure 2-5. B+ regulator.

voltage is on the order of 5 volts, think in terms of a defective Q4106 or Q4107 or a malfunction in the primary circuit of T4101.

Scan-Derived Voltages

Figure 2-6 depicts the secondary of the flyback and the scan-derived voltages. Pin 7 (via CR4705) is the -12-volt source; pin 8, the +26-volt source for vertical deflection. Pin 10 is the source for the 200 volts for the video output transistors. These sources are pretty much straightforward, except for the 200-volt source. For example, I just serviced a 35-inch TV that had washed out video and heavy retrace lines in the picture. The 200-volt source was very low. The pulse from pin 10 of T4401 was the correct value. R4720 was good; so was CR4701. The problem was an open C4716! I guess I'm just accustomed to the way RCA does things, or I would have missed the open capacitor all together.

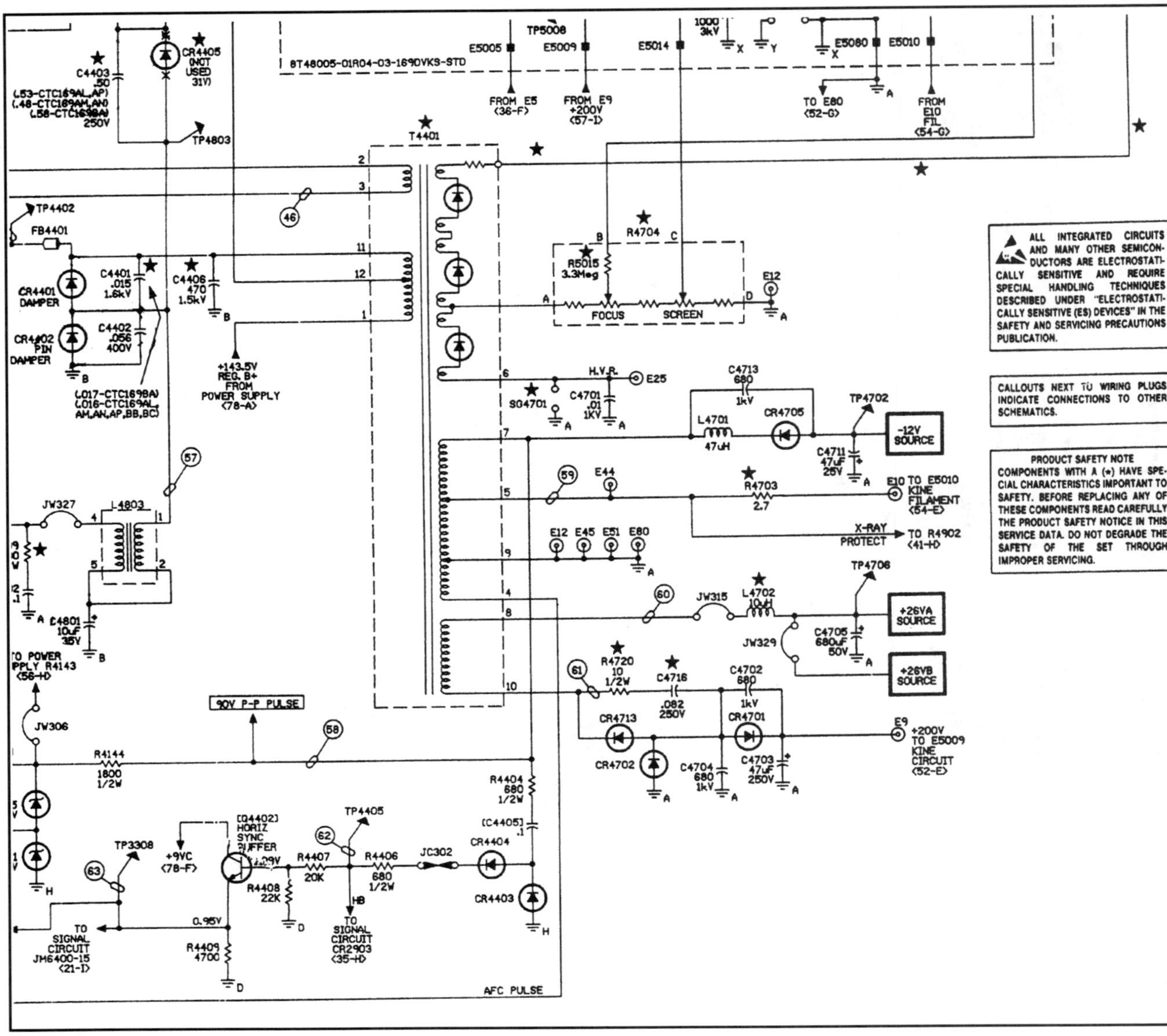

Figure 2-6. Scan-derived sources.

System Control

U3101 is the system control microprocessor. I am including for your convenience and as a reference a complete pinout diagram of this IC (*Figure 2-7*) which also gives you some basic information about the function of each pin.

As you would expect, the microprocessor controls all functions of the TV. It receives input from the front panel controls and the remote transmitter, interprets these commands, and issues appropriate instructions to the chassis. I knew this little IC handled a lot of information, but I was surprised at how much data it processed internally—digital/analog conversion, on-screen data generation, infrared signal processing, sync presence detection, AFT conversion, and analog/digital conversion. It also handles

U3101 SYSTEM CONTROL/OSD MICRO

PIN NO.	I/O	SIGNAL NAME	IN CKT RES.	DESCRIPTION
1	I	PWR-ON RESET	10K	Micro reset - Active Low.
2	I	FAULT DET	15K	When a low is sensed at this input, system control turns the set off for two seconds and turns it back on. If pin 2 goes low three times in one minute, system control keeps the set off.
3	I	IR	38K	Receives 5 Vp-p IR signal from remote receiver.
4	O	VERT KILL/ DEGAUSS	>20M	Goes high to kill vertical deflection during degaussing and the service line in direct view sets.
5	I	KS3	900K	Keyboard scan input.
6	I	KS2	900K	Keyboard scan input.
7	I	KS1	900K	Keyboard scan input.
8	I	KS0	900K	Keyboard scan input.
9	O	AUX 2	>20M	Video select control line.
10	O	AUX 1	>20M	Video select control line.
11	O	KD1	10K	Keyboard scan output.
12	I/O	KD3/ TUNING SYNC	9K	Keyboard scan output/tuning sync input.
13	O	ENABLE	>20M	Serial communications control line which goes high during data transmission and low during address transmission.
14	I/O	DATA	111K	Serial communications data line.
15	O	CLOCK	>20M	Serial communications clock line.
16	I	BLUE	16M	Blue OSD output. Active high.
17	I	GREEN	14M	Green OSD output. Active high.
18	I	RED	11M	Red OSD output. Active high.
19	-	VDD	1.4M	+5 VDC.
20	-	VSS	0	GND.
21	O	BLNK	>20M	OSD black surround out. Low = Black.
22	O	SYS RST	>20M	System reset line connected to bus devices. Goes low when set is off and high when set is on.
23	I	H-SYNC	16K	Horizontal timing input for OSD.
24	I	V-SYNC	11K	Vertical timing input for OSD.
25	-	R1	10K	OSD PLL external control pin.
26	-	VCO	>20M	OSD VCO external control pin.
27	O	SPKRS OFF	110K	Goes high to turn speakers off and low to turn them on.
28	O	RFSW/MONO	>20M	RF switch control line. Low selects Ant. A. Mono function currently not used.
29	O	TV PIX	>20M	Goes high when TV tuner is selected and low when external video is selected.
30	O	VOLUME	>20M	PWM for Volume control (currently not used).
31	O	TINT	11K	PWM for Tint control.
32	O	COLOR	19K	PWM for Color control.
33	O	CONTRAST	33K	PWM for Contrast control.
34	O	BRIGHTNESS	97K	PWM for Brightness control.
35	O	SHARPNESS	12K	PWM for Sharpness control.
36	I/O	CH CHG/AFT REF	105K	At start of channel change, voltage at pin 36 is read by micro for use in AFT A/D converter. During channel change, line goes high until channel change is executed.
37	I	AFT	73K	Automatic fine tuning input. Crossover point detected at 2.5 VDC.
38	O	TV ON	>20M	Power ON/OFF control. High = ON, Low = OFF.
39	O	OSC OUT	5M	4 MHz oscillator output.
40	I	OSC IN	4M	4 MHz oscillator input.

Figure 2-7. U3101 pinout.

channel tuning, band switching, on-screen information, and a kind of digital control over the video and audio. Moreover, it generates a varying digital output that is filtered and used to provide DC control voltages for the video and audio circuits. A 4 MHz oscillator provides the timing the microprocessor uses to execute it functions.

The EEPROM, if it is used, stores information like channel scan lists, alphanumeric channel labels which the customer selects, customer convergence settings for projection sets, VCR channel settings and RF switch options. Other feature variations such as the picture-in-picture feature are also kept in the EEPROM. Note that a TV without an EEPROM will automatically enter the autoprogram mode when it is turned on after an extended power loss. If an EEPROM is installed, the set will autoprogram only when the customer chooses the autoprogram function from the menu option.

I will not go into much detail about how U3101 works. RCA makes the information available in the books and the fiche I have already cited. If you want to know specifics, I suggest you consult those resources. Rehashing them here probably won't serve any purpose except to take up space. You can check the system control circuitry by making the following checks. Consult *Figure 2-8* as you go through the list.

(1) Check for +5 volts (VDD) at pin 19. The voltage should be very, very close to the stated value.

(2) Check for good ground connection (Vss) at pin 20. You should read 0 volts.

(3) Check for reset voltage at pin 1. Reset circuitry consists of R3114, R3115 and zener diode CR3101. The value of the zener (5.6 volts) is such that it allows the 5-volt standby to be established before pin 1 becomes high. As I have said, this circuit will also allow pin 1 to go low before the 5 volts begins to drop.

(4) Check for a good oscillator signal at pins 39 and 40. The signal should be in the 5-volt peak-to-peak range and should be at a frequency of 4 MHz.

(5) Check pins 13-15 for good activity.

The clock line (pin 15) is active only during data transfer. With no buttons pressed and the TV on, the clock line will pulse low about four times every second to request stereo or SAP presence information from the digital audio IC (U1600). The activity will vary if the customer asks the microprocessor to execute a command. You will see similar activity on pins 13 and 14. If you suspect a problem on the bus line, check for signal activity when the set is on and no buttons pressed. You should see a 0-to-5 volt peak-to-peak signal on pins 13-15. A steady, nonvarying signal between 0 and 5 volts tells you there is a defect on the bus.

A word of caution: Do not confuse pin 22 with the microprocessor reset function. Pin 22 is low when the set is off. When you turn the TV on, pin 22 will stay low for about one second and then go high and stay high until the set is turned off.

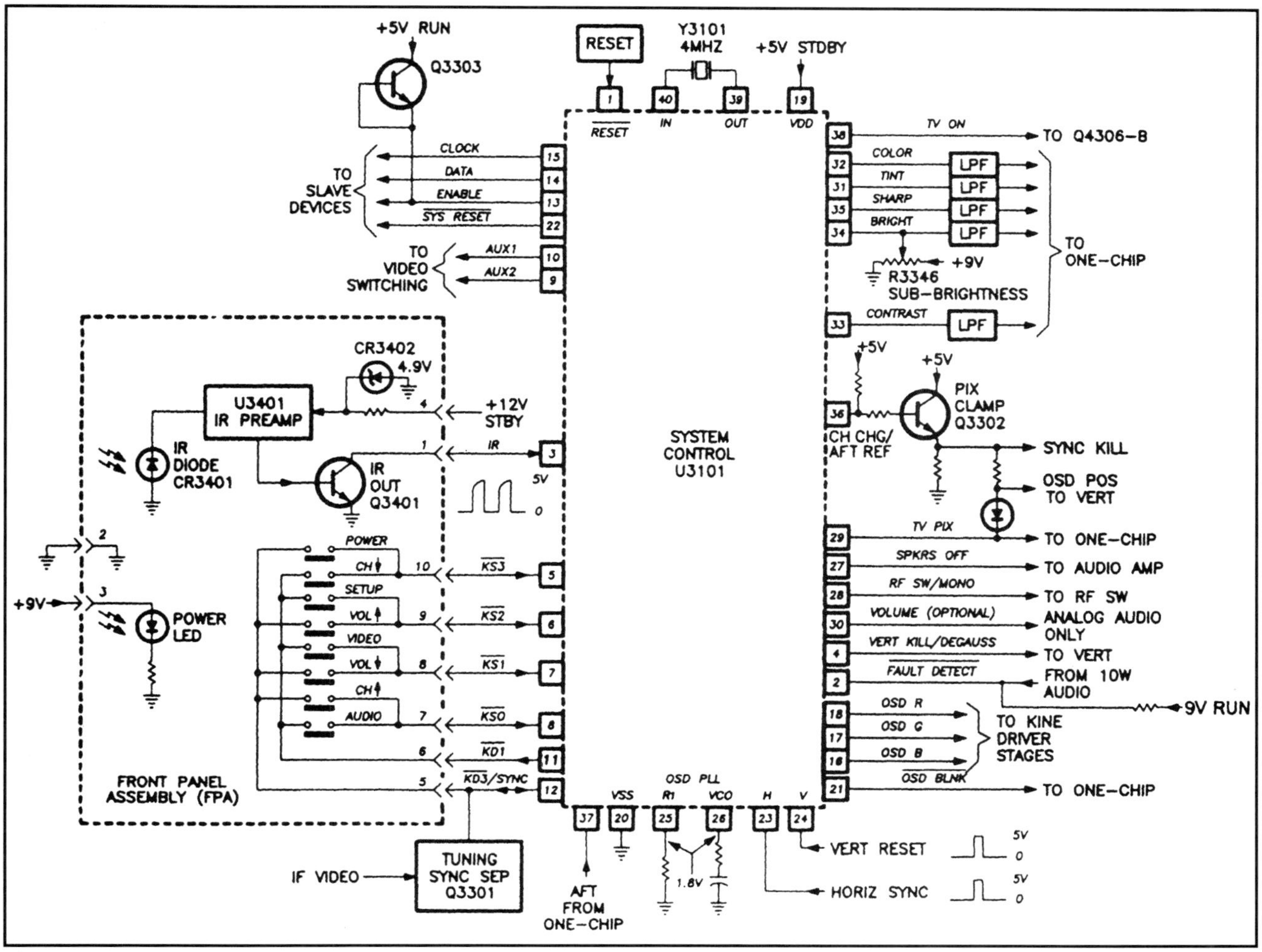

Figure 2-8. System control circuitry.

(6) Check for proper information on keyboard sense lines KSO-KS3, pins 5-8 (*Figure 2-5*). These pins should be at logic high when no key has been pressed. About once every millisecond, the microprocessor reads the KS inputs. If a key is pressed, the corresponding KS line goes low. This tells the microprocessor which of the four rows of the 4 x 2 keyboard matrix the key is in. The microprocessor will then scan the two KD lines (pins 11 and 12) to determine which of two columns the pressed key is in and then will execute the proper command. Perhaps I don't have to tell you a leaky or stuck tact switch can cause you some really weird problems!

(7) It is sometimes necessary to monitor the data out lines to find the solution to a problem. For example, suppose the TV you are working on doesn't have any color. The first thing you will check is whether you have activity on pin 32 of the microprocessor when you adjust the customer color control through its full range. If you don't have activity, you will know the no-color problem may be related to system control and not to U1001. If you do have activity, you will proceed to concentrate your efforts on the color circuits associated with U1001.

These checks should help you determine if you have a microprocessor problem or if the problem is located somewhere else in the chassis. I suggest you go through them from first to last; that is, in a systematic manner. If, for example, V$_{DD}$ is either high or low, you can expect system control to malfunction. You might even condemn the chip when the problem is its supply voltage. The moral is: Don't skip a step.

The Tuning System

I suppose this is as good a place as any to discuss the tuning system. *Figure 2-9* is a block diagram of the system, the details of which are given helter-skelter in RCA's factory literature.

The system is similar to the CTC167, but there are real differences. For example, there are two tuner options for the CTC169. The first is MTP-M-2016, like the one used and discussed in Chapter 1. It is a single-input VHF/UHF, cable-ready tuner. The second tuner is MTP-M-2030 and is identical to the first, except that it has an ANT A, an ANT B, and one cable converter input (*Figure 2-10*). The control line at pin 10 of the RF module selects the input you choose. If you have had complaints that the tuner will not switch inputs but all other functions work fine, consider doing the repair yourself. You can order just the RF switching module from, among other places, MCM. The part number is 203533, and it sells for less than $5.50 (1998 price). You don't need to be a tuner repair expert to take the tuner out and put a new RF switch on it.

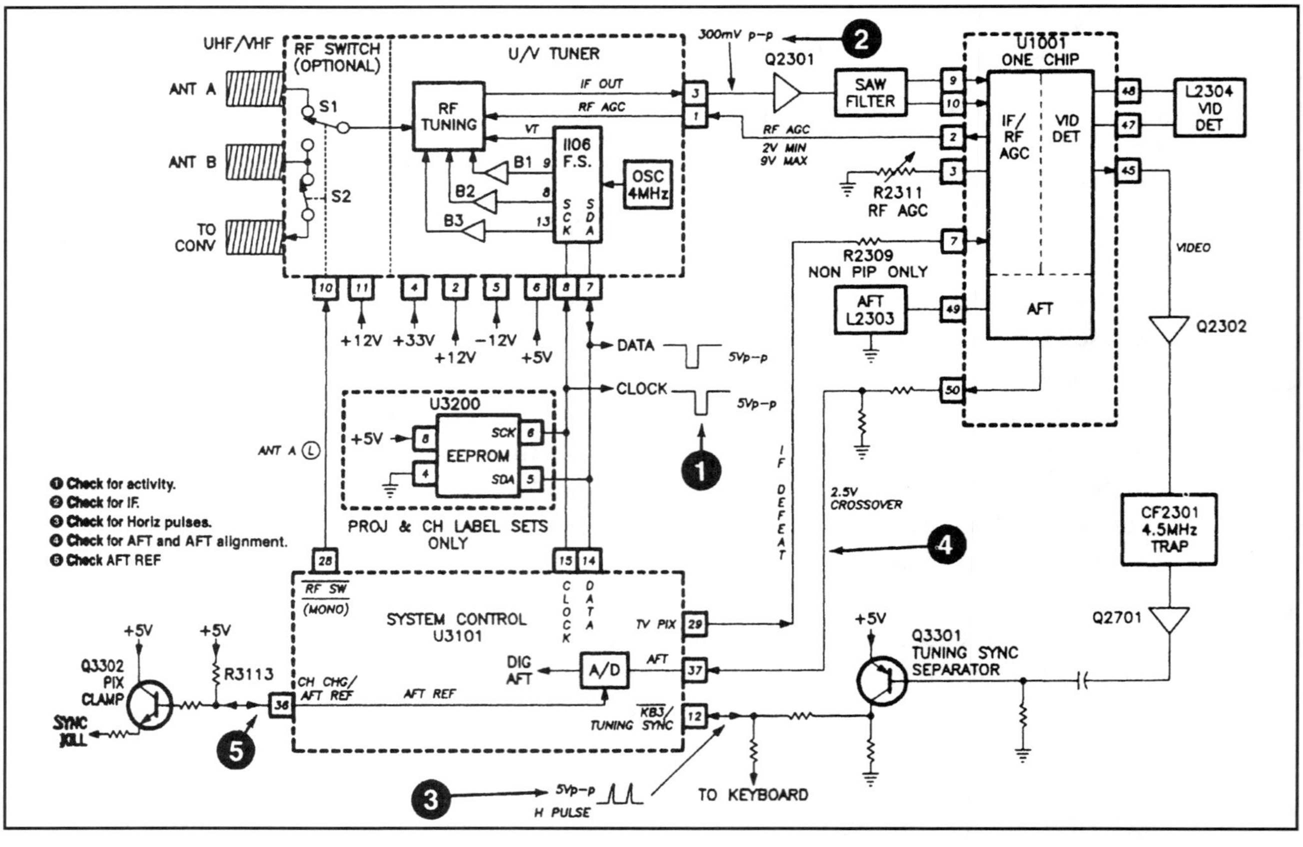

Figure 2-9. Tuning control.

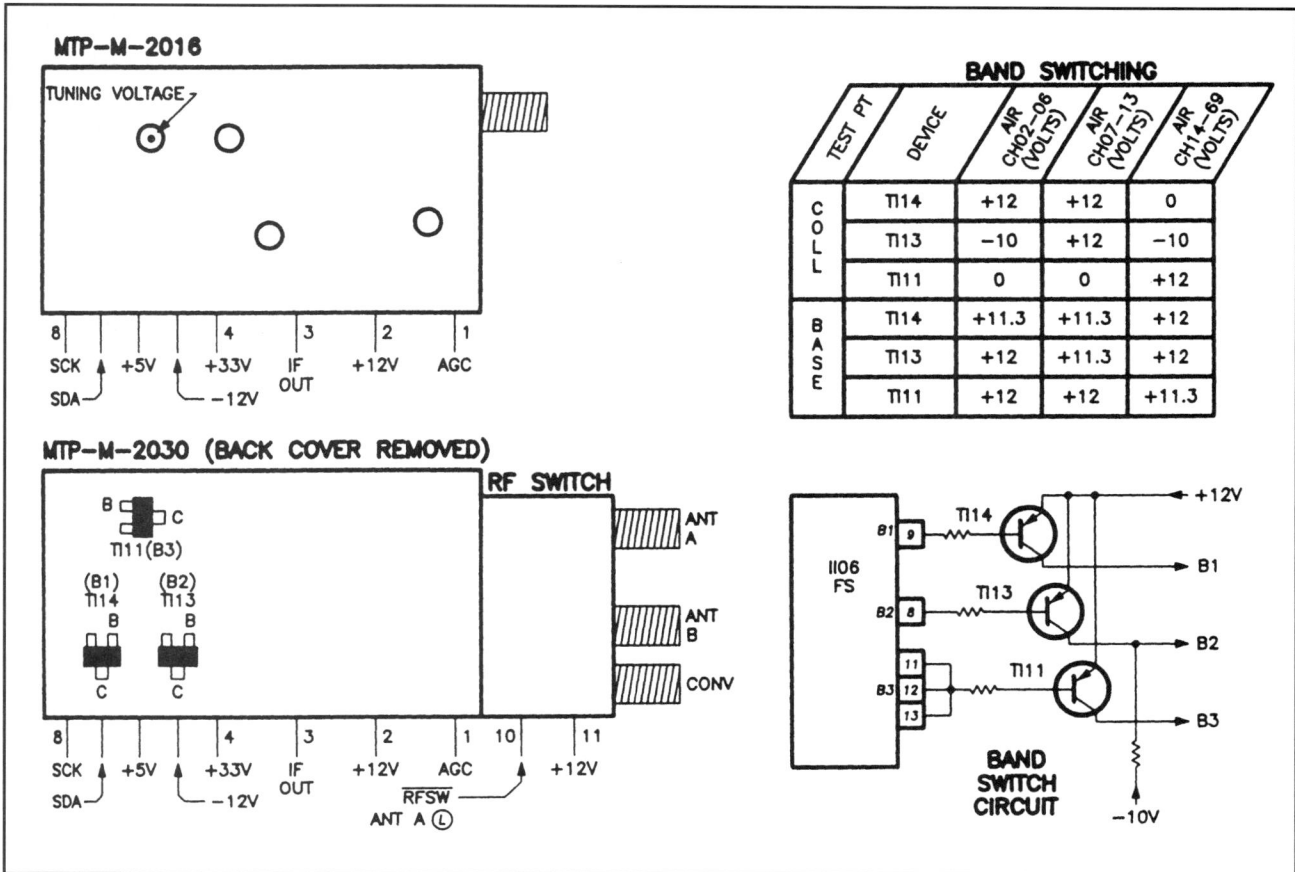

Figure 2-10. Tuners/Band Switching.

The microprocessor uses three signals to establish tuning, AFT, AFT reference, and tuning sync. The AFT voltage is a voltage representation of the video IF frequency. When the IF is at 45.75 MHz, the AFT voltage at pin 37 will be 2.5 volts. The 2.5 volts is sometimes referred to as "AFT crossover point." A voltage above or below 2.5 volts signifies the received signal is not on frequency and the local oscillator needs to make some adjustments to correct the frequency drift. If the AFT crossover voltage has not been set correctly, the picture may sometimes drift in and out of frequency as the local oscillator (in the tuner) attempts to find center IF frequency. That is the worst case scenario. What usually happens is the autoprogram function will not work.

The second signal necessary to establish tuning is AFT reference. During channel change, the DC voltage at pin 36 (AFT reference) is read by the microprocessor and used as a reference voltage for the AFT analog-to-digital converter at pin 37. This line goes high until channel change is executed.

The final input is tuning sync. It informs the microprocessor that an active channel is present and needs to be stored in channel memory. If tuning sync information is missing, the tuning system will not autoprogram because the microprocessor will have no way to determine when it sees an active channel and will therefore assume no channel information is present.

There are other details about the tuning system which I haven't mentioned because I don't think knowing them facilitates service. If you are interested, you can read about it in the literature, particularly the technical training manual for the CTC169. I can foresee certain instances when knowing the details might help you solve a problem, but I believe "the details" are at this point best left where they are.

Servicing Tuning Problems

Of course, the first thing to do when you encounter tuning problems is to CHECK THE VOLTAGES. It's a simple rule, but it's easy to forget. If you have correct OSD information but no picture or sound, check the clock and data lines going to the tuner. If they are okay, check for proper tuning voltages per the chart in *Figure 2-10*. If the tuning voltages are correct, check for about a 300 mV signal at the IF output of the tuner at pin 3. The 300 mV signal is not always easy to see even with a good scope. It therefore may be necessary to inject an IF signal from a good signal generator. If signal injection produces picture and sound, you can be reasonably certain the tuner is defective. If the on-screen information does not respond correctly—like, the channel numbers don't change as you change from one channel to another—you probably have a system control problem, not a tuner problem.

If the tuning is slow in cable but not air mode, check for tuning sync input at pin 12 of the microprocessor. You should see a high-going pulse less than 8 microseconds wide repeated at 68-microsecond intervals. If the tuning is slow in both modes, check for the proper AFT voltage at pin 36.

If a channel is momentarily tuned and then immediately drifts off frequency, think in terms of a misaligned AFT. The service literature gives specific information about AFT alignment (as 1-B1 of the fiche). I do it a bit differently. Insert a stable signal into the RF connector of the tuner. You might want to use a signal generator. Connect a voltmeter to TP2309 (pin 50 of U1001) and adjust the AFT coil (L2303) for a reading of 2.5 volts on the meter. Get the voltage as close to 2.5 volts as you can.

Horizontal Deflection

The next system to be considered is horizontal deflection. If your experience follows mine, you will have almost as much trouble with it as you do the power supply. So I will spend quite a bit of time dealing with it and its problems. I am including two illustrations of the horizontal deflection circuit. The first (*Figure 2-11*) is a basic block diagram, complete with voltages and waveforms (*The CTC 168/ 169 Technical Training Manual*, p. 18). The second (*Figure 2-12*) is the schematic from Sams PHOTOFACT.

The signal flow goes like this:

Composite video enters U1001 at pins 40 and 43. It is fed to an internal sync separator where horizontal and vertical sync are separated and routed to their respective circuits. Horizontal sync goes to the horizontal AFC stage. A pulse from pin 7 of the flyback, which is controlled by R4307 (horizontal phase control), is input to pin 24 of U1001 where it is combined with the output of the horizontal AFC circuit. The resultant signal controls the 32xH (32 times the horizontal rate) VCO oscillator. A 503 kHz

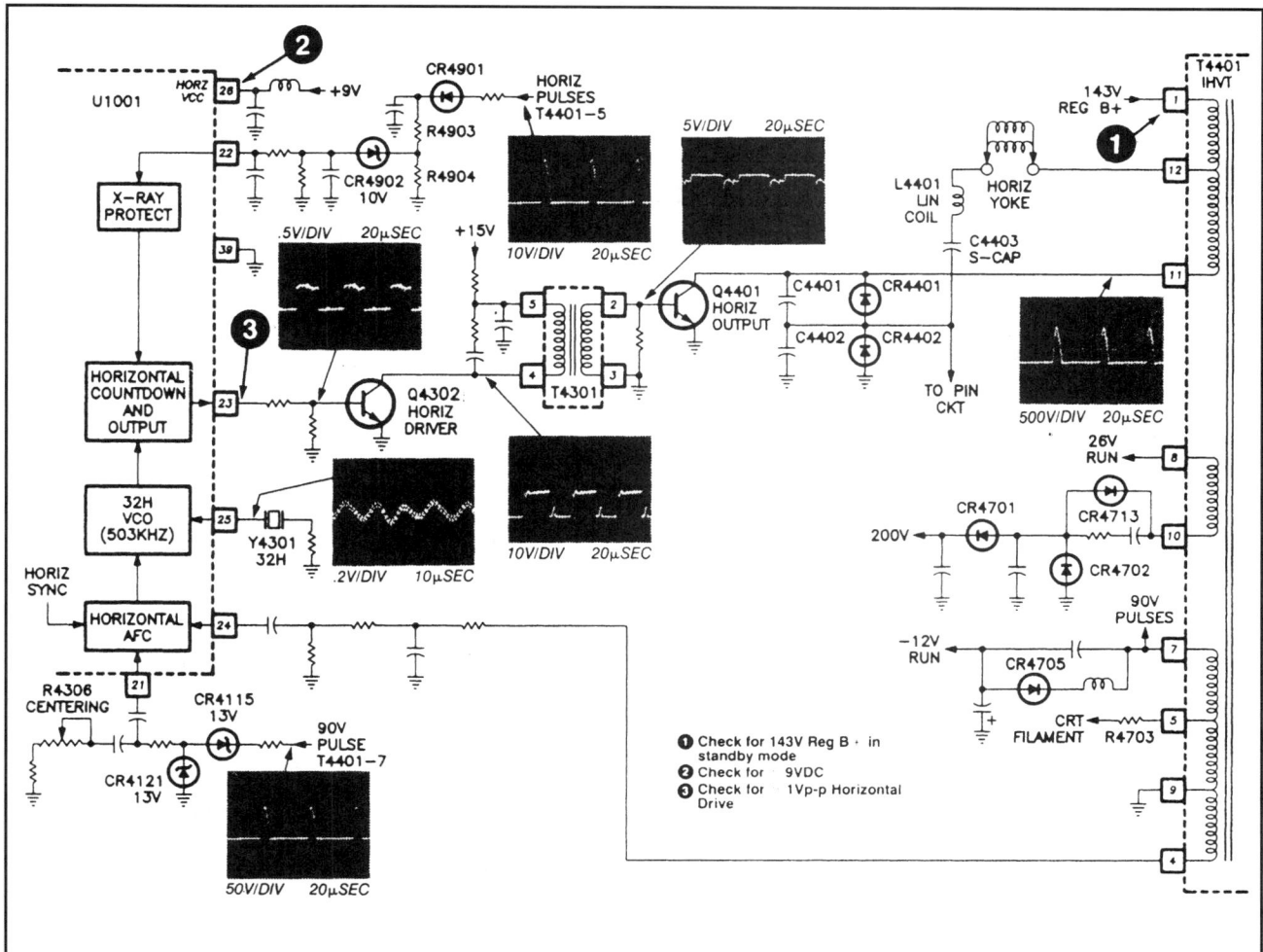

Figure 2-11. Horizontal deflection.

(32 times horizontal rate) crystal is connected to pin 25 to generate a 503 kHz signal. The output of the 32xH oscillator is applied through a horizontal countdown stage, which divides the 503 kHz signal to the proper frequency for horizontal deflection. The product is a very stable 15,732 kHz signal, which is routed to the horizontal output stage and sent out of pin 23 as horizontal drive.

Horizontal drive is routed to the horizontal driver (Q4302), which drives the horizontal driver transformer (T4301) which couples the signal to the base of the horizontal output transistor. Note that B+ for the horizontal driver is supplied by the main switching power supply (15 volts) and is present at all times, even when the TV is in standby.

The 9-volt input at pin 26 of U1001 is the B+ source for the horizontal oscillator. It is a switched supply and is controlled by the on/off circuits associated with system control. The TV comes on by supplying 9 volts to the horizontal circuits.

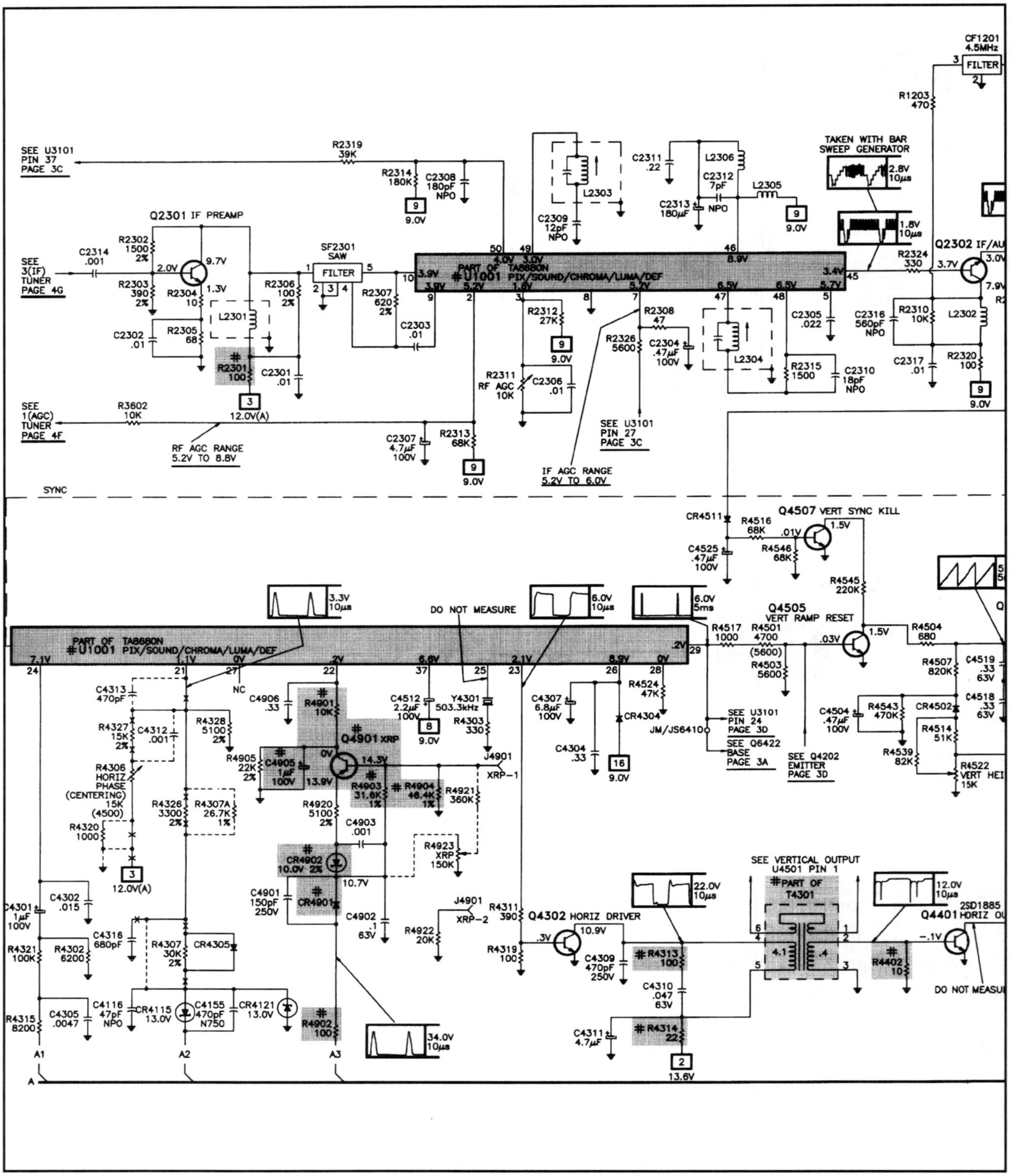

Figure 2-12. Horizontal circuits.

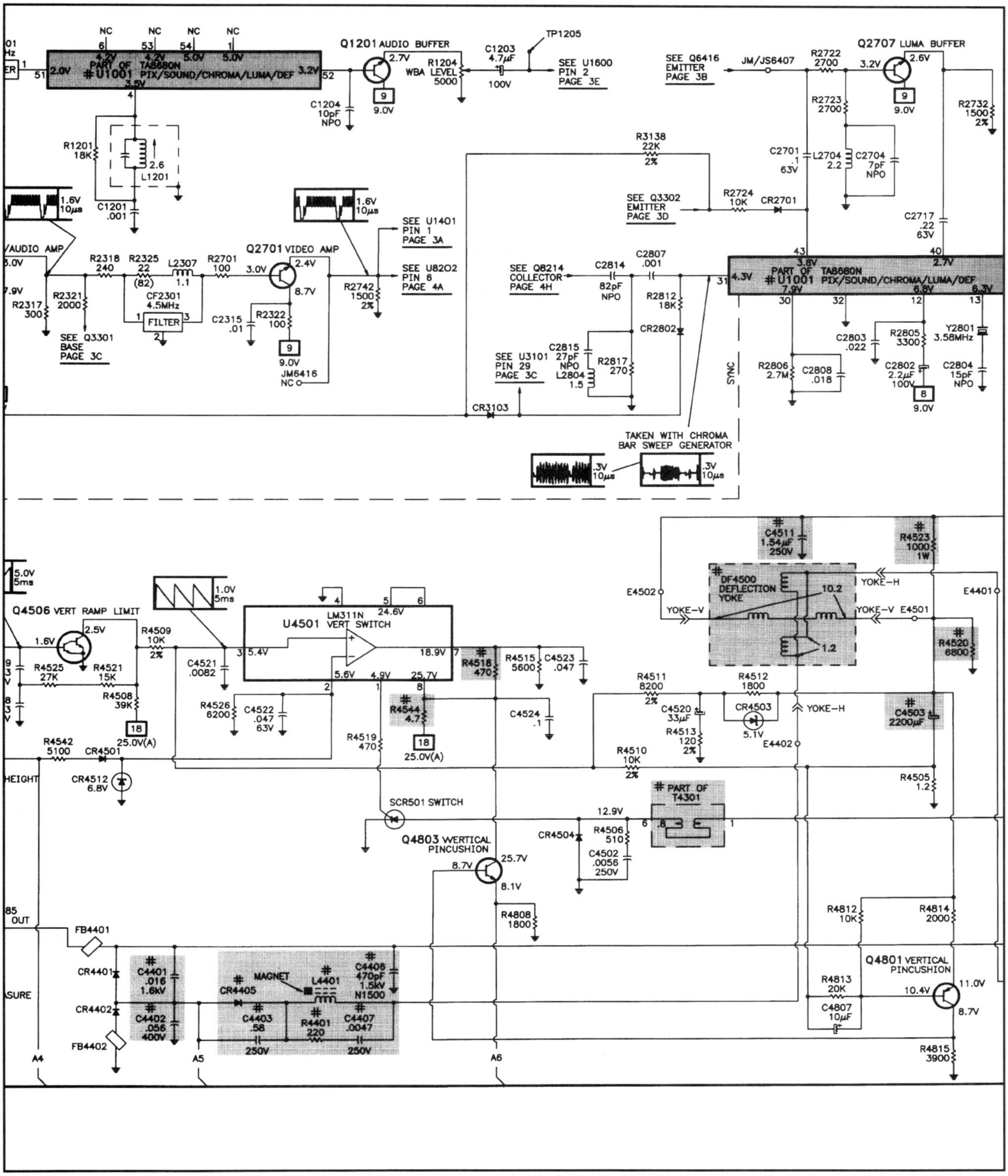

Figure 2-12. Horizontal circuits (continued).

Troubleshooting Horizontal Deflection Circuit

You need to remember if the horizontal circuit fails to operate, none of the run voltage supplies will be present. The TV will therefore be dead. Which is to say a dead-set symptom can be caused by a failure in the horizontal deflection circuit as well as the power supply, which complicates the problem for us repairpersons.

Failure of the horizontal deflection circuit will generally cause the set to go into one if its shutdown modes. For example, a shorted component in the horizontal deflection circuit will cause overcurrent shutdown. Remember the regulator IC? With 120 VAC applied, check the DC voltage at pin 16 of U4101. During normal operation the reading will be in the 10 to 11 volt range. If the voltage fluctuates between 7 and 11 volts, the TV will generally be in overcurrent shutdown. Another clue that the chassis is in overcurrent shutdown is if the power supply begins to "tick" the moment AC is applied, an indication the power supply is trying to start but cannot. The "tick" will generally cycle about three times before it stops completely. If you remove AC and wait a minute or two and reapply it, the power supply will cycle again and shut down.

Typical failures include a shorted horizontal output transistor, open capacitors in the collector circuit (C4401, C4402), shorted diodes in the collector circuit (CR4401, CR4402), and an excessive load on one of the secondary windings of the flyback (like a shorted audio output IC). These are "typical" failures. There are other possibilities as you can see when you examine the schematic.

RCA says if you suspect the horizontal section as the cause of a dead set, you need first to confirm its operation. Their engineers detail a procedure which allows you to check for proper operation while the circuits are active and before the chassis goes into shutdown. Most of the time the procedure will not be necessary because you can find the problem relatively easily. But there are those times when you need it. Since the procedure defeats regulated B+, take care to make sure the deflection circuits will not operate at higher voltages than suggested. If you operate the horizontal circuit at a voltage greater than the recommended one, you will—yes, you WILL—damage other components.

Follow these steps (*Figure 2-13*):

(1) Disable the TV's power supply. You can do it in several ways. For example, you can unsolder the collector of the chopper transistor, Q4101. It would be better to take it completely out of circuit.

(2) Force the on/off circuit into the on mode by connecting a jumper from the collector of Q4304 to cold ground.

(3) Connect hot and cold ground together by connecting a jumper between pins 12 and pin 5 of T4102.

(4) Connect a jumper between pin 10 of T4102 and pin 1 of T4401. When you install this jumper, you are connecting raw B+ to regulated B+.

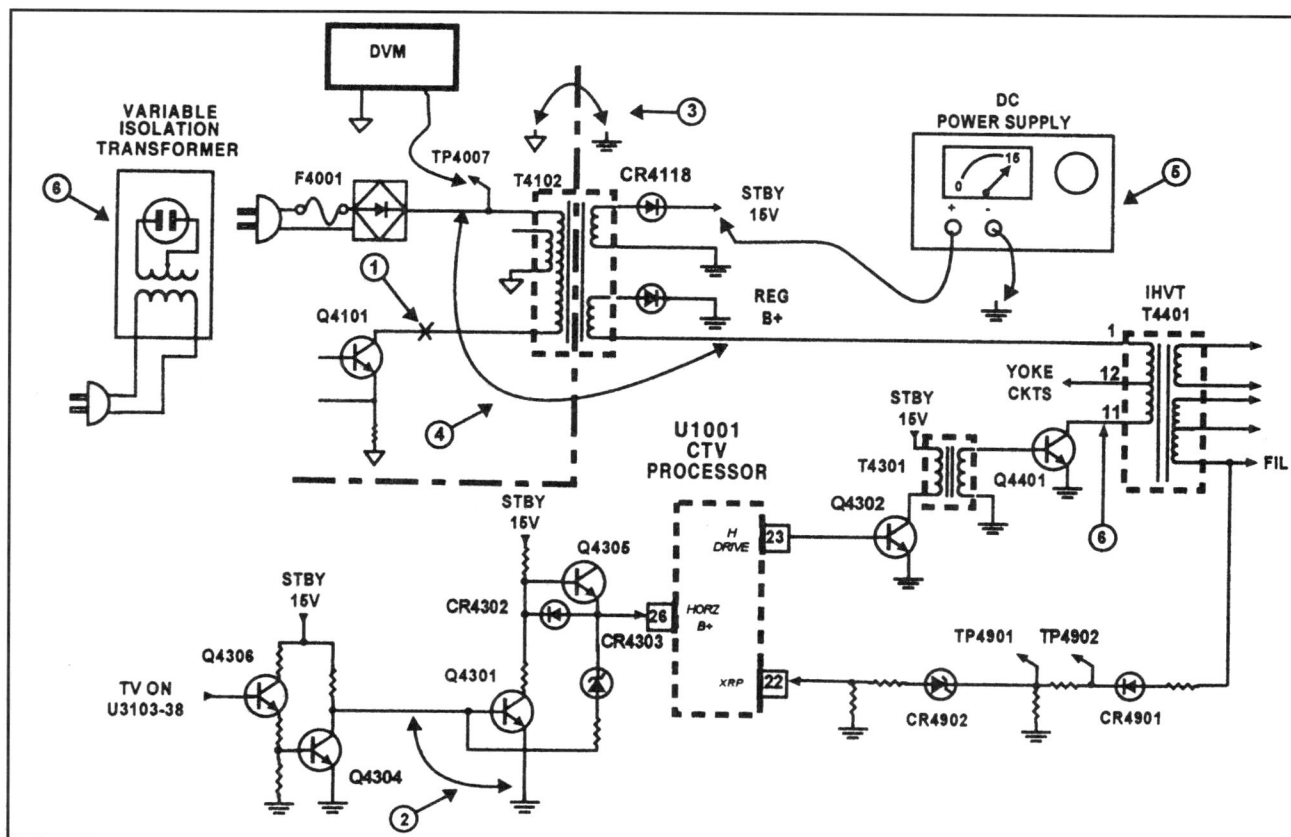

Figure 2-13. Horizontal deflection testing.

(5) Use a DC power supply to supply 15 volts to the cathode of CR4118 to provide B+ for horizontal drive to U1001. You can now confirm horizontal drive by scoping pin 23 of U1001. The waveform should be a square wave of about 5 volts p-p. If your scope will not permit automatic selection of horizontal frequency, set it for 20 µS/Div. If the signal is not present, you have a problem with the IC and/or associated components.

USE CAUTION AS YOU PERFORM THE NEXT STEP. It is a good idea to connect a scope to the collector of the horizontal output transistor to permit you to observe the retrace pulse and the DC voltage applied to it.

(6) Using a variac/isolation transformer, apply about 25 VAC to the chassis while you monitor the DC voltage at TP4007. You do not want to exceed 140 volts. Compare the waveforms to the chart in *Figure 2-14*. If the waveforms look good and the deflection circuit operates normally with the 140 volts applied, you do not have a problem with this system.

Now you have the TV up and running, and everything in the horizontal deflection circuit looks good. Take a few moments to analyze the retrace pulse because it is a good diagnostic indicator of the health of the set. Is its width about 13 microseconds? Is there any evidence of ringing? Are any secondary pulses present? If its width is correct and there is no evidence of ringing or secondary pulses, you can assume the tuned circuits in the collector of the output transistor are at their correct value and operating correctly.

If the width is narrower than 12 microseconds, suspect C4401 and/or C4402. These are timing capacitors. Their charge-discharge rate controls the current in the yoke during retrace. If the pulse narrows, high voltage increases. If high voltage increases, guess what? X-ray protection circuits are triggered. I have had it happen often enough that when I get a set that goes into x-ray shutdown and the 140 volts is correct, I automatically change both of these capacitors! Incidentally, you can check the regulated B+ while the set is in the standby mode. It should read about 143 volts at the collector of the horizontal output transistor. The 143+ volts will drop to about 140 volts when it is in the run mode.

If the pulse is wider than 13 microseconds, suspect an inductance type of problem in the collector circuit. A wider retrace pulse is often followed by ringing and/or secondary pulses. Typical causes are an excessive load on one or more of the secondary winding of the IHVT, a defective yoke (rarely), or a defective IHVT. If you want to read more on this topic, I suggest Sencore's *Tech Tips*, #s 207 and 211. The discussion is generic but applicable.

Horizontal Test Mode		
AC Line Voltage	Q4401 Collector Voltage	Q4401 Collector Waveform
25VAC	32VDC	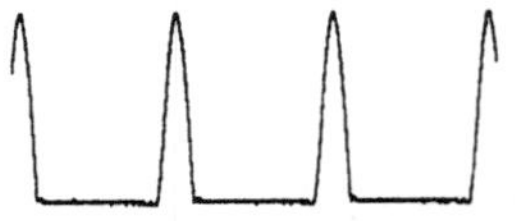 240 VP-P 20us/Div
50VAC	64VDC	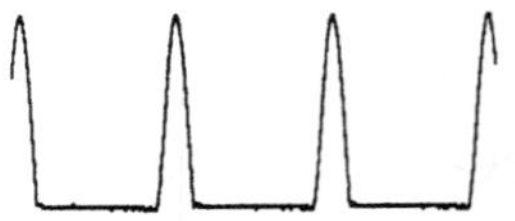 535 VP-P 20us/Div
70VAC	90VDC	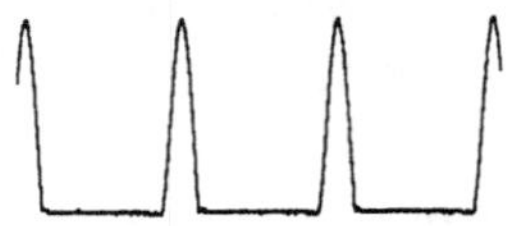 750 VP-P 20us/Div
110VAC	140VDC	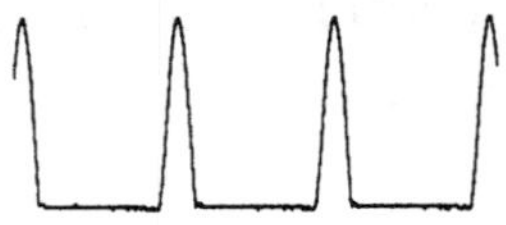 1200 VP-P 20us/Div

Figure 2-14. Horizontal deflection waveform table.

Case Histories

I will center my discussion of these case histories around *Figures 2-4* and *2-12*.

Assume a customer has brought you a CTC169 said something like this: "It was playing fine last night, but it wouldn't come on this morning." You plug it in, press the power button and nothing happens. You remove the back and reapply AC and hear a "tick-tick-tick" as the power supply comes up and then nothing as it turns itself off. You can go through an elaborate troubleshooting procedure to find the malady, but let me suggest a shortcut. Take a quick resistance-to-ground reading at the cathode of CR4116. You are checking the regulated B+ supply for a short. If you remember from a previous discussion, the reading should be 50k or greater. If it is close to zero, check the horizontal output transistor for a short. If it is shorted, replace it with a Thomson original part number and recheck the circuit for additional shorts. There is always the possibility that the damper diode has shorted if the set has one. The TV will then probably come on. While the TV is playing, monitor the output transistor heat sink. The sink should get just barely warm after many minutes. If it gets hot after about a minute, unplug the TV because you have additional problems and will need to change other components.

First, inspect the horizontal driver transformer (always a good idea when you replace a shorted horizontal output transistor in any set). Its ferrite core should be level with the top of the form, and its solder connections should be good. Next, inspect the damper diode (CR4401) and the pin damper diode (CR4402). I will usually replace them as a matter of course because they may be breaking down under load. Better to be safe now rather than sorry later on. Besides, they are relatively inexpensive. Then remove capacitors C4401 and C4402 and install new ones. Replacing these additional four components will almost always fix your problem. Don't shortcut these steps! I have learned from experience that these output transistors don't fail in and of themselves. Something invariably causes them to fail.

There is one other step a repairperson ought to take. I try to teach it to those who work with me, but they seem to forget it when they are in the middle of a repair job. Always check the horizontal output transistor after you have removed it from the circuit. In other words, confirm that it is indeed defective. It might check shorted in the CTC169 circuit because CR4116, for example, has failed. If it checks good out of circuit, you can make an extra few diagnostic checks to find the problem without wasting the time and effort of installing a new horizontal output transistor in a circuit that didn't need it.

Since I advocate original equipment parts, I'll give you the numbers for the components I have been talking about. You can get them from your local RCA distributor or from any number of mail-order companies, like MCM. Their selling price will not be that much different from the generic equivalent.

Q4401 (horizontal output transistor)	→	200167
CR4401 (damper diode)	→	198596
CR4402 (pin damper diode)	→	164589
C4401	→	206010
C4402	→	200149

While I'm on the subject of horizontal circuit failures, I will tell you about an interesting problem I serviced just last week. I turned on a 31-inch CTC169 and was greeted by a vertical line—yes a vertical line—up and down the screen, an indication that horizontal deflection was not working. The first thing I did was check the voltages and waveform at the collector of the horizontal output transistor. The regulated B+ checked okay at 140 VDC, but the retrace pulse was far too wide and had a peak-to-peak reading of 680 volts. The usually defective components were good. After a bit of thought and another look at the schematic, I saw something I had overlooked. If the usual components are good, what about the "unusual"? Maybe the horizontal yoke (a possibility) had failed, or better, something in the yoke return circuit? I found the problem in the yoke return circuit. C4407 had opened (part number 190534).

I have since learned that when it fails, C4407 will stress and destroy other parts. Those parts will usually be CR4402 and/or Q4802, both of which are in the pincushion circuit. With respect to the television under discussion, a new C4407 restored horizontal deflection, but the picture was too wide for the screen. In other words, the channel numbers and customer menus were far from where they should have been. When it failed, C4407 also caused Q4802 to fail. Replacing both components returned the TV to its properly operating condition. If the story has a moral, it would be to check for additional parts failure when you have to replace C4407. You can often tell if there is additional damage because some components may be fried. I have seen, for instance, R4401 (220 ohm, one-half watt) burned beyond recognition. But then, so was the capacitor!

I have already discussed the CTC169 that went into x-ray shutdown. It would sometimes play for five minutes and sometimes for almost an hour before it shutdown. I hooked a sample-hold DMM to pin 22 of U1001. Under normal operating conditions, the voltage here will be about 1.5 volts. If the voltage rises much beyond 1.5 volts, x-ray protection will activate and shut down horizontal drive. In this instance, the voltage would creep up to over 2 volts. C4401 in the collector circuit of the horizontal output transistor was the culprit. It had increased in value, narrowing the width of the retrace pulse, which permitted the high voltage to increase.

I have read of a report where the regulated B+ crept up to about 150 volts, which led to x-ray protection shutdown. In this instance, CR4123 (in the base-emitter circuit of Q4111, which is in the base circuit of Q4106) was not zenering at the correct value. You can spot the problem by measuring the regulated B+ and by monitoring the base voltage of the pulse-width modulator (Q4106).

Pincushion Circuitry

Since I mentioned the pincushion circuit in the previous section and since it is rather intimately related to horizontal deflection, I think this is a good place to discuss it. Please refer to *Figure 2-15* as the discussion proceeds.

As you know, the electron beam must travel farther to reach the corners of the CRT than it travels in the center. Left to itself, the results would be a raster in which each side sagged toward the center. Pincushion correction straightens the sides of the raster by modulating the horizontal yoke current at a vertical rate, so the raster more closely resembles a rectangle. Remember, such correction is needed in 27-inch and larger sets. The conventional linearity coil and "pin corrected yoke" are quite sufficient for televisions using 25-inch picture tubes.

The correction circuit develops a parabola-shaped waveform at C4801. Yoke current can be modulated here without affecting the high voltage. When Q4802 conducts, voltage across C4801 decreases. Since the voltage dropped across the yoke increases when the voltage across C4801 decreases (inverse proportion), the voltage dropped across the yoke increases. The increasing voltage drop increases the width of the picture because more B+ appears across the yoke!

As you can see, Q4802 is at the center of the pincushion circuit because it controls the charge-discharge rate of C4801. The input to the circuit is a vertical sawtooth and a signal from the return side of the vertical yoke. The vertical sawtooth is applied to the base of Q4801, while the vertical signal is applied to its emitter. Q4801 is referred to as a "subtractor" because the sawtooth waveform is subtracted from the vertical signal which results in a vertical parabola at its collector. The parabola is routed through a buffer (Q4803) to the input of the pin amp (U4801), the output of which drives Q4802.

If you want additional details about how the circuit works, you can look at the literature I have cited. I believe these details will be sufficient for our needs, but I do want to make a couple of additional comments before leaving this section.

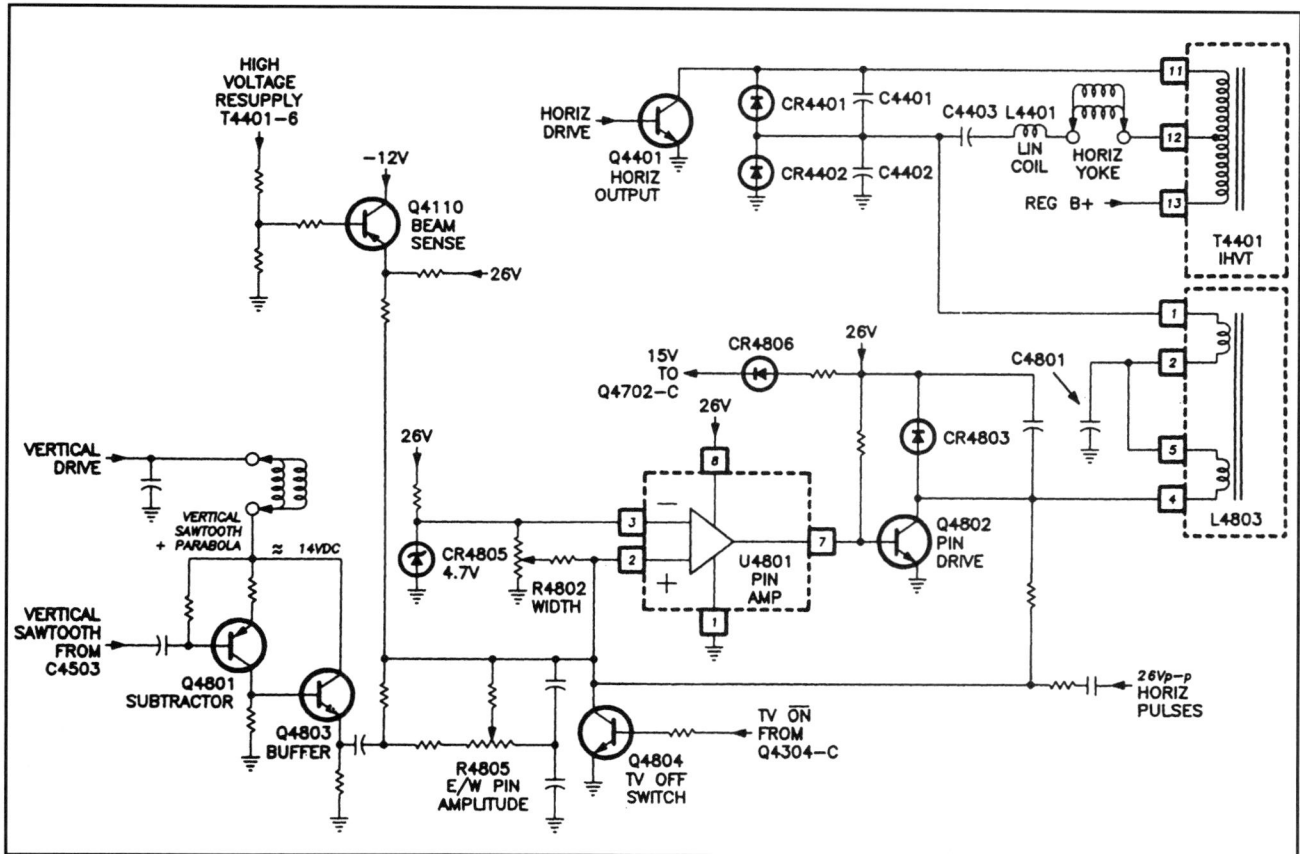

Figure 2-15. Pincushion correction circuit.

First, East/West pin amp control R4805 determines how much of the vertical parabola is input to U4801 and therefore how much East/West correction is applied to the picture. Width control R4802 sets the DC operating point of the pin amp. And a beam current input to the circuit allows U4801 to compensate for changes in width caused by variations in beam current.

Second, Q4804 keeps the pin circuit from operating when the set is turned off. It also prevents the ringing in the horizontal circuit (when the TV is first turned on) from stressing Q4802 and causing it to fail. When it turns on, Q4804 pulls pin 2 of U4801 low, which turns the IC off. When Q4804 turns off, U4801 is enabled, which is to say Q4804 is off when the set is on.

Troubleshooting the Pin Amp Circuit

Most problems with the pin amp circuit result in the same symptom—the raster goes to maximum width with significant bowing-out near the center of the picture. You can divide the circuit in half by varying R4802 (E/W amplitude control). If the picture varies in size, the problem is at the input to U4801; if it does not, check the output for the cause of the problem.

If varying R4802 causes the picture to change but does not permit the raster to be set correctly, suspect a problem with the vertical parabola. The problem could be, for example, Q4801, Q4803, or CR4805. I have some friends who have had such a problem, particularly with Q4801, but I have not. The pin amp problems I have serviced have been confined to the output circuitry, particularly Q4802, C4407, R4401, C4402, CR4402, and CR4405 (*Figure 2-12*). Remember if C4407 fails—and it does—it will take out additional components. Be especially suspicious of Q4802.

Now let's put some "flesh on these bones."

I serviced a CTC169 that exhibited a picture with a bow in it. I checked CR4402, the pin damper, and found it was leaky. The pin damper diode does not usually "just fail." Oh, it can happen, but it won't happen often. In this instance, C4402 had begun to fail, which heavily stressed CR4402 and led to its early failure. I replaced both components to cure the bowing problem. If I had replaced just the diode, I am certain I would have had a call back, a situation in which I would have worked "for nothing" for a customer who would not have appreciated the additional inconvenience.

When you troubleshoot this pincushion circuit and find the signal at pin 3 of U4801 weak, don't assume it isn't there. Wick out the pin to see if the signal reappears. There have been reports of a defective IC pulling the signal down and causing the pin correction problem.

Vertical Deflection

Our next subject is vertical deflection. I will use *Figures 2-16* and *2-12* as the basis for my comments.

I have been working on televisions for a long, long time. I cannot tell you how many schematics I have perused. I have dealt with vacuum tubes and solid-state stuff, from multivibrators to vertical count-down ICs to where we are today, and lots of things in between. My conclusion is, I have never seen a

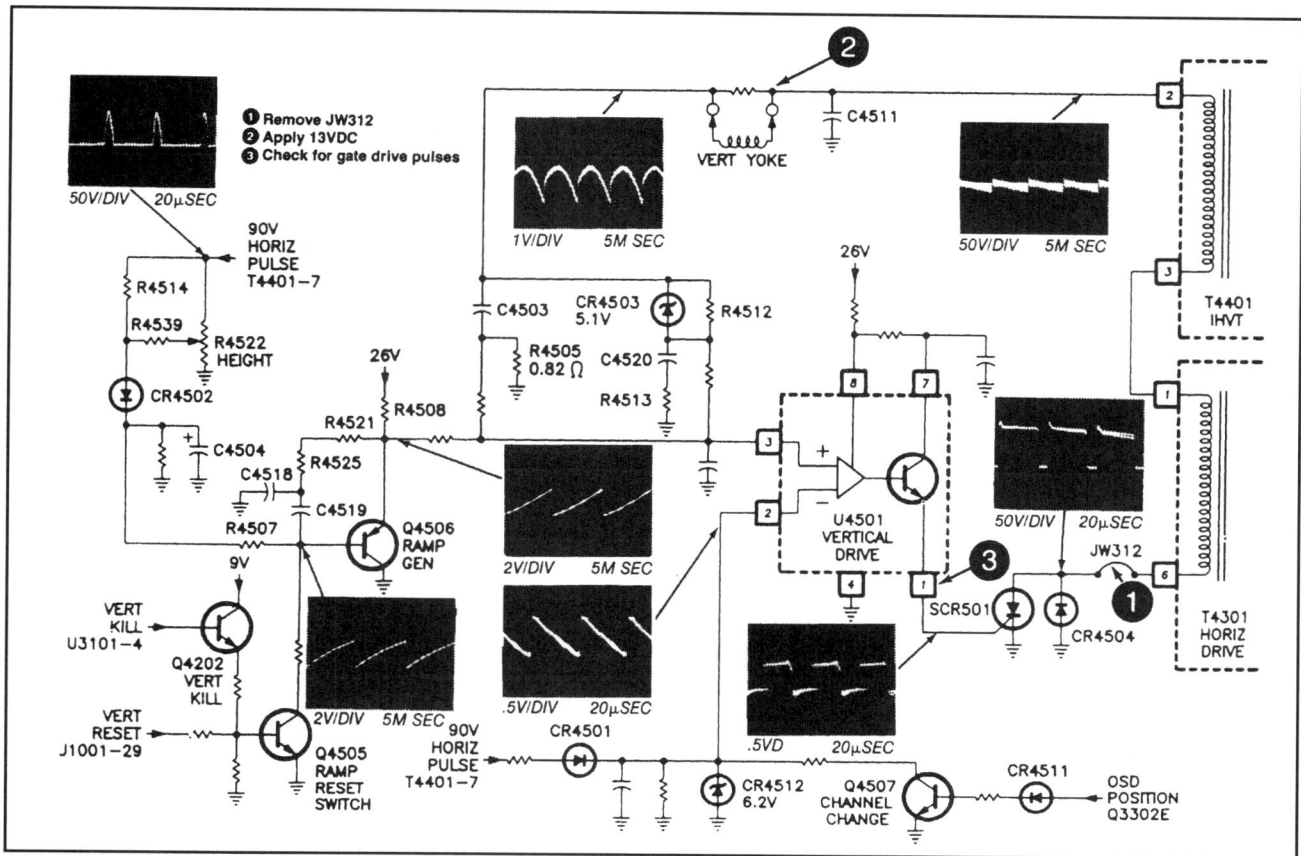

Figure 2-16. Vertical deflection circuit.

vertical circuit like the one in the CTC169 chassis! I will summarize the workings of the vertical deflection circuit to the best of my ability. It is very complicated, and the literature is not always really clear about how the components work together, as if engineers were writing for engineers. I think I can give you sufficient information to help you repair the vertical problems with which you are confronted. Having read both the fiche and technical manual, I am prepared to refer you to the technical manual as the better of the two in the event you want a more in-depth analysis of the circuit.

This vertical circuit uses a horizontal pulse from the flyback to provide current for vertical scan. A diode and SCR determine the current applied to the vertical yoke. A comparator provides the voltage-to-phase conversion used to gate the SCR. The comparator has two inputs, one from horizontal deflection and the other from the vertical ramp generator. The resulting signal determines when the SCR turns on during each horizontal line. It is turned off by the horizontal retrace pulse.

Ramp Generator Operation

U1001 develops a reset pulse at pin 29, which is about 6 volts peak-to-peak and about 10 microseconds in duration. The positive portion of the reset pulse turns on Q4505, which discharges the ramp-generating capacitors C4518 and C4519. It also places a low on the base of Q4506 ("vertical ramp limit") turning it on and limiting the peak-to-peak voltage of the pulse at its emitter to about 1 volt.

When the reset pulse goes negative, Q4505 turns off, allowing the ramp-generator capacitors to re-charge. Current to charge these capacitors comes from a power supply developed from flyback pulses rectified by CR4502 and filtered by C4504. However, these capacitors charge at different rates due to resistors R4521 and R4525. The different charge rates develop the proper waveform for the ramp pulse. Resistor R4522 sets the level for the DC supply and therefore controls vertical height.

Now pay attention to Q4507 (*Figure 2-12*), the vertical kill transistor. It turns on when the TV turns on, "killing" vertical deflection and allowing degaussing to take place, and then turns off enabling vertical deflection. I say "pay attention to it" because if it becomes leaky or shorts, it will inhibit vertical deflection. If you aren't aware of its potential failure, you certainly can spend a lot of time trying to determine why you have no vertical deflection! It doesn't fail often, but it does fail, particularly during a lightning storm.

Vertical Drive

Shift your focus now to U4501. The sum of three signals is applied to pin 3: (1) a ramp pulse which is fed through R4509; (2) a pulse developed by the current-sense resistor (R4505 via R4510), which is of the opposite polarity; and (3) the DC feedback voltage from the vertical yoke. A 90-volt horizontal pulse is applied to pin 2, the inverting input. This pulse is rectified by CR4501 and charges C4522 during horizontal retrace. Zener diode CR4512 limits the applied voltage to 6.2 volts. The comparator inside U4501 determines when to provide a gate pulse to the SCR on each horizontal line. The gating pulse exits at pin 1. It is modulated at a horizontal rate and used to gate the SCR on and off.

Now look at Q4507, the vertical sync kill transistor. Vertical sync is missing during channel change, causing the countdown circuit to go to its default setting. This causes the on-screen information to rise higher on the screen. Q4507 forces the OSD to stay in its position until the countdown circuit inside U1001 receives a valid sync signal, at which time U1001 resumes normal operation and the microprocessor turns off Q4507.

Vertical Scan

The IHVT has a dedicated winding (pins 2 and 3) which provides a 160-volt p-p to facilitate vertical deflection. The current path is through CR4504 (the vertical damper), the inductor winding of T4301, the dedicated winding of the flyback, the vertical yoke winding, the AC coupling capacitor (C4503), current-sense resistor R4505, and chassis ground. You can see the current path better if you look at *Figure 2-16*. CR4504 and the pulse from the flyback form a scan-derived power supply. With no load, the voltage on C4511 approaches 160 volts. The winding in the horizontal driver transformer limits the peak current in the vertical circuit.

During vertical trace, the horizontal retrace pulse is rectified by CR4504. A charge begins to build on C4511, permitting current to build up in the vertical yoke. The vertical rate ramp is compared to the horizontal rate signal. The vertical signal is an increasing ramp. The horizontal rate is a decreasing voltage. The SCR turns on when the vertical ramp is greater than the horizontal rate input. The voltage on pin 3 of U4501 is at its minimum at the start of vertical scan. The SCR turns on only at the very end of horizontal scan. At the middle of scan, current through the yoke is zero. When the SCR turns on, current through it is equal to the current through CR4504. Since the two currents are in opposite directions, current through the yoke will be zero. At the bottom of scan, the SCR conducts for most of horizontal scan.

Let me put it differently.

At the beginning vertical trace, the SCR gates on for short periods of time to limit the peak current building up in the vertical yoke. As vertical scan is pulled to the center of scan, the SCR stays on longer, further limiting the positive current in the yoke. When vertical trace reaches center of scan, the SCR's ON time almost equals its OFF time. Current in the yoke consists of nearly equal amounts of positive and negative current, making current flow almost zero. When current flow equals zero, the gated ON time of the SCR gradually increases to reflect a negative current flow through the vertical yoke. This pulls the vertical trace from the center of scan toward the bottom of scan.

Vertical Retrace

The vertical yoke and C4511 form a tuned circuit. During retrace, the field in the yoke collapses, causing the voltage across C4511 to increase. The SCR does not conduct, but CR4504 still conducts for short periods of time during horizontal retrace. The voltage across C4511 causes the beam to move quickly from the bottom of the screen to the top. When the electron beam reaches to the top of the screen, vertical trace begins again.

Troubleshooting Vertical Deflection Problems

The technician can divide vertical deflection problems into two categories: no vertical deflection and incorrect vertical deflection. The latter category will include insufficient height, poor linearity, and poor symmetry. This is a good place to emphasize that you <u>should not</u> power up the TV with the vertical yoke disconnected. If you do, the resistor connected across the yoke (R4523) will surely smoke.

No Vertical Scan

Note that anything that causes no gate drive to the SCR or an open SCR will cause the voltage on C4503 to rise to about 160 volts. Since it has a rating of 35 volts, the capacitor will be destroyed. If you encounter a defective C4503, the service literature suggests you proceed like this:

(1) Remove JW312 (Locate it by looking on the schematic at the anode of the SCR.) and replace C4503.

(2) Connect a DC power supply adjusted to 13 volts to the high side of the yoke (E4502).

(3) Plug the TV in, and turn it on.

(4) Monitor the gate drive pulses at the gate of the SCR. If you have your DC supply set for 13 volts, you should have a continuous string of gate pulses. Those pulses should stop as you

decrease the voltage to 10 volts. If you find this to be true, suspect a defective SCR. If not, check for defective components in the ramp generator or horizontal ramp circuit. Once you have found and corrected the malfunction, reconnect JW312.

Incorrect Vertical Deflection

Problems in vertical height can be caused by the power supply used to drive the ramp circuit, the ramp generator, or the horizontal rate ramp at U4501, pin 2. Check for a change in the DC level at the cathode of CR4502 as you rotate the vertical height control through it settings. The voltage should vary between 21 volts or so to about 33 volts. If the voltage does not vary or if it is missing, check R4522, CR4502, and C4504.

If the voltage is correct, inspect the signal at U4501, pin 3. If it is not correct or missing, suspect Q4506, C4518, and C4519. If the voltage is correct at pin 3, check pin 2 for the horizontal rate ramp If it is incorrect or missing, suspect CR4501, CR4512, or Q4507.

If either C4518 or C4519, the vertical ramp-generating capacitors, changes in value, the result will be a distorted vertical ramp. If either or both fail, the result will be no vertical deflection because there will be no vertical ramp. I have not seen these capacitors fail in a CTC169, but I have had it happen in a CTC177.

If you suspect a linearity problem, first confirm it by using a crosshatch or a circle pattern from your signal generator, then begin to look for the problem. I suggest you place C4503 (the "s-shaping" cap), C4520, and R4512 at the head of your list of suspects.

Symmetry problems is the name for a group of problems that cause the raster to be offset so that one half of it is larger than the other half. When you encounter symmetry problems, check R4513, R4521, and R4525.

Servicing Vertical Deflection Problems

I recently serviced a 27-inch CTC 169 that exhibited vertical height problems. I adjusted the vertical height control, which seemed to solve the problem, and proceeded to play the TV for several hours to check it out. Lo and behold, the problem reappeared. It is a good rule of thumb to look over the circuit board carefully. You might be surprised at what you find! In this instance, I found what appeared to be a liquid spill on the circuit board over and around CR4501. When I first checked the voltage at pin 2 of U4501, I found nothing abnormal. Evidently, the spill was causing the diode to leak when the TV heated up. I replaced the diode and cured the problem.

A second set exhibited the same symptom. In this instance, the voltages and waveforms at pin 3 of U4501 were a little off and got worse as the TV played. The problem was C4521, a very small capacitor tied to pin 3. It was leaking, bleeding off a part of the vertical ramp to the vertical switch. I have, through my reading, discovered its failure to be a common problem. So make a note to check C4521 when you have vertical height problems. Incidentally, suspect it when you encounter one of these TVs that has a vertical jitter problem or one that exhibits intermittent vertical deflection problems.

Don't let a vertical problem buffalo you. After all, it is a vertical problem. Like most vertical difficulties, a little probing with a scope and a DMM will usually put you on top of it. This vertical deflection circuit is just different. It is still a vertical deflection circuit, and techs have been fixing them since the dawn of television.

The Video Circuits

The video processing circuit is a large, complicated one. For purposes of discussion I am going to divide it into smaller, easier-to-understand segments.

Video IF Processing

The video IF signal exits the tuner at pin 3 (*Figure 2-17*) having a peak-to-peak amplitude of about 300 millivolts. You can see it with a scope if the scope is sensitive enough, but the signal will be weak. The signal goes through Q2301, the SAW preamp, to compensate for about a 10 dB insertion loss in the surface acoustical wave (SAW) filter. The signal goes from the SAW filter into pins 9 and 10 of U1001, where it is amplified and applied to the video detector. From the video detector the signal goes to the IF amplifier, the gain of which is controlled by a voltage from the AGC stage. IF AGC control voltage is filtered by the capacitors on pins 5 (for high-frequency filtering) and 7 (for low-frequency filtering). Do not confuse IF AGC with RF AGC which U1001 also develops. RF AGC voltage exits pin 1 and is applied to the tuner to control the gain of the RF amplifier.

The output of the IF amp is also applied to the AFT stage. The output of this stage varies from about 0 to 9 volts and is available at pin 50. The AFT voltage is also routed to pin 37 of the microprocessor, where a resistor network scales the voltage down to a voltage between 0 and 5 volts. The relationship between the voltage at pin 50 (U1001) and pin 37 (the microprocessor) can be illustrated like this. If pin 50 of U1001 has a reading of 4.5 volts, you should expect to see about 2.5 volts at pin 37 of the system control microprocessor. I discussed the role of the AFT voltage earlier in this chapter, and refer you to it if you want to know how the AFT voltage is used.

A typical IF problem will be "no video—audio ok." Since the audio is taken off Q2302, it follows the circuits prior to that stage will be okay. Therefore, check for video before and after CF2301, the 4.5 MHz trap. If present, look for video at the emitter of Q2701. If video is there, the problem must lie beyond the IF circuits.

Another shape an IF problem can take is "no video and no audio." In this case, check the emitter of Q2701 for video and Q1201 for audio to verify if the problem lies in this stage. If neither is present, the problem lies in this stage. If both are present, the problem lies beyond them.

If both are present, look for the 300 millivolt signal at pin 3 of the tuner. If it is present, trace the signal from there to U1001 to see where it is being degraded or lost altogether. A hint, now: Use a signal generator and not an off-the-air signal. The signal from the generator is easily recognized and does not change. It will simply make your troubleshooting go easier. Also, don't forget to check for the proper ACC voltage at pin 1 of the tuner. It should be greater than about 2 VDC. If the AGC voltage is at 2 volts or less, the gain of the tuner will be too low to process a signal. Don't forget that you can apply a voltage to pin 1 of the tuner using an external DC power supply. By varying the applied voltage, you can quickly determine if the problem is AGC related or tuner related.

As a final step, if the above check out, verify that the video detector is aligned according to the service data. If it is correctly aligned, then you most likely have a defective U1001. If you have to replace U1001, I recommend soldering a socket on the circuit board for the replacement chip. A socket costs less than an integrated circuit. It will also make replacing the new chip easier in the event it too fails. As if you didn't know, new parts can and do fail! I replaced a T-Chip in a CTC177 because of loss of red drive. The TV came back in for service about two months later with the same problem. It seems a defective picture tube was taking out the T-Chip. If you've worked on those chassis you know how delicate they are. By installing a 64-pin socket the first time around, I saved wear and tear on the circuit board and saved myself time.

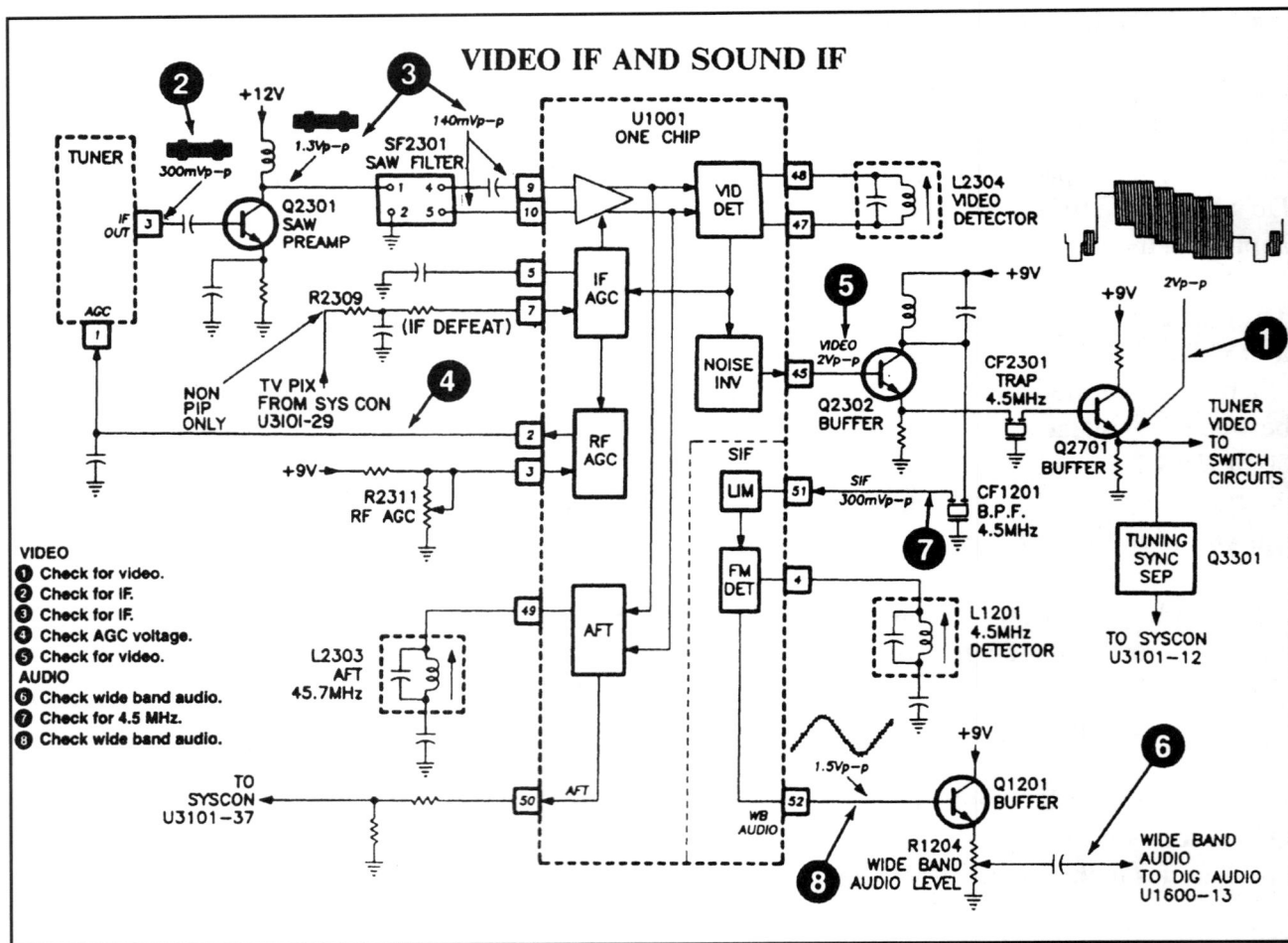

Figure 2-17. Video IF/Sound IF.

Since I brought up the subject of video detector alignment, I will include the alignment procedure as a convenience for you and hope that you never have to do it! Begin by inserting a 45.75 MHz signal into pin 3 of the tuner. The literature says the signal should be introduced through a 1000 pF capacitor, using the tuner shield as ground. If you have newer equipment, you can dispense with the capacitor. Short TP 2310 (the +lead to the -lead of C2307). Connect a meter to TP2305 (the +lead of C2304) using the tuner shield as a common reference, and adjust the video detector coil (L2304) for minimum voltage at TP2305. These alignment procedures are covered in the fiche that comes with factory service literature (frame 1-B1 and following).

IF Defeat Feature

Note a feature called "IF defeat" at pin 7 of U1001. If the set does not have the picture-in-picture feature, the TV picture line is tied to pin 7 of the microprocessor to defeat the IF when the viewer selects "external video." If the set has picture-in-picture, the IF will not be defeated when the viewer selects external video because the IF video may be used to fill the small picture when external video is the source for the big picture.

The Video Detector

The engineers at RCA call the video detector "a quasi-synchronous detector." If you want the details of its workings, consult *The CTC 168/69 Technical Training Manual*, page 38. It is sufficient for our purposes to note that the detected video exits pin 45, where it is applied to buffer transistor Q2302. The signal is picked off its emitter and sent to CF2301, a 4.5 MHz trap, to remove the 4.5 MHz audio subcarrier. Q2701 buffers the signal and provides about a 2-volt peak-to-peak signal to the video switching circuits and the sync separator, Q3301.

Luminance Processing

The luminance signal emerges at pin 45 of U1001. It proceeds to the video switching circuit (for details see *Figure 2-18*) and then goes to pin 1 of the comb filter, U2601. The luma signal is buffered and output at pin 2, where it is applied to delay line DL2601. After a bit of processing, it is reapplied to pin 5 of the comb filter. The signal is "combed" (luma separated from chroma) and sent back to pins 40 and 43 of U1001 (*Figure 2-19*).

U1001 outputs the luma at pin 38, where it is applied to the luma delay line. It is further processed and reinserted at pins 35 (low frequency) and 34 (high frequency). Luma once again exits U1001, this time at pin 16 as a nominal 2.4-volt back-to-white Y signal (*Figure 2-20*).

Several controls assist luminance processing.

The contrast control at pin 38 controls the gain of the luma amp. The DC voltage is developed from a pulse-width modulated signal generated by the microprocessor which has a range of 4.4 to 6.5 volts. A beam-current limiter is also used in conjunction with the contrast control to prevent blooming on extremely bright scenes. An increase in beam current causes the voltage at the base of Q2703 (*Figure 2-19*) to decrease, turning it on. When it turns on, Q2703 decreases the voltage on the contrast control line via

CR2705 and the voltage on the brightness control line via CR2703. A peak white limiter is used in 31-inch and larger sets in a similar manner. It works in conjunction with transistors Q2705 and Q2704.

The sharpness control is at pin 34 of U1001. At minimum setting (about 5.8 volts), little of the high-frequency luminance is added to the signal, while all of it is added at maximum sharpness ($\approx$7.9 volts).

The last control is the brightness control. Located at pin 36, it lowers the DC offset voltage of the luma signal to obtain higher brightness and raises it to obtain a darker picture. Maximum brightness will correspond to a reading of 5.6 volts; minimum brightness, about 5.2 volts. When you check the brightness control voltage, remember it can be pulled down by the beam limiter during very bright scenes. If, however, you use a signal generator, you will not have to worry about the voltage fluctuating. You can even check the circuit by increasing and decreasing the strength of the applied signal or by switching from a pattern that has lots of dark in it to one that has lots of light.

Troubleshooting Luminance Problems

You will note that some of the figures I use have troubleshooting guides on them, which is one reason I chose them. I have found them to be very helpful and pass them on in the hope that you too will find them useful.

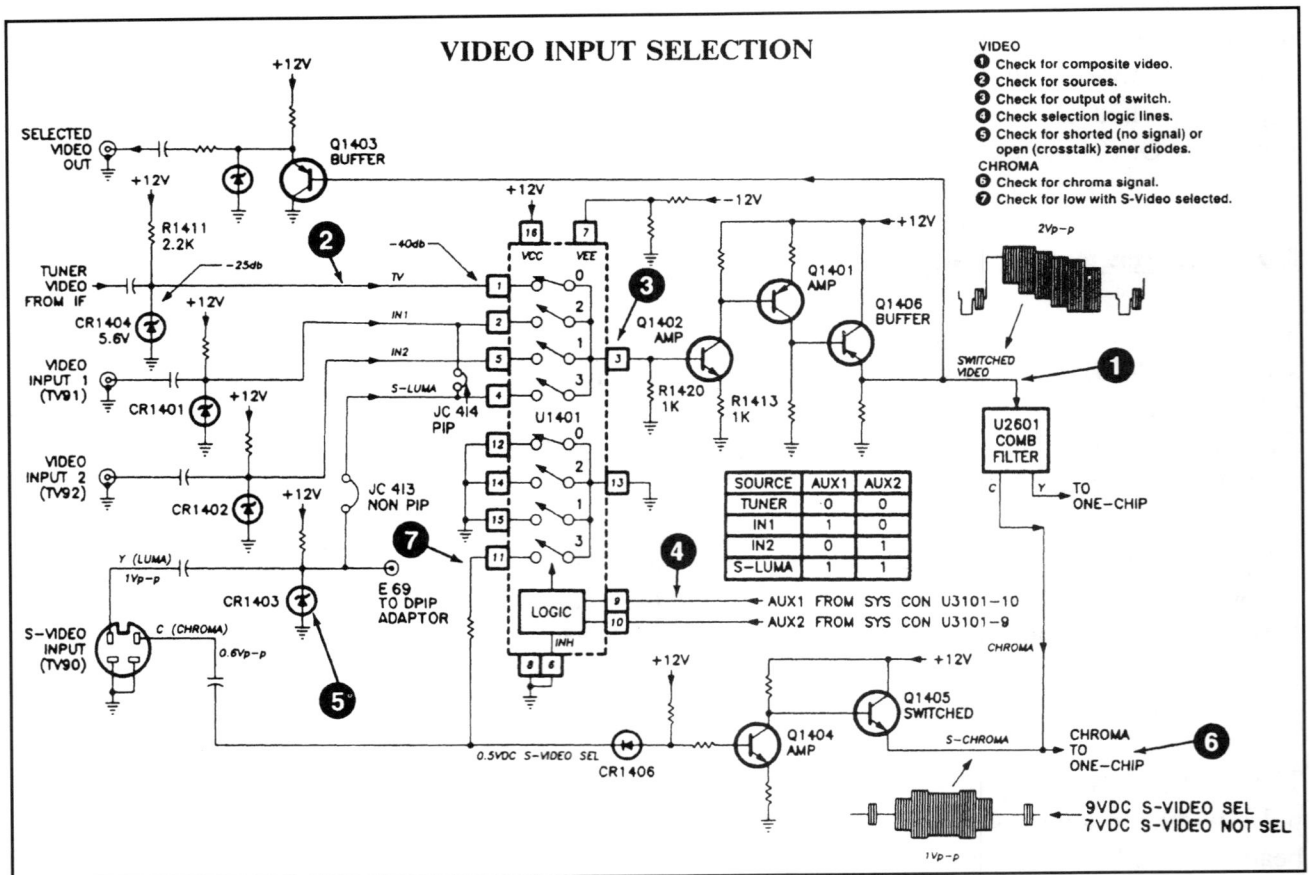

Figure 2-18. Video input selection (non-PIP).

If you have a "no video" problem, look first at the emitter of Q2906. If the signal is there and at the correct level, logically, the problem lies down the line. If it is not present at Q2906, look for it at the input to pin 1 of the comb filter. If it is missing at the comb filter, simply follow the trail back to U1001. While you are poking around U1001, double check a couple of things. For instance, make certain the sync kill line is not active except during channel change. Remember to check the picture control bias at pin 31, and be sure to check the settings of the brightness and contrast controls because if they are not set correctly, they can make you think you have a problem when you really don't. The literature also suggests you make certain pin 19 of U1001 is not stuck low. If it is low, it will account for the no-video symptom.

While I'm on the subject, I will tell you about a problem I have seen several times. The complaint that accompanies the TV is "bright screen with lines in the picture." You might even have to reduce the brightness or turn G2 down to keep the TV from going into shutdown. If you encounter such a set of symptoms, look first for the proper DC voltage and video waveforms on the collectors of the video

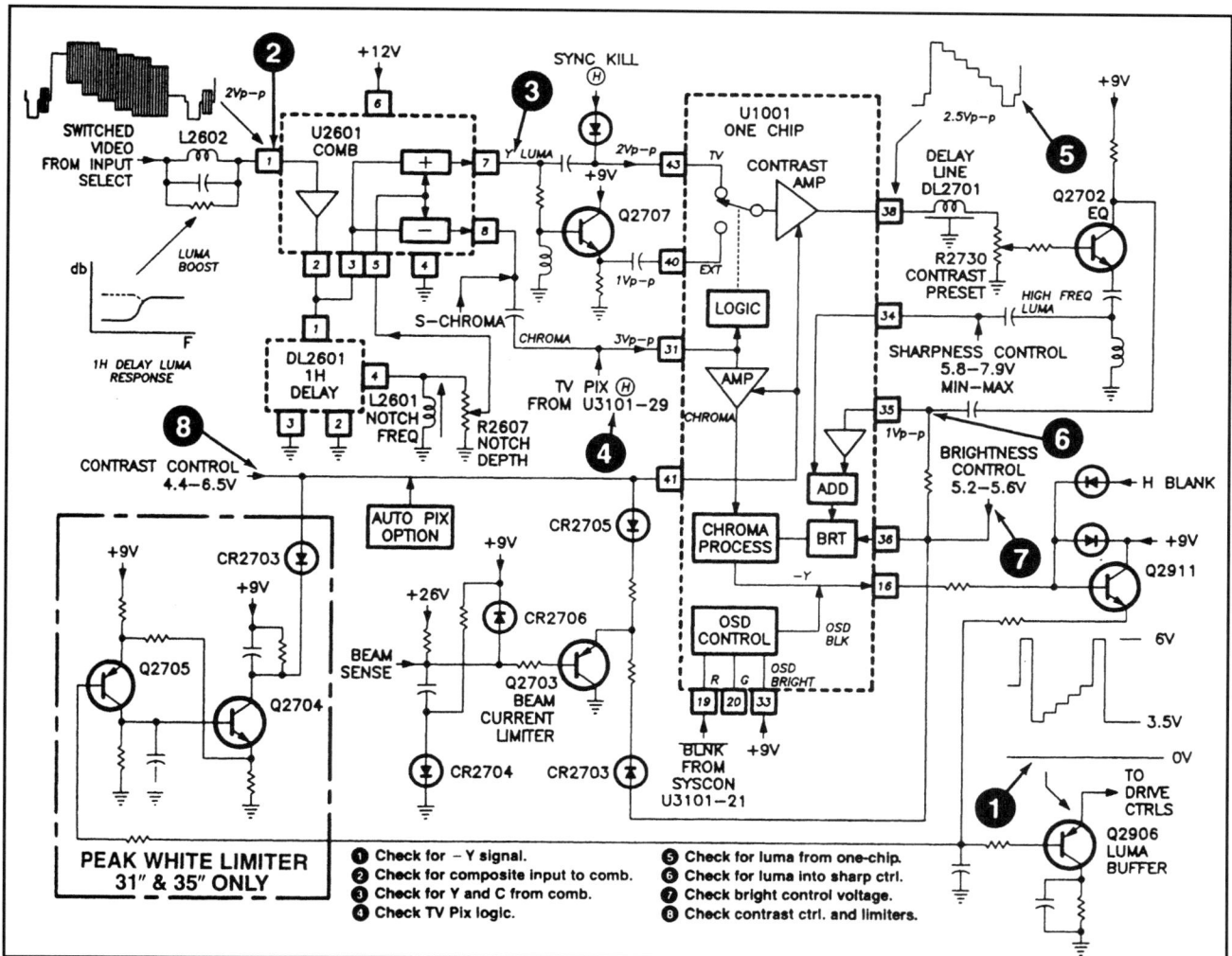

Figure 2-19. Analog comb/luma processing.

output transistors. If either is missing, the guns in the picture tube will turn fully on and cause a bright screen with heavy retrace. If the DC voltage is missing, check the 200-volt scan-derived source. If the luminance signal is missing, the troubleshooting procedure will be more involved, because any component in the luminance path can account for the missing signal. But the component most likely to fail is Q2906, the luminance buffer transistor. If it has failed, I suggest replacing it with an ECG159 which is a perfectly acceptable substitute. It usually fails by developing a collector-emitter leakage or short.

Color Processing

Please look at *Figures 2-20* and *2-21* as the discussion of the color, or chroma, signal proceeds.

The chroma signal from the comb filter enters pin 31 of U1001, where the color information is decoded and processed. U1001 contains two different sets of circuitry to process the information—one for tuner video and the other for external video. If the TV picture line is high and places about 4.5 VDC at pin 31, the chip will be set to process tuner video. If it goes low, about 1.2 volts will appear at pin 31, and the chip will be set to process external video.

Several controls "external" to U1001 facilitate the processing of the color signal.

ACC (automatic color control) monitors the chroma burst level and adjusts the gain of the chroma amplifier to maintain a constant chroma level output. The burst key, a horizontal pulse used to key the color amp and to lock the horizontal AFC stages, enters at pin 21. If you lose the burst key, you will have neither color nor horizontal AFC, which means the black-and-white picture will tear horizontally. A color level control at pin 42 varies the saturation of the demodulated color signal viewed on the screen. The tint control at pin 44 alters the hue of the chroma demodulation circuits from red at one end to green on the other. The color killer control is at pin 42. With color information on the signal, the voltage rises to about 8 VDC. If the voltage drops to about 6.5 volts, the color killer turns on to "kill" color altogether. Chroma APC (automatic phase control) is input at pin 12 and ensures accurate tracking of the color burst phase. The voltage is normally about 7 volts. The 3.58 MHz oscillator signal is at pin 13.

That's about it for the color circuit. I doubt that you will have to service many color problems. I have seen a lot of CTC169s, and just two had color problems. In one instance, the TV had no color from start to finish. The problem was loss of the 3.58 signal because the crystal was defective. In the other instance, the TV played fine until it warmed, at which time the color would gradually fade until there was nothing left but a good black-and-white picture. If I sprayed U1001 with freeze mist, the color would return and gradually fade out again. The chip was defective.

Nevertheless, here are some guidelines for servicing color problems:

(1) Check for R, G, and B outputs at pin 15, 17, and 18. If there is no output, check for chroma input as well as DC offset at pin 31.

(2) Check for the burst key at pin 21.

(3) Check the color killer voltage at pin 30.

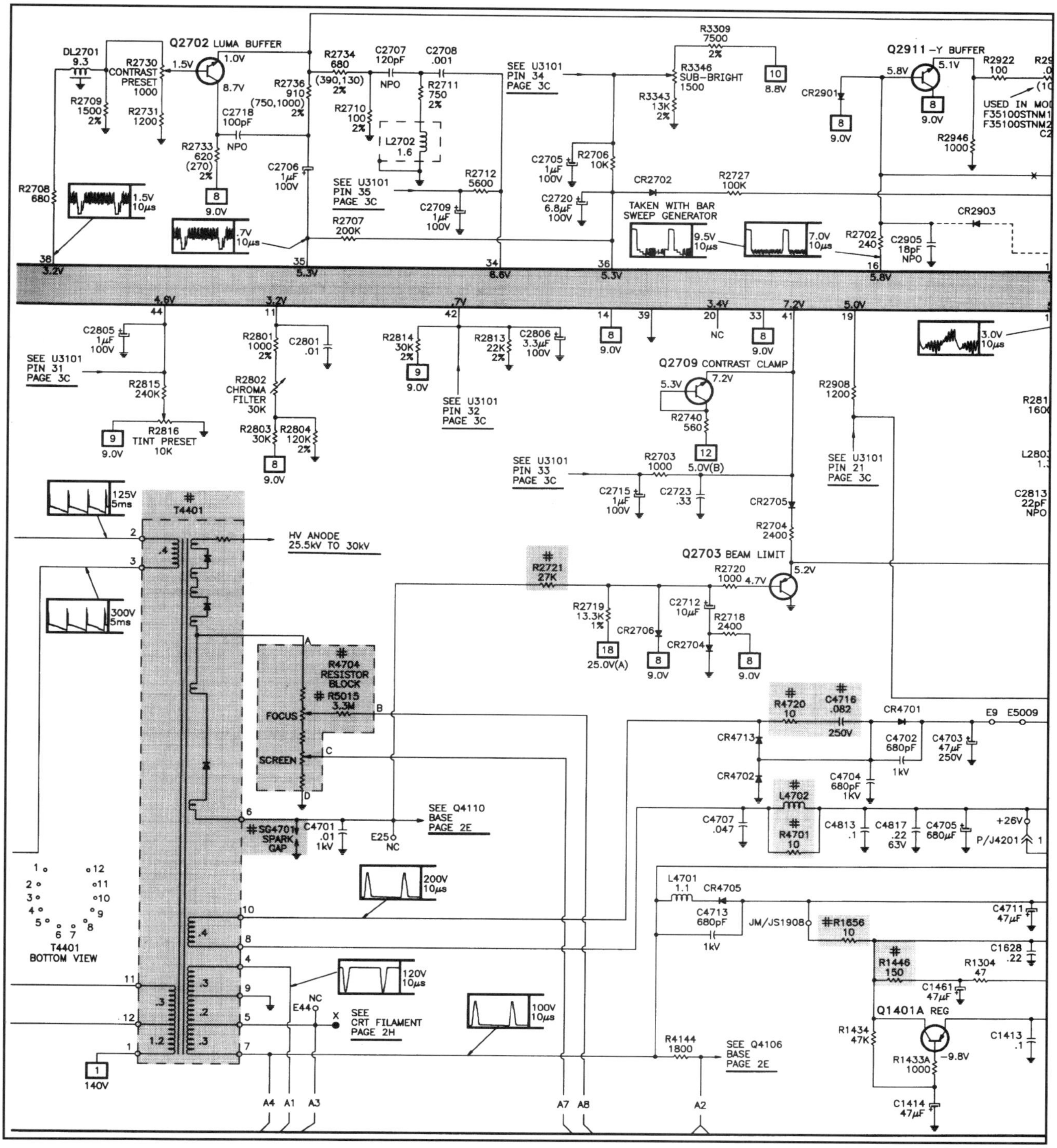

Figure 2-20. Color processing.

(4) Check for the 3.58 signal at pin 13. Check its frequency and its peak-to-peak value because both are important.

(5) As a last check, look at pin 12 for a leaky and/or open APC filter capacitor C2802.

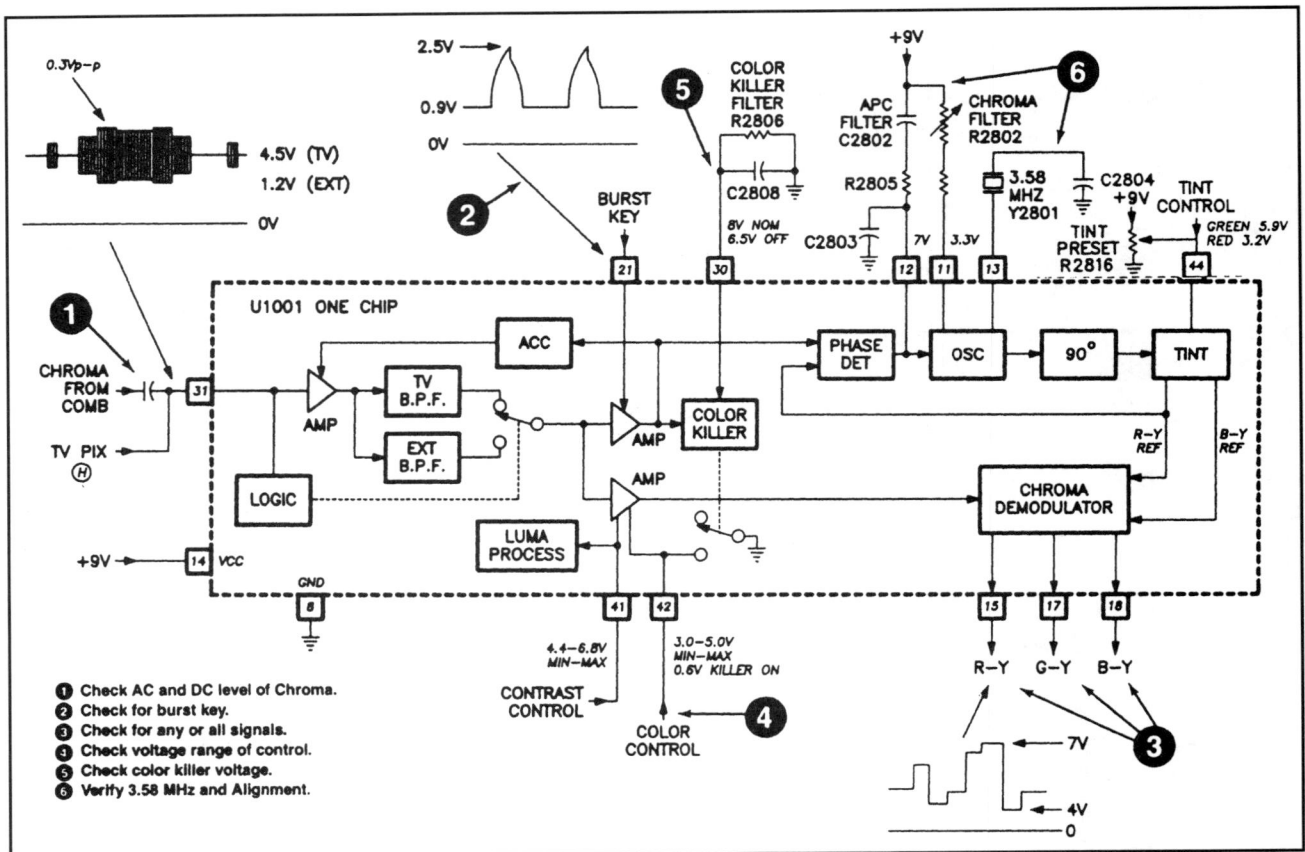

Figure 2-21. Chroma processing.

The CRT Driver Circuit

The CRT driver circuit is the last video circuit I will discuss. Use *Figure 2-22* as a reference. Since the red, green, and blue circuits are identical, I will trace the signal through the red circuit only.

The red color signal exits pin 15 of U1001 and is applied to the base of Q2903. The Y signal exits at pin 16 and is buffered by Q2906 before it is sent on to be mixed with the red signal at Q2903 to form the R-Y signal. The signal then proceeds to Q5001, the red video driver. Current changes in the emitter of Q5001 are translated into a voltage waveform at its collector. This waveform is routed to the red cathode inside the CRT to produce beam current for the CRT.

Beam current increases as the voltage at the cathode is lowered toward ground, and increases as it rises above ground. Note this, and remember it! For example, I serviced one set that had a very bright screen with heavy retrace lines. Voltage checks on the collectors of each video output transistor revealed a very low DC voltage. It is unusual for each transistor to have low voltage. If one, say Q5001, is low, you can expect to have a shorted transistor. If all three voltages are low, immediately suspect a problem in the supply voltage. In this instance the 200-volt supply was low because R4720 had opened. An open C4716 can also pull the 200 volts down.

Two other components in the video output circuit need commenting on.

CR5003 in the red cathode circuit is there to prevent grid 1 of the CRT from becoming more positive than the other two and drawing grid current which, as you know, is a real "no-no." You need to remember the beam limiter circuit won't work until the filaments heat up and the picture tube begins to draw current. If the TV were tuned to a channel that has high white content, the cathodes might be pulled very close to ground, which would send lots of electrons screaming toward the front of the picture tube. CR5003 is in the circuit to prevent just such a problem. If the voltage on grid 1 drops below 24 VDC, CR5003 will conduct to pull the grid voltage within 0.6 VDC of the cathode. The voltage difference is not enough to cause significant grid current, but it is enough to protect the picture tube or tubes.

Q2908 at the emitter of Q2903 is used to produce on-screen-display information. To produce a red character, the red OSD line from the microprocessor goes high to turn on Q2908. This lowers the voltage on the collector of Q5001, permitting current to flow through the red cathode to produce the character on the screen. Notice the bias controls for the three electron guns affects the OSD color temperature.

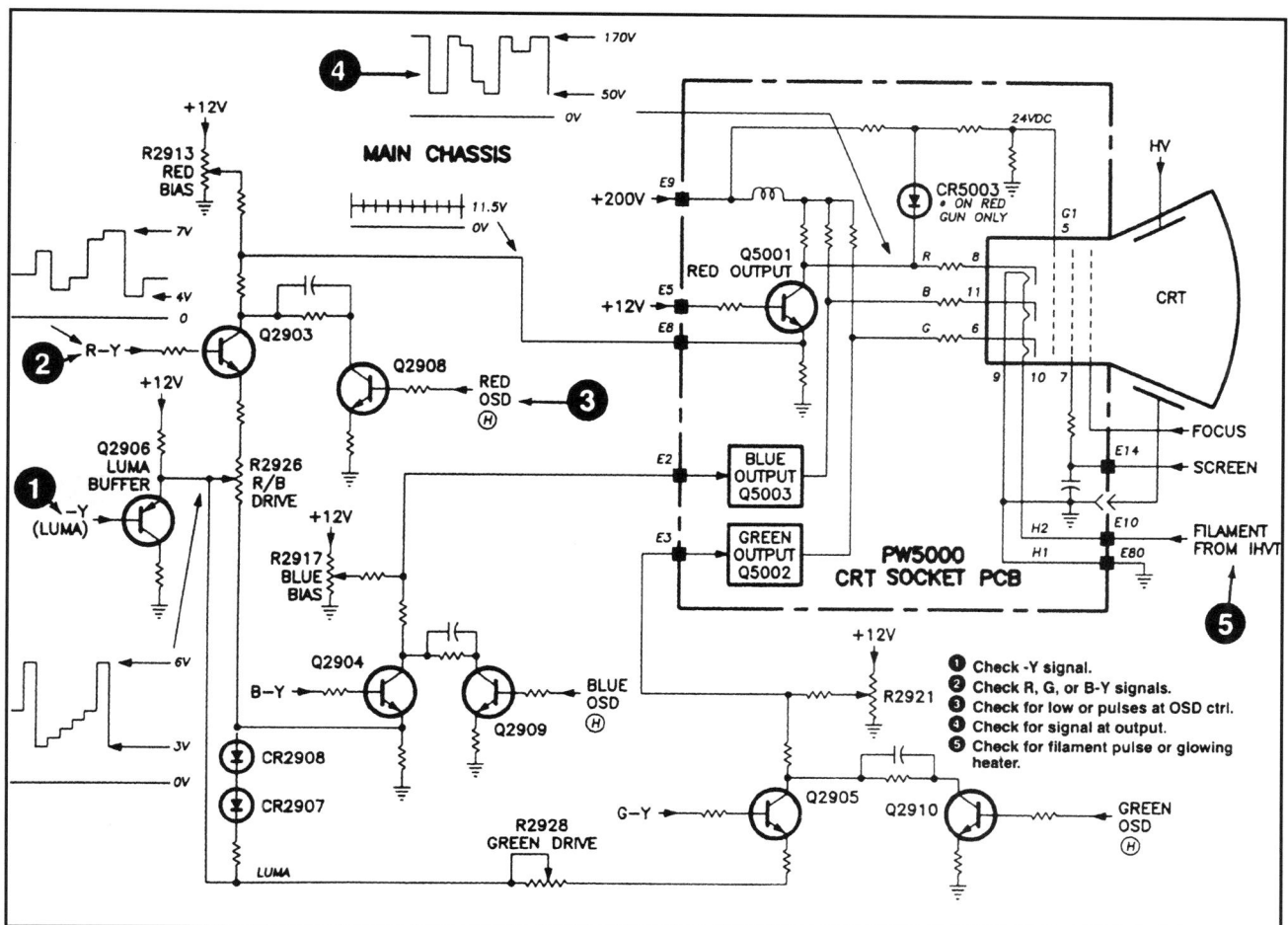

Figure 2-22. CRT driver circuit.

Video Options

Depending on which version of the CTC169 you service, you will encounter one or more special features. I will not spend much time dealing with these features, because you can read about them in the technical sections of the references I have cited.

Video Features SIP

The features contained on this board are digital comb filter, white and black stretch, auto picture and PIP interface (*Figure 2-23*). Each feature can cause its own problem. For example, "white stretch" is supposed to increase the gain of the luminance signal in the low to mid-light regions of the picture. If the circuit fails, the symptom could be "no video." The literature tells you how to troubleshoot the problem.

Picture-in-Picture Options

There are three PIP options: DPIP, D-SPIP, and FF-SPIP. A DPIP module produces a small picture which can be frozen, moved to one of four positions, and swapped with the big picture. D-SPIP (defeatured SPIP) modules perform the same functions and can be used as a DPIP substitute in early

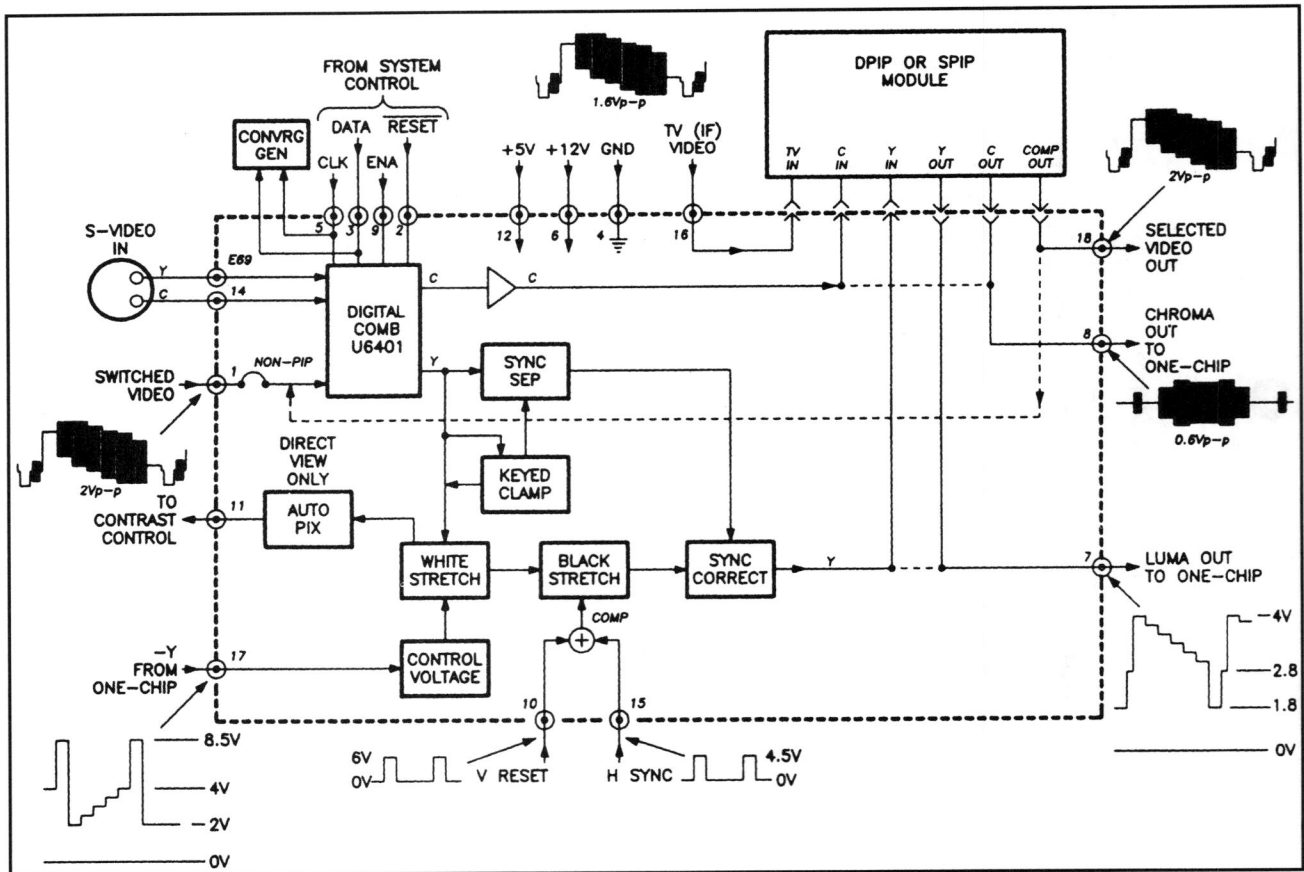

Figure 2-23. Video features SIP.

production sets. The FF-SPIP has the same features, plus some features similar to the CPIP module used in the CTC140 chassis. The FF-SPIP module permits the viewer to move the small picture to any position on the screen, zoom in/out, and pan. It also permits the viewer to view several channels at one time. The actual modules are shown in *Figure 2-24*.

The Audio Processing Circuits

All audio functions except power amplification are performed by U1600, the digital audio IC. It performs audio source selection, MTS/SAP decoding, expanded stereo operation, volume, tone, and balance control. It operates as a slave to the system microprocessor and will respond only to commands from the system microprocessor. These commands are communicated to U1600 via clock and data lines at pins 1 and 2. A third line, the "enable" line from system control, connects to pin 3.

Digital Audio Processing

Wideband audio is developed by U1001 and enters pin 13 of U1600. According to the literature, the peak-to-peak level of the signal is about 400 mV, which in my experience is generally true. The IC converts the incoming analog signal to a digital format for processing and reconverts it to analog before

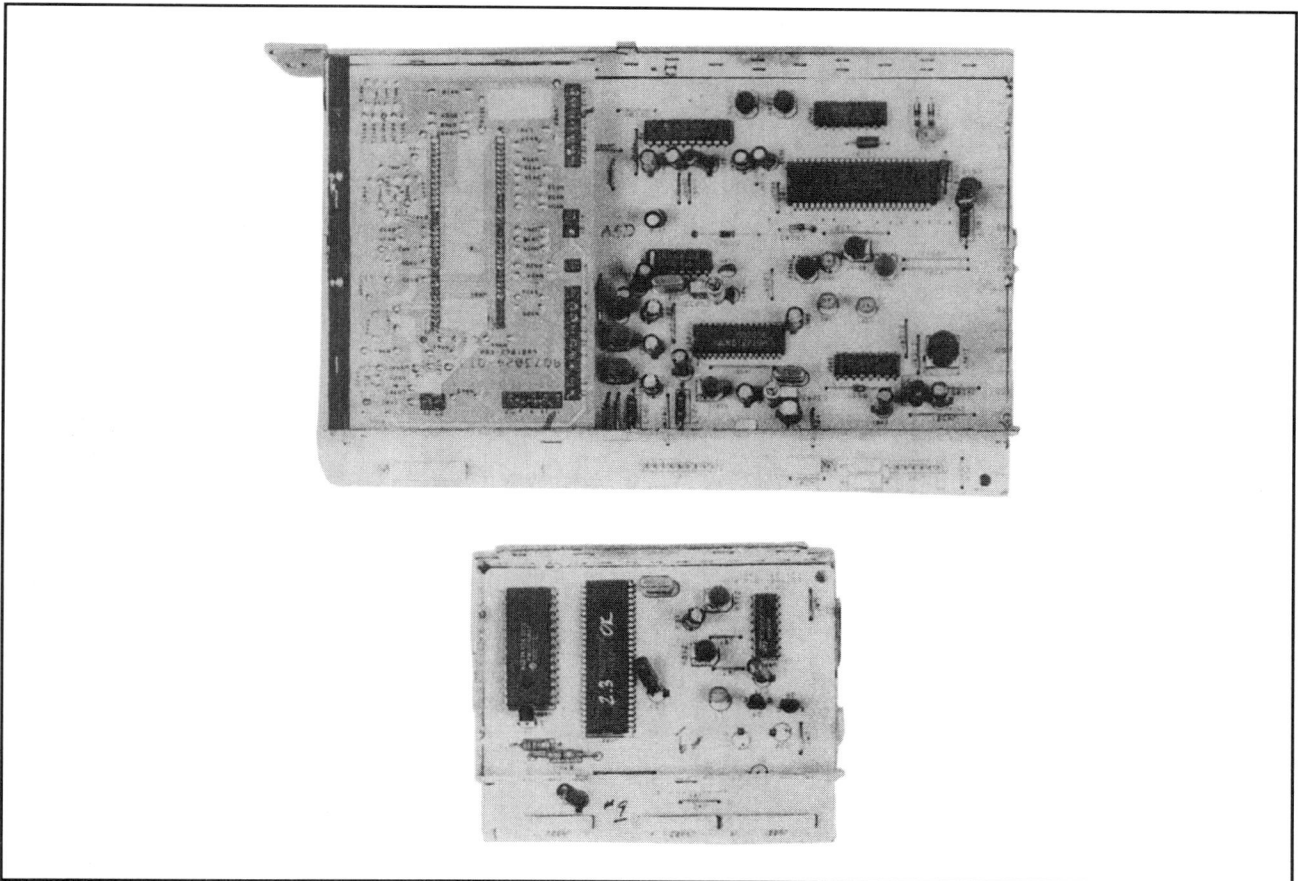

Figure 2-24. FF-SPIP modules.

sending it on to the power amp and the speakers. U1600 can select one of three audio inputs, wideband audio from U1001 or audio from one of two audio inputs (line-in audio from the jack pack). Remember, selection is performed via clock and data lines from the microprocessor.

Before we get to the specifics of how the audio circuit works, I need to remind you that the exact configuration of the audio circuits will depend on the model with which you are working. For example, I will be using factory literature for the 5-watt amp used in the CTC169L, AS, BD, BE, CA, CB, CD, CF, and JA models. Different models will have different configurations. The block diagrams I use are based on information found in the Technical Training Manual. If you compare the block diagrams to the schematic, you will immediately see the difference. Keep these comments in mind as we proceed with the discussion.

U1600 is powered by the 5-volt run source (*Figure 2-25*). Five volts is applied to pin 19 for analog processing and pin 7 for digital processing. Y1601 at pins 15 and 17 sets the operating frequency for the digital processing circuitry. The free-run frequency of the oscillator at pin 15 should be 11,014,000 Hz ±500 Hz when the TV is tuned to a mono station and mono selected from the customer menu. The VXO

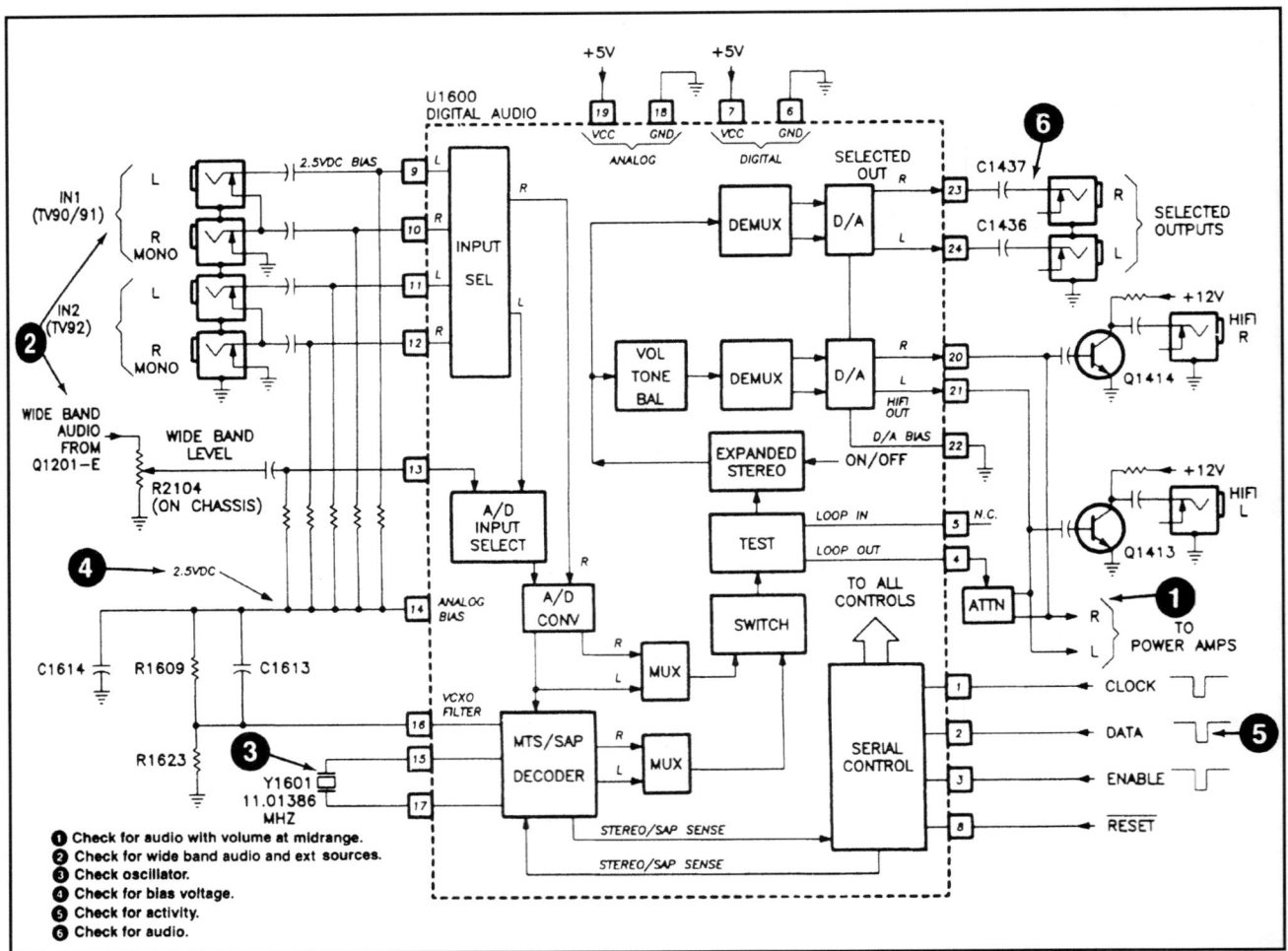

Figure 2-25. Digital audio IC.

will not lock onto a pilot signal if the free-run frequency is too far off. Remember, system control checks for stereo availability and SAP about four times a second via the bus. If stereo or SAP is present and has been selected by the viewer, U1600 is given instructions to select stereo or SAP.

The right and left outputs of the stereo decoder are multiplexed and applied to a switch that selects between tuner audio or external audio (*Figure 2-26*) and is then sent to an expanded stereo stage to increase stereo separation. Expanded stereo can be turned on or off from the audio menu.

Audio from the expanded stereo stage follows two paths:

The first path is through the volume, tone, and balance stage. Volume, tone, and balance are controlled via the serial bus instead of individual voltage controlled inputs we are accustomed to seeing. After volume, tone, and balance control, the audio signal is demultiplexed to produce right and left channel digital information. The digital-to-analog converter then produces right and left analog signals at pins 20 and 21. The signals are buffered by Q1413 and Q1414 before being sent to the AUDIO OUT jacks. The customer has AUDIO OUT, which can be routed to an external amp but controlled via the television. The outputs at pins 20 and 21 are also applied to the internal audio amp in the TV.

The second path is to a demultiplexer stage and a digital-to-analog converter. The right and left outputs of this converter are at pins 23 and 24 and are capacitively coupled to the selected output jacks. The difference between pins 23-24 and 20-21 is that the former is not affected by volume, tone, and balance processing. The customer, therefore, has a fixed audio level output.

Digital Audio IC Troubleshooting

One of the first symptoms you might encounter is "picture but no sound." Don't do as I have done more than once, begin troubleshooting without checking the basics. What do I mean? Make certain the volume is <u>not</u> at minimum, the speakers are turned <u>on</u>, and the customer has <u>not</u> turned on an external audio input source!

Once you have made these checks, listen for audio. If you can hear anything, the problem will more than likely lie in the wideband audio from the tuner or in the MTS/SAP decoder section. You should try both tuner and external audio. If you hear nothing, check for audio at pins 20 and 21 of U1600. If audio is present, the problem may just be in the power amp. If it is not present at pins 20 and 21 but is present at pins 23 and 24, the problem will more than likely lie in the volume control section of U1600 or in the serial communications bus.

Another likely problem is that the stereo indicator never comes on, even when the TV is tuned to a stereo signal. First, select "mono" in the audio menu. Then check for the proper signal at pin 15 of U1001. The frequency should be within ±500 Hz of the stated frequency. If there is no oscillation at all, the chip will not work. I have experienced this only once. If the frequency is correct, check for 2.5 VDC at pin 14. The audio signals are biased at 2.5 volts before entering the IC. If these checks are okay, check for wideband audio level at pin 13.

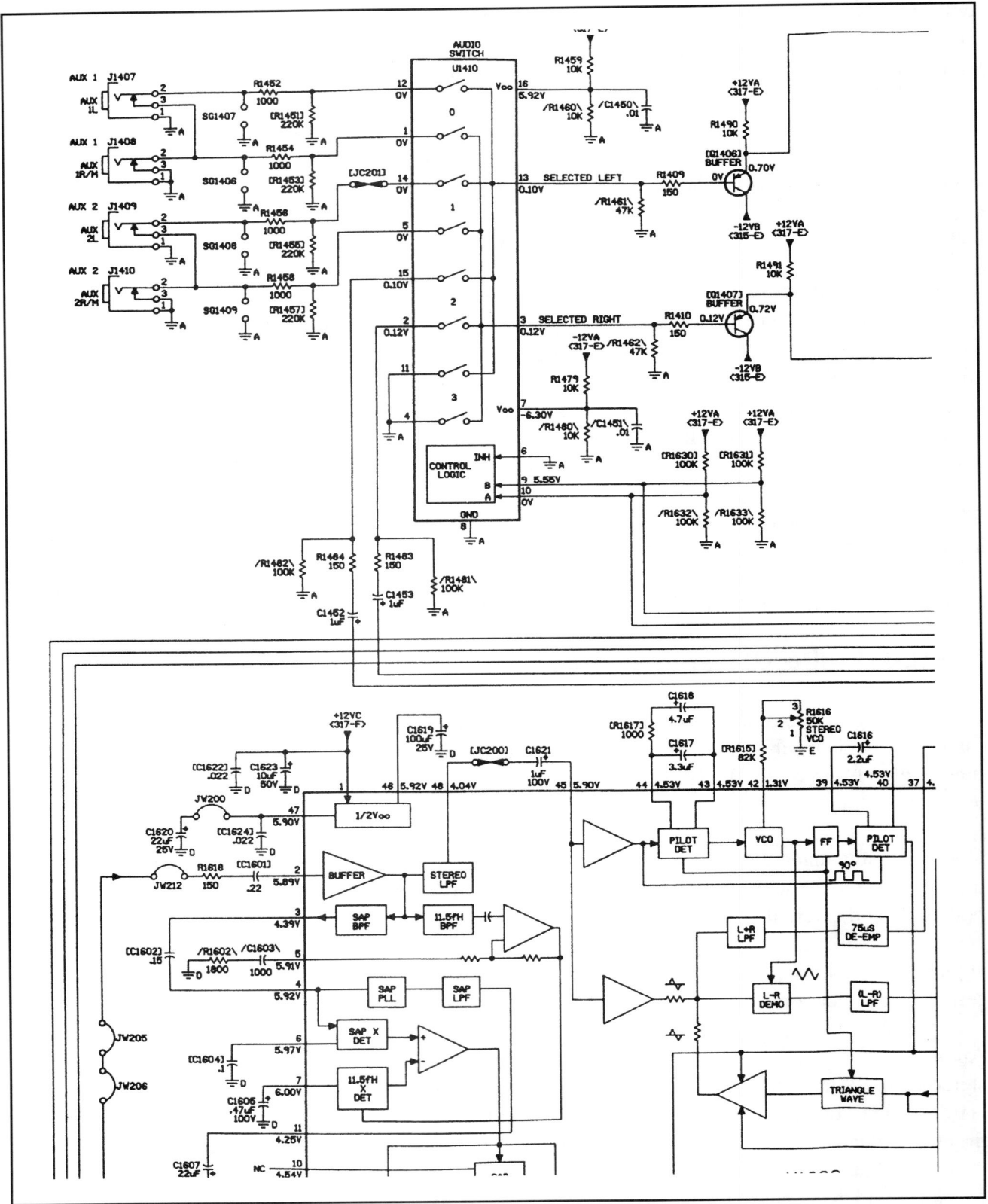

Figure 2-26. Audio circuits.

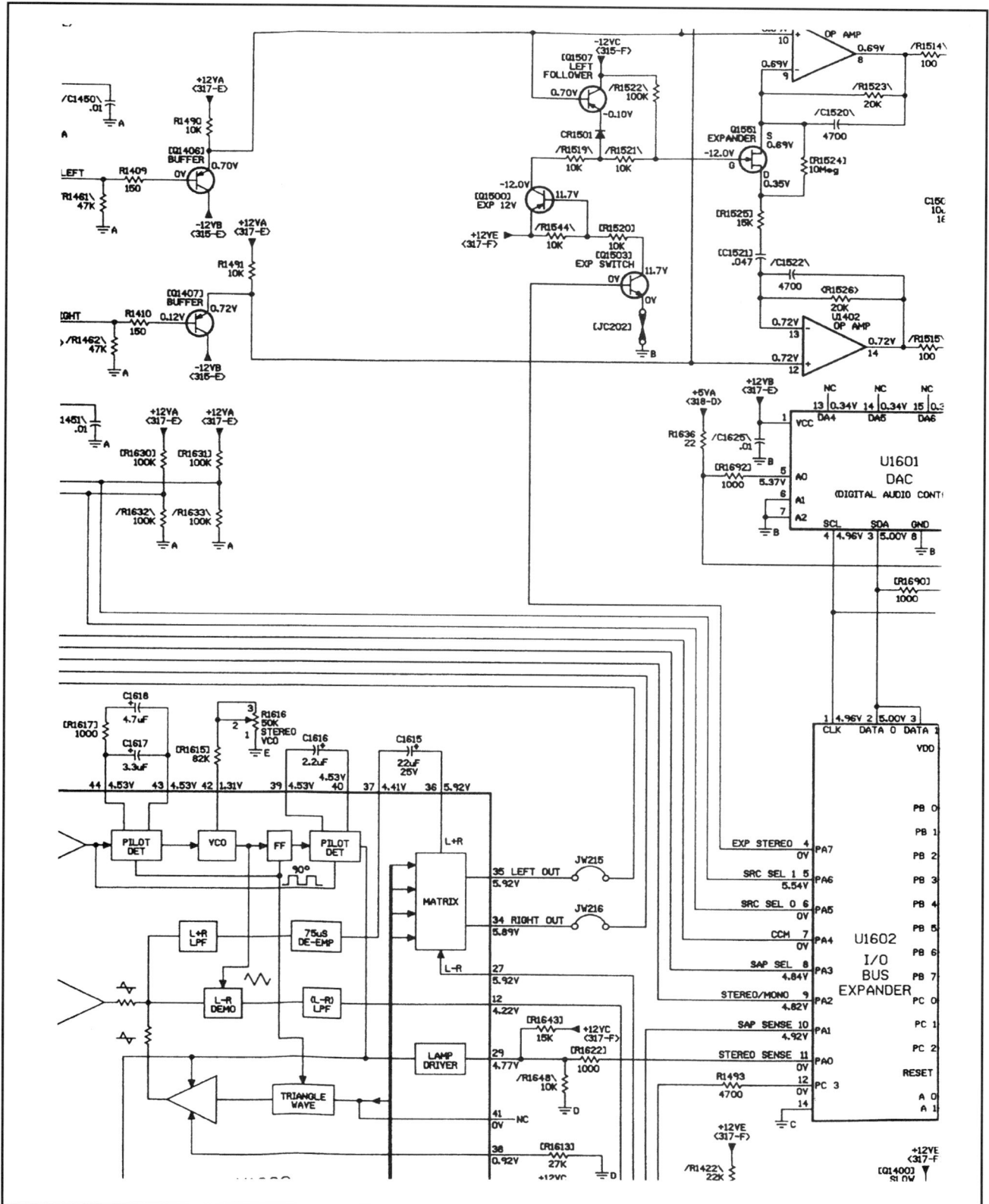

Figure 2-26. Audio circuits (continued).

Power On/OFF Mute Circuit

The CTC169 has a circuit which mutes the audio when the TV is turned on or turned off to prevent pops from being heard in the internal speakers (*Figure 2-27*). Depending on the model you are servicing, the circuit may be mounted on an adapter board or may be on the audio SIP board. It works the same wherever it is mounted.

When the TV is off, C1111 charges from the 26-volt line. System reset is low, keeping Q1111 off. Q1110 is turned on via the 26-volt line, keeping the other transistors on. These transistors pull the audio lines low. The result is no sound from the speakers. When the TV comes on, system reset goes high, turning Q1111 on, grounding the anode of CR1110 and removing the 26 volts to Q1110, which turns off, turning off the other transistors. With these transistors off, the audio lines are allowed to go high.

If you suspect a problem in the audio mute circuit, check Q1110. You should find no voltage at the emitter with the set turned on. If there is voltage on the emitter, Q1112-1115 will be turned on, muting the audio. It's a quick and easy check and will tell you fairly quickly where the problem lies, but you must make sure the set has been turned on for several seconds before you make the check.

The Audio Power Amplifier

Depending on the model, you will encounter one of three power amplifier configurations. The direct-view models use the 1-watt or 5-watt amplifiers; the projection models use a 10-watt amp (*Figures 2-28*

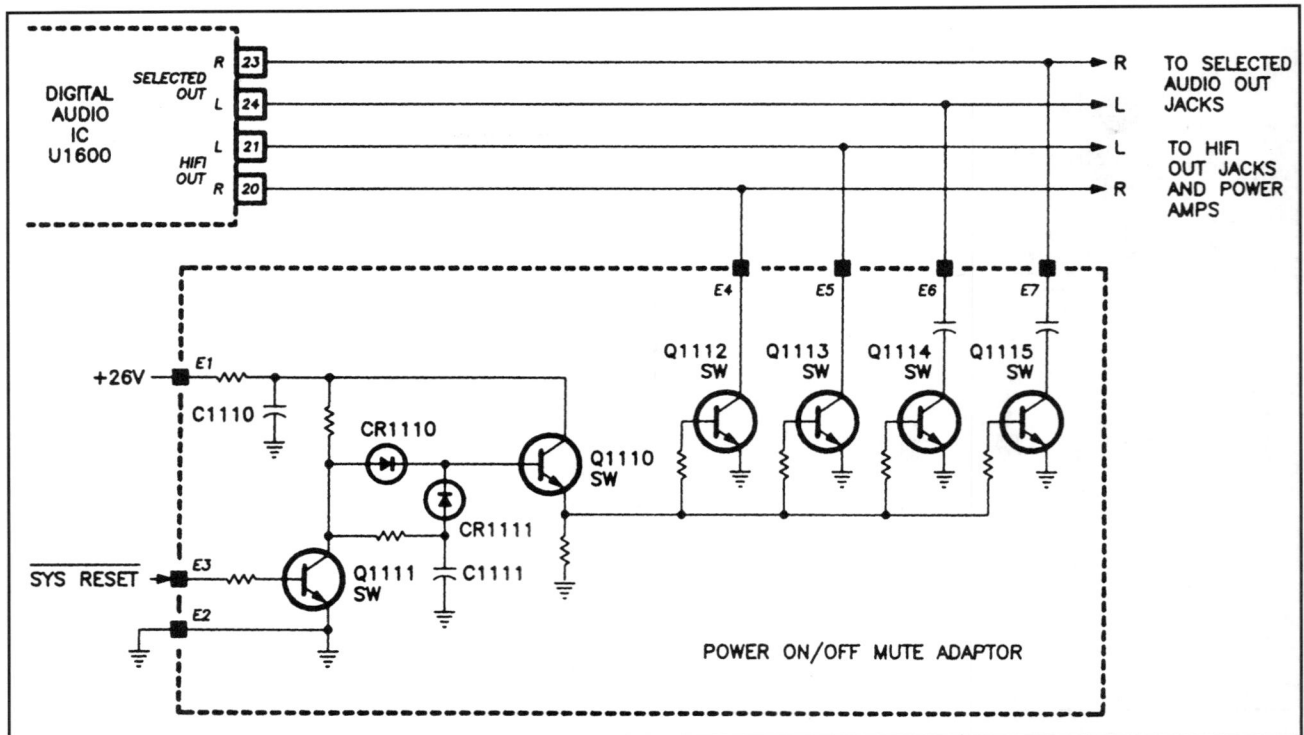

Figure 2-27. Power on/off audio mute.

and *2-29*). The amplifier for the direct-view televisions is designated U1900 and can be configured to output 1 watt or 5 watts. The difference will be in the size of the heatsink, the 5-watt amp requiring a bigger sink.

If you suspect a problem with the power amp, first check to make sure the speakers are turned on! Then supply a signal from a signal generator and get set to make a few checks. I encourage the use of a signal generator as opposed to an off-the-air signal because, as I have said, a signal generator outputs an easily recognized signal that does not change in amplitude. I usually set my generator for a 1000 Hz sine wave.

Begin troubleshooting by confirming 26 volts B+ and audio input at pins 1 and 5. Be sure to watch the heatsink as you troubleshoot. If it gets hot, turn the TV off because you probably have an audio output IC that is shorted internally. There is no need to add to the trouble you already have! Of course, if you have proper voltages and waveforms at the input to the amp but no output, you will most likely have a defective IC. Before you turn on the TV after replacing the audio output IC, check the speakers for shorts or opens.

The 10-watt amplifier is configured differently and can present the troubleshooter with a different set of problems. As you can see, the amp is direct-coupled to the speakers. To prevent damage to the speak-

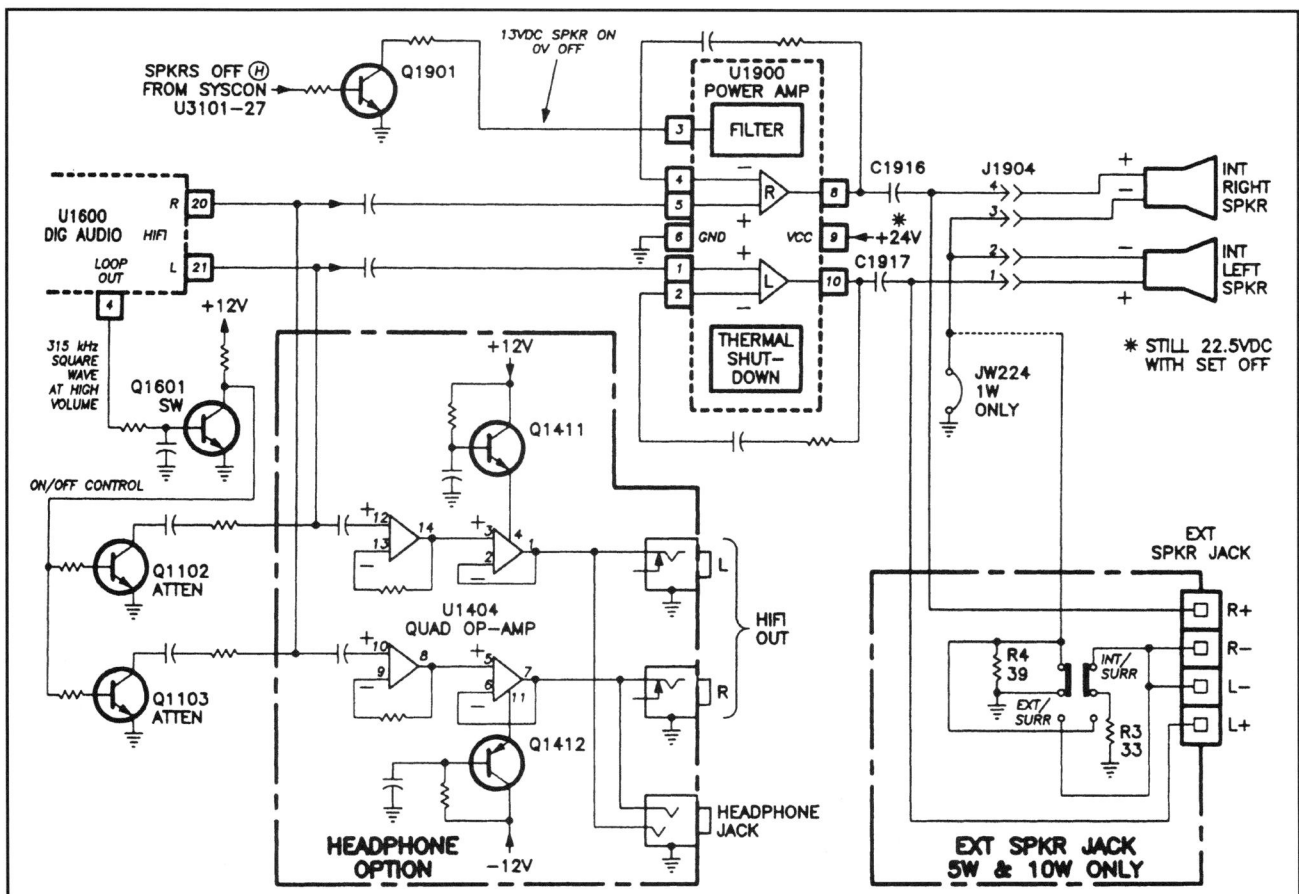

Figure 2-28. 1W/5W audio output.

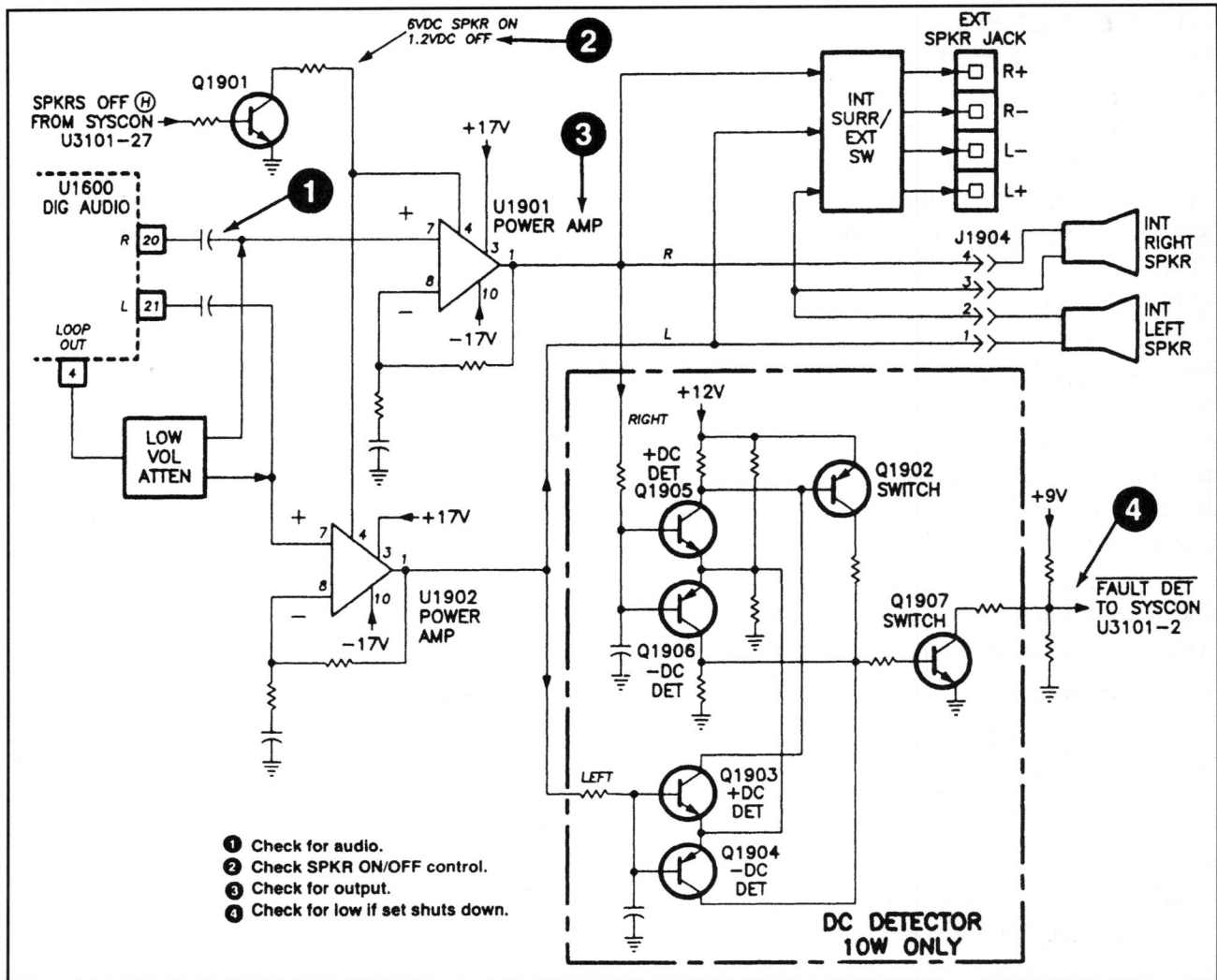

Figure 2-29. 10W audio output with fault detect.

ers, the projection sets have a DC fault detect circuit which will turn the TV off if it detects a high positive or negative DC voltage on the audio output lines.

If you suspect the TV is shutting down because the fault detect circuit is activating, confirm your suspicion by doing the following:

(1) Move the Int/Surr switch to the Ext/Surr position, disconnecting the internal speakers.

(2) Ground the base of Q1907 to make sure it will not turn on and send a fault detect signal to the microprocessor.

(3) Apply AC power and turn the set back on. If the set comes on, turn it off immediately and check the audio output stage for shorts and/or defective components. If the TV still does not come on, check for a loss of the 9-volt run voltage supply.

The Rest of the Story

I wish I could quote Bugs Bunny with, "That's all, folks." There is, unfortunately, more. I have just covered the basics and not really gotten into the fine details. What you find in the audio circuit will depend on the model you are servicing. If you have major problems, you can always order the audio SIP board as a unit and replace it. However, it is relatively expensive (about $110.00 if I remember correctly) and to our advantage to repair it if we can.

Let's take a look at the audio flow from U1410 to U1900, the audio switch to the power amplifier (*Figure 2-30*). Note the differences between the schematic and the technical training manual, and remember it. Remember also to use the correct supplement for the model you are servicing, because one schematic will not cover all models.

There are lots of goodies to cause trouble here. For example, I clearly remember the television the owner said had perfect picture but no sound. He said it happened while he was watching TV during a thunderstorm. I confirmed his complaint and checked to make sure the speakers were on, the audio level was set correctly, and tuner audio was selected. Then I injected a signal from a signal generator into the tuner, turned on my scope, and began probing the audio circuit at U1410. I had audio input but no output. The audio switch was defective. A new chip passed audio, but I still had the same symptom. I traced the signal to the input of U1402 but found nothing at its output. The op-amp was also defective. A new op-amp sent the signal to U1501, which did not pass the signal. It was also defective. I had to replace it and the power amp to get audio out to the internal speakers. Incidentally, I checked to see if audio could be injected at the audio input jacks. Since it could not, I started with the audio switch instead of the stereo decoder.

Did the set work? Well, after a fashion, because the audio was terribly distorted. After some hair-pulling and a call to the tech line, I isolated the problem to Q1400, one of the "slow-start" transistors. There are five of these little fellows (*Figure 2-31*). They all checked good except for this one. It is a surface-mount job and might be a little difficult to find because the legend on the circuit board is difficult to read. I won't say Q1400 fails often, but I will say I have had to replace it for causing distorted audio several times. So, look out for it when you encounter the "distorted sound" symptom. One of the keys the slow-start transistors may be the problem is incorrect voltage readings around U1411. The voltages may be just slightly off, and you may be tempted to ignore it. Those "slightly" off readings may be pointing you to one or more leaky slow-start transistors.

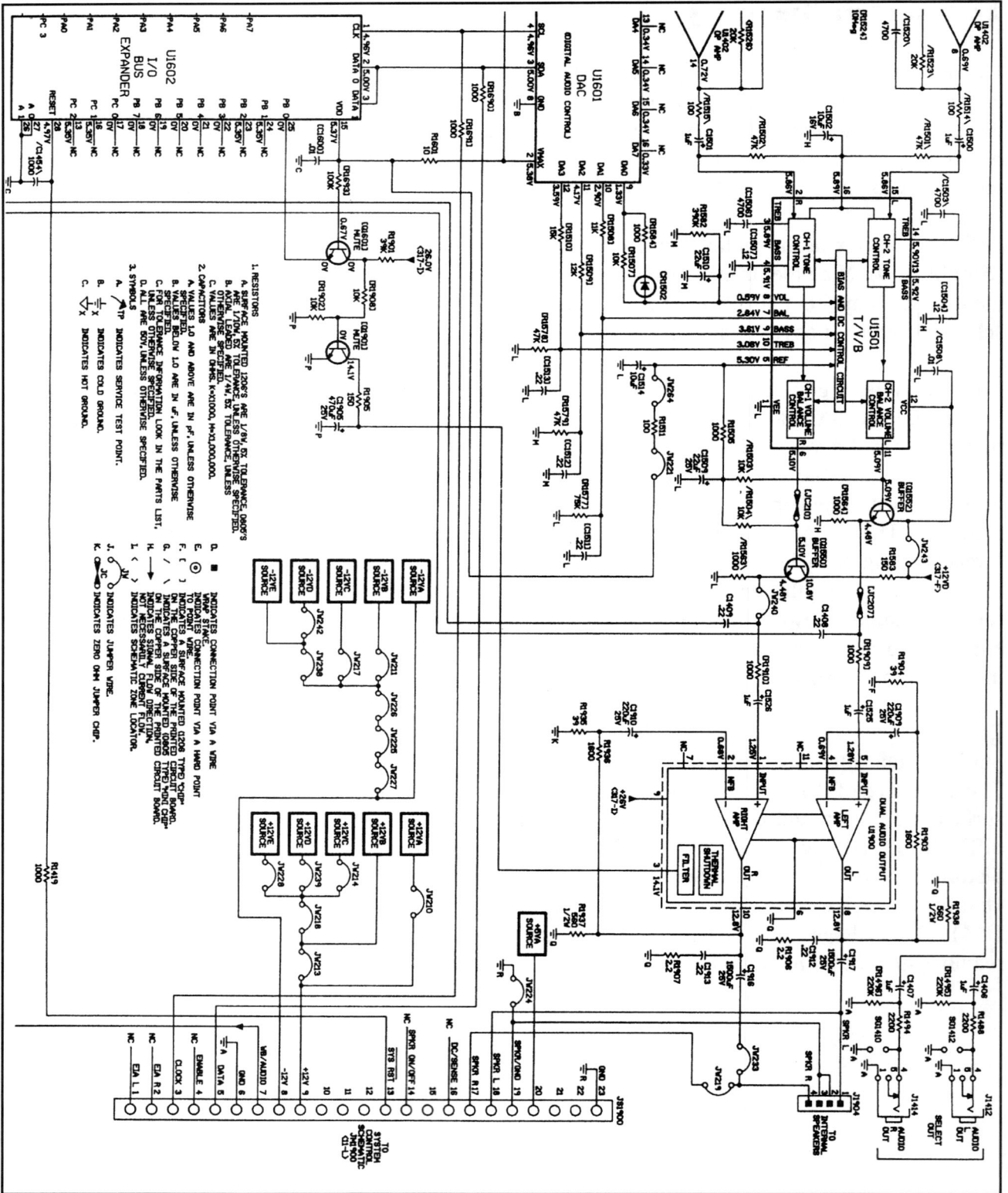

Figure 2-30. Audio flow.

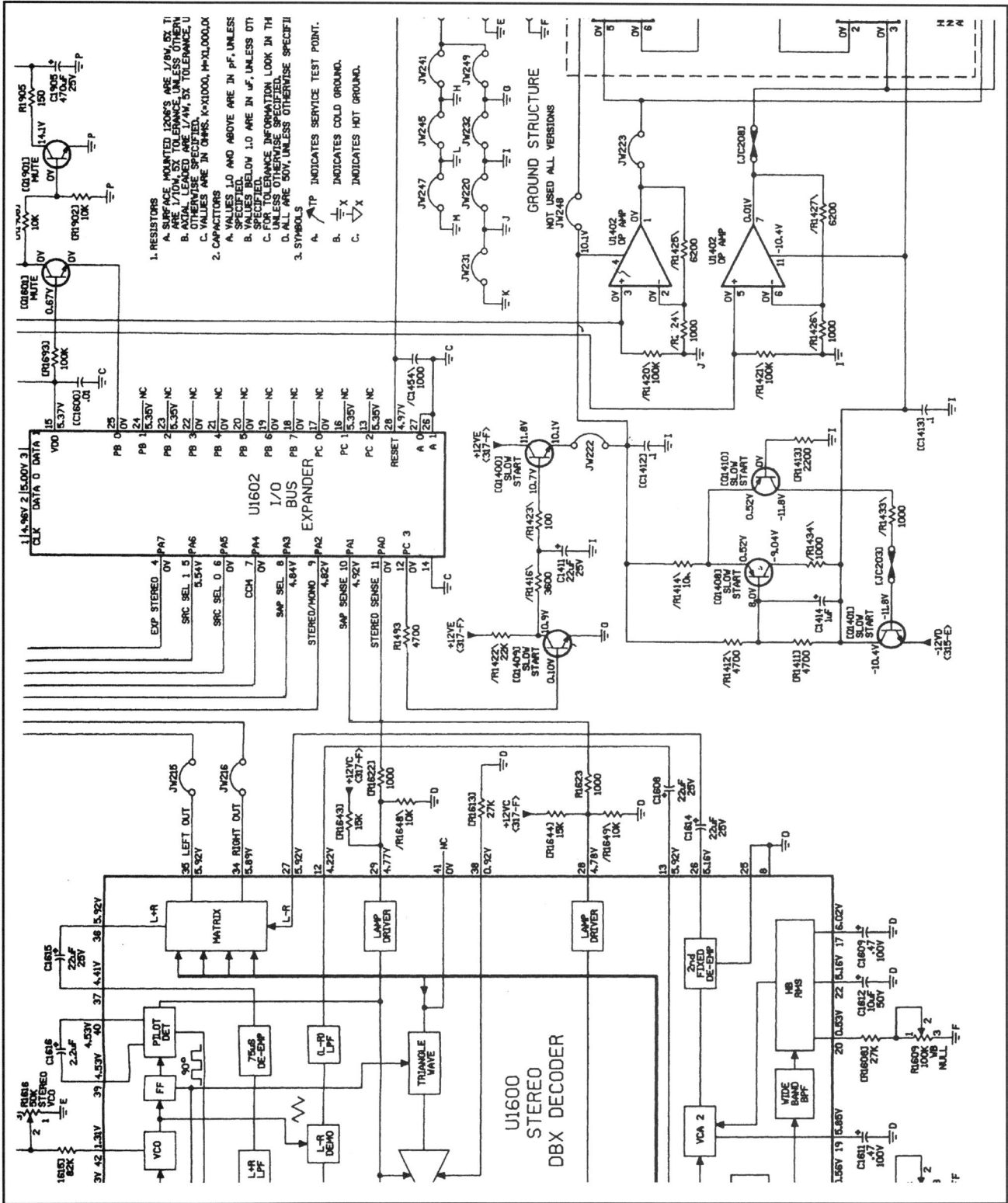

Figure 2-31. Slow-start transistors.

CHAPTER 3

CTC175, CTC176, CTC177, and CTC187

I first became aware of the CTC175/177 family when I attended a Thomson service school in Memphis in 1992 taught by the now-deceased Ken McDaniels. The technology was so new it left many of us speechless—the use of an EEPROM to store chassis data, thus eliminating the familiar pots, variable caps and inductors, and the infamous TOB (tuner-on-board) that we service centers had to learn to service. I remember shaking my head and wondering how we small, independent servicers would ever learn our way around something this new and innovative. But Thomson's new technology merely paved the way for future developments, because many manufacturers simply followed suit. Moreover, and much to my surprise, I discovered I actually like the new way of doing things.

As I have done in preceding chapters, I will begin by giving you a small bibliography, acting on the assumption the more you know the better equipped you are to service the product. I have two articles in *Electronic Servicing and Technology*. The first is "Servicing the RCA CTC 175/177 Color Television Set" in the November 1996 issue; the second is "Servicing EEPROM Problems in RCA Televisons" in the July 1997 issue. Glen Kropuenske has two articles in *Sencore News*, one in the July/August 1995 issue (#170) entitled "Quality Repairs For 'On The Board' Tuners," and the other in the Sept/Oct 1995 issue (#171), "Service Alignment of the RCA/GE CTC175/176/177 Chassis." RCA has two publications I consider a must for the serious servicer: *CTC 175/176/177 Technical Training Manual* and *CTC 177/187 Troubleshooting Guide*. You can obtain the Thomson publications from your local RCA/GE distributor or by writing to the address I gave you in the introduction.

An Overview of the Thomson Televisions

These chassis are very similar, the differences being the power supply and the features each model offers the consumer. The CTC187 utilizes a newly-designed microprocessor, which necessitated slight changes in the circuitry and the addition of U1600, a stereo decoder IC. With these exceptions, the CTC187's system control is functionally the same as the CTC177. The CTC175 is a hot chassis which uses a linear regulator in the power supply to develop B+ for horizontal deflection. Therefore, it does not come equipped with audio or video input/output jacks. The CTC 176/177 and 187 have a switching power supply that permits a cold ground. They also have a fully-featured jack pack.

Thomson incorporated four innovations into these TVs: (1) All alignments are performed digitally using a remote control. Digital alignment does away with certain mechanically adjustable parts, like potentiometers, variable capacitors, and variable inductors; (2) The tuner is an integral part of the motherboard. Since the motherboard or main chassis is not available as a separate module, the tuner requires troubleshooting to the component level; (3) Full-feature models contain a new PIP which is located on the main board. It has just two ICs, the PIP processor (U2901) and the SRAM memory (U2902); (4) Later versions follow the mandated FCC regulation and come equipped with a closed-caption decoder. The decoder is primarily contained within the system microprocessor and requires very little, if any, service.

Of course, different models support different screen sizes. Picture tubes are avaliable in 19", 20", 25", 27" and 31" sizes. Various options accompany the different sets. Full-feature sets will have the new picture-in-picture circuit with S-video input and the standard video/audio in/out jacks. And, as you should expect, the new circuits work in ways that require some getting accustomed to. I will point out the new way of working as we go through the different circuits.

There is one other aspect of the new technology I want to discuss. The tuner-on-board circuitry necessitated the use of a circuit board with less capacitance than the traditional one. RCA developed a board made of a glass-steel composition as a way to accommodate it. The new circuit board requires a slightly different approach to service because it can be easily damaged by heat. If you use too much heat, you will cause the circuit traces to peel off the board, especially the small traces, for which RCA is "famous." So be careful not to apply too much heat when you work with it. The key is using a "hot iron" (about 700 °F) and getting it on and off the circuit board quickly. The hot iron melts solder rapidly, permitting you to do the job without lingering. It also permits you to desolder components using solder wick without overheating the connection and damaging the delicate traces. Of course, a desoldering tool is better, but we don't always have access to one, do we?

The Tuner Wrap Problem

Everyone has to know about the tuner wrap problem by now. It cropped up as soon as these televisions appeared on the market because of a thermal mismatch between the tuner wrap (or tuner shield) and the printed circuit board. These two components heated and cooled at separate rates, causing cracks in the solder that mated the two together. The cracks in the solder caused a variety of symptoms: loss of picture quality, loss of vertical deflection, loss of audio, reduced vertical height, and EEPROM corruption.

The tuner-on-board (TOB) has caused more problems for Thomson than anything I can remember. If reports are reliable—and I believe they are, it also cost them quite a lot of money. It was also a problem nobody seemed to anticipate—except us servicers, that is! The engineers took a rectangular piece of metal, soldered it to a circuit board in a spate of places, and expected it to work. How many times have we had to resolder large components to a PCB after the product had been in use for a few years? Ever fix the older MSC/MST and MTT modules by removing the covers and resoldering the wirewrap stake around the periphery? If you have, you know what I mean.

Thomson has spent quite a lot of money and tried a variety of fixes to solve the problem. I have, I think, three service bulletins which discuss it and offer a fix. Service bulletin TV 96-005 detailed "the final fix." I guess the problem has been solved, because I don't see it very often in the newer sets, but I see a variation of it every now and then. The cracks in the solder are no longer present, but solder is lacking. I have repaired several of the tuners in the newer sets by soldering what the manufacturing process left unsoldered.

The "final fix" involved several changes in the original design:

First, the tuner wrap itself was changed. The original wrap was made of steel. It was changed first to copper and later to zinc, the material found in all the newer sets. Thomson warns us that all three wraps are tin-plated and look the same. You can identify the copper wrap by removing the top cover of the tuner and looking for copper visible along the top edge. You can distinguish zinc from steel by using a small magnet which will adhere to the steel but not to copper or zinc. It is important to know which wrap you are working with because zinc melts at about 800 degrees! Therefore, keep the temperature of your soldering iron at about 775 degrees when you work on it. If you don't, you will melt the zinc wrap and possibly have to replace it! The copper and stainless steel don't require the gentle handling. Again, know which wrap you are working on before you proceed with repairs.

Second, the engineers at Thomson came up with a new solder for the servicer to use when he/she repaired a tuner wrap, particularly if the job was covered under warranty. Preparatory to applying the new solder, the tech was instructed to remove completely all solder from the interior tabs of the wrap and from the copper pads. The kit (called S-kit-1) contained solder, paste flux, and a solder template (reproduced in *Figure 3-1*). The new solder was then applied using the paste flux and the enclosed template as a guide. When it cooled, the solder remained elastic and would expand and contract as the PCB and the tuner wrap expanded and contracted.

Incidentally, the template (*Figure 3-1*) is a handy tool to have around. It is easy to miss some of the smaller solder connections inside the tuner area unless you go over your work several times before you reinstall the bottom shield. The template makes missing these small areas difficult. Make a copy of it, and try it. You might be surprised how handy it really is.

Third, the tech who repaired the wrap was expected to install six jumper wires at specific locations. *Figure 3-2* gives you the particulars. The jumpers were then to be secured to the PCB by hot glue! I confess to a certain reluctance in putting jumpers and hot glue on a new television, especially one as nice as these televisions are. But I did it anyway if the warranty was in effect.

Fourth, the printed circuit board was also changed, but the change involved just a few of the copper pads in the tuner area. The new chassis built on the revised board is called "chassis version II" and is not supposed to need the extensive repairs to the tuner area "chassis version I" requires.

I installed the jumpers and used the new solder only if the job was covered under warranty. Like you, I discovered the problem before Thomson ever mentioned it, and I solved it by using rosin-core solder. I don't know how many hundreds-upon-hundreds of these sets I have repaired. I do know we have never had a single recall for a "tuner shield job." Based on my experience, I saw no need to alter what we were doing by changing our procedure for out-of-warranty repairs. So, S-kit-1 (what remains of the ones I ordered) lies unused in a storage bin.

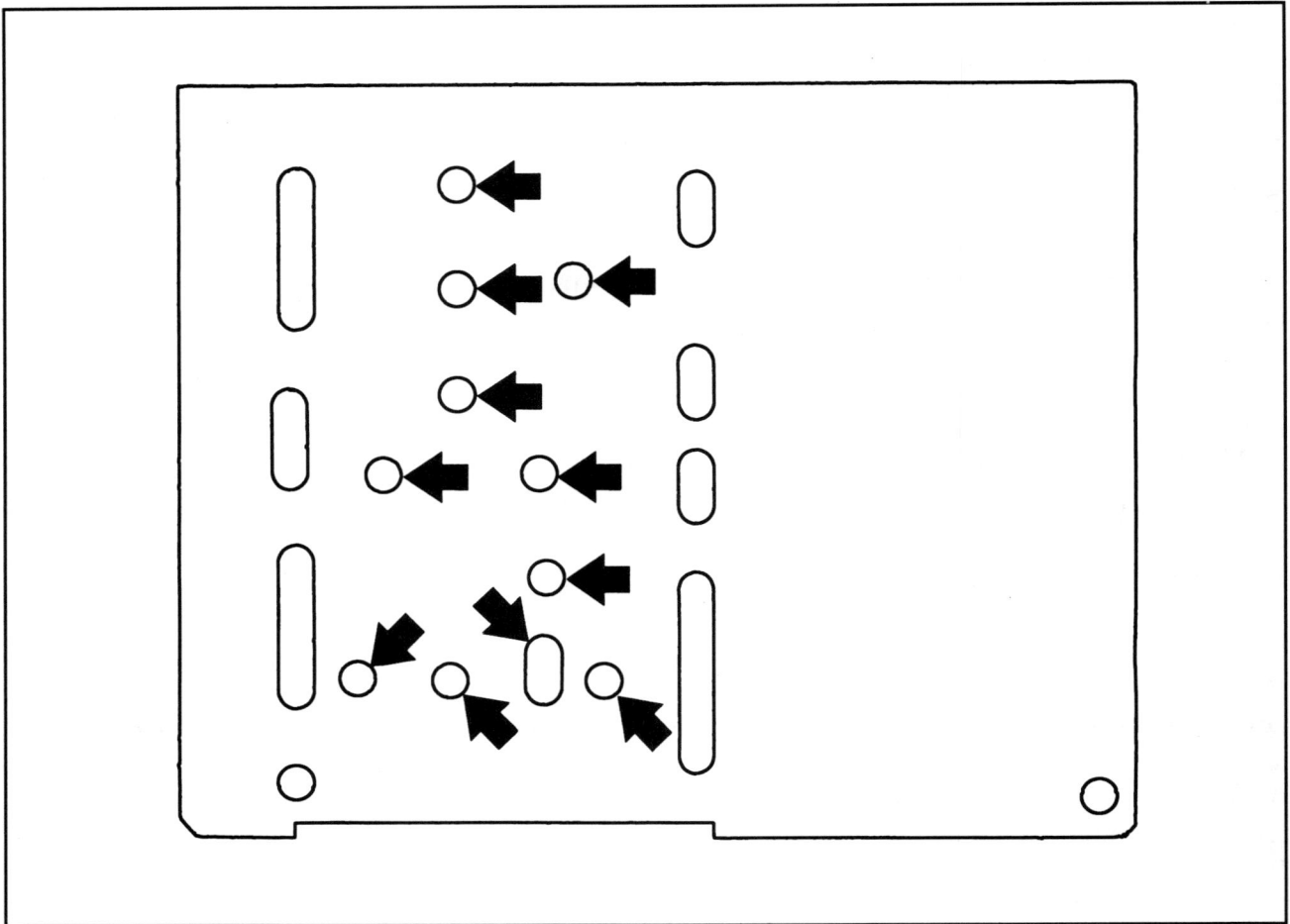

Figure 3-1. S-Kit-1 template.

The microprocessor wrap or shield suffered a similar fate, but it did not give nearly the scope of problems the tuner shield did. I haven't seen more than a dozen or so microprocessor shield problems in the last five years. The loose shield around the microcomputer caused a variety of symptoms, ranging from loss of audio and video to intermittent shutdown. So, along came service bulletin TV 93-001, reproduced in part in *Figure 3-3*. You can see for yourself where the problem areas were and what the resulting symptoms were.

I don't want to use this book as a soap box or a pulpit, but I do wonder how much time and money Thomson would have saved if their engineers had followed the conventional way of doing things. I remember asking our field service engineer a couple of years ago if Thomson had plans to revert to the traditional modular tuner. He said there were no such plans. End of discussion! Price shouldn't be the issue now. For example, I just purchased three tuners from Zenith for their new sets. The price for each was less than $29.00! Makes you think, doesn't it? However, I will admit that the TOB has turned out to be one of the most reliable, if not *the* most reliable, tuner in the many televisions I service. It rarely gives trouble, and the trouble it causes is often relatively easy to fix.

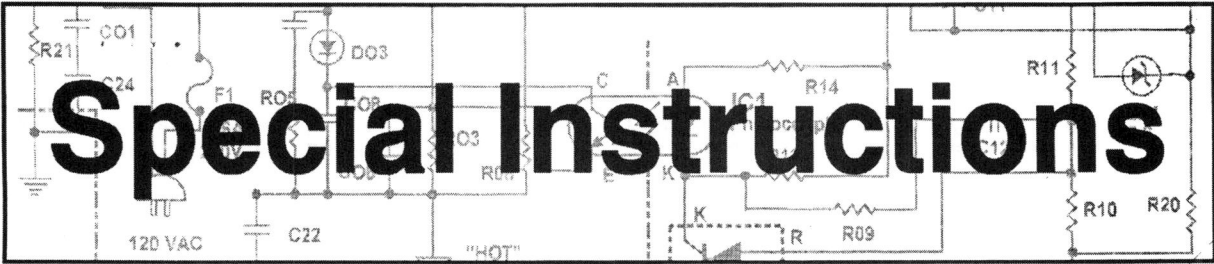

Special Instructions

PROSCAN RCA GE

SPS 4382-1

Stock Number S-Kit-1

Contents: Solder and Paste Flux (enough for 10 repairs) and Template

Procedure: The special solder supplied in the kit is <u>not</u> a rosin-core solder. <u>Paste flux must be used</u> to get the solder to flow. The solder included in the kit remains elastic when cool to prevent joints from breaking due to thermal expansion. It is necessary to remove the tuner bottom cover to gain access to the ground connections on the tuner wrap. *Use caution when removing the solder from the ground tabs; excessive heat can lift/damage the copper traces.*

Note: The tuner wrap was changed from steel to copper, and then to zinc. All three materials are tin-plated, and therefore look the same. The copper wrap can be identified by removing the top cover of the tuner (the cover that is not soldered on). Copper will be visible at the top edges of the wrap. Copper is an excellent conductor of heat, therefore, it is necessary to apply more heat to remove the bottom cover of the tuner.   The zinc wrap will look identical to the steel tuner wrap.   However, a small magnet will stick to the steel tuner wrap but will not stick to either the copper or zinc wraps. The zinc wrap will melt at approximately 800 degrees, so it is necessary to keep the tip temperature of the soldering iron below 775 degrees when working on a zinc wrap.

1. **Remove Solder From Interior Tabs:** (Figure 2)
    Completely remove all solder from the interior tabs of the tuner wrap and from the copper pads. Refer to figure 2 for tuner wrap tab locations. Use the template supplied in the kit as an aid in locating the tabs. Placing the solder braid flat against the wrap tab and perpendicular to the board will minimize the risk of damaging the copper or lifting surface mounted devices.

Figure 1 Install Template

Figure 3-1A. S-Kit-1 instructions.

119

2. Re-solder Interior Tabs:

Apply the paste flux generously to each interior ground connection of the tuner wrap. Use the supplied solder to re-connect all of the interior wrap grounds. The solder has a dull appearance when cool. This is normal and does not indicate a cold solder joint. The solder joint should be smooth in appearance. A rough finish does indicate a bad solder connection. A tip temperature of 750-800 degrees Fahrenheit works well with the solder.

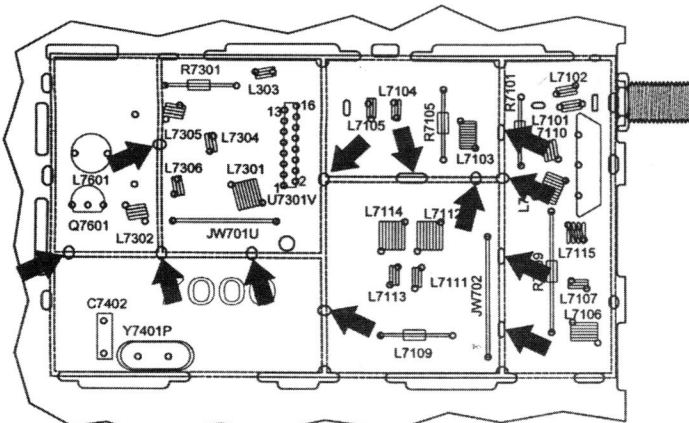

Figure 2 Interior Solder Connections

3. Install 6 Jumper Wires: (Figure 3)

Refer to Service Information Notice TV96-005 to determine if jumpers are needed. Install jumper wires at the locations shown in figure 3. Use 24 or 26 gauge solid hook-up wire for all jumpers. It is important that all jumpers be connected to the printed circuit board and not to the wrap, and that jumpers A, B, C, and D are connected outside of the tuner wrap. *Jumper C must be long enough to fit over the outside of the tuner bottom cover. Jumpers E and F inside the tuner wrap must be kept as short as possible and dressed against the board in order to prevent oscillations.* It is not necessary to use the supplied solder to attach the jumper wires, standard 60/40 rosin-core solder should be used.

4. Re-Solder External Wrap Connections:

Re-solder all of the external wrap connections. Use standard 60/40 rosin-core solder for all external wrap connections.

5. Remove Excess Solder Flux and Secure Jumper Wires:

Remove excess solder flux before attaching the bottom cover of the tuner. Use hot melt glue to secure the jumper wires to the printed circuit board. The jumper wires inside the tuner wrap do not need to be glued in place. The jumper wires on the outside of the tuner should be glued on the insulation near each end.

Figure 3 Jumper Wire Locations

Figure 3-1B. S-Kit-1 instructions.

3. **Install 6 Jumper Wires:** (Figure 3)

Install jumper wires **(on Version I chassis only)** at the locations shown in figure 3. Use 24 or 26 gauge solid hook-up wire for all jumpers. It is important that all jumpers be connected to the printed circuit board and not to the wrap, and that jumpers A, B, C, and D are connected outside of the tuner wrap. Jumpers E and F inside the tuner wrap must be kept as short as possible and dressed against the board in order to prevent oscillations.

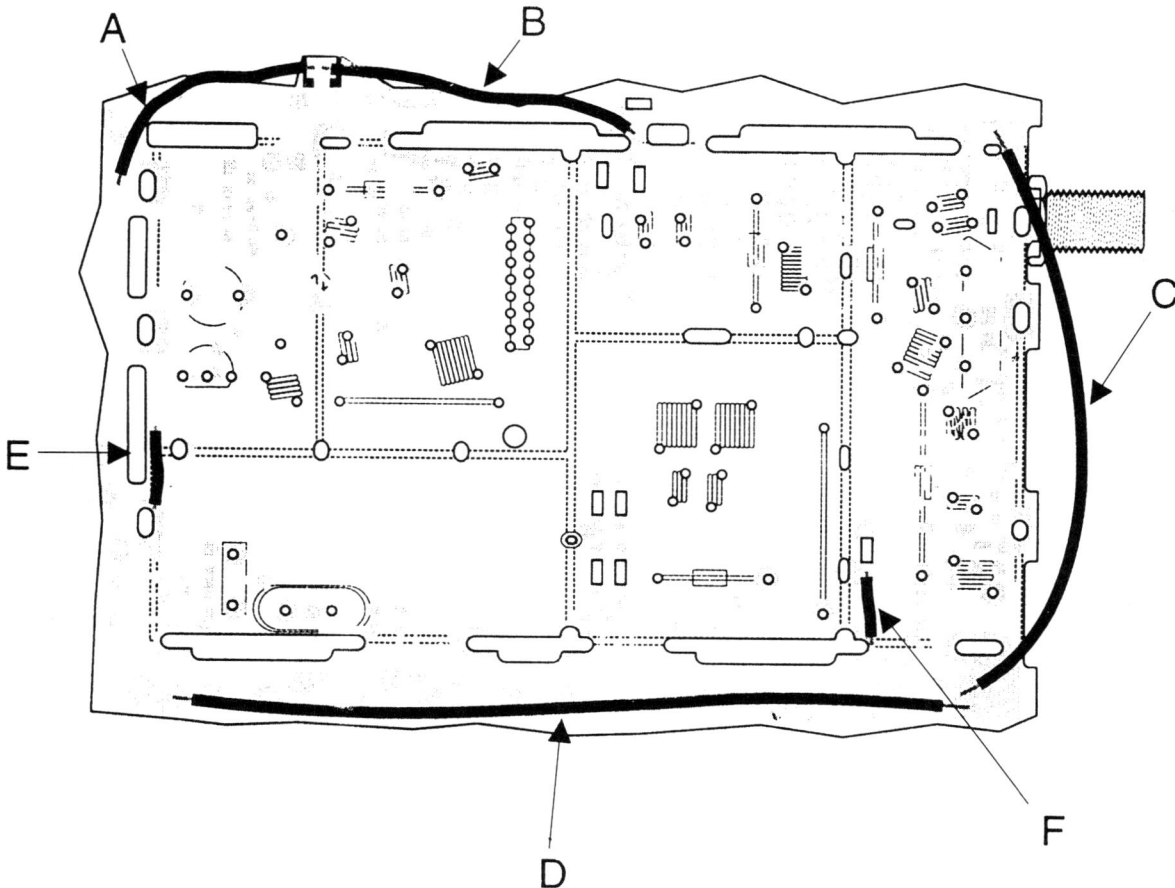

Figure 3   Jumper Wire Locations
(Bottom View)

4. **Re-Solder External Wrap Connections:**

Re-solder all of the external wrap connections. Use standard 60/40 rosin-core solder for all external wrap and jumper wire connections.

Figure 3-2. Jumper wire locations.

RCA **GE** PROSCAN

Product Support • Mail Stop I-400
P.O. Box 1976 • Indianapolis, IN 46206-1976

Television Service Bulletin

TV 93-001

File in Appropriate Section of Your Thomson Technical Bulletin Binder

DATE: June 15, 1993

CHASSIS: CTC175, CTC176, CTC177

Warranty Labor Rate: Intermediate

TOPIC: INTERMITTENT GROUND CONNECTIONS ON MICROCOMPUTER SHIELD

SYMPTOM: A variety of symptoms have been observed ranging from loss of audio and video to intermittent shutdown.

CAUSE: There have been reports of intermittent ground connections on the shield of the microcomputer U3101. The intermittent tab is normally caused by the shield not being seated prior to soldering. The open connections can result in some unusual symptoms.

Testing was conducted to determine the effects of the open ground tabs. The following symptoms result when each ground is opened independently of the others. The effects of multiple intermittent grounds were not evaluated.

OPEN TAB	SYMPTOM
J1	NONE
J2	NONE
J3	LOSS OF AUDIO AND VIDEO
J4	NONE
J5	NOISY PIX
	UNABLE TO TUNE ABOVE CHAN 4
	(INTERMITTENT SHUTDOWN)
J6	NONE
J7	NONE
J8	NONE
J9	NONE

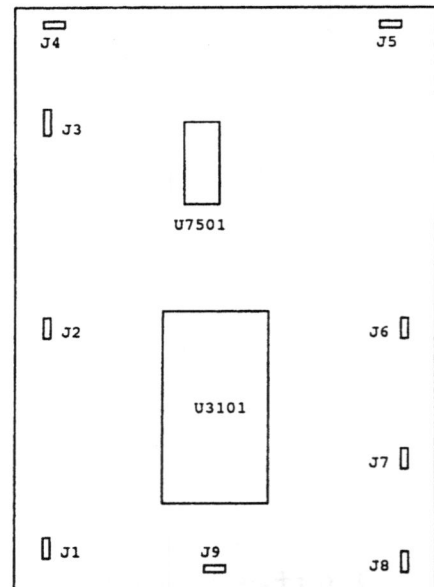

Grounds J6,7, and 8 are used for RFI capacitor grounds. Grounds J4 and J9 have redundant ground jumper wires which maintain circuit continuity.

Cure: Carefully inspect the ground connections, resolder if necessary.

Product Safety Information

Product safety information is contained in the appropriate TCE Service Data covering models/chassis referenced in this bulletin. All specified Product Safety requirements and testing shall be complied with prior to returning equipment to the customer. Servicers who defeat safety features or fail to perform safety checks may be liable for any resulting damages and may expose themselves and others to possible injury.

Figure 3-3. Tuner wrap bulletin.

Repairing the Tuner-on-Board

A common problem with the TOB, with the exception of the tuner wrap, is the RF connector breaking off. I am not sure the service literature even mentions it, but you can buy an exact replacement for the RF connector by ordering part number 2155543. MCM lists it for $1.83. It's far easier to install than one that you've ripped off a VCR RF converter, for example, or torn out of a defective tuner. I make it a policy to keep several in stock, especially with the advent of the CTC185 chassis. Thomson changed things again when it started selling the CTC185, producing, in my opinion, a mechanically less stable shield than the CTC175 family has, because the RF connector pulls rather easily out of the zinc wrap.

If you are like me, you find it economically feasible to send a defective tuner to a repair depot or replace it with a new one. I repair tuners only if the problem is minor and I am reasonably certain of the cause of the failure. You see, I am hampered by not having a way to check tuners out-of-circuit, which means I have to reinstall the tuner in the offending TV to check the fix. Taking the tuner out and putting it back is time-consuming and risky because I will eventually damage the circuit board by soldering-unsoldering-resoldering. I don't have a choice when I face a TOB problem. Of course, I can send the whole board to a repair depot like PTS, but I have to pay about $140.00 for the repair. Not many of my customers will pay such a price to have a 1990s TV fixed, especially if it is smaller than 25". I am, therefore, left with the necessity of fixing the tuner or losing the repair job.

Tuner Basics

Time to review tuner basics, as a way of getting ready to fix tuner-on-board problems.

A tuner does three things: (1) It selects a band of frequencies from all the available frequencies at its input. For example, if my television at home is instructed to tune channel 3, it will select just the band of frequencies associated with channel 3, as opposed to the thirty-odd channels it could select; (2) It amplifies the frequency band (channel) you have selected; (3) It downconverts the frequency band (channel) to a 41.25 MHz carrier for the audio and a 45.75 MHz carrier for the video. As a final act, it sends the audio and video IF frequency on to the IF section of the television for processing.

The tuner-on-board is composed of the basic elements shown in *Figure 3-4*. As a matter of fact, many of the tuners with which we deal will have these same elements. What I am about to say will therefore apply to most of the tuners we see as television repairpersons.

The "front end" has a filter network that will not pass unwanted FM and IF frequencies, which are invariably present at the antenna. It also contains a filter tuned to the frequency you want to receive, and inputs the selected frequency into the RF amplifier. This filter is one of three bandpass filters in the tuner. A bandpass filter, as you know, will pass a "band" of frequencies while rejecting frequencies that fall above or below the band to which it is tuned. It is composed of inductors and capacitors that form a parallel resonant circuit and is "tuned" so that it is resonant only at the desired frequency. The tuning is accomplished in electronic tuners by electronically switching capacitors and inductors in and out of circuit. When the resonant frequency matches the selected frequency, the inductor-capacitor (LC) filter passes the RF signal on to the RF amplifier (*Figure 3-5*).

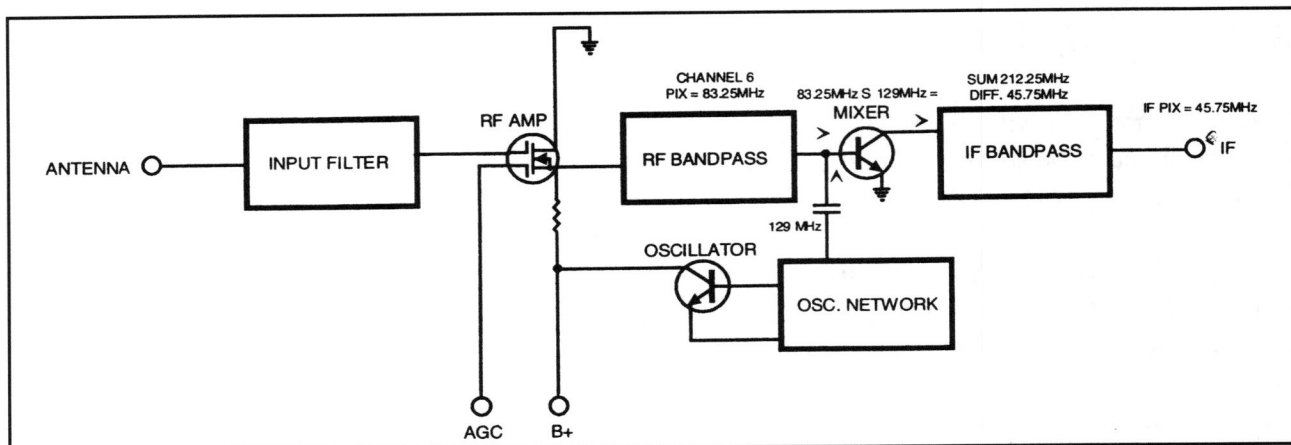

Figure 3-4. Tuner-on-board basic elements.

The RF amplifiers in the TOB circuits are dual-gate depletion MOSFETs (metal-oxide-semiconductor field-effect transistor). These transistors have an extremely high input impedance and are voltage controlled, much like vacuum tubes. They are also normally "on" without gate bias (as opposed to enhancement MOSFETs which are on only when a bias voltage is present). The MOSFETs Thomson uses are N-channel devices, which means a negative voltage applied to the gate with respect to the source will reduce drain current flow. A sufficiently high negative voltage will completely halt drain current flow. A positive voltage on the gate with respect to the source will increase drain current to a point.

We are dealing with dual-gate MOSFETs, which means both gates affect drain current. Gate 1 is configured to receive the RF signal from the input filter while gate 2 receives an AGC voltage. If the AGC voltage rises, drain current also rises, increasing the amplification of the RF signal applied to gate 1. The converse is also true—If AGC voltage decreases, the less the RF signal is amplified. During normal operation, .5 to 5 mV of RF signal is input to gate 1. The transistor will increase or decrease gain, following the amplitude of the RF envelope.

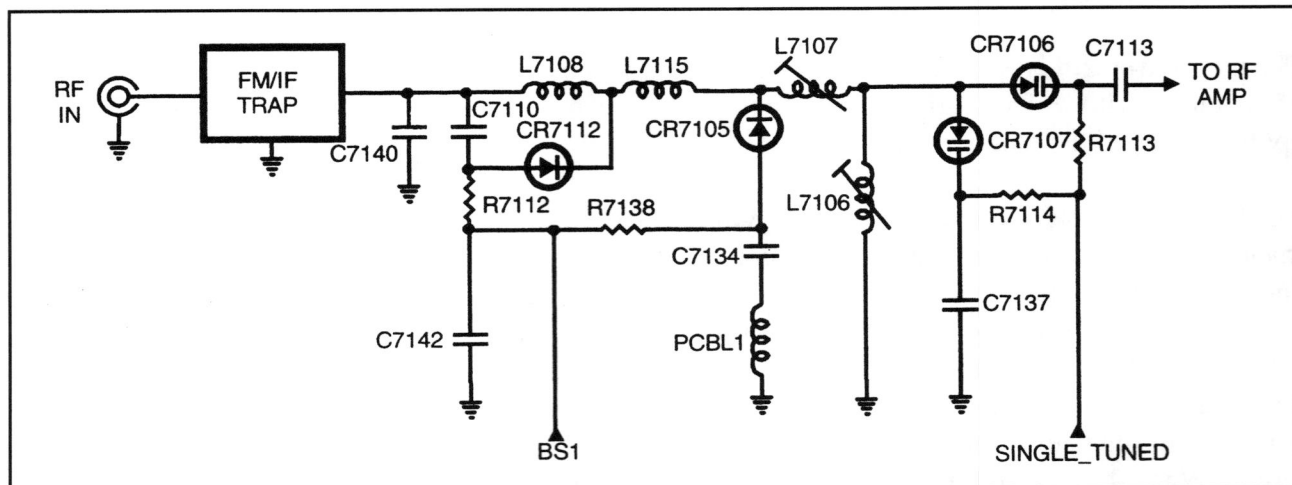

Figure 3-5. Tuner front end.

AGC voltage at gate 2 also controls the gain of the amp. A rising voltage increases gain, while a decreasing voltage lowers gain. At full gain, tuners with depletion MOSFETs will have an AGC voltage ranging from approximately 6 volts upwards to about 9 volts. If the RF signal increases in strength, the AGC voltage drops to lower the gain. This action keeps a large signal from producing interference in the mixer stage. Let me put it differently: AGC voltage ensures the amplified RF signal is of relatively uniform amplitude, neither too small to be of use nor so large that it overdrives the mixer stage. Just to remind you, AGC (automatic gain control) voltage is developed by the T-Chip.

The TOB has two RF amplifiers—one for VHF and the other for UHF—which are switched into the circuit by bandswitching transistors Q7403 and 7405. When a VHF channel is selected, the switch turns on, providing a ground path for the source of Q7102. When a UHF channel is selected, the switch is turned off, removing the ground path for Q7102, which turns it off while turning on the UHF RF amplifier (*Figure 3-6*).

The next block is labeled "RF bandpass." It is a double-tuned filter that receives the amplified frequency and basically retunes it; that is, it performs the sharp tuning necessary to obtain the needed selectivity. Good selectivity is necessary to ensure that the receiver processes just the channel it has been instructed to tune and no other. In addition to improving selectivity, the double-tuned filter also matches the impedance of the RF amplifier to the next stage (*Figure 3-6A*). Note that there are two of these filters following the RF amplifier. Both are bandpass filters like the one I just discussed.

The next stage consists of an oscillator network, a mixer and IF bandpass filter. The oscillator network is comprised of a local oscillator and its control circuitry. It produces a frequency 45.75 MHz higher

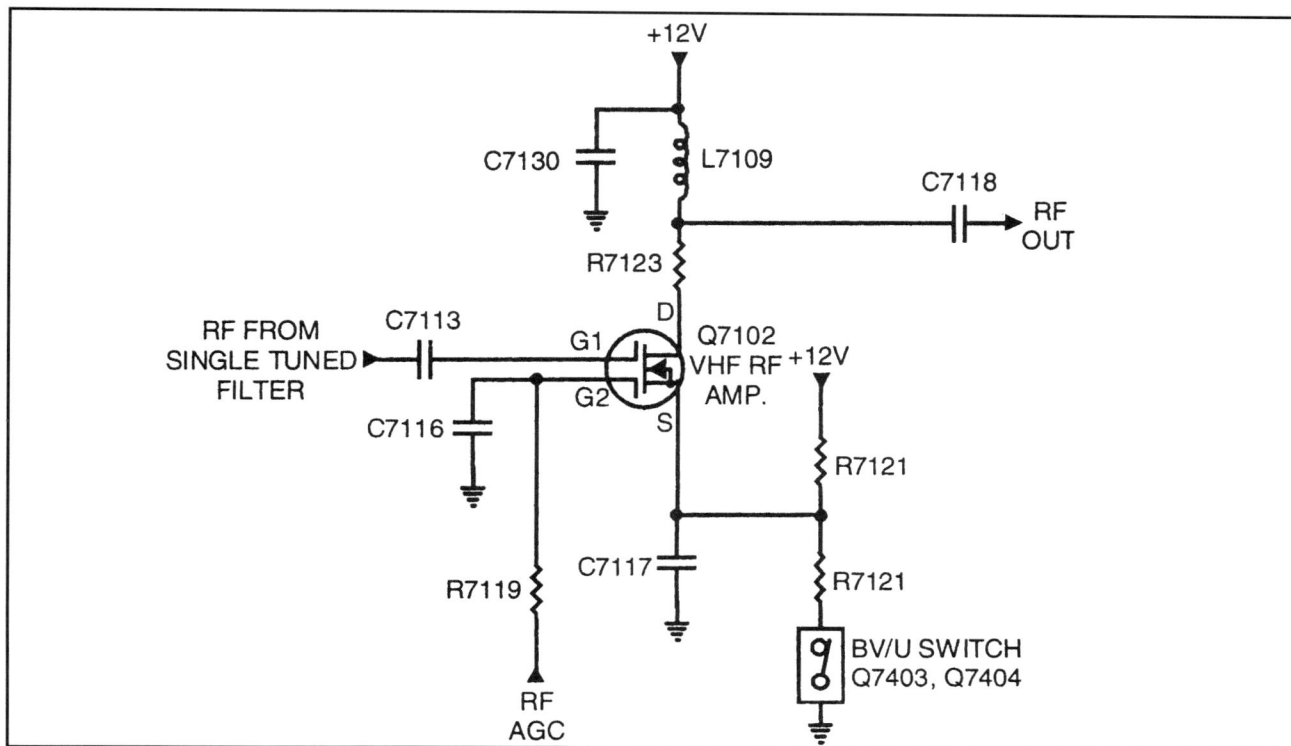

Figure 3-6. TOB RF amp.

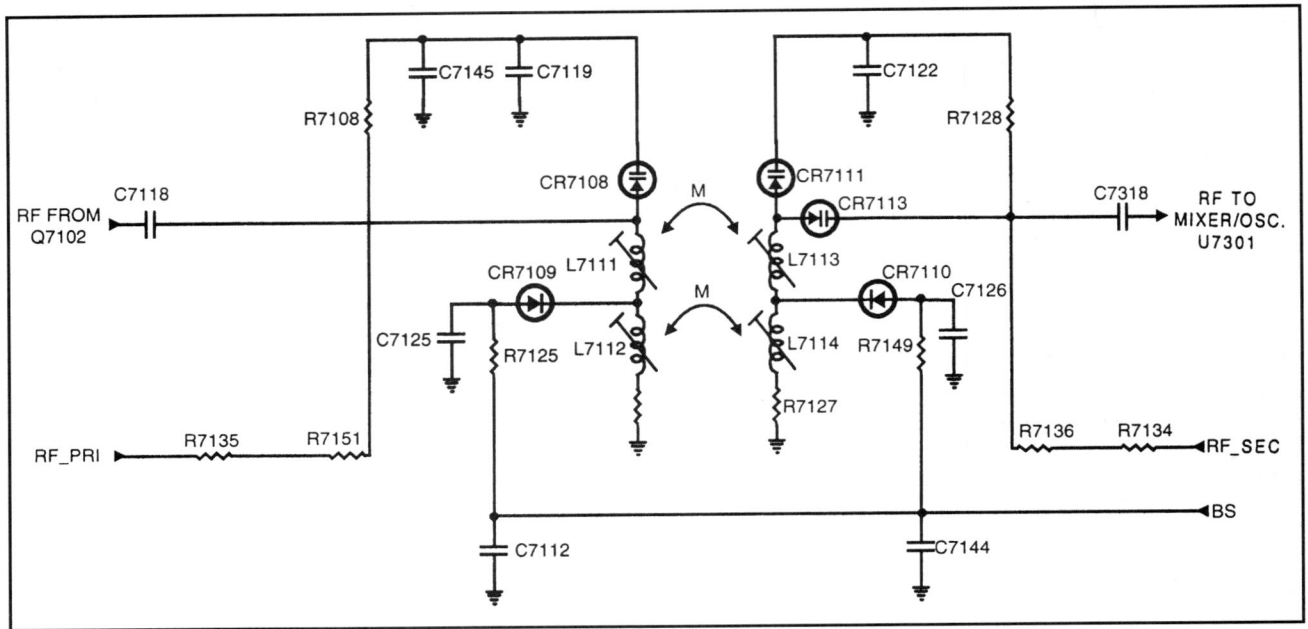

Figure 3-6A. RF bandpass.

than the selected channel frequency and is beat (heterodyned) against the incoming RF signal, resulting in an output from the mixer of four different frequencies: the original RF frequency; the local oscillator signal; the sum of these two; and the difference of these two, which will always be 45.75 MHz (the video carrier IF frequency). The IF bandpass extracts the signal we want, the difference between the RF signal and the local oscillator signal, and passes it on to the IF section of the TV (*Figure 3-7*).

The Tuner-On-Board Versus The Traditional Tuner

The TOB is a tuner, and therefore has much in common with a traditional tuner module, but it is also different. For example, the CTC169 tuner uses a microprocessor-controlled frequency synthesis scheme to tune channels. The microprocessor controls a divider inside a phase-lock loop (PLL) IC which produces a DC voltage that controls the frequency of the local oscillator. The same tuning voltage is sent to the single- and double-tuned circuits that select the correct band of frequencies and serve as impedance matching devices. The local oscillator also operates 45.75 MHz higher than the carrier of the received frequency. *Figure 3-8* is a block diagram of a traditional tuner.

The traditional tuner can be called a "track" tuner because one voltage controls the bandpass circuits and local oscillator; that is, all stages of the tuner "track" together as the tuning voltage changes (*CTC 175/76/177 Technical Training Manual*, p. 40). It is obvious that the same circuit will act and react differently as the frequency of the received signal gets higher and higher. Since the same voltage controls all stages, different stages of the tuner cannot be adjusted independently of each other. Therefore, the design of the track tuner introduces a compromise signal performance on some channels. This is a phenomenon we servicers have seen time and again. Ever listen to a customer complain that some channels he/she watches are not as clear as others? The track tuner works well, but its performance cannot be optimized to the particular channel the viewer has selected.

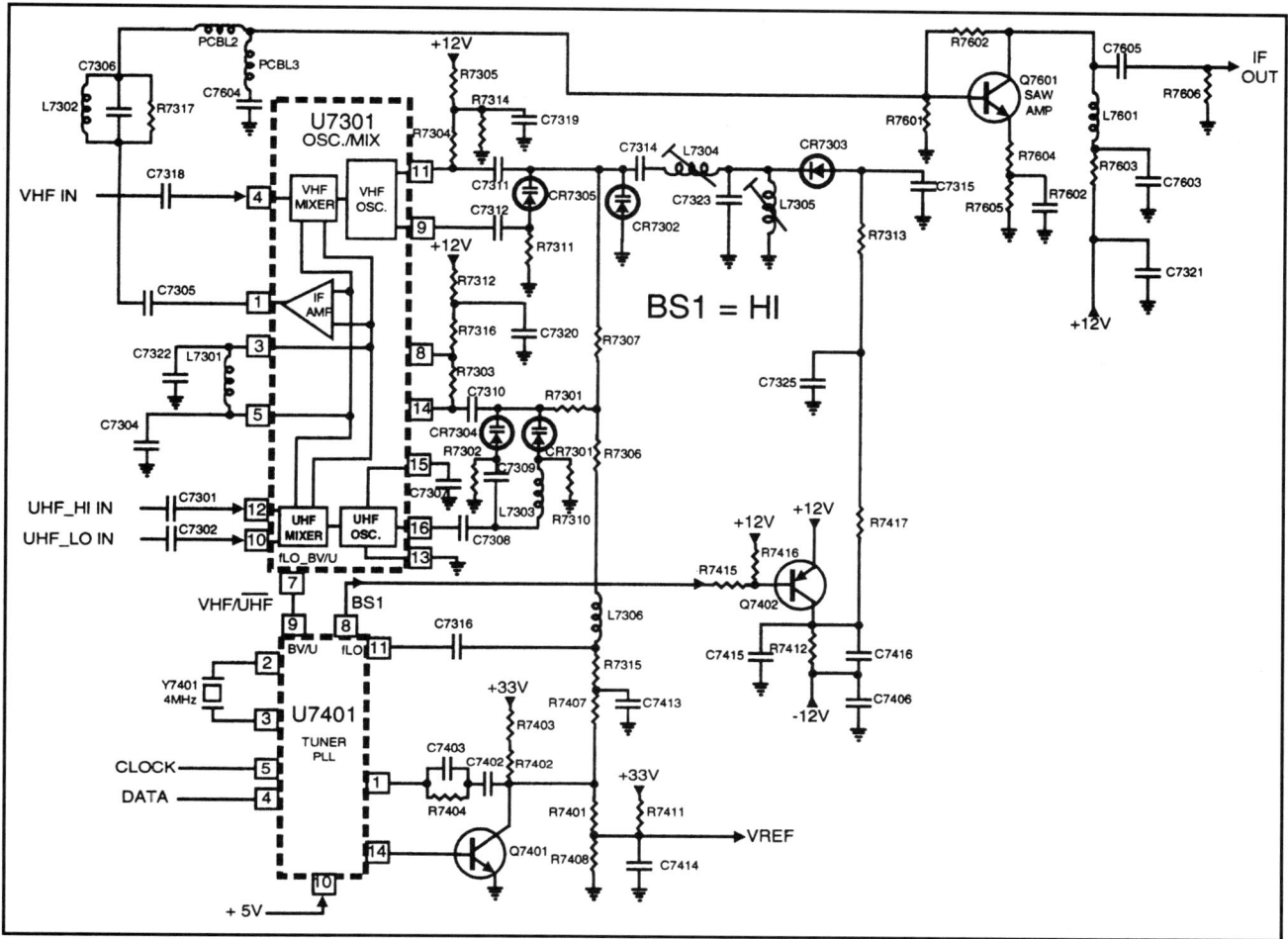

Figure 3-7. Oscillator, mixer, and IF bandpass.

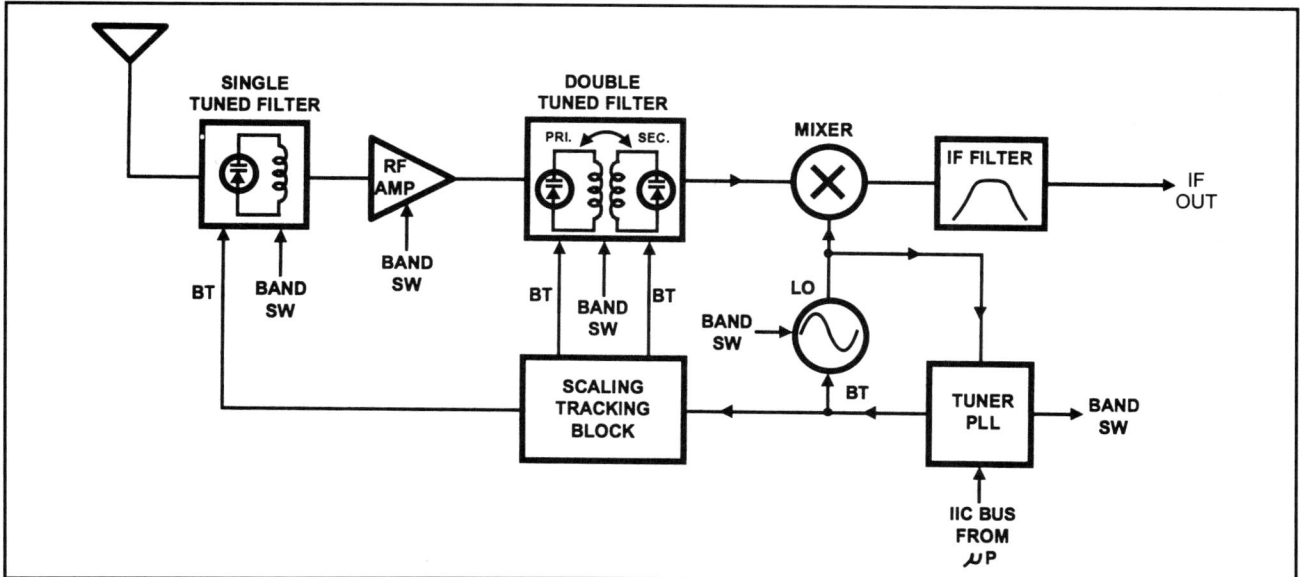

Figure 3-8. Traditional tuner.

How does the TOB differ? It is digitally aligned, which means the local oscillator is controlled by one voltage and each bandpass filter is controlled by its own voltage. Each of the three bandpass filters (single-tuned, double-tuned primary, and double-tuned secondary) is supplied a tuning voltage developed by the system microprocessor (U3101) from values established by the EEPROM and sent to the filter via a low-pass filter/summing amp (U7501). *Figure 3-9* is a block diagram which should make the TOB configuration a little clearer. Such a scheme improves overall tuner response.

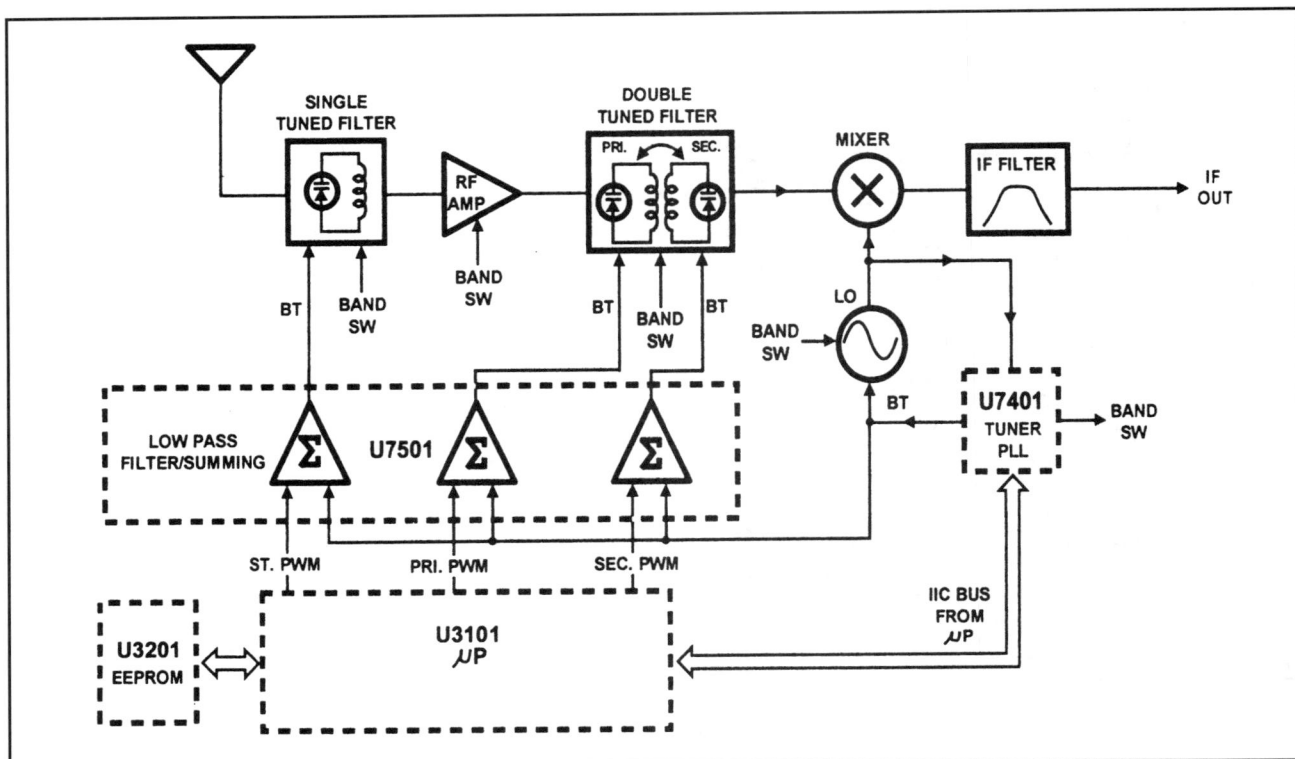

Figure 3-9. TOB configuration block diagram.

Look at *Figure 3-10* while the discussion of the TOB proceeds. The viewer selects a channel. U3101 sends clock and data information to the tuner PLL (U7401), telling it what band and frequency to synthesize. The output of pins 1 and 14 set the tuning voltage (VT) for the local oscillator and the reference tuning voltage (VREF) for the bandpass filters. The tuning voltage adjusts the frequency of the local oscillator to produce the IF frequency for the selected channel. A sample of the signal from the local oscillator is sent back to U7401 as feedback for the PLL, and to U3101 for AFT information. Bandswitching voltage from U7401 pins 8 and 9 tells the filters which band to tune. The VREF voltage is then summed with the digital-to-analog voltages from U3101 and sent to the bandpass filters to tune the desired RF carrier. The digital-to-analog voltage levels are set via digital information stored in the EEPROM.

In addition to the microprocessor, there are four integrated circuits integral to the TOB: the summing amp, the EEPROM, the oscillator/mixer, and the PLL or tuner control. This is a good a place to take an in-depth look at each.

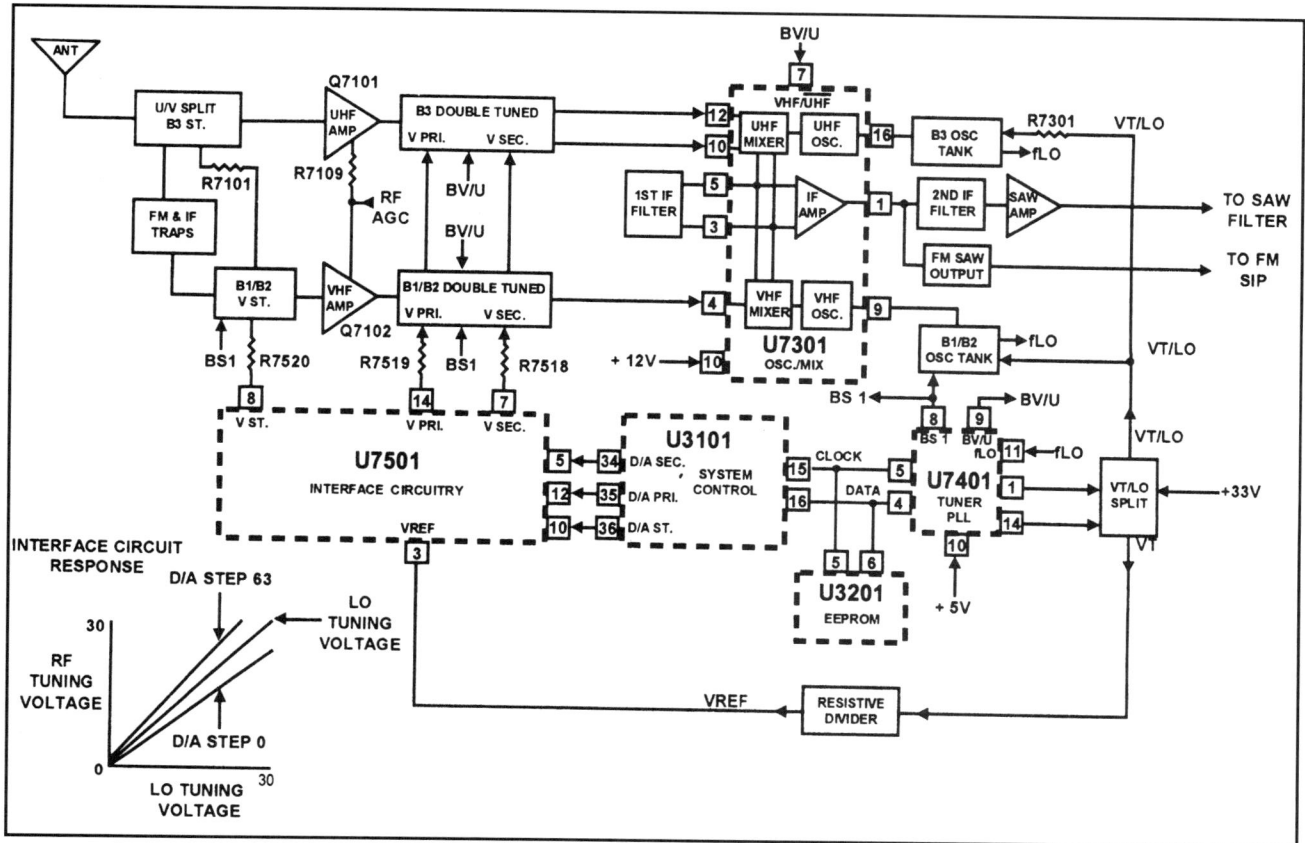

Figure 3-10. TOB expanded diagram.

The first is the microprocessor, which you can call "the brains of the outfit." It interprets customer commands, sends the information to the tuner PLL (U7401) via clock and data lines telling it what channel to tune, pulls pertinent data from the EEPROM to tune the bandpass filters, and sends the information to U7501 via pulse-width modulation.

The second is the low-pass filter/summing amp, U7501. It forms the interface circuit that low-pass filters a PWM signal from U3101 (the microprocessor) and sums it with the VREF tuning voltage from U7401. The interface circuit response is shown in the bottom left of *Figure 3-10*. The higher the tuning voltage, the greater the offset voltage has to be the from digital-to-analog. This is necessary because varactor diodes require more voltage across them to get the same change in capacitance at higher tuning voltages than the lower tuning voltage.

The third is the EEPROM—the little integrated circuit that causes problems.

The tuner is aligned on nineteen "data channels." If you need to know what these channels are, you will find them in *Figure 3-11*. Obviously, these televisions tune more than these nineteen channels. The TOB uses what RCA calls "linear interpolation" to determine the correct settings for the channels that fall between the data channels. The reason given is "it reduces the number of alignments and saves

TEST CH.	TV CH.	BAND	PIX FREQ	MID FREQ	CHROMA FREQ	SOUND FREQ	LO FREQ
1	2	1	55.25	57	58.83	59.75	101
2	6	1	83.25	85	86.83	87.75	129
3	14	1	121.25	123	124.83	125.75	167
4	17	1	139.25	141	142.83	143.75	185
5	18	2	145.25	147	148.83	149.75	191
6	13	2	211.25	213	214.83	215.75	257
7	34	2	283.25	285	286.83	287.75	329
8	37	2	301.25	303	304.83	305.75	347
9	48	2	367.25	369	370.83	371.75	413
10	50	2	379.25	381	382.83	383.75	425
11	51	3	385.25	387	388.83	389.75	431
12	57	3	421.25	423	424.83	425.75	467
13	63	3	457.25	459	460.83	461.75	503
14	76	3	535.25	537	538.83	539.75	581
15	83	3	577.25	579	580.83	581.75	623
16	93	3	637.25	639	640.83	641.75	683
17	110	3	709.25	711	712.83	713.75	755
18	117	3	751.25	753	754.83	755.75	797
19	125	3	799.25	801	802.83	803.75	845

Figure 3-11. Nineteen "data channels".

space in the EEPROM" (*CTC 175/176/177 Technical Training Manual*, p.42). As you know, there are three alignments per data channel: single-tuned, double-tuned primary, and double-tuned secondary. The three alignments per channel correspond to parameters 100 through 156 in the tuner alignment section of the service menu. If you change the parameters of one data channel, be prepared to adjust the parameters of each succeeding channel because the settings for the one channel will affect the linear interpolation curve.

The fourth integrated circuit employed by the TOB is the oscillator-mixer (*Figure 3-12*). It receives RF from the VHF or UHF amplifier, mixes it with a signal from the local oscillator, and outputs the 45.75 MHz difference signal to the IF amplifier. The signal passes out of pin 1, where it is routed to Q7601 (the SAW amp) and on to the T-chip where it is demodulated and processed for video, audio, and sync.

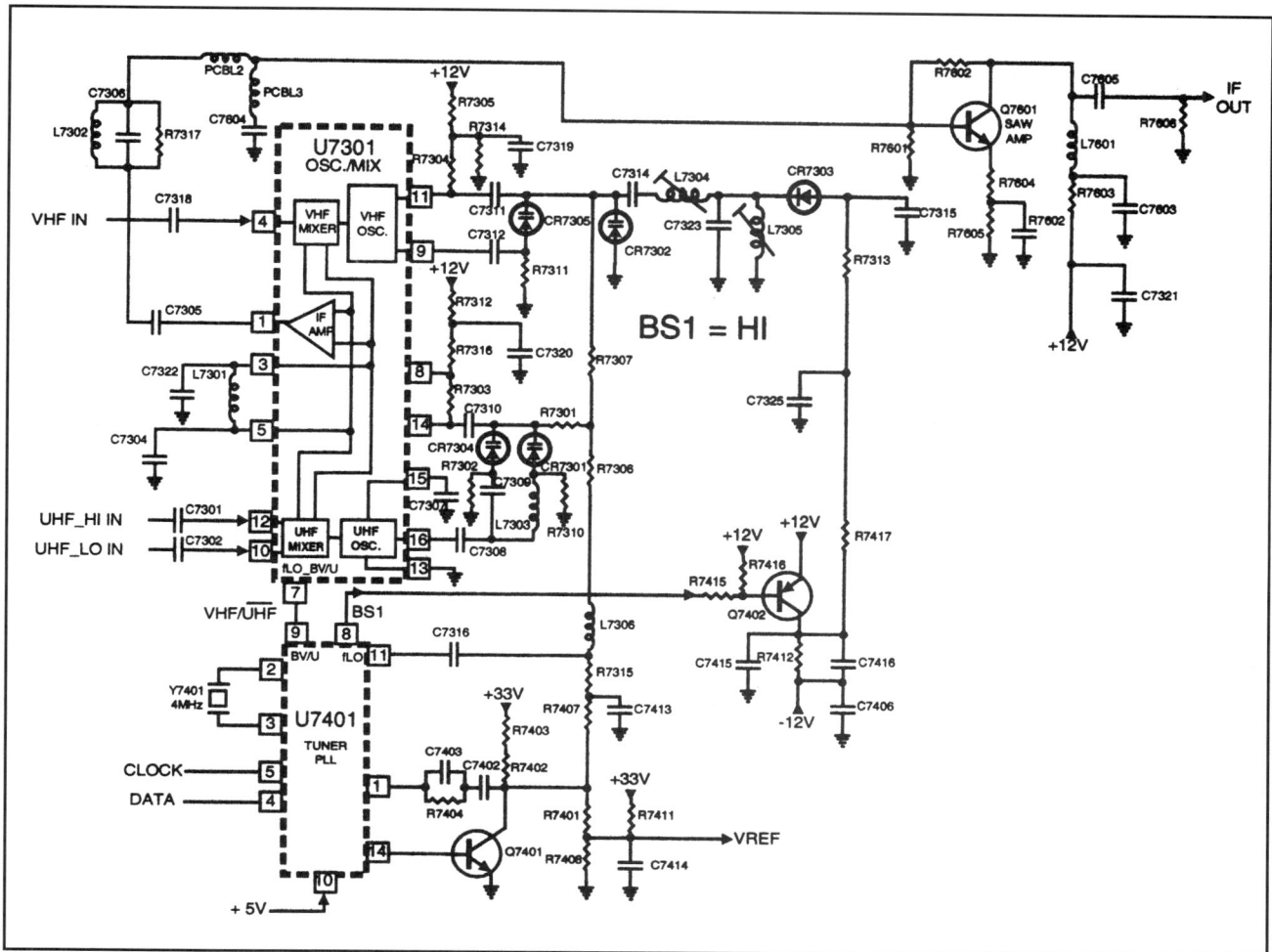

Figure 3-12. Local oscillator.

The local oscillator is controlled by a voltage generated in the PLL integrated circuit, the tuner controller IC (U7401). The reference oscillator frequency is generated by a 4 MHz crystal at pins 2 and 3. The frequency of the local oscillator is sampled at pin 11, divided down, and compared to an internal reference frequency. Pins 1 and 14 output a series of pulses that are low-pass filtered and used to sink current through Q7401 to lower the tuning voltage at its collector. As it conducts less, the +33-volt pull-up supply pulls the tuning voltage higher. The resulting tuning voltage is split and sent to the summing amp to tune the filters, and to the local oscillator to set its frequency.

Well, enough of this. If you want to read more—and there is a lot more that could be said—I suggest you consult the literature I have mentioned. My purpose is to give you an overview of the TOB that will assist you in your troubleshooting, not to bore you with theory. However, a certain amount is theory is an absolute must if you are going to repair these new tuners.

Aligning The Tuner-On-Board

Tuner alignment parameters are found in Level 3 of the service menu. When we discuss the service menu, I will tell you how to enter the service mode and access each of the three levels. I will also give you set of values that in many instances will spare you the necessity of complete tuner alignment. But for now let's do it Thomson's way.

First of all, do the tuner wrap repair if it is necessary and replace the top and bottom shields. Do not attempt tuner alignment with the shields removed! These chores completed, you are ready to do the alignment.

Thomson suggests you use a piece of equipment they manufacture and sell. It is called a "tuner alignment generator" and is given the designation "TAG001" (*Figure 3-13*). The TAG001 requires an external 5 volts and an external video and audio signal. It then takes the video and audio signals and converts them to whatever channel you choose. It also has a built-in step attenuator and is controlled by the same remote control that operates the TV. The TAG001 is a very handy piece of equipment, but it costs about

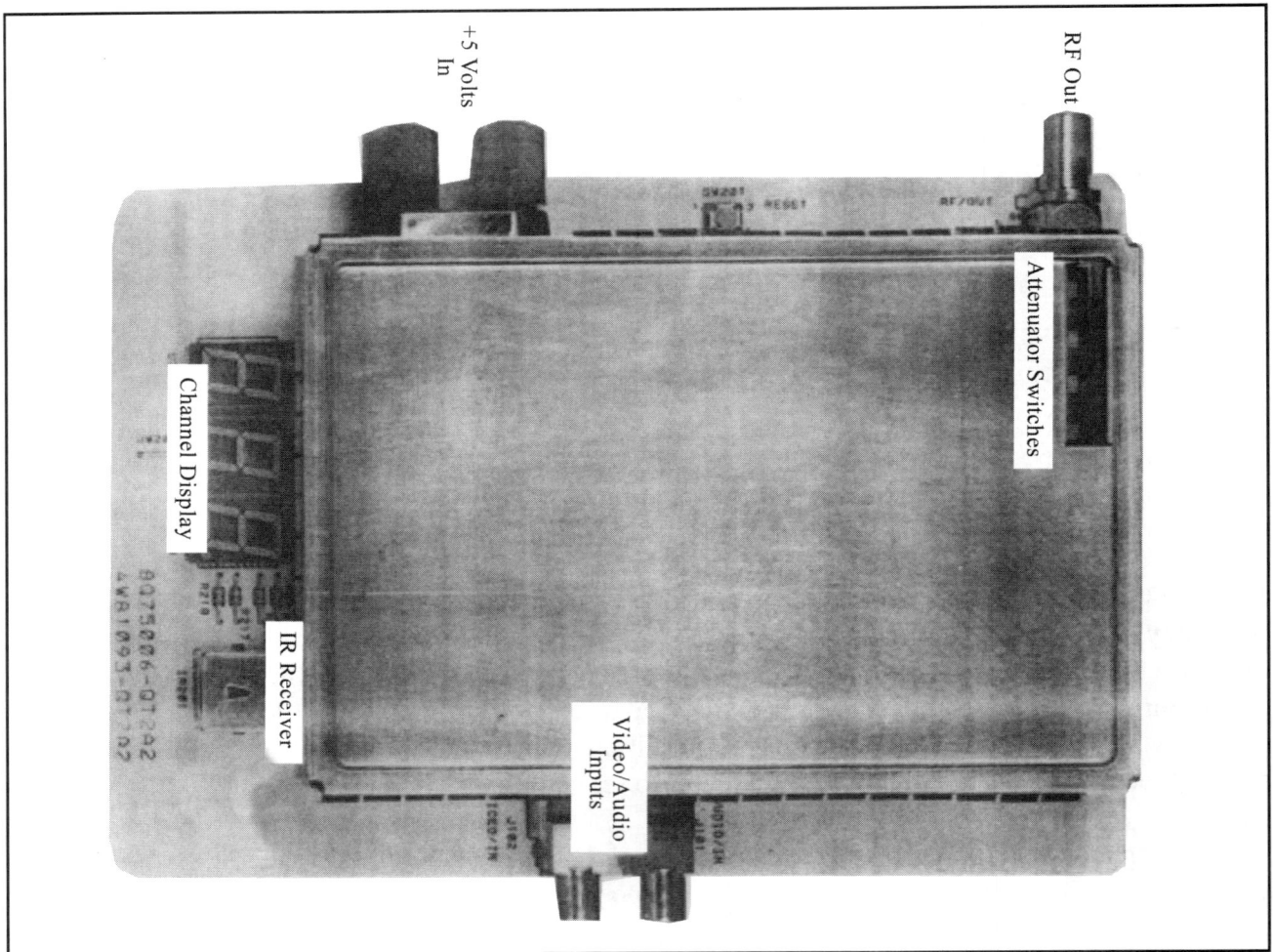

Figure 3-13. Thomson's tuner alignment generator.

$130.00. The good news is you can use any signal generator capable of generating a minimum of 100 channels and which has a signal attenuator. I have already mentioned an article in *Sencore News,* which tells you how to use one of their signal generators to align the TOB. Perhaps I need to underscore the fact that you <u>must not</u> attempt alignment using an off-the-air signal. So, use what you have, but use what you have!

Thomson tells you to monitor the RF AGC voltage at pin 12 of the T-Chip (U1001). That's okay, I guess, but I prefer to monitor it at pin 14 because I can see the voltage change there more easily than at pin 12. Also, use a good DMM. Don't try it with an analog meter. As you monitor the voltage, adjust each parameter for <u>minimum</u> voltage while you attenuate the test signal. In other words, use the minimum signal necessary to obtain a reasonably good picture. Be sure to change the channel on the TV to the correct data channel and to change the channel on the signal generator if you are not using the TAG001. You can change the channel on the TV while it is in the service mode just one way, with the number pad on the remote control. Remember, the channel up/down arrows let you scroll through the parameters in the service mode and do not control channel change.

Troubleshooting Tuner Problems

If you don't have it, I suggest you spend a few dollars and buy the *CTC177/87 Troubleshooting Guide.* It will come in handy as you work with this tuner. In case you don't have it or access to it, I am including a top view of the tuner (*Figure 3-14A*), a component layout guide for the bottom of the tuner (*Figure 3-14B*), and a troubleshooting flowchart (*Figure 3-15*), which is also found in the CTC187 factory service literature.

If you have to change a varactor diode in the VHF (CR106, 107, 108, 111, 113, 302 and 305) or the UHF section (CR101, 102, 103, 114, 301, 304), you must change all the diodes in the section because the diodes are matched for capacitance characteristics and come as a set. Use stock number 215494 for the diode kit. If you do not change all the diodes, the tuner will not work correctly.

Before I get to troubleshooting procedures, I will take a little time to discuss some of the common problems I have encountered. If my information is correct and if your experience parallels mine, these are problems you can expect to encounter.

First, the loss of the +33-volt tuning supply.

The +33 volts is a scan-derived voltage developed at pin 7 of the flyback by CR4113, filtered by C4130, and regulated by CR4108, a 33-volt zener diode (*Figure 3-16*). The voltage can be checked at the cathode of CR4108, which is located on the top of the PCB near the front right edge of the tuner wrap. It can also be checked at R7411, a one megohm resistor, and at the collector of Q7401. The voltage at the collector of Q7401 should change as you change from one channel to another.

Lightning has a way of damaging CR4108, which removes the tuning voltage and renders the tuner inoperative. If I face a TOB repair, the first check I make is on the +33 volt line. I have also heard reports of R7411 failing, shutting the tuner down. A friend in Middle Tennessee says he has serviced

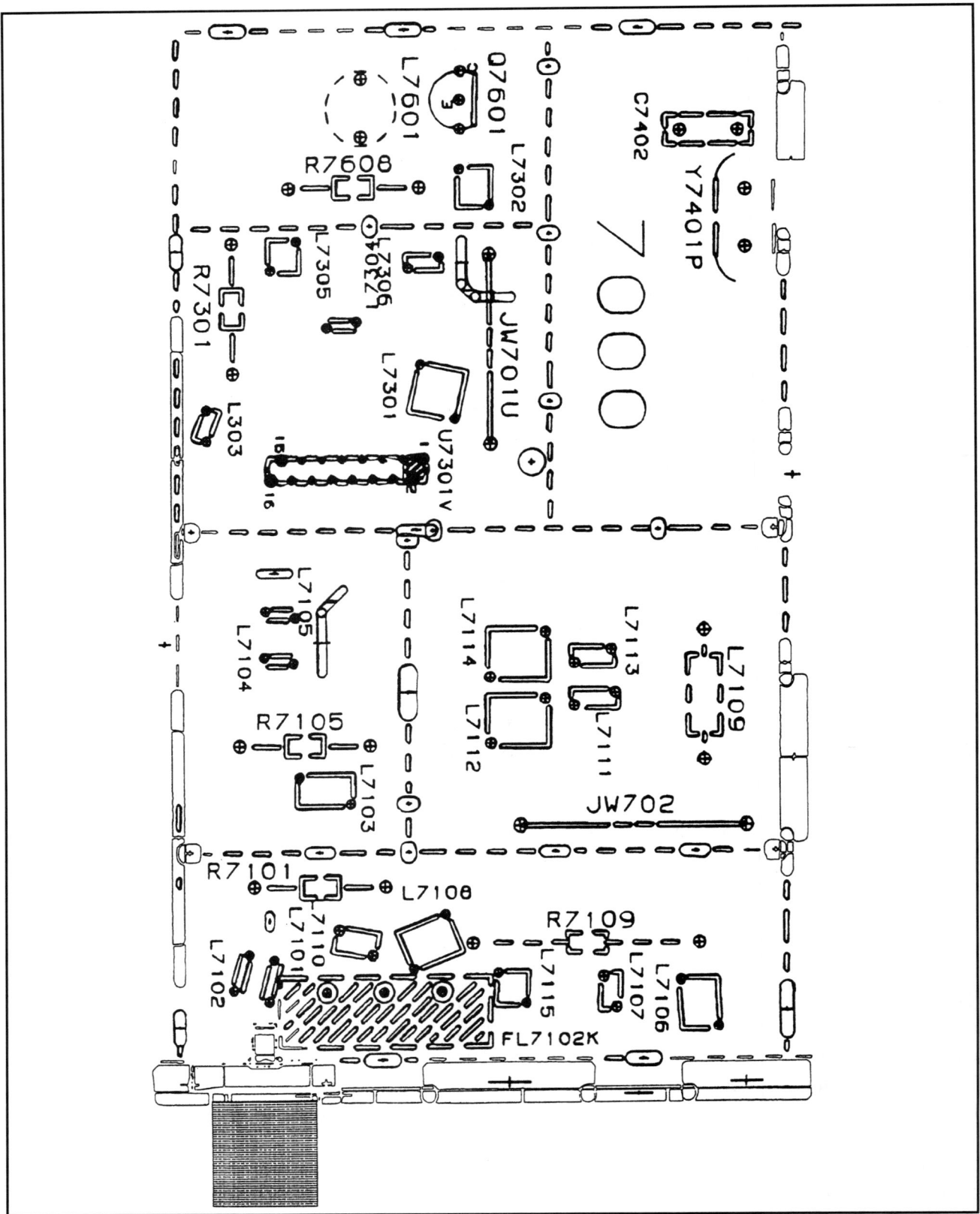

Figure 3-14A. Tuner, top view.

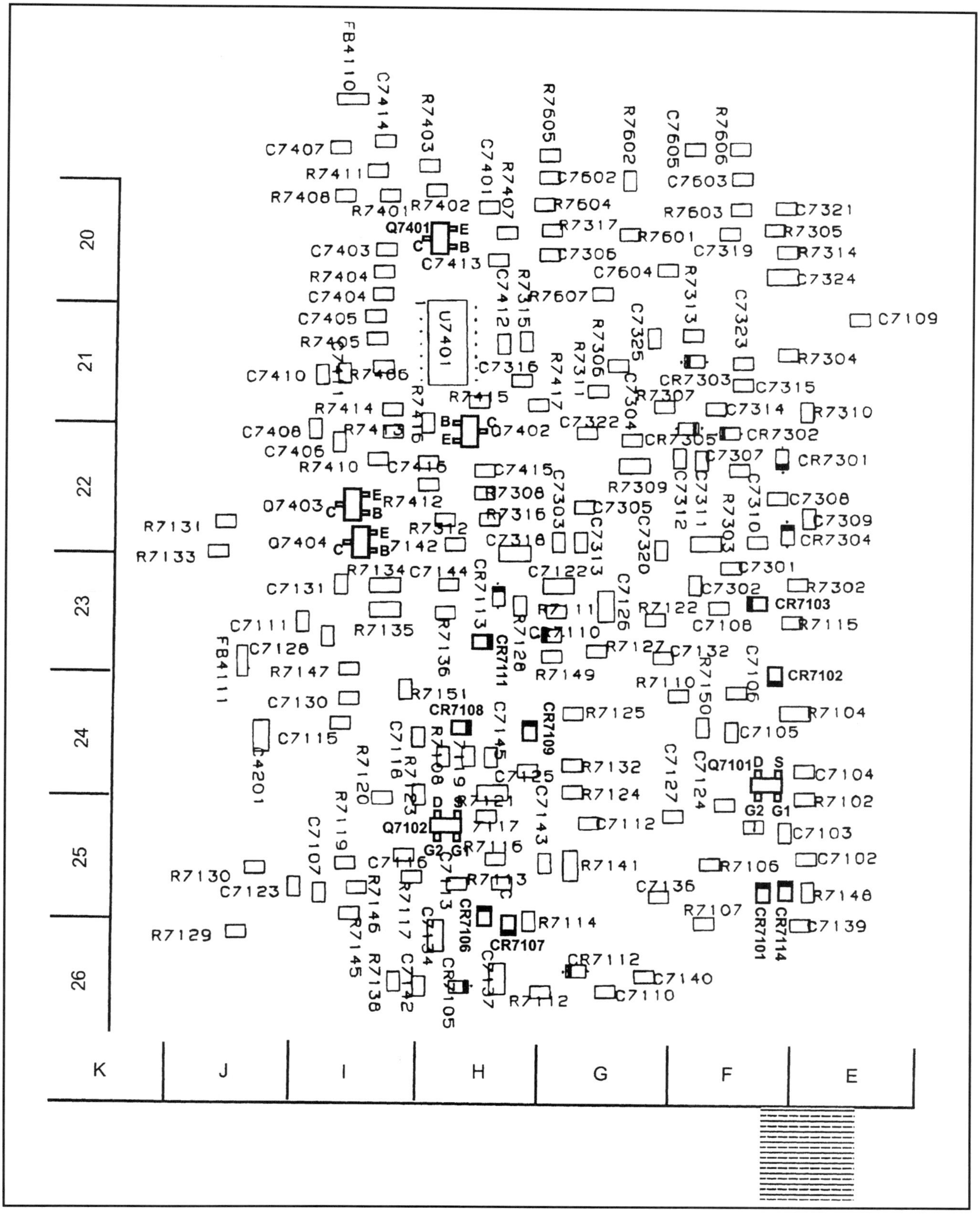

Figure 3-14B. Tuner, bottom view.

CONNECT SERVICE
GENERATOR TO
ANTENNA CONNECTOR
NO ATTENUATION

CHECK
OPERATION
ALL BANDS
(SEE NOTE)

NO
PICTURE

NOTE: IF ONLY ONE (1) OR TWO (2) BANDS
ARE NOT FUNCTIONING LIMIT
TROUBLESHOOTING CHECKS TO
THOSE BANDS.

CHECK
VOLTAGE
SUPPLIES TO
TUNER

+5V - RUN
+12V - RUN E
+12V - RUN F
-12V - RUN
+33V - RUN C

OK

CHECK
VREF AT
TP7401

B1-CH17- +4.40~4.60V
B2-CH50- +4.80~5.00V
B3-CH125- +4.55~4.75V
(SEE TUNER COIL
ALIGNMENT)

IF VOLTAGE MISSING
OR INCORRECT,
CHECK R7401, R7411

REMOVE TUNER
COVER(S)

OK

CHECK
LO VOLTS
AT R7301

NOT OK

0V-CHECK 33V LINE
CHECK BANDSWITCH
VOLTS AT R7313
CHECK TUNING
VOLTS AT R7407,
R7315 & R7307

OK

PICTURE
PRESENT BUT
NOT GOOD

CHECK SINGLE
TUNED, PRIMARY,
ANDSECONDARY TUNING
VOLTAGE

CHECK FOR ACTIVITY
ON TUNER-DATA,
TUNER-CLOCK LINES.
CHECK FOR OP AMP
VOLTS U7501.
CHECK FOR CORRECT
EEPROM VALUES BY
CHECKING FOR
IMPROVEMENT USING
TUNER COIL
ALIGNMENT
PROCEDURE.

OK

CHECK
TUNER IC
BIAS (U7301)
↑ LO_B V/U

CHECK BV/U
SWITCHING LOGIC
PIN 9 U7401

OK

A

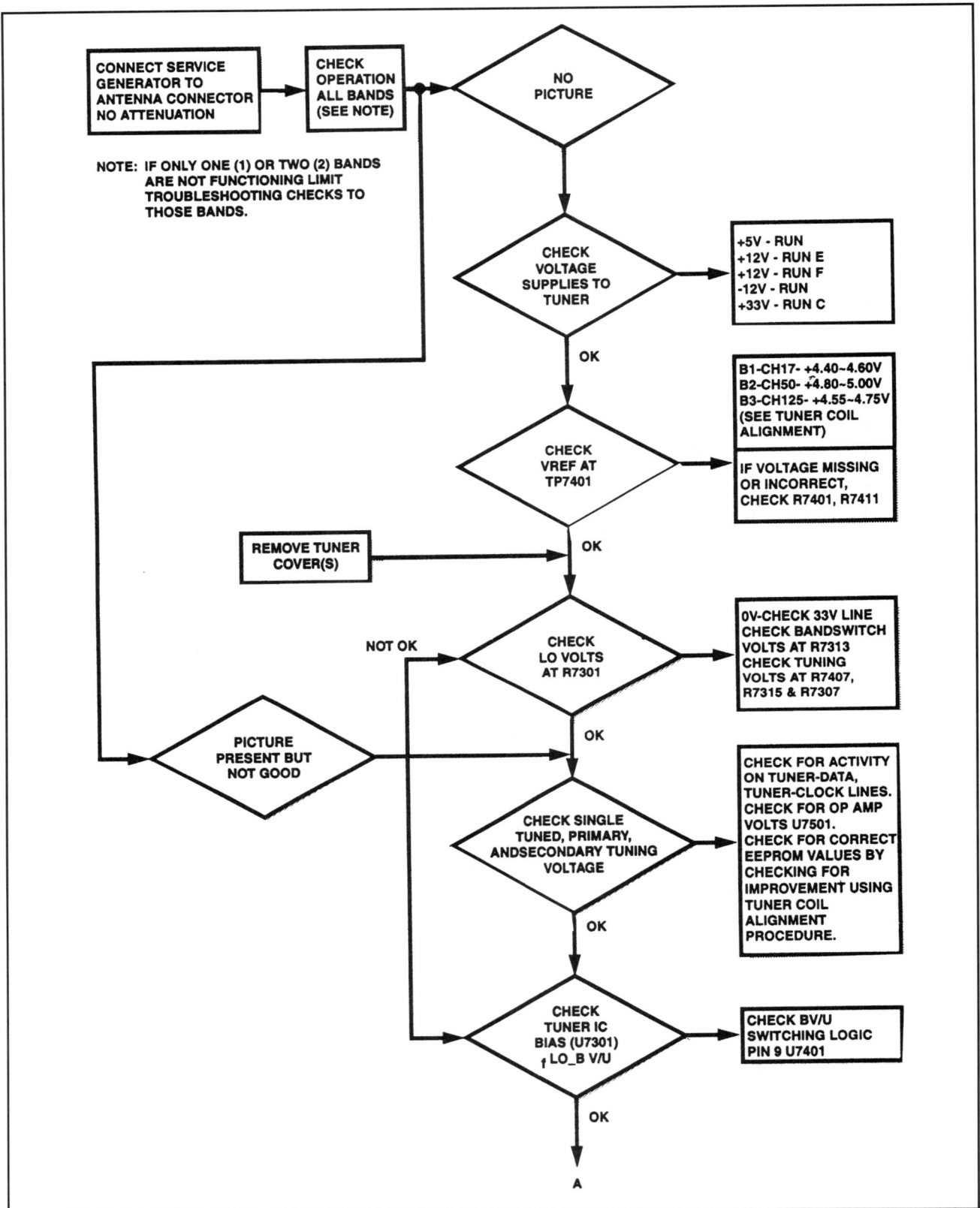

Figure 3-15. TOB troubleshooting flowchart.

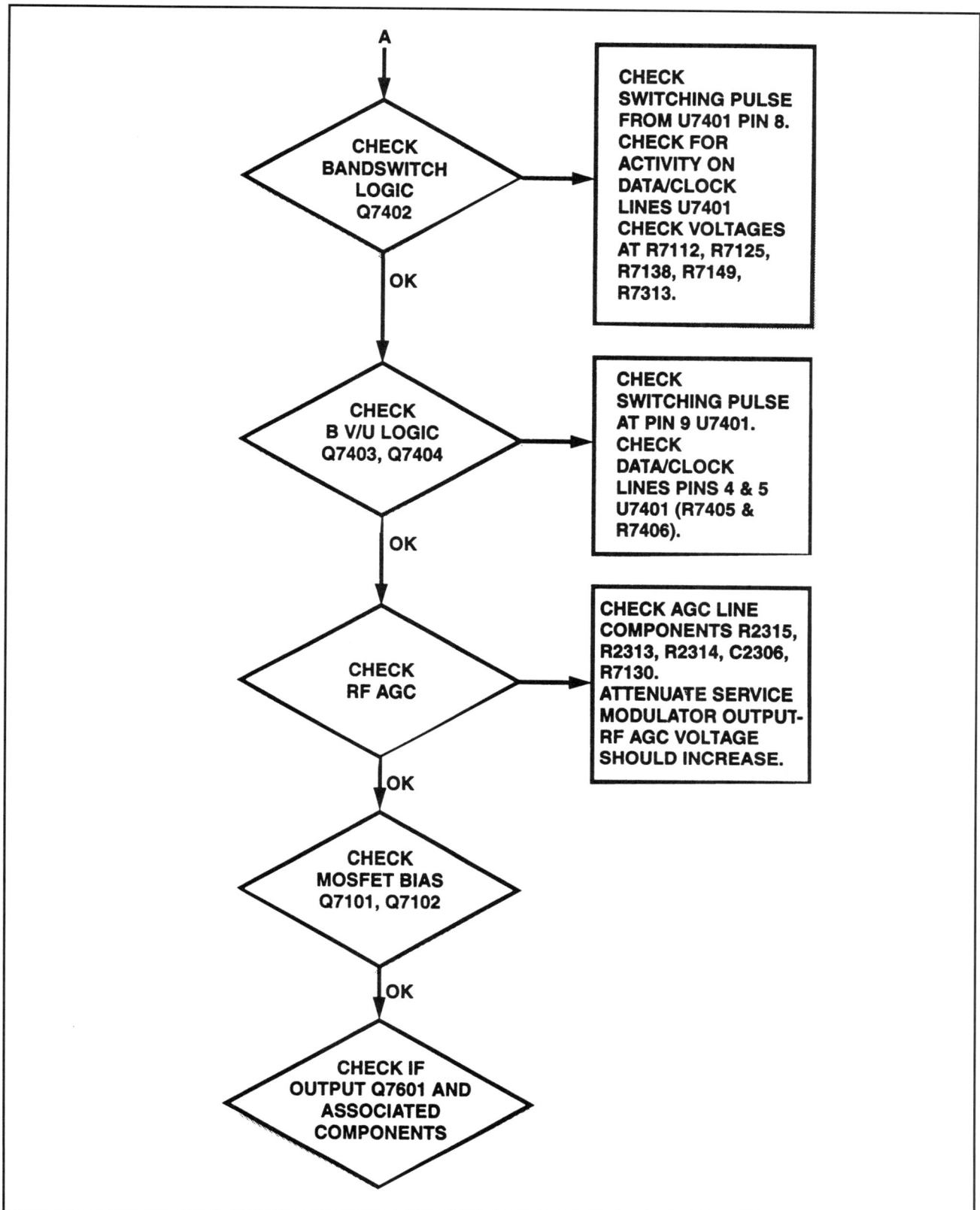

Figure 3-15. TOB troubleshooting flowchart (continued).

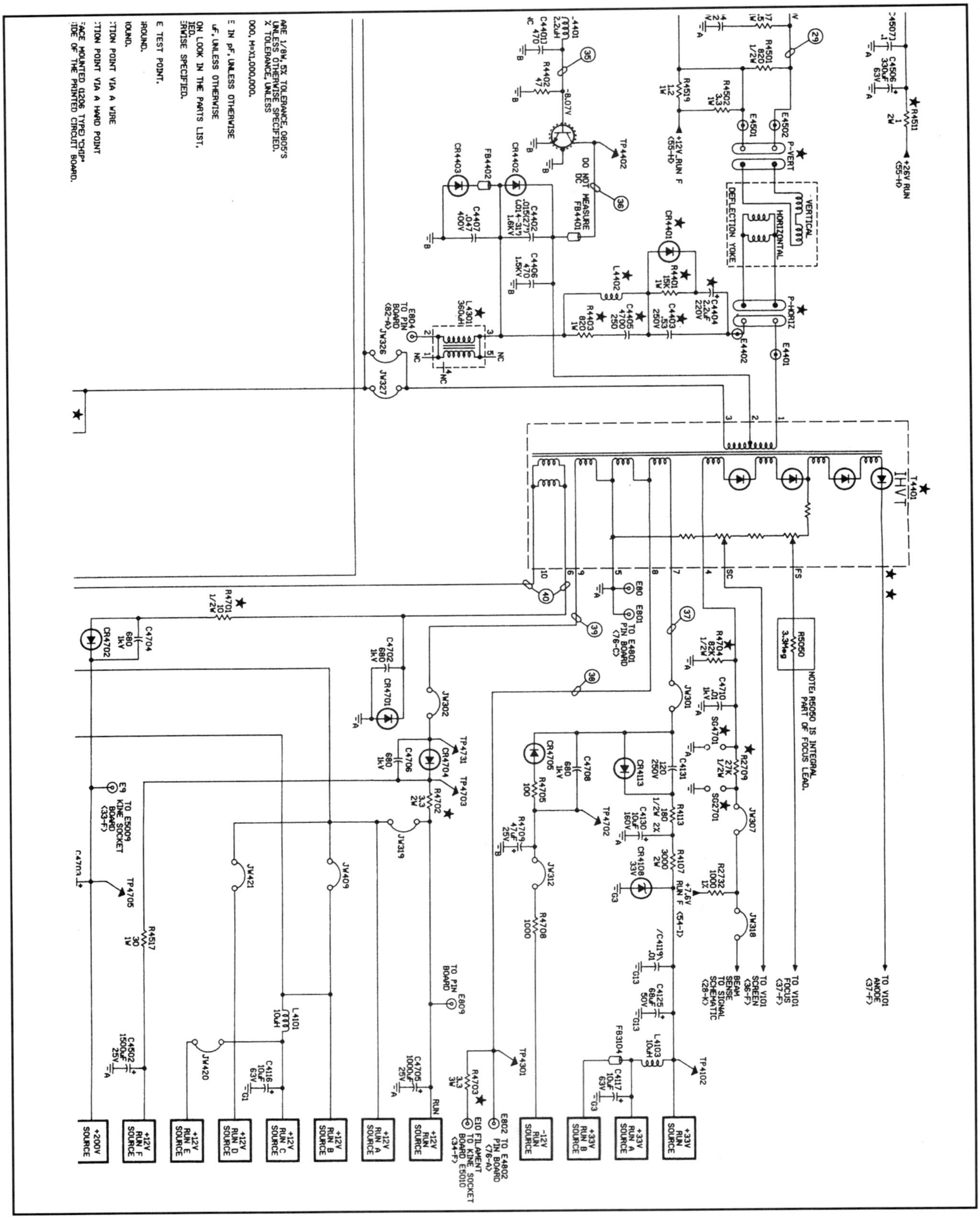

Figure 3-16. Tuning supply circuits.

several CTC176 tuners that had an open R7411 (*Figure 3-17*). The only tuning voltage problem I have repaired has been on the +33-volt line, but as you know, anything that can go wrong probably will.

Second, a defective Y7401.

The 4 MHz crystal at pins 1 and 2 of U7401 develops the reference signal for the tuner controller IC. You can check its performance by scoping the signal on either of its pins or at U7401. The signal should be about a volt peak-to-peak and should be at the stated frequency. A bad crystal will likely cause one of two problems: a completely dead tuner, or one that drifts off frequency. I serviced a set just last week that had an inoperative tuner because the crystal was defective. I also serviced one that would drift off channel after the chassis warmed up. At first, I thought the automatic fine tuning was causing the tuner to drift off frequency, but when I applied freeze spray to the crystal, the tuner returned to its original frequency. New crystals fixed both sets.

Third, the mixer/oscillator, U7301.

Let's use as an example a TV that has raster but not video or audio, a typical tuner related problem. You should always begin by checking all VHF channels and all cable channels. Then you should place the TV in the "off air" mode to see if the UHF section is also dead. This is a step we often overlook because we just "know" what the problem is! I ought to know because "I've been there and done that." Assume we checked the VHF and cable channels and got nothing. We then instruct the microprocessor to receive the UHF band, and, lo and behold, we can tune the UHF channels.

What have we learned? That at least part of the tuner is working and that the problem will be either in the VHF circuitry, the oscillator/mixer IC, or the tuner control IC. Let's set the TV to channel 3 and inject a channel 3 signal of the proper strength into U7301 pin 4, using tuner shield as ground (*Figure 3-18*). We get nothing. Next, check the bandswitching voltage at pin 7, which should be about 3 volts for VHF and 0 volts for UHF. If the voltage is correct for VHF reception (and it probably will be), you have isolated the problem to U7301 and its associated circuitry. Make a few more checks to confirm your suspicion. If the voltage on pin 7 is correct, check for the appropriate bandswitching voltage at pin 9 of the tuner control IC. If that voltage is good, get ready to change the oscillator/mixer IC. I have never had to change the controller IC for this problem, but I have had to change the oscillator/mixer many, many times.

Fourth, the tuner control chip, U7401.

I fixed a CTC175 that had an interesting problem. It would "sort of" receive channel 4 when it was instructed to tune channel 4. It would also receive channel 4 on any low-band VHF setting and would lose picture and sound completely on high-band VHF channels. It would pick up some cable channels but would quickly drift off frequency.

These symptoms pointed directly to U7401, which was in fact the problem.

If you face the necessity of changing this surface-mount IC, don't get excited because it is an easy chip to change. Methods will vary from tech to tech. If you have it, a desoldering tool is the instrument of

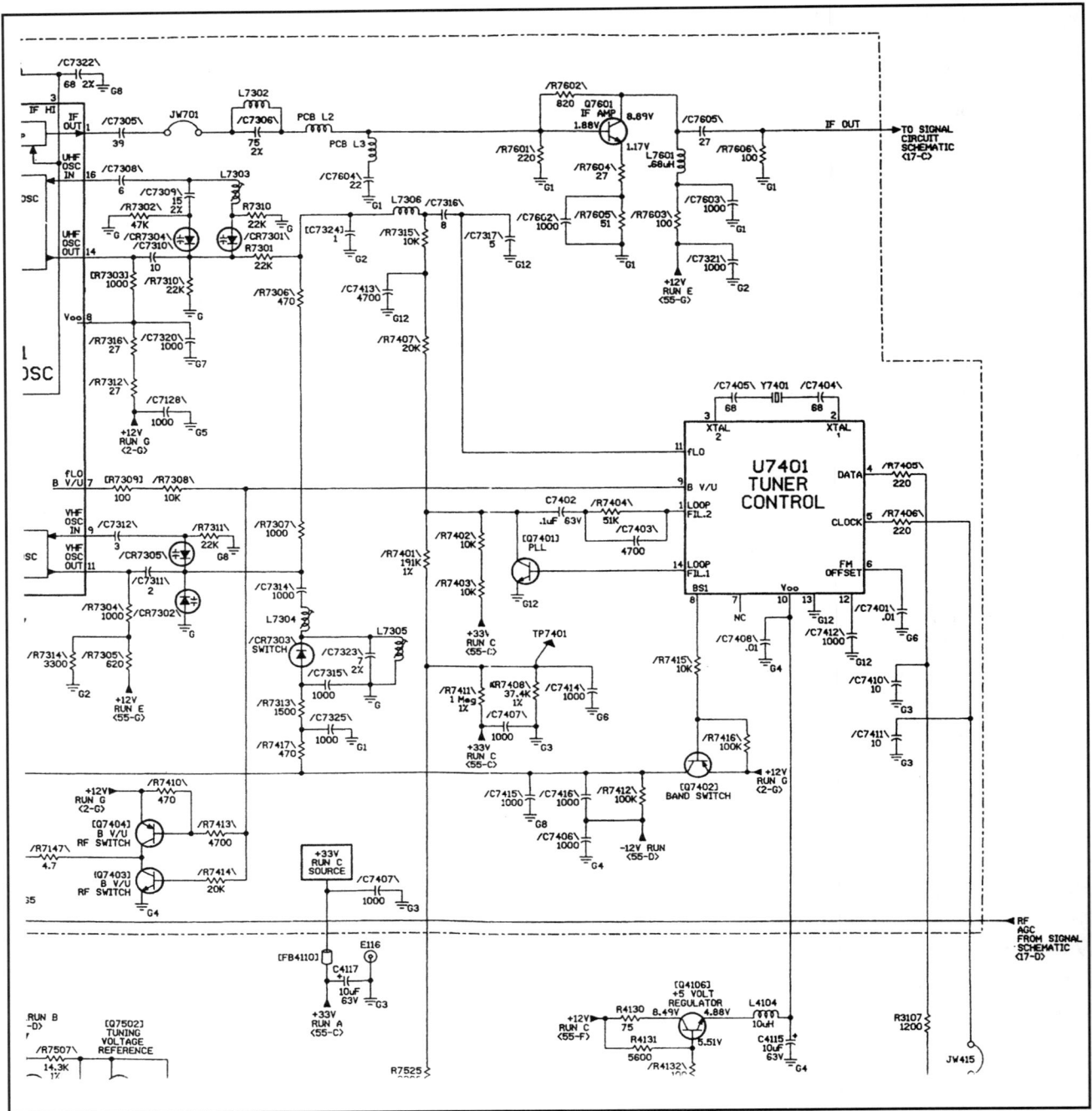

Figure 3-17. R7411 location on 33-volt line.

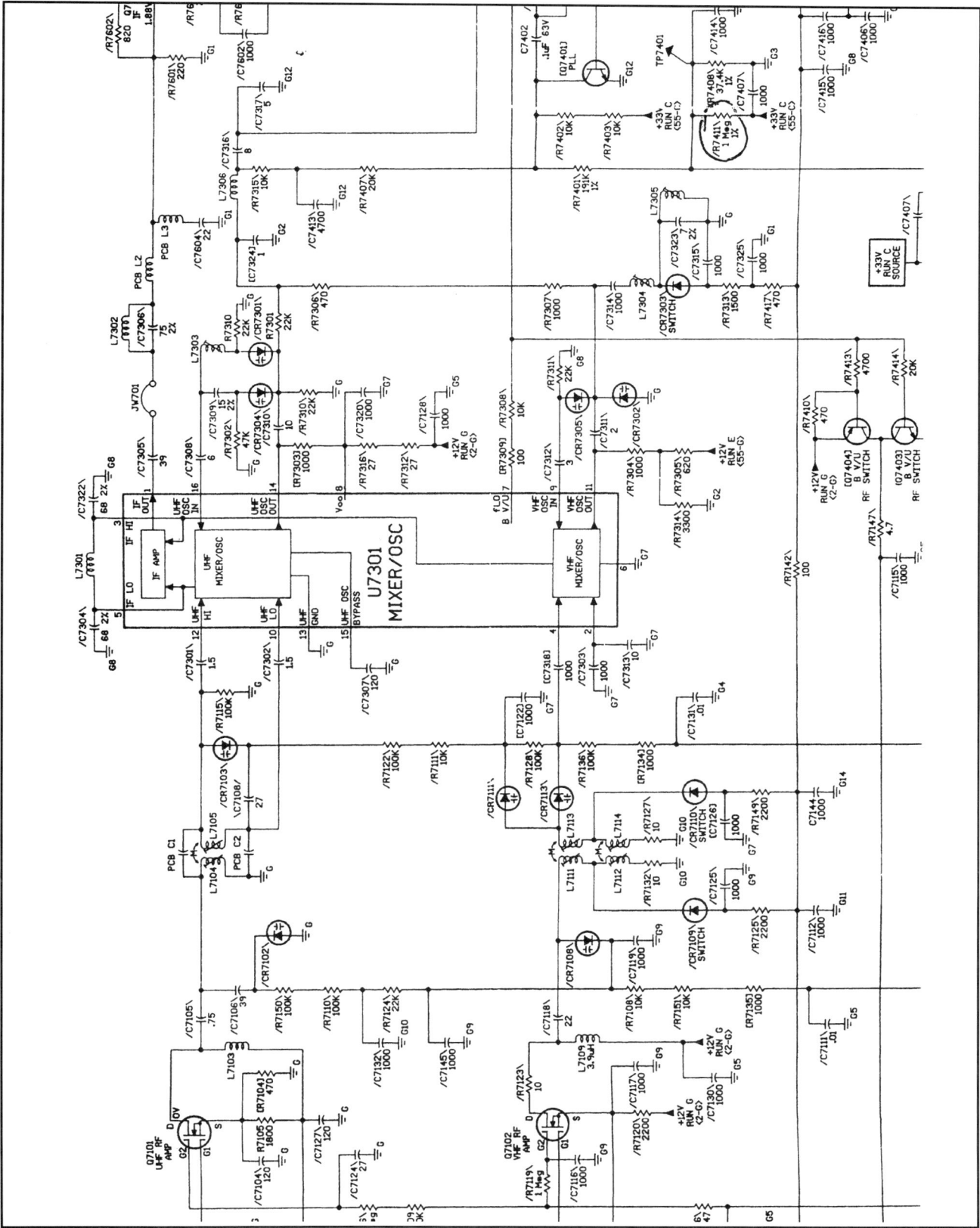

Figure 3-18. U7301 mixer/oscillator circuits.

choice. A new product on the market sold under the label "Chip Quik" makes changing this IC or any surface-mount IC relatively easy. You can order it from just about anyplace, and instructions come in the package. However, I use an Exacto knife and solder wick to remove a chip as small this one. I cut the pins close to the body of the IC, being careful not to damage the circuit board, remove the body of the chip with the point of the knife, and use solder wick to clean the pins off their pads. The whole procedure takes about a minute and has never failed me.

Now, let's proceed with our troubleshooting. If you like to use flowcharts, remember *Figure 3-15*. It's for the CTC187 but will work fine for the CTC175/176/177 chassis.

The very first step is a thorough performance test. Use a good signal generator with a known input level, say about 1000 microvolts. Such a signal level will permit the RF amplifiers to operate at full gain. If the TV is working properly, you will see a noise-free signal at the CRT. A thorough performance test will include testing VHF, UHF and cable channels. The results of the test will give you important clues with respect to possible problems. Will all channels tune but are noisy? Then you most likely have a signal gain problem. Look at the RF amplifiers, the AGC voltage, or IF stage as likely culprits. Do you get VHF but not UHF or vice versa? Are you missing channels 2-6 but receive 7-13?

Using the performance test as a starting point, we will now try to find the solution to some real-life problems.

One Band is Inoperative

If the tuner will tune channels on all bands but one, limit your troubleshooting to the band it will not receive. If one band works, U7501, U3201, U7401 and a part of U7301 works. Assume you are missing VHF reception.

If you can inject a VHF signal into U7301 and get good results, obviously U7301 is also working. Therefore, check -12 and +12-volt supplies, and check the biasing on the appropriate RF amplifiers. If you are missing, say, channels 2-6 or 7-13, make sure the bandswitching voltage from the collector of Q7402 turns on/off CR7112, CR7105, CR7109, and CR7110.

Picture Present but not Good

Check AGC voltage. Then check the supply voltages to the tuner: +5, +12, -12, and +33 volts. Check the bandpass filters (single-tuned, primary- and secondary-tuned) per the enclosed voltage chart, then the voltages at U7501 per the chart. Check the EEPROM values by trying to improve one channel by realigning it. (Remember to write down the original values so you can restore them if necessary.) If these steps fail to isolate the culprit, go to the next step.

The Tuner will not Tune any Channel

First, verify that the channel numbers change on screen. If the OSD does not respond to channel change requests, the problem will be in system control and not the tuner.

Second, check your supply voltages. I have already indicated there are four separate supply voltages.

Third, check for correct bandswitching voltages on pins 8 and 9 of U7401, pin 7 of U7301, the collector of Q7402, and the collectors of Q7403 and 7404. These voltages are given in the appropriate voltage chart.

Fourth, check the tuning voltage on the collector of Q7401. (See the voltage chart, *Figure 3-19*.) If the tuning voltage is stuck high or low, the problem will be in the PLL loop.

Monitor the voltage while you scroll through several channels. If it doesn't change, look for the 4 MHz signal on Y7401. It should be about 1 volt peak-to-peak. In my experience, Y7401 fails about as often as any part in the tuner, and more often than most.

Fifth, check the local oscillator voltage at R7301 (*Figure 3-18*). It should increase as the channel numbers increase and decrease as the channel numbers decrease.

If the voltage is missing, follow the circuit path to see why. Remember to check for a leaky or shorted CR7301, CR7302, CR7304, and CR7305.

Sixth, check the bandpass varactor diode tuning voltage per the chart.

Seventh, check bandswitch logic (BS1) on Q7402.

Eighth, check the V/U logic on Q7403 and Q7404.

Ninth, check RF AGC. If you attenuate the signal applied to the RF connector, the AGC voltage should increase.

Tenth, check MOSFET bias on Q7101 and Q7102.

Eleventh, check the IF output, Q7601, and associated components.

Now, that is the troubleshooting procedure as RCA has outlined it in their literature and service meetings. You might want to modify it to suit your taste. For example, why not inject an IF signal into pin 1 of U7301 right at first? If you get a good picture and audio, you can be reasonably certain the tuner is at fault. I am suggesting a procedure and inviting you to use it or to make up your own. The key to successful troubleshooting is proceeding logically and methodically.

I shall conclude the discussion of the TOB with two comments:

First, if you find a voltage missing or improper, you can use an external DC power supply to substitute for the missing voltage or to correct one that is improper. By subbing a missing voltage or shifting one that is present, you can determine if a stage is working as it should.

Second, don't forget how helpful signal injection can be. For example, suppose you want to see if the local oscillator is the cause of a dead tuner. Feed a 45.75 MHz IF test signal into the mixer input and vary its strength. If you get a good picture on the CRT, you know the mixer and IF stages are working but the local oscillator is not. You can then find out why the oscillator is not working. On the other hand, if you don't see improvement, move your injection point to the IF amp (See *Sencore News*, issue 170, p. 26).

Off-Air Signal	U7301 Pin No.	Lo V Chan.		Hi V Chan.		UHF Chan.		
		2	6	7	13	14	40	69
	1.	5.43V	5.42V	5.38V	5.38V	5.30V	5.22V	5.30V
	2.	2.93V	2.98V	2.95V	2.95V	3.18V	3.16V	3.18V
	3.	7.80V	7.81V	7.75V	7.69V	7.56V	7.51V	7.56V
	4.	2.99V	2.99V	2.96V	2.97V	3.18V	3.16V	3.18V
	5.	7.82V	7.81V	7.75V	7.71V	7.56V	7.51V	7.56V
	6.	0V	0V	0V	0V	0V	0V	0V
	7.	3.06V	3.06V	3.04V	3.01V	0V	0V	0V
	8.	9.06V	9.03V	8.97V	8.89V	8.84V	8.80V	8.84V
	9.	3.02V	3.02V	2.98V	2.98V	3.36V	3.33V	3.36V
	10.	3.25V	3.25V	3.22V	3.19V	2.88V	2.87V	2.88V
	11.	4.94V	5.00V	5.06V	5.02V	9.62V	9.58V	9.62V
	12.	3.25V	3.23V	3.22V	3.19V	2.87V	2.87V	2.88V
	13.	0V	0V	0V	0V	0V	0V	0V
	14.	.9.05V	9.04V	8.97V	8.90V	5.43V	5.39V	5.46V
	15.	3.43V	3.43V	3.40V	3.37V	2.88V	2.84V	2.88V
	16.	3.44V	3.43V	3.40V	3.38V	2.89V	2.89V	2.89V

U7401 Pin No.	Lo V	Hi V	UHF
1.	1.75V	2.11V	1.72V
2.	2.11V	2.11V	2.11V
3.	2.11V	2.11V	2.11V
4.	4.78V	4.78V	4.78V
5.	4.71V	4.71V	4.71V
6.	0V	0V	0V
7.	1.32V	1.32V	1.34V
8.	11.5V	0V	0V
9.	7.41V	7.41V	0V
10.	4.85V	4.85V	4.85V
11.	2.30V	2.30V	2.30V
12.	2.30V	2.30V	2.30V
13.	0V	0V	0V
14.	0.60V	0.60V	0.60V

Figure 3-19. Tuner voltage charts.

U7501

Pin	Lo V		Hi V		UHF		
No.	2	6	7	13	14	40	69
1.	1.36V	1.96V	1.58V	1.93V	1.74V	2.90V	4.84V
2.	1.36V	1.96V	1.58V	1.93V	1.74V	2.90V	4.84V
3.	1.35V	1.95V	1.57V	1.92V	1.73V	2.89V	4.83V
4.	33.0V	33.0V	33.0V	33.0V	33.0V	33.0V	33.0V
5.	1.06V	1.74V	1.47V	1.87V	1.46V	2.53V	4.08V
6.	1.06V	1.74V	1.47V	1.87V	1.46V	2.54V	4.09V
7.	1.06V	6.05V	4.09V	7.03V	3.95V	11.8V	23.2V
8.	0.68V	4.57V	3.46V	6.19V	4.41V	12.3V	24.0V
9.	1.01V	1.54V	1.39V	1.76V	1.52V	2.60V	4.19V
10.	1.01V	1.54V	1.39V	1.76V	1.52V	2.60V	4.19V
11.	0V	0V	0V	0V	0V	0V	0V
12.	1.05V	1.75V	1.36V	1.72V	1.44V	2.51V	4.15V
13.	1.05V	1.75V	1.36V	1.72V	1.44V	2.51V	4.15V
14.	1.01V	6.10V	3.24V	5.92V	3.78V	11.6V	23.6V

Q7101

	Lo V	Hi V	UHF
	2	7	14
G1	0V	0V	4.84V
G2	5.03V	6.54V	7.19V
D	0.16V	0.20V	11.3V
S	0.19V	0.17V	4.83V

Q7102

	2	7	14
G1	4.64V	4.58V	4.61V
G2	5.32V	6.85V	7.19V
D	11.3V	11.2V	11.4V
S	4.08V	4.20V	11.3V

Q7401

	Lo V	Hi V	UHF
E	0V	0V	0V
B	0.60V	0.60V	0.60V
C	2.06V	3.85V	17.8V

Q7402

E	11.4V	11.2V	11.3V
B	11.3V	10.5V	10.6V
C	-14.9V	11.1V	11.2V

Q7403

E	0V	0V	0V
B	0.70V	0.70V	0V
C	0.10V	0.10V	11.3V

Q7404

E	11.4V	11.2V	11.3V
B	11.0V	10.9V	10.6V
C	0.11V	0.11V	11.3V

Figure 3-19. Tuner voltage charts (continued next page).

Tuner Voltage Charts Cable Signal

U7301

Pin No.	Band 1 2	17	18	Band 2 50	51	Band 3 75	99
1.	5.44V	5.40V	5.41V	5.40V	5.30V	5.28V	5.48V
2.	2.99V	2.96V	2.96V	2.96V	3.18V	3.17V	3.00V
3.	7.80V	7.78V	7.77V	7.75V	7.57V	7.57V	7.89V
4.	2.99V	2.96V	2.96V	2.96V	3.18V	3.17V	3.00V
5.	7.82V	7.78V	7.77V	7.77V	7.57V	7.57V	7.89V
6.	0V	0V	0V	0V	0V	0V	0V
7.	3.06V	3.04V	3.05V	3.05V	0V	0V	3.08V
8.	9.02V	9.01V	8.98V	9 00V	8.84V	8.83V	9.14V
9.	3.01V	3.01V	2.98V	2.98V	3.36V	3.35V	3.01V
10.	3.26V	3.23V	3.22V	3.23V	2.88V	2.88V	3.28V
11.	4.96V	5.04V	5.07V	5.16V	9.62V	9.60V	5.14V
12.	3.26V	3.23V	3.22V	3.23V	2.88V	2.87V	3.28V
13.	0V	0V	0V	0V	0V	0V	0V
14.	9.05V	9.00V	8.98V	9.00V	5.43V	5.42V	9.13V
15.	3.43V	3.41V	3.42V	3.41V	2.88V	2.87V	3.46V
16.	3.41V	3.41V	3.41V	3.41V	2.87V	2.89V	3.47V

U7401

Pin No.	Band 1 2	Band 2 18	Band 3 51
1.	1.74V	1.74V	1.74V
2.	2.11V	2.11V	2.11V
3.	2.11V	2.11V	2.11V
4.	4.78V	4.78V	4.78V
5.	4.71V	4.71V	4.71V
6.	0V	0V	0V
7.	NC	NC	NC
8.	11.5V	0V	0V
9.	7.47V	7.42V	0V
10.	4.85V	4.85V	4.85V
11.	2.31V	2.31V	2.31V
12.	2.31V	2.31V	2.31V
13.	0V	0V	0V
14.	0.60V	0.60V	0.60V

U7501

Pin No.	Band 1 2	17	18	Band 2 50	51	Band 3 75	99
1.	1.36V	4.63V	1.30V	5.52V	1.20V	2.21V	2.86V
2.	1.36V	4.63V	1.30V	5.52V	1.20V	2.21V	2.86V
3.	1.36V	4.63V	1.30V	5.52V	1.20V	2.21V	2.86V
4.	33.0V	33.0V	33.0V	33.0V	33.0V	33.0V	33.0V

Figure 3-19. Tuner voltage charts (continued).

5.	1.06V	3.62V	1.16V	4.68V	0.99V	1.90V	2.54V
6.	1.06V	3.62V	1.16V	4.68V	0.99V	1.90V	2.54V
7.	1.07V	19.8V	1.78V	27.6V	0.54V	7.20V	11.9V
8.	0.69V	24.2V	1.30V	25.3V	0.86V	7.63V	12.5V
9.	1.01V	4.22V	1.10V	4.38V	1.01V	1.96V	2.62V
10.	1.01V	4.22V	1.09V	4.38V	1.04V	1.96V	2.62V
11.	0V	0V	0V	0V	0V	0V	0V
12.	1.06V	4.36V	1.07V	4.76V	0.99V	1.86V	2.66V
13.	1.06V	4.36V	1.07V	4.76V	0.99V	1.86V	2.66V
14.	1.02V	25.2V	1.12V	28.1V	0.54V	6.91V	12.7V

Q7101

	Band 1 **2**	**Band 2** **18**	**Band 3** **51**
G1	0V	0V	4.82V
G2	5.05V	4.17V	7.19V
D	0V	0.10V	11.3V
S	0.10V	0.10V	4.83V

Q7102

G1	4.59V	4.65V	4.61V
G2	5.44V	4.50V	7.20V
D	11.3V	11.2V	11.1V
S	4.08V	3.71V	11.0V

Q7401

E	0V	0V	0V
B	0.60V	0.60V	0.60V
C	2.05V	1.65V	1.00V

Q7402

E	11.3V	11.4V	11.4V
B	11.4V	10.6V	10.6V
C	-14.5V	11.2V	11.2V

Q7403

E	0V	0V	0V
B	0.71V	0.71V	0V
C	0.11V	0.11V	11.3V

Q7404

E	11.4V	11.4V	11.4V
B	11.0V	11.0V	10.7V
C	0.11V	0.11V	11.3V

Q7401 Collector	
CABLE CHANNEL	**VOLTAGE**
2	1.6
6	5
14	12.7
17	22.7
18	1.8
13	5.6
34	10.8
37	12.7
48	26.2
50	31
51	0.7
57	2
63	3.8
76	8
83	10.1
93	12.6
110	15.9
117	18.3
125	23.5

Figure 3-19. Tuner voltage charts (continued next page).

Varactor Diode Tuning Voltage Chart			
CABLE CHANNEL	**CR7106, CR7107**	**CR7108**	**CR7111**
2	0.69	1	1.1
6	4.37	5.7	6.3
14	14.92	14.5	14.9
17	26	25.5	23.5
18	1.6	1.3	1.9
13	5.7	5.5	6.7
34	11	10.5	12.8
37	12.8	12.2	14.7
48	22.8	23.1	26.2
50	24.1	24.3	26.6
	CR7101, CR7114	**CR7102**	**CR7103**
51	0.6	0.64	0.69
57	1.7	1.81	1.98
63	3.06	3.28	3.68
76	6.88	7.09	7.76
83	8.92	9.08	9.83
93	11.45	11.63	12.3
110	14.43	14.73	15.61
117	16.54	17	18.2
125	19.6	21.6	26.5

Note: Voltages are approximate cathode voltages only and will vary from set to set. This chart is supplied as a basic guide for typical voltages on the alignment channels. DO NOT USE THESE VOLTAGES AS A BASIS FOR TUNER ALIGNMENT.

Figure 3-19. Tuner voltage charts (continued).

The EEPROM and the Problems it Causes

An EEPROM is an electrically erasable programmable read-only memory integrated circuit. In the real world, it is used to store all kinds of information, such as configuration parameters, data files, and even full programs. Some are capable of storing data well into the megabyte range! In consumer electronics, the EEPROM is usually used to store manufacturer's setup data and/or user settings. If you recall the discussion of the CTC169, you will remember its EEPROM stores information about the RF input, channel scan data, and customer control settings. In the new family of RCA/GE televisions, the EEPROM stores almost all operating parameters.

There are several different types of EEPROMs on the market today. The type that concerns us is the "serial EEPROM" because it is used in modern consumer electronics. Most of these are of the 12C type, represented by the 24C02 chip. It typically has a small memory capacity, about 1 kbyte, and uses a serial bus to interface it with the microcomputer. According to some manufacturer's specs, it has a nonvolatile memory (does not require battery backup) which is good for about 40 years!

The 24C02 identifies a family, which is how EEPROMs are grouped. The 24C01, 24C02, 24C04, 24C08, etc., make up the family in which we are interested. These chips are all identical except for memory capacity. For example, the 24C01 has a memory of about 1k while the 24C02 has about 2k, or one page of the hex editor's screen. If you want more information, I suggest you look up an article in the January/February (1998) edition of *ProService Magazine* by Jerry Rains, "The How, Why, and Which of EEPROM Programmers." You might also like to peruse a web site or two. I particularly like National Semiconductor's web site, *http:\\www.nsc.com/pf/NM*. It has lots of useful information, the only cost being what the telephone company charges for your phone service.

Thomson configures the EEPROM according to *Figure 3-20*. U3101 is the master; U3201 the slave. When the TV is plugged in, U3201 causes clock and data lines to go high, seeking acknowledgment (a "handshake") from U1001, U7401, U3201, and U2901 if it is present. After it receives the handshake, U3101 causes clock and data activity to cease. When the TV is turned on, U3101 polls the EEPROM (U3201) for the necessary operating data—horizontal frequency, vertical height setting, gray scale settings, etc. It also pulls data out of the EEPROM that permits it to tune the channel the user has requested. The microprocessor is the "brains" of the outfit; the EEPROM is, I guess you could say, "the gray matter." It is the place where all operating parameters for the set are stored, and the set will not operate without a working EEPROM.

The Service Menu

The operating parameters are in what RCA calls "the service menu." To access it, place the TV in the service mode by depressing and holding the menu button and pressing and releasing the power ON button and the volume UP button, and then releasing the menu button. You now have the set in the service mode. A "P 00" will appear in the lower left part of the picture, and a "V 00" will appear on the lower right side (*Figure 3-21*). The P stands for "parameter"; the V for "value." The parameter is what you adjust; the value is the adjustment itself.

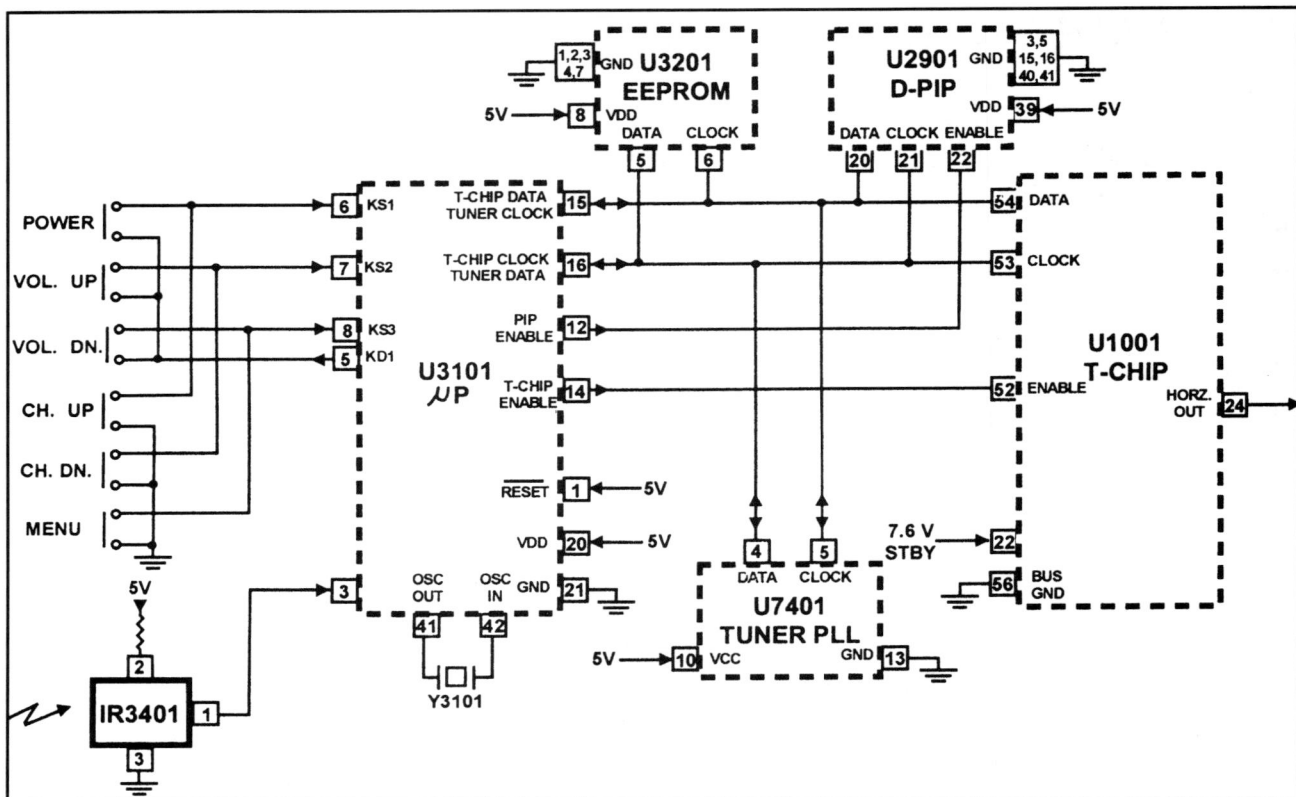

Figure 3-20. Thomson EEPROM configuration.

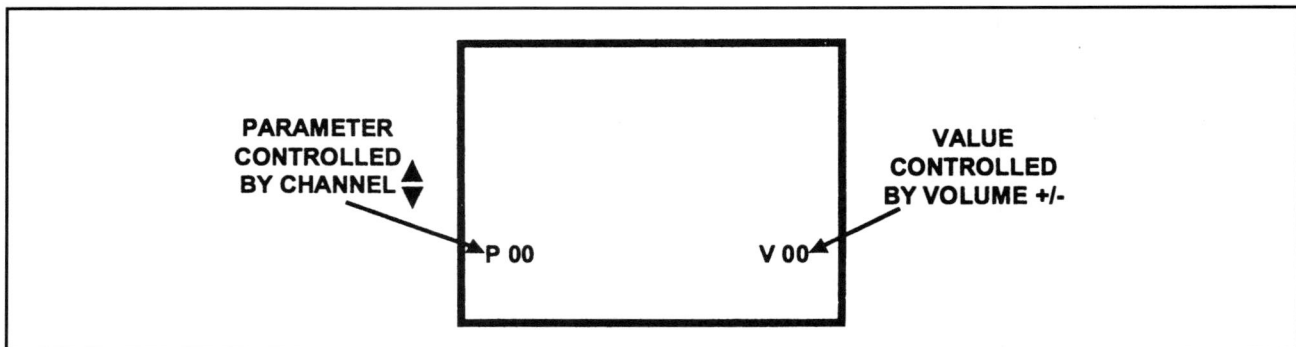

Figure 3-21. Service menu view.

There are three pages, or levels, in the service menu: Service Adjustment Parameters, Chassis Alignment Parameters, and Electronic Tuner Alignment Parameters. Each level has its own pass number. For example, the pass number for page 1 (level 1) is 76. Depress either the volume UP control on the front of the TV or on the remote control until the reading on the right side of the screen is "V 76." Then use either the channel UP button on the TV set or the remote control to set the "P" to the setting you want to check or adjust. Suppose you want to adjust horizontal frequency. Use the channel UP control to select "P01." Then use the volume UP/volume DOWN control to set the value ("V") to the desired setting. To access level 2, first scroll through all the parameters in level 1 and then set the value to 77. To access level 3, first scroll through all the parameters in level 2 and then set the value to 78.

Typical EEPROM Problems

We can now turn our attention to the problems a defective EEPROM can cause. I must emphasize that you should never replace an EEPROM without first doing the tuner shield. Ground faults at the tuner wrap are responsible for almost all EEPROM problems, something I learned the hard way. I replaced and programmed an EEPROM for a no-sound problem. The customer took the TV home and called me right back. He said it still didn't have any sound. I had to do the tuner wrap, replace the EEPROM, and reprogram it to fix the TV. Never service one of these sets without first fixing the tuner wrap!

No Start-Up

Several things can cause a no-start condition. The EEPROM stands at the head of the list and can easily be checked. First, check pin 8 for about 5 volts. The voltage in real life will be about 4.9 volts. If it is there, the power supply is up and running. Second, check pins 5 and 6 with a scope for activity. If you find activity on these pins, assuming you have not touched any control and that the TV has been plugged in for a few minutes, chances are excellent the EEPROM is bad.

Remember what happens when the TV is first plugged in? The system control requests acknowledgment from the various ICs. When it gets the handshake, it ceases clock and data activity. If it does not get acknowledgment, clock and data activity continue. Of course, if you find activity on pins 5 and 6 after the TV has been plugged in for a while, the problem *could be* elsewhere, like the tuner controller integrated circuit. But based on lots of experience and to save time, I usually replace the EEPROM. Since I use an 8-pin socket when replacing an EEPROM, if I make a mistake, I don't have to go to extraordinary lengths to correct it.

In the world of electronics, there are always exceptions. For example, I just serviced a dead set CTC175. As I teach our techs, when you encounter a dead set CTC175/176/177/187, check the standby voltage first at pin 8 of the EEPROM. Then scope pins 5 and 6 for activity. The voltage at pin 8 checked fine, and clock and data activity ceased as it should have, which indicated a successful handshake with the microprocessor. If you pressed the power button, clock and data activity reappeared at about a 5-volt level. However, the TV would not start. Was the EEPROM, therefore, good? If it was good, where did the problem lie?

I told the tech to replace the EEPROM anyway. He did, and the TV came to life. You see, parameter 01, the horizontal oscillator frequency adjustment, had been corrupted. We later found out the value had been reset to 00, probably by a AC power line spike. The 00 in parameter 01 kept the television from starting because the horizontal oscillator could not start.

There are two ways to restart a TV when the horizontal oscillator is so far off frequency it won't permit the TV to fire up. You can replace the EEPROM, or you can temporarily add additional capacitance across the retrace capacitor to detune the horizontal deflection circuit. The added capacitance will lower the high voltage, allowing the set to stay on while the horizontal oscillator frequency alignment is changed.

You must exercise precaution when you add the additional capacitance, to prevent damage to the horizontal circuit. The literature says to add an extra 1000 to 1500 picofarads to 20" versions and about 3000 picofarads to 25" sets. These sets have no pincushion circuits, so doubling the capacitance will not be dangerous. You can parallel an identical capacitor to the one in the circuit, in other words, without damaging the TV. But the 27" and 31" models are a bit more tricky because they employ diode modulator pin-correction circuits. If the balance is disturbed too much, one or both of the damper diodes will more than likely be destroyed. The engineers suggest that you keep the added capacitance under 2000 picofarads if possible, and be prepared to parallel extra capacitance across the pin retrace cap (C4407) to maintain the balance. The literature suggests as a rule of thumb adding four times as much capacity to C4407 as you add to the retrace cap. Use capacitors rated at least 1.6 kilovolts, and remember to remove them once you have correctly adjusted the horizontal oscillator frequency. That's the suggested procedure.

Wouldn't it be simpler, and safer, just to remove the old EEPROM, install a socket, and at least temporarily insert a new EEPROM? Once you get the set up and running, while the TV is on, you can remove the new EEPROM, install the old one, and make whatever adjustments you need to make.

No Audio

A customer complains of no audio or a very low level of audio when the volume is turned up as high as it will go. You confirm the problem, but you are not sure where to begin troubleshooting. Logically, you should begin with the audio output IC. But there is one check you should make before you do that. Measure the voltage on pin 29 of the microprocessor. If the voltage is high, about 5 volts, wick it out and let it float. If the sound returns when you plug the TV back in and turn it on, you have a corrupted EEPROM. Instead of unsoldering it, you can take pin 29 to ground using a small value resistor if you prefer.

The EEPROM contains data that mutes the speakers to prevent an audible pop when the set is turned on and off. A ground fault created by a poor tuner-wrap ground corrupts this data bit and causes the EEPROM to instruct the microprocessor to mute the speakers at all times. The microprocessor performs the mute function by toggling pin 29 high (mute) and low (no mute). The corrupted data bit causes the microprocessor to hold pin 29 high, resulting in no audio.

Our friends at RCA tell you to use one of two procedures to correct the problem in lieu of replacing the EEPROM. First, use the Chipper Check to reinitialize the EEPROM. (Let me confess I have not been able to reinitialize a dead EEPROM with the Chipper Check, and I have given up trying!) Second, with AC disconnected from the chassis, wick out pin 8 of the EEPROM. Use a pick to short the pin to the pad to temporarily apply B+ to the EEPROM. Plug the TV in and turn it on, and then remove the pick to allow pin 8 to float free of the pad. Use the menu function to select the speaker on/off feature and turn the speakers back on. Finally, unplug the TV and resolder the pin.

On the other hand—there is another way—clip R1915 out of the circuit and throw it away! By removing this resistor, you are disabling the audio mute function. You will hear a slight pop when you turn the TV on or turn it off, but the pop is not objectionable in the least. The big savings is you don't have to install and program a new EEPROM and charge the owner for it. Other techs have come up with different solutions, like grounding pin 29 of the microprocessor (which I don't like) or removing the surface-mount transistor which mutes the audio output IC. Choose whichever method you like. Clipping R1915 is the easiest for me because it is easy to access. Perhaps I should note that this is not "a factory approved repair" and will be frowned upon by the engineers at Thomson. There is one occasion when you should not use it, namely if you are effecting a repair covered under warranty. Always, <u>repeat always</u>, follow the correct "factory approved" procedures when you do warranty work.

No Vertical Deflection

The customer complains the television has a line across the center of the screen. You confirm the lack of vertical deflection, but you can find no reason for it except no vertical drive out of the T-Chip. Before you change that 64-pin IC, do yourself a favor and change the EEPROM. If you are wrong and the problem really is the T-Chip, you can always reinstall the original EEPROM. As I have said, I recommend putting the new EEPROM into a socket. You can buy 8-pin IC sockets from almost any parts house for just a few pennies, and it is guaranteed to save you time.

If you look at parameter 06 in the service menu, you will see that vertical size is set by it. A fault that resets the parameter to 00 will result in no vertical deflection. I have on occasion been able to restore the parameter in the original EEPROM. The first time was rather difficult since I couldn't see a picture. I entered the service mode, pressed the volume UP button 76 times, pressed channel UP 6 times, and toggled the value of parameter 06 up and down and restored full vertical deflection. I do it quicker and easier now by using the 8-pin socket. Use a new EEPROM to get the TV up and running. Unplug it and install the old EEPROM. Access the necessary parameters in the service menu and reset them. At the very least, this procedure is worth a try.

No Horizontal Phase

My customer said the picture on his TV looked like some kind of fancy "latticework." When I turned the CTC175 on, I observed a picture that floated across the screen from left to right—in other words, a horizontal oscillator that was off phase. It always pays to check service menus on these new televisions before you drag out the service literature and turn on the test equipment. For instance, I serviced a Zenith television because it was stuck on just one channel. It would not autoprogram and would not budge off channel 69. I checked the service menu and found the "channel lock" function had somehow been turned on. I turned it off and solved the problem. With respect to the CTC175, I entered the service mode and accessed parameter 01 and noted it had a full range of adjustments. The horizontal frequency changed, in other words, as I adjusted the "V" numbers from 05 to 20. I accessed parameter 02 and noted that the picture did not change as I ran the parameter through its full range of adjustments. A new EEPROM solved the "no-horizontal-phase" problem.

No Video

The usual scenario is no audio and a very dim raster without even a hint of video. If you troubleshoot such a chassis, you will find all voltages basically correct except you won't have video coming out of the T-Chip. Before you replace U1001, try a new EEPROM.

Installing And Programming A New EEPROM

Anticipating your needs, I am including a list (*Figure 3-23*) that has EEPROM part numbers for most of the RCA chassis in circulation. I am also going to suggest two ways to program a new chip. You can use the parameters in *Figure 3-22*, or you can do it according to the service literature. I arrived at the parameters in *Figure 3-22* by accessing the service mode in several new TVs, writing down the values in each level, and averaging them. They will not work in every set, but they will work in most. If they won't work, the suggested values will get you to a very good starting point. As a matter of fact, I have had just two TVs in three years that these values wouldn't fix. In both instances, I had to do the tuner alignment with a TAG001 to get the tuner up and working.

You can get Thomson's suggested alignment procedures from the literature I have mentioned, but I will summarize it here. One of the best "non-Thomson" discussions of the alignment procedure is found in issue 171 of *Sencore News*. I recommend that you read that article carefully because it has some very helpful insights into the alignment process. I will not discuss tuner alignment because I have already been over it.

(1) Begin by preparing the following test terminals on the bottom of the circuit board: (a) IF input at pin 1 of the SAW filter, SF230; (b) TP1201, which is pin 55 of U1001; (c) TP2305, which is pin 14 of U1001 (AGC); (d) TP7102, which is at the junction of R7130 and R2312 (RF AGC); (e) TP2302, which is pin 63 of U1001; (f) TP1202, which is pin 3 of U1001; (g) and ground.

(2) Turn on the set and make the following adjustments to the customer control menu: set the color to minimum, the contrast to about three-fourths of maximum, and the brightness to mid-range. Remember to use a good signal generator instead of an off-air signal.

(3) Enter page 1 of the service menu, the service adjustment parameters. Adjust the horizontal frequency until the picture just drifts across screen. Make adjustment in single increments, because if you run the value too high or too low, the TV will shut off, and you will have to try a new chip. P02 and P06 concern horizontal centering and vertical height and centering. These are easy to do.

P07-P12 set the bias and drive controls to ensure proper white balance and color temperature. First, preset the drive controls (P10-P12) to 32, which is their mid-range setting. Then set the bias adjustments (P07-09) so that you get about 170 VDC on the red, blue, and green video output transistors. You might want to collapse the raster to a service line when you make these adjustments. To get a service line, merely press the menu button on the front of the TV. Press it again to get full raster. Now check the raster to make sure no color other than black, gray or white exists in any of the low-light or high-light bars. It may be necessary to tweak the bias adjustments to remove the coloring from the low-light areas. If color appears in the high-light areas, remove it by adjusting the drive controls.

(4) Enter page 2 of the service menu, the chassis alignments.

TABLE ONE
Service Adjustment Paramenters

Parameter Number	Parameter	Value Range	Suggested Value
01	Horiz. Frequency	00-31	Approx. 11
02	Horiz. Phase	00-15	Approx. 11
03	EW DC (pincushion)	00-15	Approx. 07
04	EW Amplitude	00-07	Approx. 03
05	Vert. DC (centering)	00-15	Approx. 08
06	Vertical Size	00-31	Approx. 24
07	Red Bias	00-127	Begin at 23
08	Green Bias	00-127	Begin at 20
09	Blue Bias	00-127	Begin at 27
10	Red Drive	00-63	Begin at 36
11	Green Drive	00-63	Begin at 32
12	Blue Drive	00-63	Begin at 27

TABLE TWO
Chassis Alignment Parameters

Parameter Number	Parameter	Value Range	Suggested Value
14	PLL Tuning	00-63	31
15	4.5 MHz Trap	00-07	04
16	Video Level	00-07	03 to 05
17	FM Level	00-15	07
18	B+ Trim	00-15	07
19	RF AGC Ch. 6	00-31	Begin at 00
20	RF AGC Band 1	00-31	Begin at 00
21	RF AGC Band 2	00-31	Begin at 00
22	RF AGC Band 3	00-31	Begin at 00
23	D-PIP Chroma		
24	D-PIP Tint		
25	D-PIP Bright		
26	D-PIP Contrast		
27	Factory Tint	00-63	

Figure 3-22. Alignment parameters (continued next page).

TABLE THREE
Electronic Tuner Alignment Parameters

Parameter Number	Parameter	Value Range	Suggested Value
100	Ch.2 Secondary	00-63	32
101	Primary	00-63	27
102	Single	00-63	17
103	Ch. 6 Secondary	00-63	63
104	Primary	00-63	53
105	Single	00-63	40
106	Ch. 14 Secondary	00-63	63
107	Primary	00-63	62
108	Single	00-63	63
109	Ch. 17 Secondary	00-63	48
110	Primary	00-63	62
111	Single	00-63	62
112	Ch. 18 Secondary	00-63	47
113	Primary	00-63	44
114	Single	00-63	45
115	Ch. 13 Secondary	00-63	63
116	Primary	00-63	44
117	Single	00-63	56
118	Ch. 34 Secondary	00-63	63
119	Primary	00-63	44
120	Single	00-63	57
121	Ch. 37 Secondary	00-63	63
122	Primary	00-63	44
123	Single	00-63	55
124	Ch. 48 Secondary	00-63	51
125	Primary	00-63	51
126	Single	00-63	37
127	Ch. 50 Secondary	00-63	43
128	Primary	00-63	38
129	Single	00-63	23
130	Ch. 51 Secondary	00-63	30
131	Primary;	00-63	33
132	Single	00-63	44
133	Ch. 57 Secondary	00-63	28
134	Primary	00-63	30
135	Single	00-63	38
136	Ch. 63 Secondary	00-63	26
137	Primary	00-63	26
138	Single	00-63	34

Figure 3-22. Alignment parameters (continued).

RCA EEPROM LIST

24-Oct-95

CHASSIS #	PART #	CHASSI #.	PART #.
CTC175A	217321	CTC175A2	218397
CTC175C	221047	CTC175C2	218397
CTC175K	217323	CTC175K2	218397
CTC175L	217323	CTC175L2	218397
CTC176C	218540	CTC176D2	218541
CTC176E	217657	CTC176F	217658
CTC176F2	223923	CTC176G2	224836
CTC176H2	221097	CTC176J2	221097
CTC176K2	221098	CTC176L2	221099
CTC176N2	223932	CTC176P	224752
CTC176P2	224836		
CTC177AA	218400	CTC177AA2	224279
CTC177AA3	227715	CTC177AC	218542
CTC177AD	218401	CTC177AE	218403
CTC177AF	218402	CTC177AF2	223935
CTC177AF3	228287	CTC177AG	218401
CTC177AH	218403	CTC177AJ2	221101
CTC177AK	218400	CTC177AK2	224279
CTC177AM2	223935	CTC177AR	227716
CTC177BB	215861	CTC177BD	218397
CTC177BD	218397	CTC177BE	218398
CTC177BF	218397	CTC177BG	218398
CTC177BH	218399	CTC177BH2	224281
CTC177BH3	224281	CTC177BP2	227714
CTC177CC	218404	CTC177CD	218405
CTC179BJ	229779	CTC179BJ	229780
CTC179CJ	229779	CTC179CJ	229780
CTC179CK	229781	CTC179CK	229782
CTC179CM	229781	CTC179CM	229782
CTC179DJ	229779	CTC179DJ	229780
CTC179DK	229781	CTC179DK	229782
CTC179DL	229781	CTC179DL	229782
CTC186D	225767		
CTC187AA	223912	CTC187AB	223940
CTC187AC	223912	CTC187AD	223940
CTC187AF	228907	CTC187BC	225688
CTC187BD	225688	CTC187BD2	225688
CTC187BF	225688	CTC187BF2	225688
CTC187BG	223941	CTC187BH	225689
CTC187BJ	225689	CTC187CH	228905
CTC187CJ	223957	CTC187CK	223954
CTC187CL	223957	CTC187CL2	223957
CTC187CL3	223957	CTC187CM	223957
CTC187CN	228905	CTC187CN2	228905
CTC187CN3	228905		
CTC189DA	230052	CTC189DA	230055
CTC189DB	230052	CTC189DB	230055

Figure 3-23. EEPROM list.

To align the IF VCO, connect a jumper from TP7102 to ground. Then apply a 45.75 MHz IF signal to pin 1 of the SAW filter. Connect a DC power supply to pin 14 of U1001 and apply 4 VDC. Connect a scope to TP1201 with the time base set at 0.1 microseconds and the volts/div. to .02 mV. If you don't see a waveform, adjust L2303 to produce the waveform. Adjust parameter 14 until there is minimum noise and interference on the 4.5 MHz signal. Then remove the DC power supply.

Proceed to P15 (the 4.5 MHz trap). Attach the scope to pin 63 of U1001. Inject a mono audio signal into the SAW filter. You should see a low-level 4.5 MHz signal. Starting at "V07," reduce the value until you see the first reduction in the amplitude of the signal.

Move to P16, the video output level of the T-Chip. Disconnect the signal generator from the SAW filter and connect it to the tuner input. Select a channel and signal level needed to produce a good picture on the CRT. Be certain to remove all jumpers and other test instruments. Connect the scope to pin 63 of U1001 and adjust the value of the parameter until the amplitude of the waveform is 2 Vp-p from sync tip to peak white.

The last parameter to adjust is P17, the audio level output. Connect a jumper from TP2305 to ground and remove the signal generator from the RF input. Inject a 4.5 MHz signal between pin 55 of U1001 and ground. While looking at the signal at pin 3 of U1001, adjust the value for a signal of about 1.2 Vp-p. Use a 1000 Hz audio signal for the adjustment.

I have obviously not covered all the alignment procedures—just the ones that are more-or-less critical to the operation of the television. You can nevertheless see what you are up against if you do a complete alignment per the service literature. I have not even mentioned how to do the stereo alignment for the CTC187, which is a world unto itself! The complete alignment for the dbx stereo decoder is covered in pages 5 through 9 of the *CTC 177/187 Troubleshooting Guide*.

System Control Problems

Late production CTC175/176/177 chassis have a different microprocessor (U3101) than early production models. The newer chassis can be identified by the presence of a "2" after the alpha suffix on the barcode label. For example, a CTC177AA having the newer microprocessor will be labeled CTC177AA2. The pinouts of the two chips are <u>not</u> the same and therefore are not interchangeable. If you need to replace a U3101, make certain you have the correct part number. Incidentally, the CTC187 utilizes the newly designed chip, which is designated the "ST-9 microprocessor." Pinout information for the ST-9 system control microprocessor is given in *Figure 3-24*.

Five integrated circuits make up the system control circuitry: U3101, the main microprocessor; U1001, the T-Chip; U2901, the D-PIP microprocessor; U3201, the EEPROM; and U7401, the tuner PLL (*Figure 3-30*). Add U1600, the stereo decoder, if you are working on the CTC187.

These chips communicate via the serial bus. The communication format is called "bus protocol." Would you believe Thomson uses three protocols: IM Bus, I-squared-C bus, and the T-Bus! We don't need to understand the intricacies of the protocol system to service these televisions. Knowing what type of

U3101 - Thomson ST-9 SYSTEM CONTROL MICROPROCESSOR

PIN NO.	NAME	VOLTAGE	IN CKT. RES.	DESCRIPTION
1	IR IN	4.5	>300K	IR input signal from remote control.
2	RESET	5	>200K	Micro reset - active LO.
3	NC	-	-	
4	DEGAUSS	0	>20M	Activates degaussing relay.
5	KD1	0	>20M	Keyboard drive line.
6	KS1	4.7	>200K	Keyboard scan input.
7	KS2	4.7	>200K	Keyboard scan input.
8	KS3	4.7	>200K	Keyboard scan input.
9	DATA OUT	0.3	>20M	Data out for commercial television.
10	ATE ENABLE	0	10K	Used for factory testing.
11	CC VIDEO	2	>50K	Closed caption video input.
12	PIP ENABLE	4.7	>20M	Serial communication line used to control data between the micro. and PIP.
13	NC	-	-	
14	T-CHIP ENBL	4.6	>20M	Serial comm. line used to control communication between the micro and T-Chip.
15	BLUE OSD	0	1K	Blue OSD output.
16	GRN OSD	0	1K	Green OSD output.
17	RED OSD	0	1K	Red OSD output.
18	FSW	0	2K	Fast switch - controls OSD and video switching in T-chip.
19	T-CHIP DATA TUNER/ST CLK	4.5	>20M	Serial communication - data/clock.
20	T-CHIP CLK TUNER/ST DATA	4.7	>200K	Serial communication - data/clock
21	VDD	4.7	>20M	Power supply input for microprocessor.
22	VSS	0	0	Ground for microprocessor.
23	PLL FILTER	2.6	>20M	PLL filter network.
24	PLL BIAS	2.2	6K	PLL bias resistor Connection.
25	PLL VCC	4.7	>20M	Power supply for PLL.
26	H	0.7	8K	Horizontal timing input for OSD.
27	V	0.2	1.8K	Vertical timing input for OSD.
28	EXP ST SW ST / MONO	11 / 0	36K	Expanded stereo, NWS-gain and mono /stereo control line.
29	SPK MUTE	0	>100K	Goes HI to mute speakers.
30	TONE	0 / 5	>100K	Goes HI for low tone and LO for high tone.
31	SCLOCK/FM ON/OFF	0	>20M	Serial clock for DBX stereo decoder. FM on control for commercial TV.
32	ST SENSE	0 / 5	>20M	Stereo Detect line for non-DBX stereo sets.
33	RF SEC	* VARIES	>20M	PWM output - Controls the secondary of the double tuned filter in the tuner.
34	RF PRI	* VARIES	>20M	PWM output - Controls the primary of the double tuned filter in the tuner.
35	SINGLE TUN	* VARIES	>20M	PWM output - Controls the single tuned filter in the tuner.
36	OSC OUT	2.3	>20M	8MHz crystal connection.
37	OSC IN	2.3	>20M	8MHz crystal connection.
38	TUN SYNC	2.3	>20M	Sync input to detect the presence of an active channel when tuning.
39	FM TUN	* VARIES	>20M	Input to detect an active FM station on commercial TV.
40	IF DEFEAT	0	>20M	Output to defeat IF circuit via AGC.
41	STBY SW	-	>20M	
42	DATA IN	4.8	>200K	Data input for commercial television.

Figure 3-24. ST-9 pinout information.

information is exchanged and what IC pins are involved is important (*Figure 3-20*). The IM Bus uses three wires to communicate with the D-PIP microprocessor. The I-squared-C Bus, the serial bus with which we are most familiar, uses two wires to communicate with the tuner PLL and the EEPROM. The T-Bus also uses three wires, which the microprocessor uses to talk to the T-Chip. Fortunately, Thomson has moved away from the three-bus arrangement to just one bus in its later-model televisions. If you just have to know the details of the three-bus protocol, I will refer you to the Thomson service literature I have already cited.

Power-On Sequence

These chassis are unlike anything RCA/GE has ever manufactured. For example, U3101 does not have a single pin to control the on/off function of the television. The T-Chip, where horizontal drive is generated, is controlled entirely over the serial bus. When it receives an ON command, the microprocessor sends information over the bus to the T-Chip (U1001), telling it to turn on horizontal drive. U1001 has its own standby 7.6-volt regulator to keep communications alive and the horizontal stage operative while the set is off; that is, in standby mode. (There really is no "off" state. The TV is either "on" or in standby.) When the ON command is received, horizontal drive comes on, bringing up the run supplies which biases the rest of U1001 on, and the TV comes alive. An OFF command causes U1001 to turn horizontal drive off, shutting the TV down and placing it in standby.

The "Must Haves"

The microprocessor must have certain inputs to work. I will not give you pinout information because we are dealing with two different microprocessors. You will have to consult the appropriate service material to get the pin numbers.

(1) It must have 5 volts. The voltage in the CTC177 will be slightly less than 5 volts. It should be stable and without ripple. Not long ago, I was called to service a CTC177 that was dead. I measured the 5 volts, and it checked slightly low. I turned the TV off to wick out the VCC pin at the micro. The voltage measured lower! I resoldered the pin and checked it again. The voltage measured less than 3 volts! Dummy me! I finally got around to checking the standby regulator, Q4103, which was leaky collector-to-emitter. Incidentally, an ECG123AP substitutes just fine for it.

(2) It must have a good ground. One way to check the ground connection is to connect the negative DMM probe to the ground pin when you check for 5 volts. You "kill two birds with one stone" by making the voltage check like this.

(3) It must have reset (*Figure 3-25*). Reset starts the microprocessor at a known point in its program. When AC is first applied, reset goes high after about 55 milliseconds, giving the oscillator time to stabilize before the microprocessor starts. As you can see, the reset circuit also monitors the 12-volt standby line. If it drops below 10 volts, reset will activate and put the micro in a low-power mode. A 5.6-volt reference is applied to the emitter of Q3102. A voltage divider network reduces the 12 volts to about 6 and applies it to the base of

Q3102. The collector of Q3102 is connected to the base of Q3101. Under normal conditions, the voltage at the base of Q3102 is high enough (at about 6 volts) to keep it turned off.

If the 12-volt line drops far enough to allow the base of Q3102 to get to about 5 volts, Q3102 will turn on, turning on Q3101 which pulls the 5-volt reset to ground, initiating reset. Q3101 also disables the oscillator by grounding it through R3191 and CR3101. The microprocessor then goes into its low-power mode to maintain memory.

I have spent some time detailing the workings of the reset circuit because it can give trouble. I don't see reset problems often, but I do service them. The usual symptom is a dead microprocessor; that is, one that will not initiate the start sequence. In every instance the problem has been a leaky Q3101. Note that it is a surface-mount device.

(4) The oscillator must be working. Look for the signal to be about 5 volts p-p and very close to the stated frequency. If the signal is not present, the usual culprits will be either the crystal or the microprocessor. I have never had to change a defective crystal in one of these sets, but as you know, in the world of consumer electronics, anything can happen.

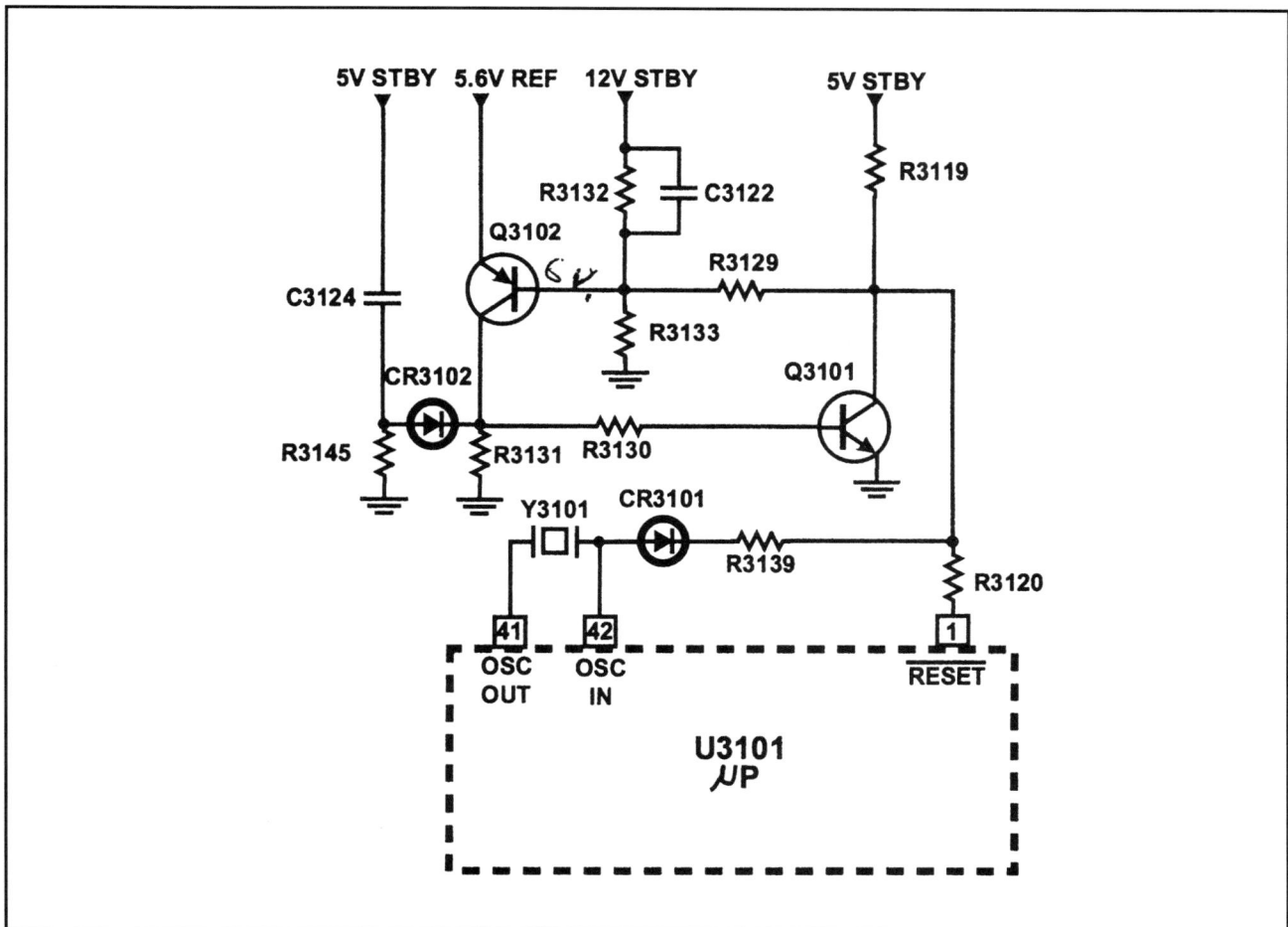

Figure 3-25. Reset circuit.

Servicing System Control Problems

The system control circuit controls <u>every</u> function of the television. If a part of it fails, the television will not work. Because U3101 and U1001 are interrelated, troubleshooting procedures will necessarily overlap. As a matter of fact, a failure of any IC in the system control network (U3101, U1001, U3201, U7401, etc.) will not permit the TV to work. The typical symptom will be a "dead set." Because anything can cause a dead set, we techs must be efficient in localizing the problem. Time is money!

Where, then, do we begin?

(1) It's an old adage but a true one: Check the standby voltages. If you are like me, you can get to the really weird stuff quickly. It's the small stuff, the easy things, that buffalo you. Therefore, verify the presence of +12 volts, +7.6 volts, and +5 volts. *Figure 3-26* is a partial schematic for the CTC175 chassis; *Figure 3-27* for the CTC176/77/87 chassis.

(2) Use a scope to check for horizontal drive at pin 24 of U1001 while you issue an ON command. If the pulses are there even for a split second, system control is working, and you need to troubleshoot horizontal deflection. If the pulses do not appear, check for 7.6 volts at pin 22 of U1001. If this voltage is not present, wick out the pin to see if it comes up. If it does, the chip is bad. If it doesn't, proceed to the power supply.

(3) I am assuming you have checked U3101 for its "must haves"—VCC, ground, reset, and oscillator. If, for example, the oscillator signal is not present, you will most likely have either a defective crystal or IC or both.

(4) Check for the presence of clock and data activity on the appropriate pins of the microprocessor. According to *Figure 3-30* (We are dealing with two different microprocessors.), the pins are 15 and 16. These pins should have <u>no</u> signal present in the standby mode after the

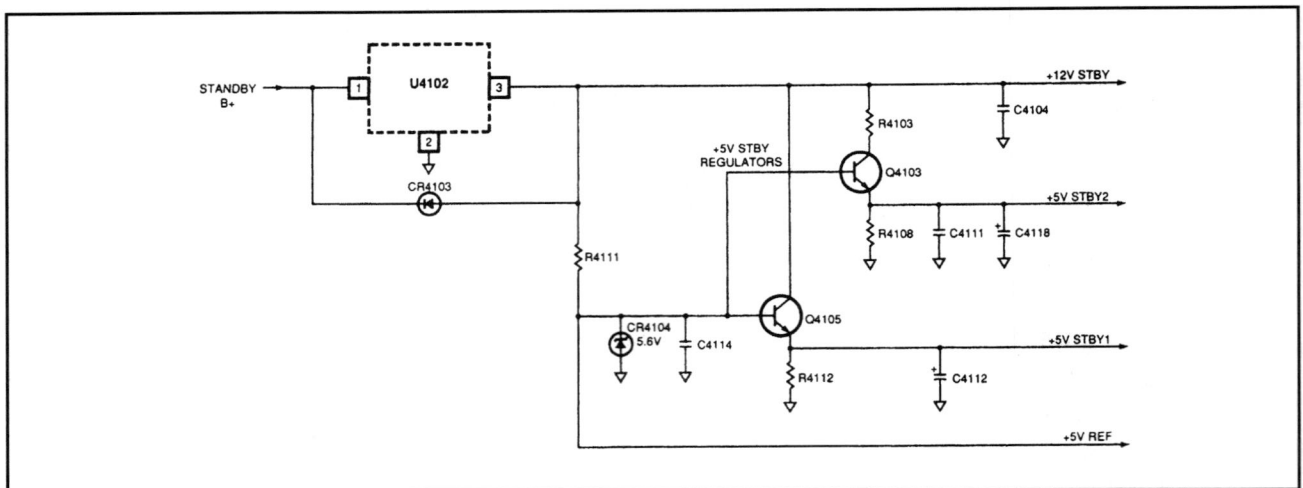

Figure 3-26. CTC175 system control (continued next page).

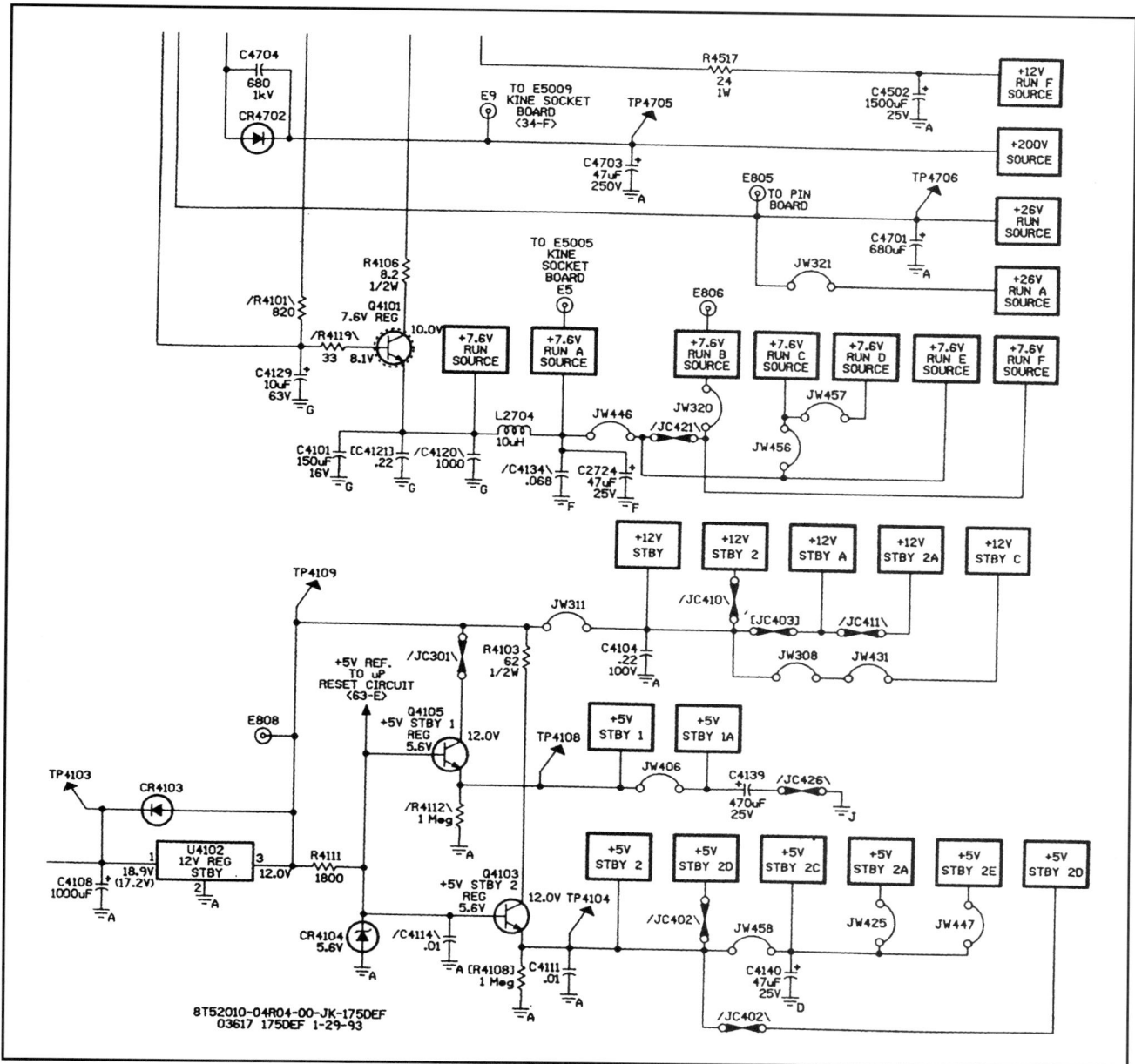

Figure 3-26. CTC175 system control (continued).

circuit has initialized. To be safe, check for clock and data activity on these pins after the TV has been plugged in for a minute or so.

If you issue an ON command, you should see a 5 Vp-p data train present on both pins. If it is not there, wick out the pins to see if data appears. If it does, the problem will be external to the microprocessor. It could be the EEPROM (most likely), the tuner PLL, or the T-Chip, for example. Use common sense to locate the fault by unsoldering the appropriate pins on each of the system control ICs. If activity resumes after a particular set of pins has been unsoldered, you have found the culprit.

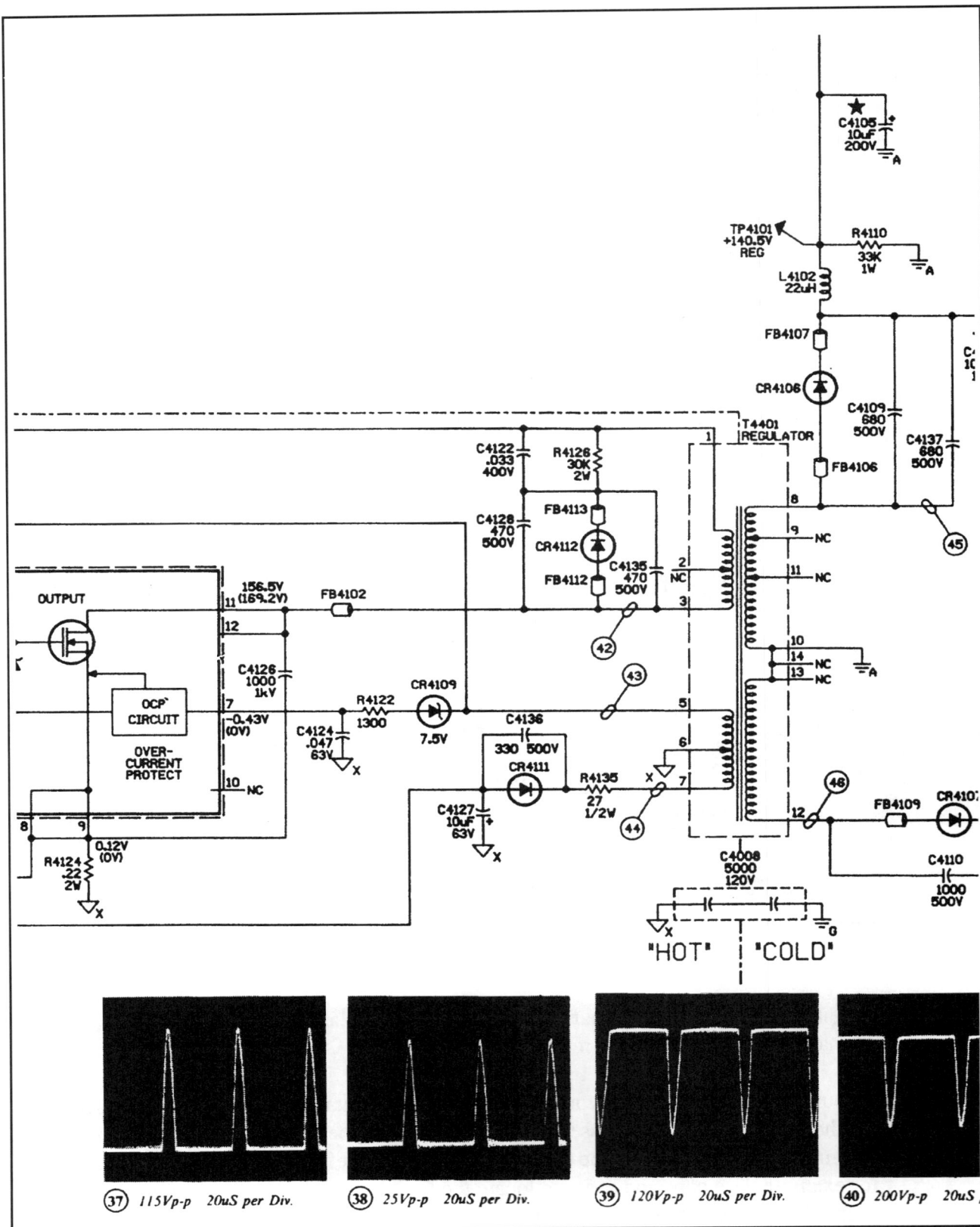

Figure 3-27. CTC176/77/87 system control.

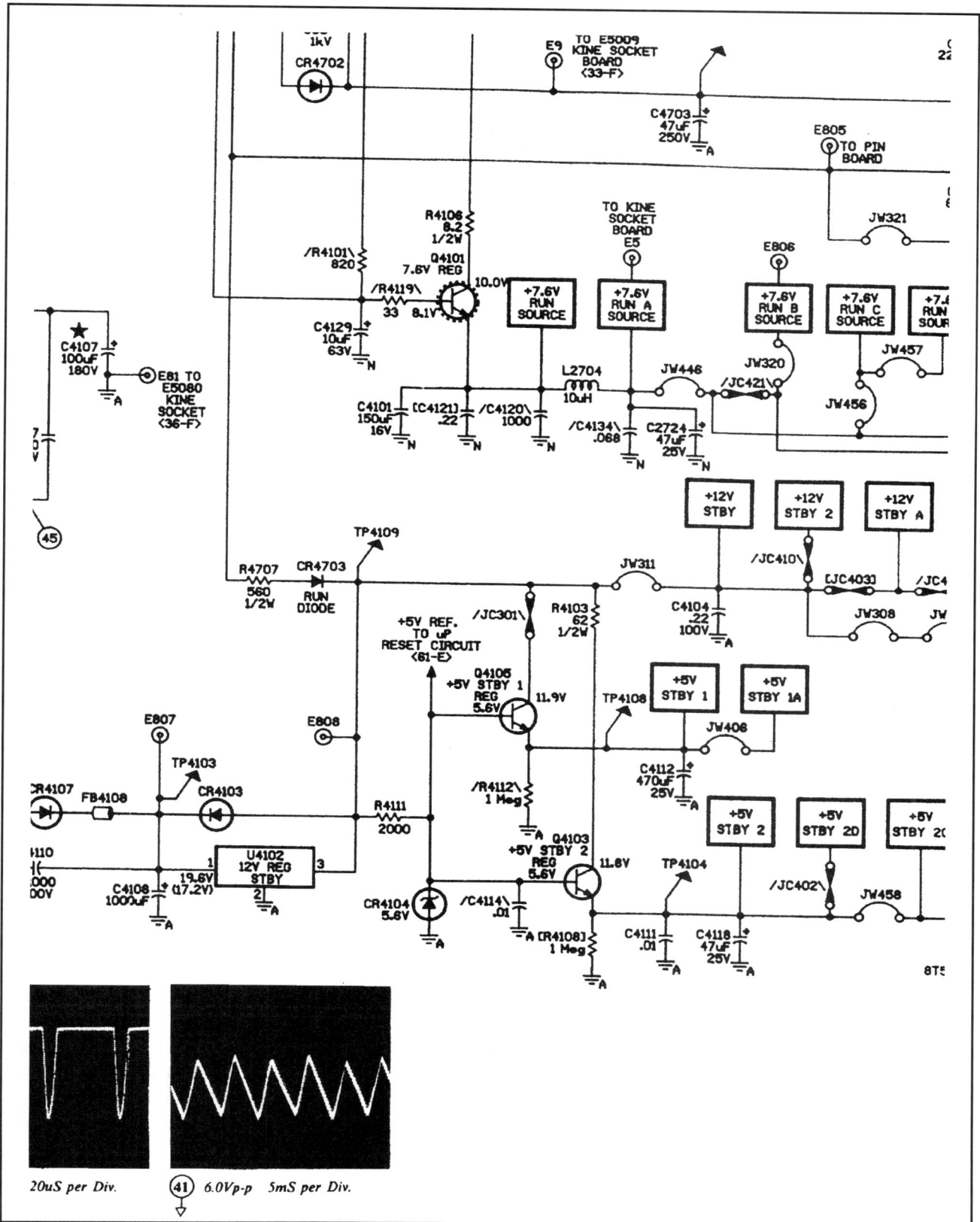

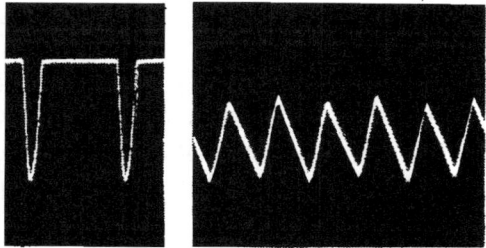

Figure 3-27. CTC176/77/87 system control (continued).

Because the EEPROM is almost always at fault, I suggest you use a slightly different approach to a dead-set problem. Check pin 8 of the EEPROM for 5 volts. If it is there, the power supply is working. Check pins 5 and 6 for activity after the set has been plugged in for a minute or so and before you have pressed a front-panel control. If the activity is there, the EEPROM becomes a likely suspect.

I have seen a couple of these TVs when the "must haves" were present, but there was no activity on either clock or data line (one or both lines might be involved) when the TV was first plugged in or after I issued an ON command. The microprocessor was defective in both instances. Be careful about condemning this chip because it is a low failure component. I suggest you rule out as best you can all alternatives before you order a new one.

(5) There is one other problem I want to mention which the literature does not cover. Somewhere in the troubleshooting process, check the data-in lines of the microprocessor for inappropriate activity. Look again at *Figure 3-22*, paying attention to pins 5, 6, 7 and 8 of U3101, the data input lines. We have seen such a configuration before, but there is an important difference. There is only one key drive line, KD1, which drives the power, volume up, and volume down keys. When one of these keys is pressed, KD1 pulses the corresponding sense line low. U3101 detects which button is being pressed by monitoring the sense lines for pulses. The other three switches pull KS1, KS2 or KS3 to ground. When it sees a constant low instead of a string of pulses, U3101 knows one of the other three buttons has been pressed and initiates the appropriate response.

Why mention the keyboard? Let me put it like this: I recently accepted a CTC177 for repair. The customer said it began to channel up by itself, turned itself off, and would not come back on. The problem almost had to be in system control, but where? I checked the "must haves," and they were good. I then checked pins 5-8 for any activity. Normally these lines will be 0 volts. Sure enough, one pin was high. I had a leaky channel-up tact switch. It won't happen often, but it does happen. The moral is, make your four "must have" checks first, and then check to see if there is inappropriate activity (activity when there shouldn't be any) on the data-in lines to the microprocessor.

The Power Supply

This family of televisions uses two different power supplies. The CTC175 uses one type, while the CTC176/177/187 chassis use another. Because the two supplies are different, the servicer will have to vary his/her approach to troubleshooting them.

As you can see from the block diagram in *Figure 3-28*, the CTC175 employs a basic series-pass regulator. It differs from a run-of-the-mill series-pass regulator because it is microprocessor-controlled. RCA employs its usual scheme to generate the raw B+ (about 150 volts) which is applied to the collector of Q4150 through F4150. R4155 lends the regulator a hand by passing some of the current the horizontal section needs. C4150 filters the regulated voltage.

Every regulator needs feedback. R4157 and R4158 comprise a voltage-divider network to feed the input of a comparator inside U1001. U1001 responds by outputting a PWM signal at pin 29. The PWM

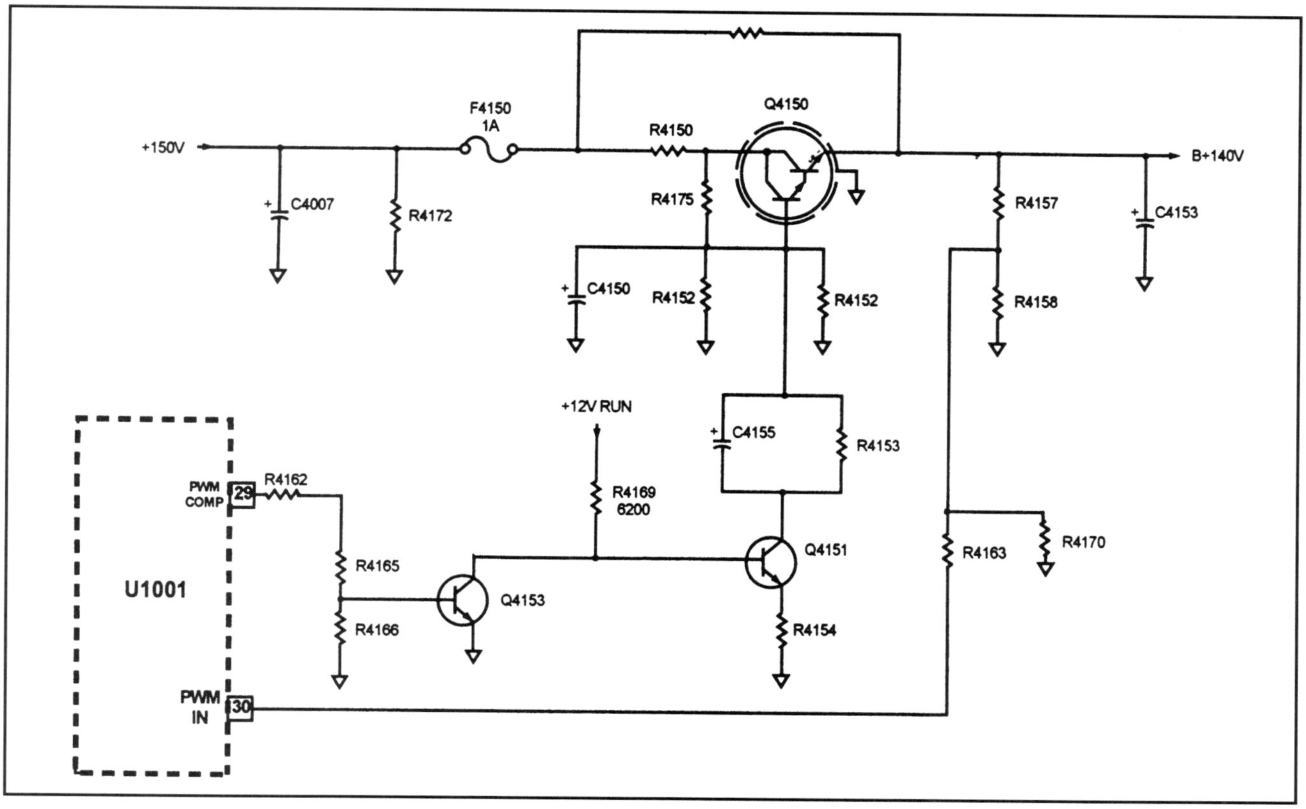

Figure 3-28. CTC175 power supply.

signal is inverted by Q4153 (common-emitter configuration) and controls Q4151, which in turn controls Q4150. For instance, if the load on the 140-volt line increases, the voltage at pin 30 of U1001 drops. The voltage drop at pin 30 causes the PWM at pin 29 to increase its output. The increased output biases on Q4153 harder, which reduces the bias on Q4151, increasing the B+ output of the power supply.

One other interesting feature about the regulator circuit. Beam current which is sensed at pin 16 of U1001 also affects the PWM output at pin 29. The purpose of the circuit inside the T-Chip (a compensation circuit) is to minimize blooming by helping to stabilize the regulator during heavy beam current transitions. If the beam current increases, regulator output decreases. The result is a raster that does not bloom or shrink.

If you suspect regulator problems, you will need to employ a variac/isolation transformer. Begin by applying between 105 to 130 VAC while you check for 140 volts out. If the B+ is not at 140 volts, enter the service mode and select parameter 18 (the B+ adjust). If you can set B+ at 140 volts by varying its setting, the circuit is probably working as it should. If you can't set the regulated B+ at 140 volts, suspect either Q4150 (the most likely suspect) or a component in the control circuit.

Note: A shorted horizontal output transistor (Q4401) will almost always cause Q4150 to fail. Replace Q4150 with part number 217309, and Q4401 with part number 17791. You may have success with generic parts, but I don't. Wherever possible, especially in critical circuits, I opt for exact replacement parts. It should go without saying that you use genuine RCA parts when you perform warranty service.

The CTC176/177/187 chassis use a switching regulator power supply. *Figure 3-29* shows you the basic circuit minus the raw B+ supply. I have found this switching power supply to be among the most reliable on the market. Aside from the usual damage caused by lighting/power surges, I seldom see one that needs repair.

The supply utilizes a variable frequency/variable pulse-width hybrid integrated circuit. The IC itself contains most of the components the power supply needs to operate, including a power FET. The FET turns on and off, inducing a voltage in the secondary of the transformer. The induced AC-type voltage is rectified and filtered and used to operate the chassis. Since it is a variable-frequency switching supply, the lower the frequency the more energy the IC transfers to the secondary. The converse is also true. As the frequency rises, the IC delivers less energy to the secondary windings.

The IC works like this:

Raw B+ enters at pins 11 and 12, the drain connection, of the IC. The source connection is at pins 8 and 9 and is taken to ground through R4124. R4104, a 1.5 megohm resistor, provides start-up voltage for the FET. Current flowing through the primary of T4101 induces a voltage between pins 5 and 6. This voltage is coupled from pin 5 through R4125 and C4123 to pin 4 of the IC. The polarity of the voltage at pin 4 is such that it turns on the FET harder. The increase in current through the FET causes more voltage to be dropped across R4124, the source resistor. This voltage will soon become large enough to turn on the overcurrent protection circuit inside U4101, which shuts down the FET. When the FET turns off, energy transfers from the primary winding of T4101 to the secondary, charging C4107 and C4108. The cycle repeats itself before stable oscillation starts. The frequency of the oscillation will vary from about 100 kHz in standby when power demands are minimal, to about 38 kHz with a full load.

Look again at *Figure 3-29*, paying attention to pins 5 and 7 of the switching transformer. As you can see, this winding develops feedback voltage for the regulator. It is tightly coupled to the secondary windings, which means the induced voltage closely follows the voltage developed in the secondary. The voltage is rectified by CR4111, filtered by C4127, and applied to pin 1 of U4101, where it is input to a precision voltage reference, trimmed to -40.5 ±0.5 volts. The error amp works at keeping the voltage at pin 1 equal to the reference voltage. If the load on the secondaries increases and the voltage drops, the voltage at pin 7 will become less negative (will decrease). The decreased voltage lets the FET stay on longer, increasing the output voltage. The IC is able to hold its output constant regardless of the line voltage and/or the load.

If the resistors tied to the error amp open for any reason, be assured the secondary voltages will rise! I serviced a CTC177 last month that had been "in another shop." I plugged it in and heard a high-pitched squeal indicative of a heavy current demand. I have learned a few shortcuts in my years of service. A heavy current demand means a short, and a short makes me think of the horizontal output transistor. Sure enough, the horizontal output transistor had shorted. I always get suspicious when one of these transistors blows and wonder what causes it. These TVs don't just blow them. Oh, it does happen from

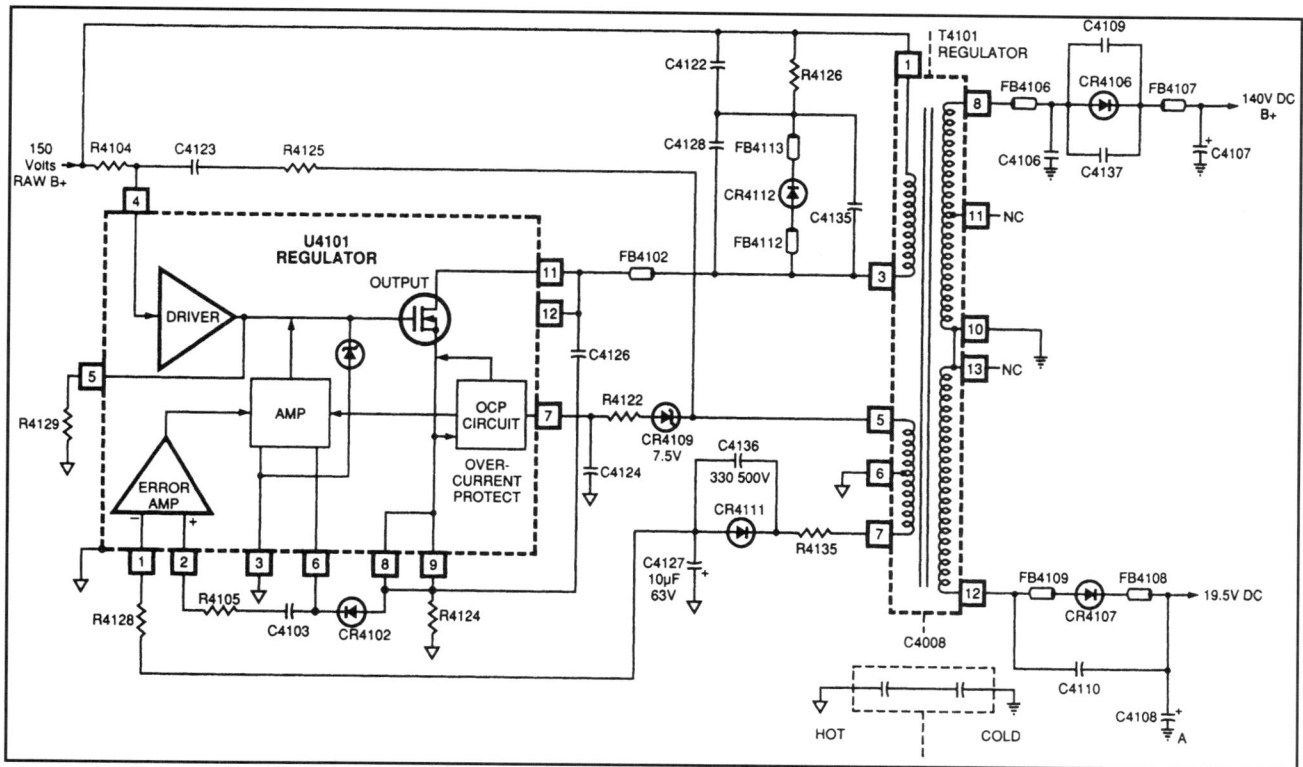

Figure 3-29. CTC176/177 power supply.

time to time, but not often. I did a few checks and reapplied AC and heard a sizzle and a pop. C4108 had blown. Why? I replaced the cap and monitored the 140-volt line as the TV powered up. The voltage read about 440 volts! The regulation circuit was obviously not working. This time I bothered to inspect the printed-circuit board around U4101 and immediately saw that R4128 and R4105 had been fried. Replacing both resistors, the capacitor, and the horizontal output transistor fixed the TV.

There is, of course, a moral to my little tale: It is easy for anybody to forget the basics. Using your senses—seeing, smelling, and hearing—is basic to effective troubleshooting. My hearing led me to the shorted horizontal output transistor and defective capacitor. But I forgot to look. If I had inspected the circuit board—if I had looked carefully—I would have seen the burned resistors. So, use your senses as well as your test equipment.

If an excessive load is placed on the power supply, the ON time of the FET increases, resulting in more current through the FET and its source resistor, R4124, which increases the charge on C4124. At some point the overcurrent protection circuit turns on, causing the FET to turn off. The value of C4124 determines the trip point for the protection circuit. The result will be a power supply that tries to start and shuts down, a cycle which repeats itself thousands of times a second. The key that something is placing an excessive load on the supply will be the audible, high-pitched squeal.

Before I leave this discussion, I need to make a few comments about some peripheral components. The peripheral components don't often give trouble. When they do, though, they create some bizarre symptoms. Knowing their function makes finding the problems they cause a little easier.

C4122, C4228, R4126, and CR4112 comprise a snubber network used to reduce high-voltage spikes developed when the FET turns off. C4102 and R4105 (at pin 2) stabilize U4101 by keeping parasitic oscillations at a minimum. R4129 at pin 5 does duty as an electrostatic protection resistor for the gate of the FET. R4129 and CR4109 (pin 7) provide some protection against varying line voltage. The ferrite beads in the circuit reduce radio frequency interference (RFI) emissions. C4107, L4102, and C4105 reduce ripple in the regulated B+ and any high-frequency switching noise.

Troubleshooting the Switching Power Supply

What sort of trouble does the switching power supply cause? As I have said, I have found it to be relatively trouble-free and among the most reliable on the market. Fortunately for us repairpersons, it does fail now and then.

The most common problem is caused by R4104, a 1.5 meg resistor in the start-up circuit. If the resistor opens, U4101 cannot start, and the complaint is a "dead set." If you are confronted by a dead set, check the voltage at pin 4 of U4101. If it is 0 volts, check R4104 first. It is rated at a quarter-watt and is located on the top of the circuit board. It was initially such a problem that RCA issued a service bulletin calling attention to the problem and the fix. Since it is in circuit with a capacitor, you will either have to discharge the capacitor or lift one end of the resistor to get an accurate reading.

If you happen to replace the regulator, make a couple of checks before you apply AC. First, check CR4106 in the 140-volt secondary source for a short. If it checks shorted or leaky, confirm that the problem lies with it or the horizontal output transistor. You really do need to check for shorts in the secondary because if, for example, CR4106 is shorted, you will probably blow the new integrated circuit when you apply power. (Don't even ask how I know!) Second, check R4129, a 100 ohm resistor tied to pin 5. The literature says if R4129 is open, the new chip will instantly destruct. Since I have never had a problem with R4129, I assume the literature is correct.

I conclude the troubleshooting section with a few miscellaneous remarks.

(1) Check pin 1 of U4101 for -40.5 ±0.5 volts. If the voltage is correct, the power supply is probably working as it should. If pin 1 shorts, the regulated B+ will be low (about 30 volts). If pin 1 opens, as when R4128 fails, the regulated B+ will rise. Remember the television with 400 volts on the B+ line?

(2) If it is under a heavy load, like a shorted horizontal output transistor, the frequency of the power supply will be in the audible range. Once you have heard it, you will never forget it! It is your clue to see what is loading it down. Incidentally, a heavy load will not normally cause it to self-destruct, one of the nice features about switching supplies. If it is not under a load, the B+ will rise and the supply will go into "a burst mode," which means it produces a series of burst pulses.

(3) If F4001 opens and the bridge diodes are good, replace the chip. The RCA part number is 215530.

(4) In the event that the supply will not oscillate, suspect U4101, R4104, R4125, C4123, or T4101. I have never had to replace a transformer in one these sets, but all things are possible. As a matter of fact, the only transformer problems I have ever serviced were in the CTC140 chassis. If its transformer failed, it would "eat" chopper transistors at a rather regular rate. Some transistors would last for a few days before they failed. The key was a transformer with a funky yellow color. If you saw the color and found the chopper transistor shorted, you just replaced both. I recall, I believe, having to replace a couple of capacitors in the secondary windings too. The transformers in the CTC177 chassis aren't like that, but in our business all things are likely! Therefore, "Never say 'never!'"

(5) If the B+ will not regulate or regulates poorly, think in terms of the components tied to pins 1 and 2 (R4128, R4105, and C4103), the IC itself, and as a last resort, the transformer. If the B+ is low and there is no load on the output like a shorted or leaky component, check R4135, CR4111, and C4127. These parts are in the feedback loop to the IC.

Scan-Derived Voltages

The scan-derived power supply is detailed in *Figure 3-30*. You can see it is basically straightforward and rather more simple than the CTC167 or CTC168. I will note two problems I have had to deal with. The first is the loss of the +33 volts to the tuner. In two instances the loss of the voltage was due to a shorted CR4108. In a third instance, the loss was due to a crack in the PCB.

The biggest problem has been the loss of the +12-volt run source. I will use as an example a set that comes into the shop with the complaint, "The picture just went black, and the sound stopped as if I had turned it off." You plug the set into an AC outlet and confirm the symptom. Confronted with a no-raster situation, you turn G-2 up and note a horizontal white line, indicative of no vertical deflection. Two problems in a television cause you think what the two circuits have in common. Vertical deflection and audio depend on the +12-volt run supply. A good place to check the voltage is at R4702 (*Figure 3-30*). It is a 2 watt resistor mounted about half an inch off the PCB, about midway between the flyback and front of the chassis. Chances are the resistor has opened because CR4704 has become leaky. The problem was fairly common early on because Thomson had gotten hold of a batch of inferior diodes. I don't see it often now, but I still have to correct the problem. Don't replace the resistor without replacing the diode. The part number for the resistor is 133393; the part number for the diode is 207878. If you don't use a genuine RCA part for the resistor, be certain to use a 2 watt, flameproof resistor because it is in a critical, high-current carrying circuit. It's better to be safe than sorry. I used an ECG part to replace the diode on one occasion because the customer was in a devil of a hurry for his television. The generic part seemed to work.

Horizontal Deflection and Pincushion Circuits

I will begin discussion of the horizontal deflection circuit by emphasizing the fact that these TVs utilize two different horizontal output transistors. In typical RCA fashion, the 27" and 31" sets employ diode modulator pin correction circuits and a horizontal output transistor without the built-in damper diode. The part number is 215539. The part number for the output transistors with the built-in damper diode

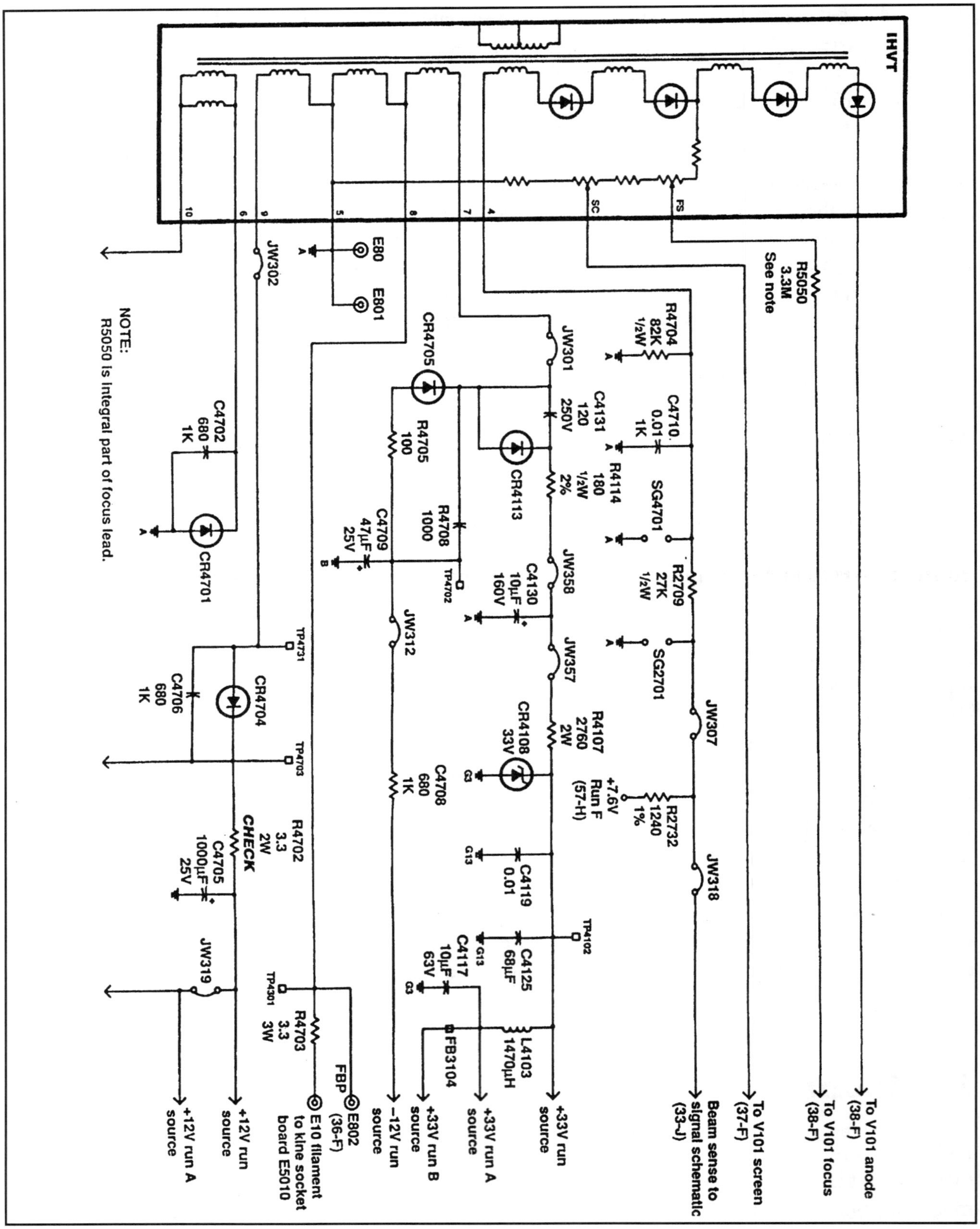

Figure 3-30. Scan-derived power supply.

for the 19", 20" and 25" sets is 177791. In the event you replace Q4401, remember to put the correct transistor into the TV!

Thomson confines low-level horizontal processing to the T-Chip, as they have in most of their chassis. These functions include automatic frequency control (AFC), automatic phase control (APC), horizontal drive, east-west (EW) pincushion correction, x-ray protection, and horizontal V_cc standby regulation. The difference between these chassis and former ones is the functions are now controlled via the serial bus.

There really is nothing new or different about the design of the horizontal drive and output circuits, as you can see by studying *Figure 3-31*. The output at pin 24 of U1001 is an open collector, that is low when horizontal drive is on. It might interest you to know the pulse width is adjustable from 32 to 36 microseconds via serial bus commands to the T-Chip's "horizontal duty" register. It is factory-set and cannot be field adjusted. Q4302 is the horizontal drive buffer, the output of which is capacitively coupled to the horizontal driver (Q4301). Q4301 drives the primary of the horizontal driver transformer, which drives the base of the horizontal output transistor, which drives the integrated high-voltage transformer. As I said, there is nothing new or different here—just straightforward circuitry. *Figure 3-32* gives you the particulars for the 19" through 25" sets, while *Figure 3-33* gives them to you for the 27" and 31" screens.

Horizontal AFC and APC (*Figure 3-34*) maintain proper synchronization between horizontal scan and the incoming video signal. AFC phase-locks the horizontal oscillator to the incoming sync signal, and APC locks the phase of the horizontal output to the phase of the horizontal oscillator. Most of the adjustments for APC and AFC are performed via serial bus commands at the factory and are not field adjustable, except for parameters 01 and 02 in the service menu.

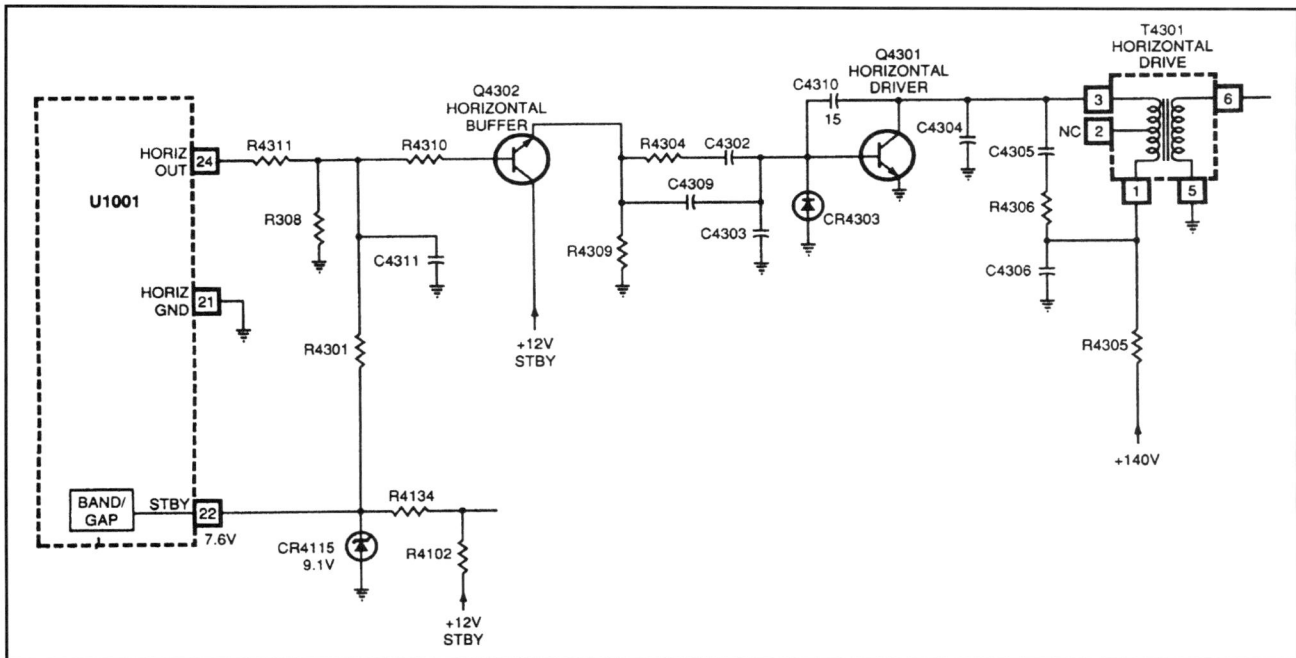

Figure 3-31. Horizontal drive and output circuit.

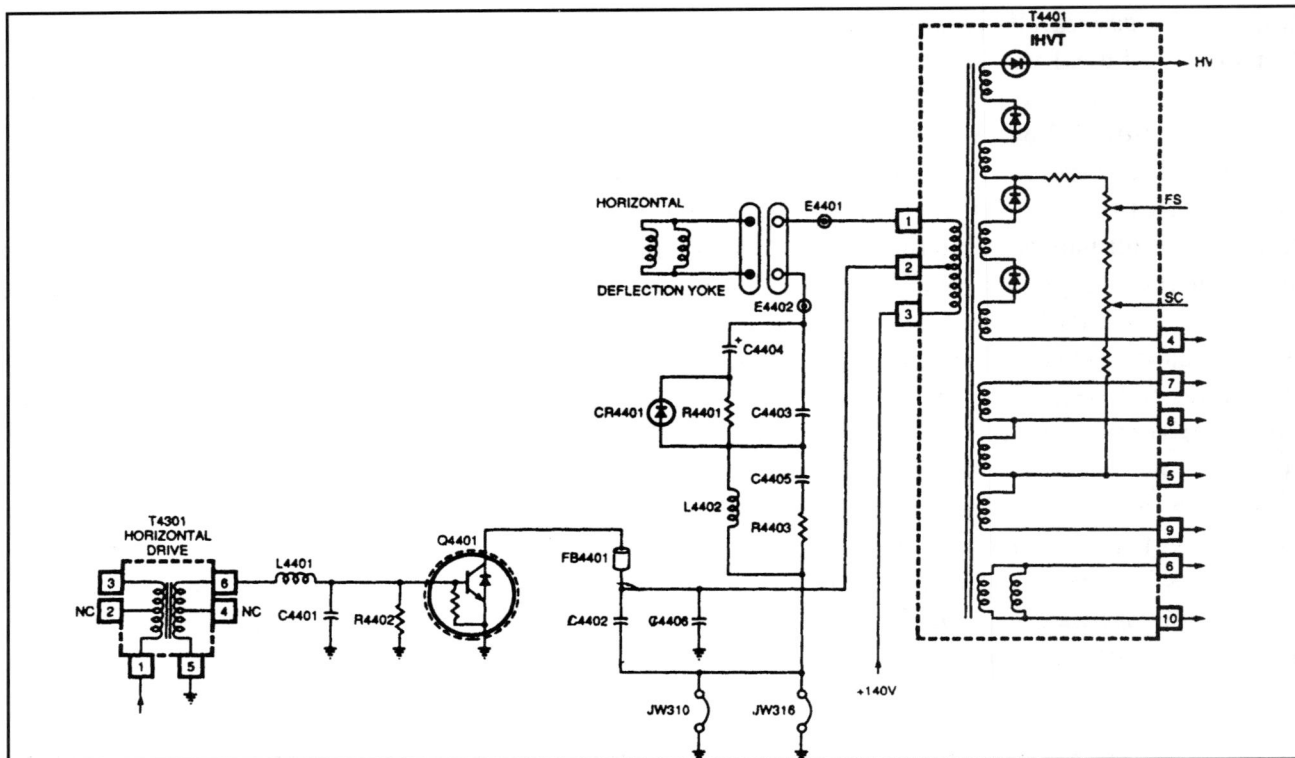

Figure 3-32. 19" through 25" circuitry.

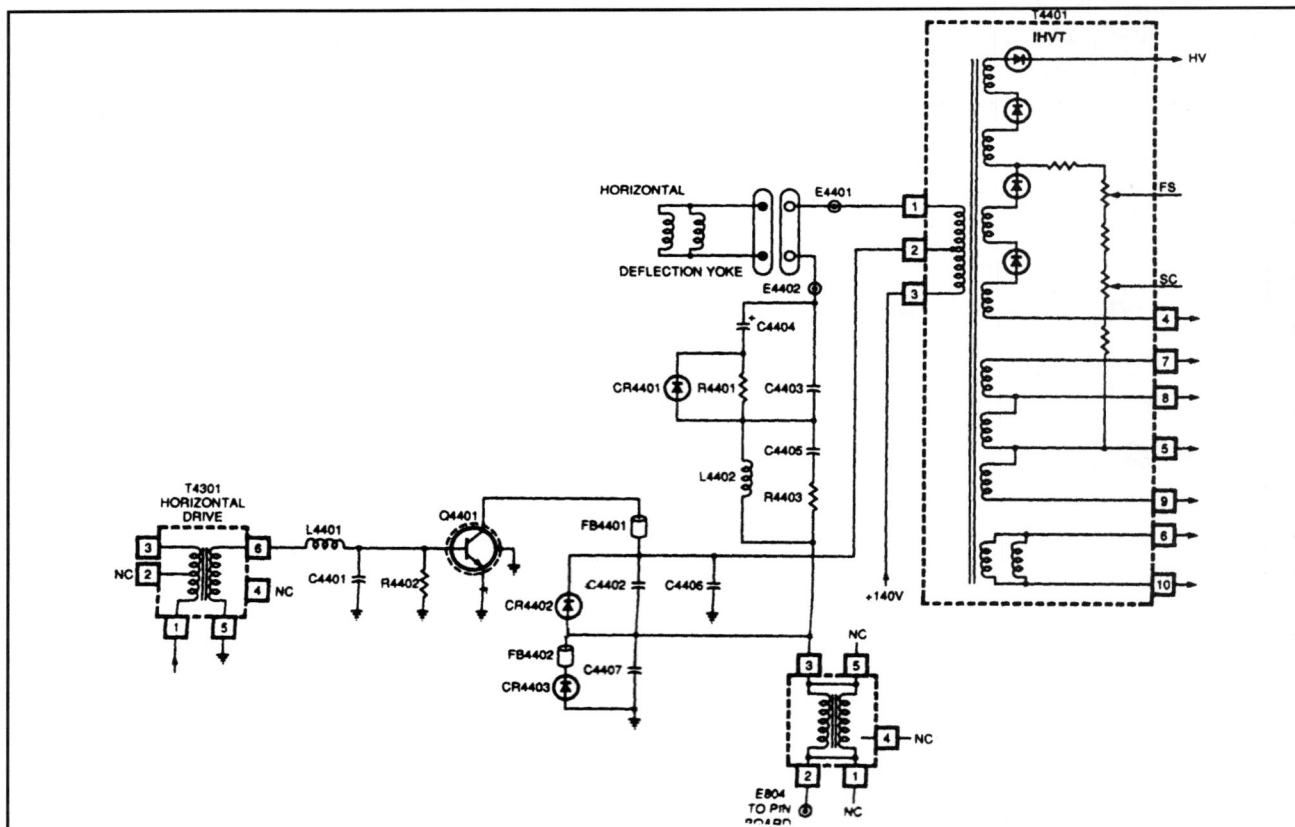

Figure 3-33. 27" and 31" circuitry.

I personally have not had a problem with any of the T-Chip's low-level horizontal processing functions, but I know of a situation which might interest you. The customer's complaint was a slight bending of the picture at the top of the raster. We call it "flag-waving." The tech who fought the problem found the solution almost by accident. He said the voltage at pin 23 of U1001 was slightly off. He replaced C4307, a 1 µF capacitor rated at 100 volts, and cured the problem.

I say I haven't had a problem with these functions, but that's not quite correct. The x-ray protection circuit sometimes acts up because of an interesting set of circumstances. I have not to my knowledge had to replace a faulty component, but I have had to replace R4903 and R4902 on occasion. There was one instance in which R4902 was left out of circuit during production and an instance in which R4903 has come loose because of contraction-expansion of the circuit board. These problems are easy to spot because the base voltage of Q4901 (*Figure 3-34*) will be significantly off.

There is a test for x-ray protection shutdown. If it tries to start three times and then stays off, you are dealing with x-ray protection shutdown. How can you tell? Listen to the clicking of the degauss relay! You can also monitor the voltage at pin 26 of U1001 with a scope while you press the power button. If DC voltage appears momentarily just before the TV turns off, you have an x-ray protection problem. If a voltage doesn't appear, you have another problem.

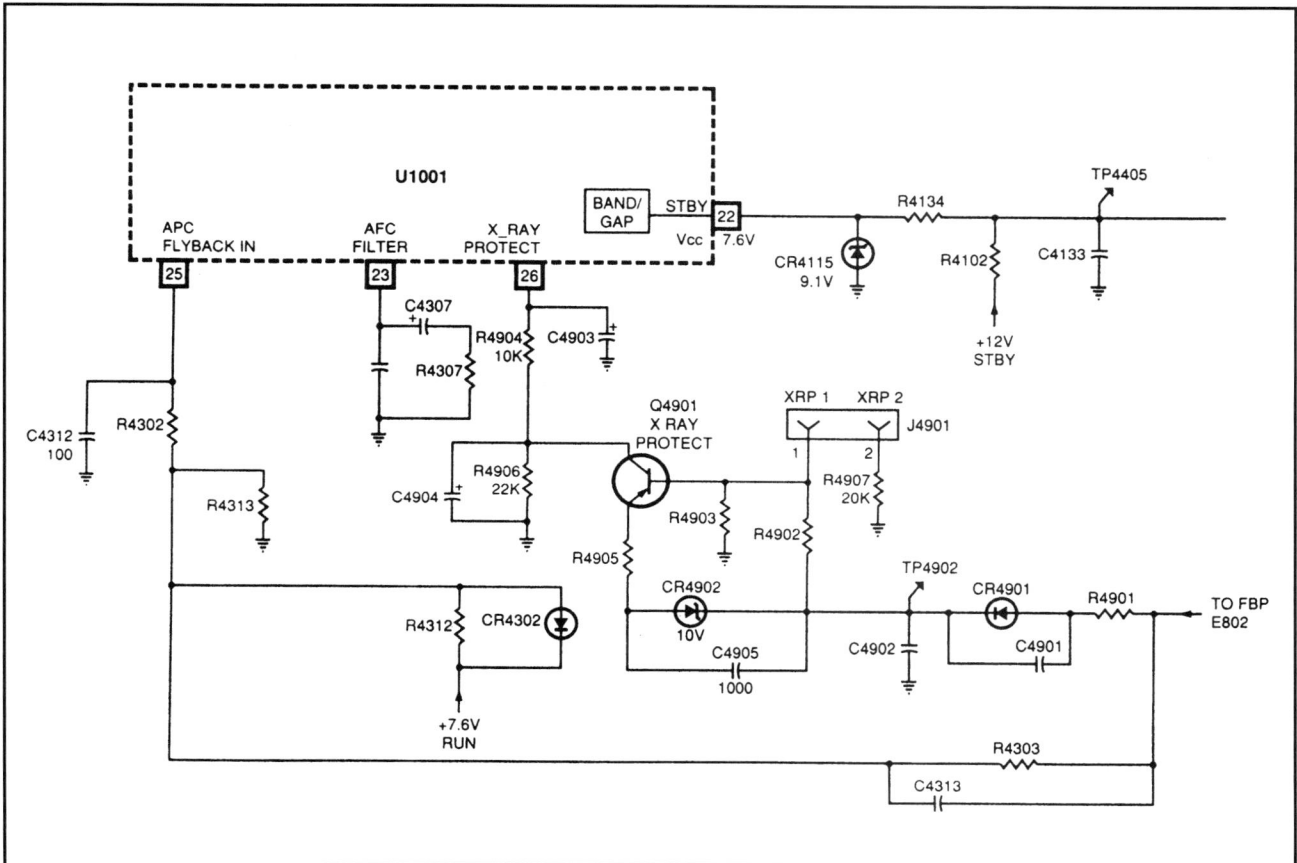

Figure 3-34. Horizontal AFC and APC.

Troubleshooting Horizontal Deflection Problems

Troubleshooting horizontal deflection problems is really straightforward. Since I try to do things systematically, I will give you the troubleshooting procedure in a 1-2-3 order. Since it isn't carved in stone, modify it to suit your particular quirks.

(1) Check for B+ at the collector of the horizontal output transistor. If it is not there, fix the power supply.

(2) Check for 7.6 volts at pin 22 of the T-Chip. If it is not there, troubleshoot the 12-volt standby source. If the 12-volt standby source is good, check for an internal short or a leaky pin 22 to ground. For example, if the voltage is less than 7.3 volts and goes up when you wick out the pin, you will need to replace U1001.

(3) Check for horizontal drive pulses at pin 24 of the T-Chip as you issue an ON command. If you don't see the pulses, confirm that you don't have a system control problem, like a corrupted EEPROM, before you think about replacing the T-Chip. The T-Chip is a reliable part, which means it doesn't fail often. It is also more difficult to replace than an 8-pin EEPROM. So replace the EEPROM before you even think about replacing the T-Chip.

(4) Check drive on the emitter of Q4302 and the collector of Q4301. If drive is missing, check the corresponding stages.

(5) Check the drive signal at the base of the horizontal output transistor. If it is present, suspect a problem with Q4401. If it is not present, check the driver transformer.

Pincushion Correction

The T-Chip outputs a vertical rate parabola at pin 19, which drives the pincushion correction circuitry (*Figure 3-35*). The output has bus-controlled AC and DC components, which are set via serial bus commands to registers inside the IC. The East-West DC control affects picture width, while the East-West AC control affects the amount of pin or barrel distortion in the raster.

The vertical parabola from pin 19 is combined with a filament pulse at pin 2 of U4851. U4851 turns on at the horizontal rate, while the amplitude of the signal is modulated at the vertical rate. When U4851 turns Q4851 on, current is pulled from the horizontal yoke through L4853 to ground, which lowers the voltage on C4851, increasing current flow through the yoke. Increased current flow through the yoke naturally widens the picture. When Q4851 turns off, the energy stored in L4853 is released into the 26-volt supply when it is high enough to forward bias CR4851 (about 26.6 volts), which prevents excessive voltage from damaging Q4851.

A little service note here: If the pincushion circuit fails and permits the 26-volt line to go high, you will likely face a failure of the vertical output chip, IC4501. If you replace IC4501, check the 26-volt supply. If it is high, suspect a problem with the pincushion circuit.

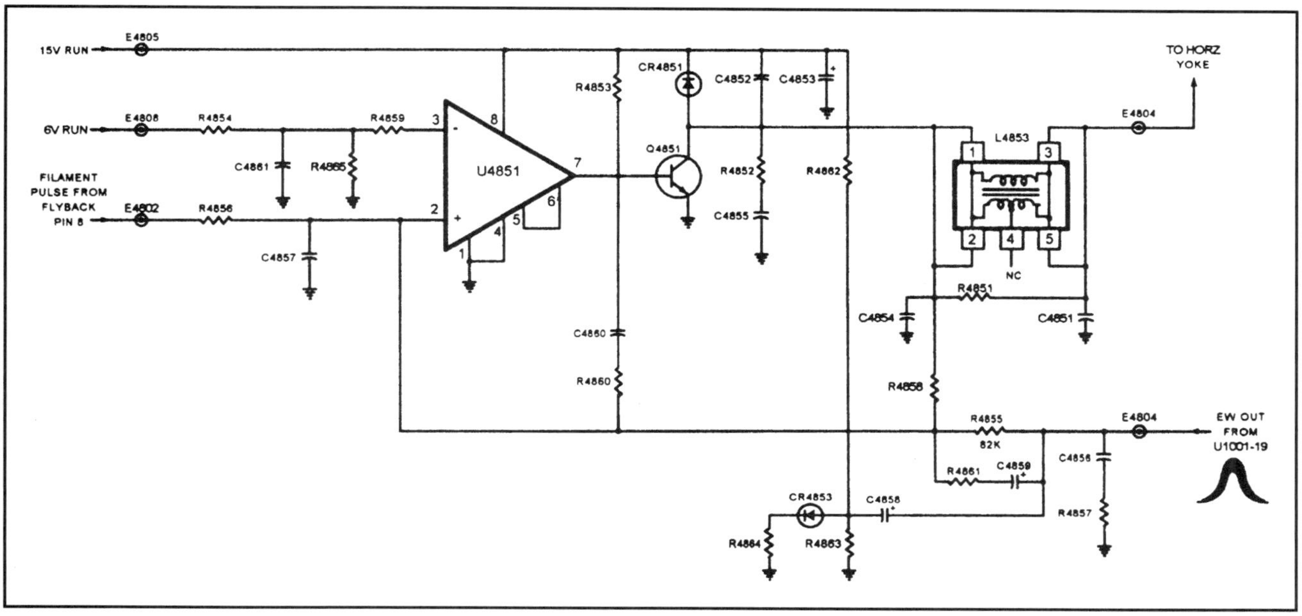

Figure 3-35. Pincushion correction circuitry.

If the pincushion circuit fails, you will face one of two problems: a raster that is too large for the picture tube, or one that has pulled in from the sides (an hourglass configuration). Obviously, any component in the circuit can cause the problem. Some are more likely than others, for example Q4851. But the likely culprit will more than likely be the pin modulation circuit. To confirm pin modulation problems, check the 200-volt scan-derived voltage source. If it is low, suspect an open C4407. If it has opened, CR4403 has been significantly stressed. Replace both to cure the problem.

Again, I recommend using genuine RCA parts because those parts are specifically engineered for this critical circuit. All of us will use generics from time to time, usually because the customer is in a hurry for his television. If you elect to use a generic part, double check its value because the part is in a critical circuit. The value of the capacitor will depend on the chassis. There are, I believe, four different part numbers for it. The part number for the diode is 164589.

The Vertical Deflection Circuit

The vertical deflection circuit is relatively straightforward (*Figure 3-36*), but it does have a quirk or two. If you examine the circuit you will note that it differs from other RCA circuits in that it is DC-coupled as opposed to capacitively-coupled. Such a circuit has certain advantages over its traditional capacitively-coupled relation: fewer parts, lower cost, and less dependence on linearity-correcting electrolytic capacitors that lose tolerance as they age. Linearity correction is facilitated inside the T-Chip.

DC coupling also means the DC level of the vertical ramp exiting the T-Chip affects vertical centering. The average DC level of the ramp is about half the vertical Vcc applied to pin 32 of U1001 (about 3.8 volts DC). The ramp can be adjusted ±150 millivolts via the data bus. Remember page 1 of the service menu? Parameter 05, the vertical DC adjustment, adjusts vertical centering, while parameter 06 adjusts vertical size. When you replace the EEPROM, you can quickly adjust these parameters by choosing parameter 06 and slightly collapsing the raster. Choose parameter 05 and adjust it so the raster is centered vertical, and then select parameter 06 and adjust it for full vertical deflection.

U4501 (part number 215531) is really a voltage-to-current converter. It takes the vertical DC ramp from pin 17 of U1001 and converts it to a current ramp before routing it through the vertical yoke. It is also an inverting amplifier; that is, it sinks current at pin 5 when pin 1 is low.

Two DC voltages are necessary for the vertical circuit to work correctly: the 26-volt scan-derived supply, and a so-called "half-supply" (one-half of the 26 volts) developed from the 12-volt run supply. The 26-volt supply is the main B+ voltage; the half-supply is employed to deflect the electron beam from the bottom to the top of the screen at the end of vertical scan.

Now let's take a look at some of the major components. I will key my comments to *Figure 3-36*. R4517 limits current in the yoke to keep the beam from deflecting off the screen if U4501 shorts to ground or to the 26-volt supply. R4518 is used to adjust the circuit for various screen sizes and is basically used only in the 20" TVs. C4502 provides additional filtering for the 12-volt supply, and in conjunction with R4518 assists in reducing vertical ripple current on that supply. R4519 and R4502 are current sensing resistors that develop a voltage drop across them proportional to yoke current. Some of this voltage is routed to pins 4 and 5 of resistor pack RN4501 and fed to pin 7 of U4501.

How does the circuit work?

The vertical sawtooth waveform exits U1001 at pin 17 and goes immediately to pins 1 and 2 of resistor pack RN4501. The average DC level of the ramp is about half the vertical Vcc supplied to pin 32 of U1001 (about 3.8 volts). The vertical ramp and error signal riding on the half-supply from R4519 and R4502 (the current-sense resistors) are added and input to pin 1 of the vertical output IC. The 7.6-volt supply is input to pin 7 of the resistor pack, where it is divided to half the Vcc, added to the error signal riding on the half-supply from the current-sense resistors, and routed to pin 7 of U4501. The average voltage at pin 7 should be about 9 volts.

When the vertical ramp is at the bottom of the slope, pin 5 of U4501 sources current from the +26-volt supply through the yoke to the +12-volt (the half-supply) supply, deflecting the electron beam to the top of the screen. As the ramp increases the voltage on pin 1, the current source from pin 5 decreases, which lowers the voltage across the yoke. The beam moves toward the center of the screen. When the voltage on pin 1 equals the voltage on pin 7, pin 5 is half the +26-volt supply, and current ceases to flow in the yoke. The electron beam is now at the center of the screen. As the voltage on pin 1 rises higher than pin 7, pin 5 begins to sink current, causing voltage to flow from the half-supply through the yoke to pin 5. The current flow is now reversed, and the beam is deflected downward. During retrace, the ramp resets causing pin 5 to go high, deflecting the beam upwards.

The extra current needed during retrace is produced by C4505. Its negative lead is grounded through pin 3 of U4501 during scan time, permitting the it to charge to 26 volts. During retrace, the flyback generator inside the IC connects pin 3 to pin 2, applying 26 volts to the negative side of C4505. The charge stored in the capacitor plus the 26 volts on the negative lead produces 52 volts on pin 6. The increased B+ quickly retraces the beam.

Troubleshooting Vertical Deflection Problems

Before we proceed, I will repeat a previous observation: <u>A failure in the pincushion circuitry can lead to vertical problems. Always check the +26-volt line for a higher-than-normal voltage in the event U4501 fails.</u>

The vertical deflection circuit has proved to be highly reliable. I seldom have to service it for problems other than those caused by the EEPROM. Of course when you are confronted by a failure of the circuit, you should first check the DC voltages at U4501. If they are present and at the correct level, check for the vertical ramp on pin 1. It is really is a vertical parabola and should be at about 2 Vp-p. If it is absent, check for 7.6 volts on pin 32 of U1001. If the voltage is correct, check for the vertical ramp at pin 18. If the ramp is not present, change the EEPROM before you even think about changing the T-Chip!

I have serviced a couple of these sets where the raster was pulled in from the bottom but slowly filled out within seconds of channel change. Checking the voltages at RN4501, I found no voltage on pin 4 because R4519, one of the current-sense resistors, had opened.

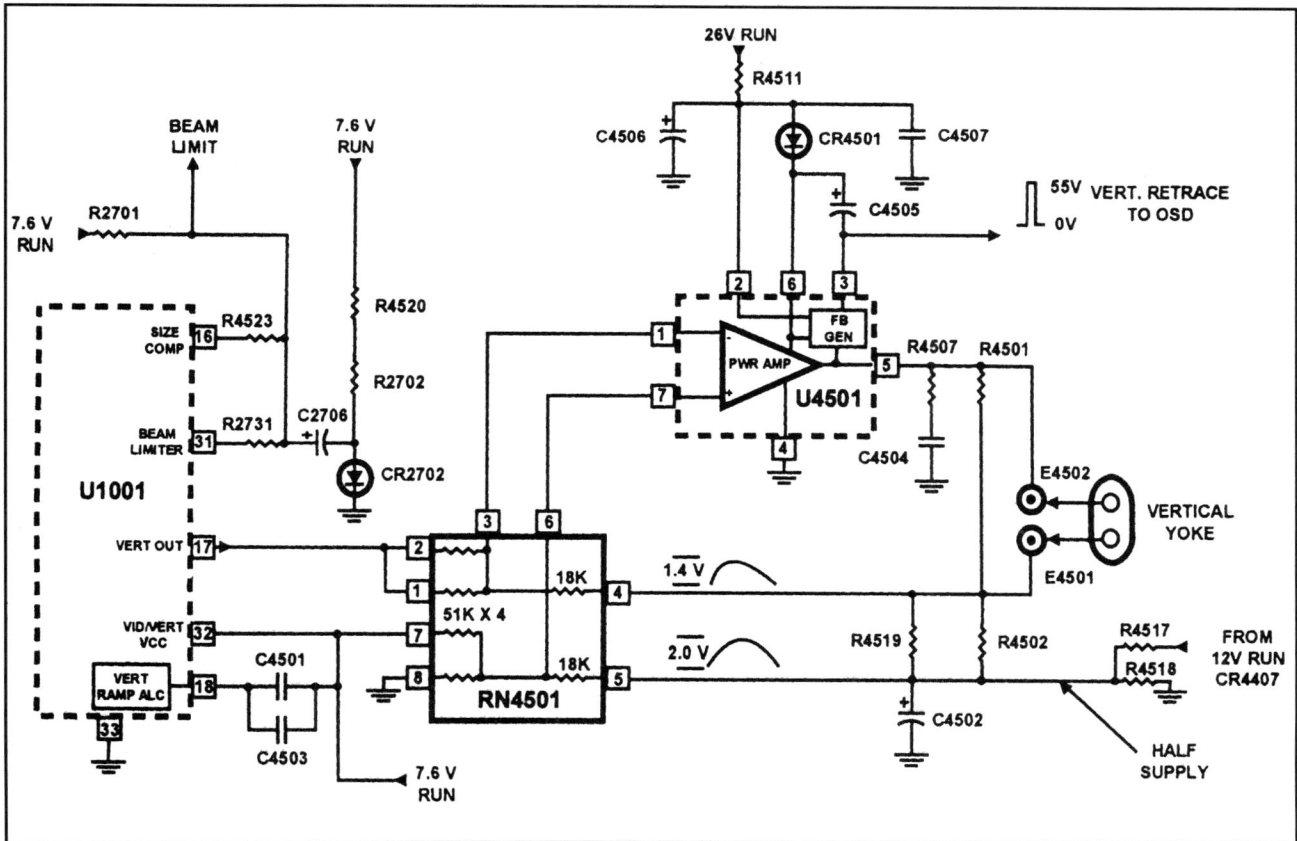

Figure 3-36. Vertical deflection.

I have also had problems with R4511, a 1 ohm, 2 watt resistor in the +26-volt line. If it opens, it defeats vertical deflection. In one instance it had increased in value, causing vertical foldover at the top of the screen.

Resistor pack RN4501 can also cause problems. If one or more resistors in the pack opens, the result may lead to no vertical deflection. On one occasion it was the cause of intermittent vertical deflection. I located the problem by pecking the circuit board with the handle of a nut driver. When I tapped directly on the resistor pack or wriggled it with my fingers, I could make vertical deflection appear and/or disappear.

I have serviced one case of nonlinear vertical deflection and am aware of a few other instances of it. Each case was caused by the same problem. If it is your problem, check for above normal voltages and/or noise on pin 18 of the T-Chip. If you confirm these conditions, check C4501 and C4503 in the vertical ramp automatic level control circuit. These surface-mount capacitors help to keep the vertical ramp at a constant level. If they are correctly soldered, then one or both will be defective.

There have also been reports of poor solder around the switching IC in the switch-mode power supply causing vertical deflection to fail. I have not experienced it, but I have seen enough poor solder on these pins to motivate me to resolder them in every chassis I service—a procedure I heartily recommend.

The Video Processing Circuits

As I was thinking over the video path for these chassis and consulting my notes, I realized I have had only three video problems out of all the chassis I have serviced! It is incredible, but true. Two of the problems involved loss of red drive out of the T-Chip and were caused by a defective CRT. I solved the other by consulting a service bulletin. It is this one I want to talk about now. The TV came into the shop with the complaint, "The sound is fine, but the screen is black." If you encounter these symptoms, check first to see if the filaments in the CRT are glowing. If they are, see if the OSD function is working. If it is not, turn up the G2 setting and check again. Chances are you will then have very dim video and very dim OSD information.

In these chassis, OSD information is inserted before picture and brightness information. If picture and brightness controls are reduced to minimum, the result is no picture and no OSD. The cure is to access the menu with the G2 control set high and restore the picture and brightness settings as you reduce the G2 setting! That's the way I did it the first time I encountered the problem. There is another way. If the menu is not visible, push "reset" on the remote control to return menu functions to a preset level. I owe this little tidbit to one of the techs on RCA's technical assistance phone line.

Luminance Processing

Now I will very briefly review the video path using *Figure 3-37A* (a rather full schematic) and *Figure 3-37B* (a block diagram).

Selected video from pin 51 of U1001 is separated into its luma and chroma components. Luma is input into pin 48 where it is processed by a "black stretch" circuit. Pay attention to the DC voltage on pin 43 because it must be above 1.2 VDC to enable the luminance circuit. If it is less than 1.2 volts, the TV will switch to an RGB mode by turning off the luminance path. The signal then goes to the internal/ external switch which selects between OSD and video. You might also note that R-Y and B-Y from the chroma circuits are also routed to this switch.

As I have indicated, OSD information is added ahead of brightness and picture controls, which is a departure from the way RCA usually does it. Pins 40, 41 and 42 are the OSD inputs from system control. Pin 39 is the control point for another switch, the "fast switch input" control. If it is high, U1001 switches external video (OSD) out. If it is low, it will permit internal video to be switched out.

The internal/external circuitry produces Y, R-Y, and B-Y signals. They are sent to their respective contrast and clamp circuits which are controlled by the serial bus. Luminance is sent to the brightness control; the other two signals are matrixed and summed with the luminance. The summed red, green and blue signal are amplified and sent to the CRT through pins 36, 37 and 38.

Pin 31 is the beam-sense input. It is used to reduce brightness and contrast during high beam-current scenes to keep the picture tube from "doming." The circuit is active below 6.2 volts.

The suggested troubleshooting procedure for a "no luminance" problem goes something like this:

(1) Begin by pushing the reset button on the remote control to ensure the problem is not caused by incorrect customer settings.

(2) Check the input signal at "Y1" and "Y2" inputs (*Figure 3-36*). It should be about 1 Vp-p.

(3) Check for the signal at the selected Y output on pin 44. It should be the same as Y1 and Y2.

(4) Check the DC voltage at pin 31. The circuit is active below 6.2 volts.

(5) Measure the voltage at pin 39, the fast switch input. A DC level greater than 1.7 volts will blank the video to let OSD information through.

(6) Measure the voltage at pin 43. Luminance will be enabled only if it is above 1.2 VDC.

Chroma Processing

Chroma is processed inside U1001 (*Figure 3-38*). Pin 49 is "selected chroma" that can be input from the TV or an aux input source. Pin 46 is for S-Video input on sets that support S-Video. *Figure 3-38*, a block diagram of the insides of U1001, makes the signal path so clear that I will resist the urge to comment on it. Please note that pin 47 is 3.8 volts or greater with a color signal present, and low when the color burst is not detected. The color killer at pin 47 can be defeated by applying 3.8 volts to it.

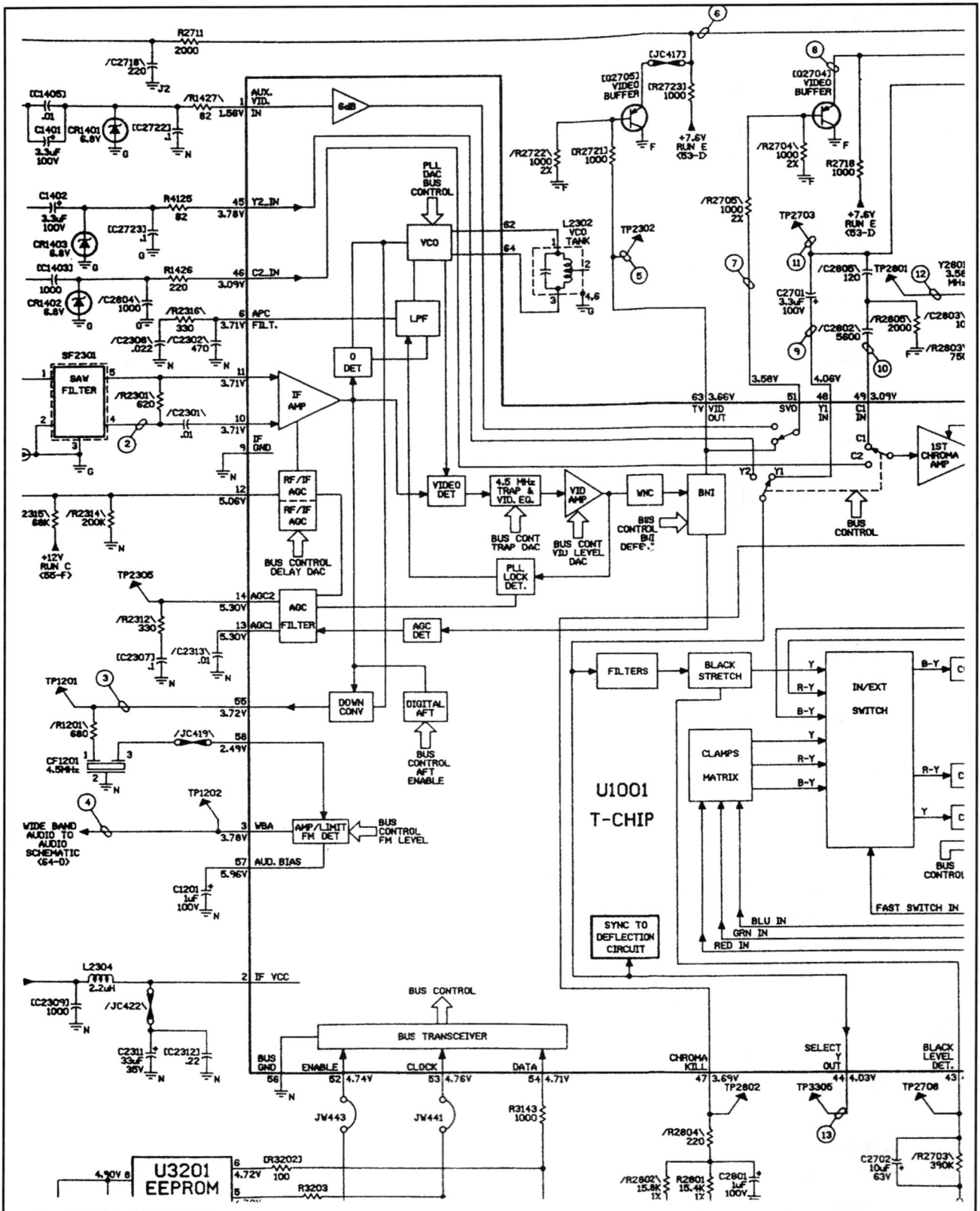

Figure 3-37A. Luminance processing.

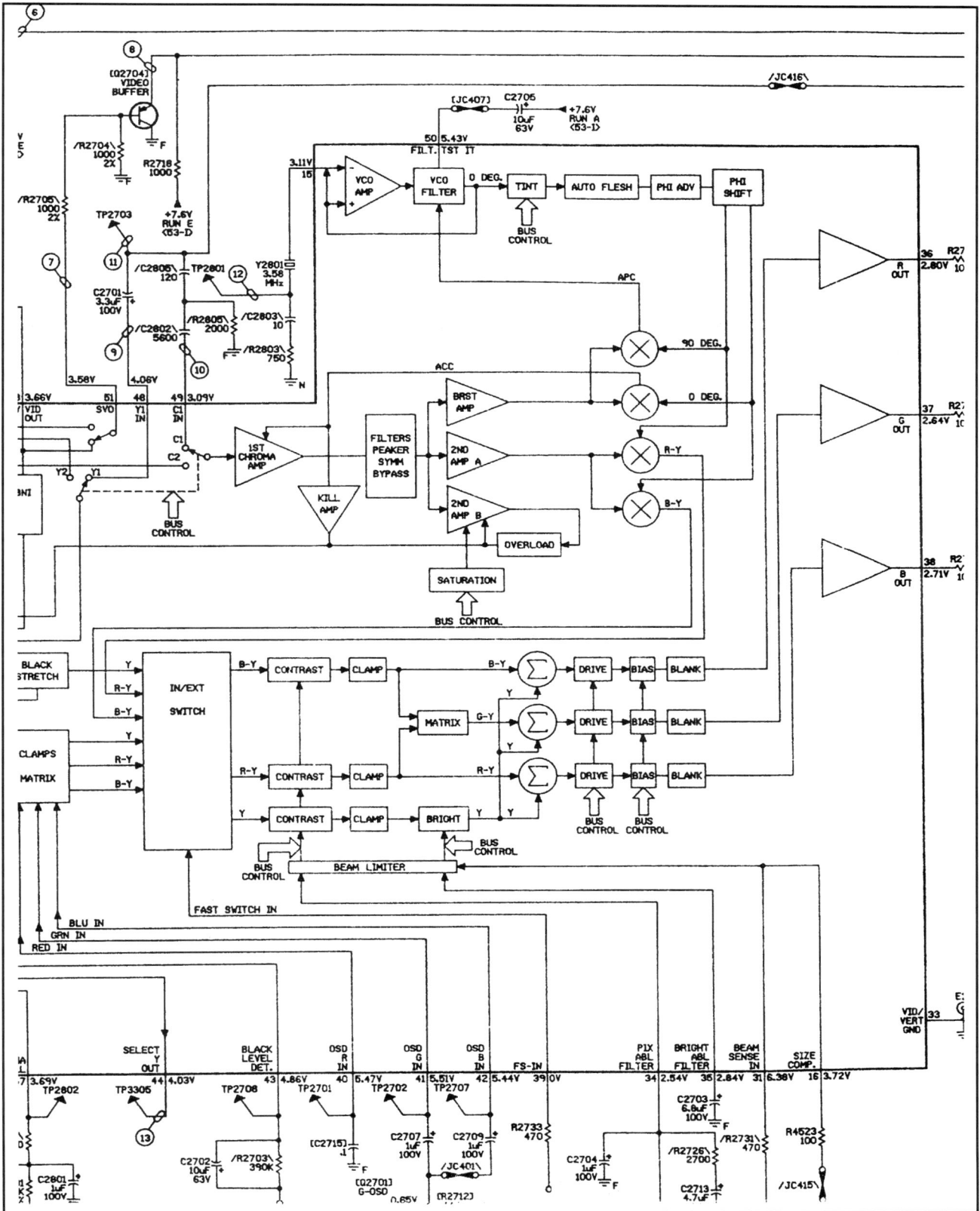

Figure 3-37A. Luminance processing (continued).

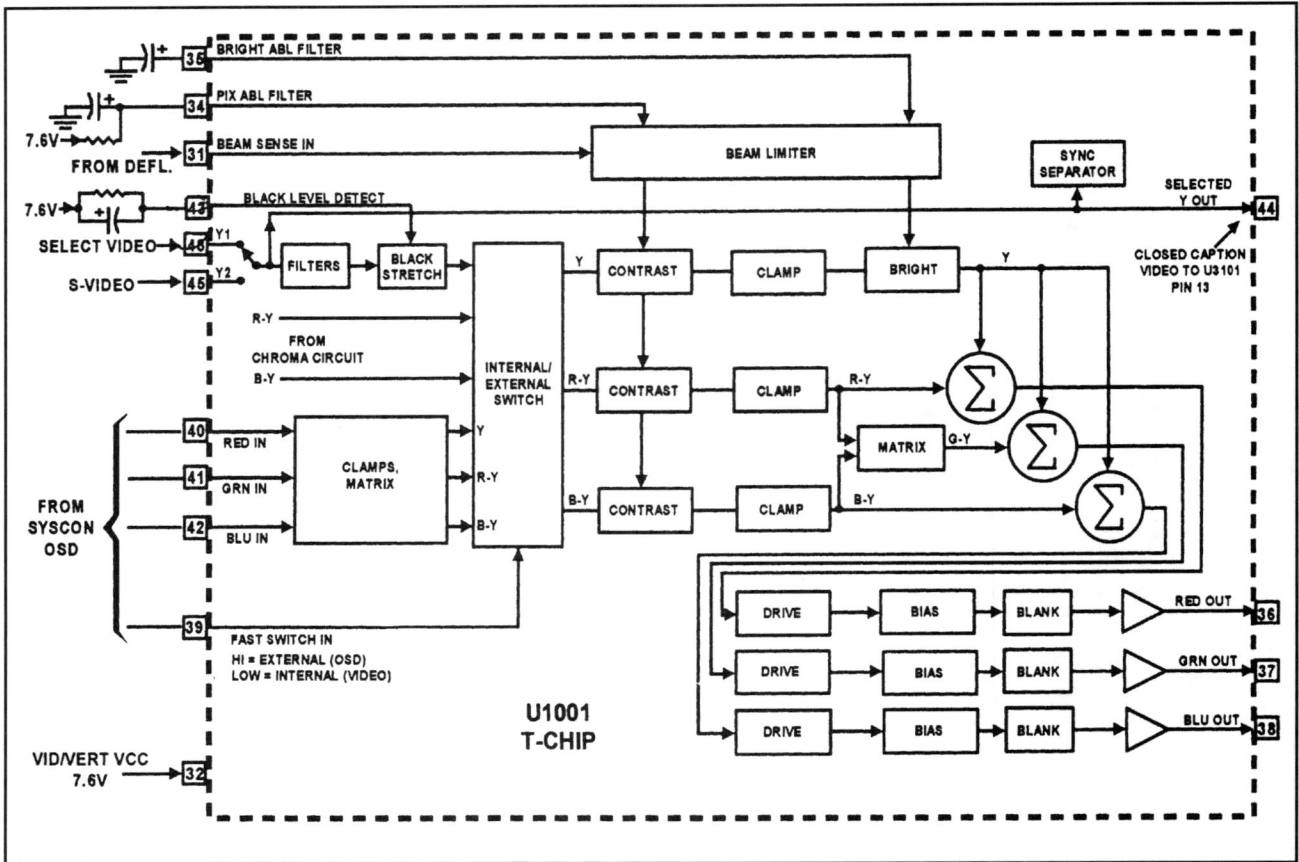

Figure 3-37B. Luminance processing block diagram.

The biggest difference between this circuit and similar chroma circuits has to do with the 3.58 oscillator input at pin 15. The crystal is in series with the comparator circuit. When the oscillator is locked on frequency, no oscillator voltage will be present at pin 15. What does this mean? It means that you cannot view the oscillator signal outside the IC with a scope!

After it has been processed, the chroma information is sent to the internal/external switch, where it is switched with OSD information and matrixed to demodulate the G-Y signal.

Troubleshooting Color Problems

The literature says color problems can best be diagnosed using a good scope and digital multimeter. It suggests, first of all, that you check the setting of the color and tint controls in the customer menu. If they are at a sufficient level, check the chroma input at pins 49 and 50 of the T-Chip. The signal should be in the neighborhood of 300 millivolts. If you need to defeat the color killer, apply about 4 VDC to pin 47. With 4 volts applied, you should see free-running chroma on the screen (the barber-pole effect) if the 3.58 oscillator is running. As far as I know, this is the only test we can use to determine if the 3.58 oscillator is up and running.

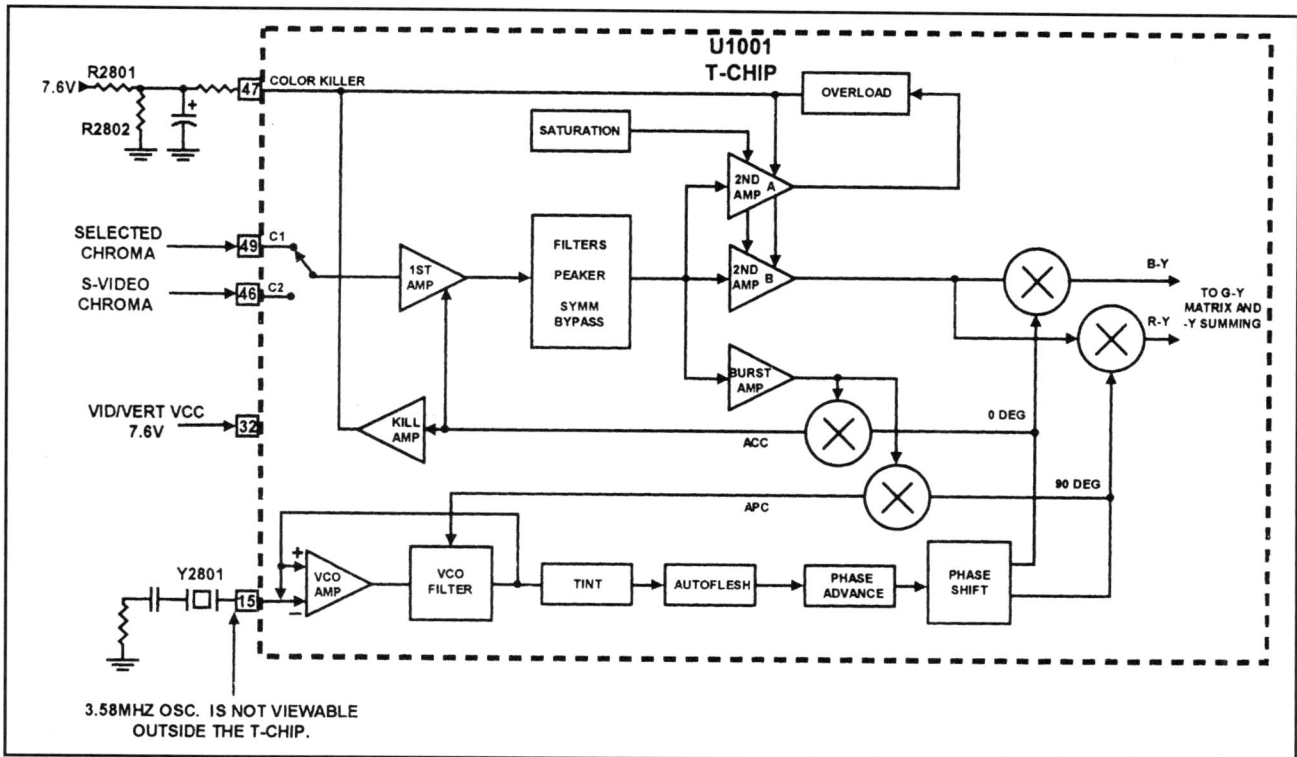

Figure 3-38. Chroma processing.

The Audio Processing Circuits

First a note about the audio circuits used the CTC175/176/177/187 chassis. If the TV does not have stereo, the audio output circuit will look like that in *Figure 3-39*. The stereo models will look like the partial diagram in *Figure 3-40*. The block diagram to which I will refer in this discussion is *Figure 3-41*.

In the mono sets, wideband audio from the IF block is output at pin 3 of U1001 and coupled back into it at pin 5, the right channel serving as the mono channel. The audio signal is then routed out of pin 59 to the push-pull amps. Note that the mono sets do not have aux jacks.

In stereo sets, wideband audio from the IF block is output from U1001 pin 3 and sent to pin 5 of the stereo decoder IC, U1701 (*Figure 3-42*). U1701 decodes the stereo information and outputs left and right channel signals at pins 13 and 14. The left and right channel information is sent back to U1001 pins 4 and 5, where they are switch-selected. The selected audio, either external or internal, then exits the T-Chip at pins 59 and 60 and sent on to the audio amp, U1901.

As you can see, audio selection and volume control take place inside the T-Chip, eliminating the external switching ICs which Thomson has used in the past. Stereo is detected by U1701, which notifies the microprocessor that stereo is present by pulling pin 20 low. U3101, the microprocessor, can then select stereo or mono with a low or a high from pin 23. Normal audio mute takes place inside the T-Chip when the microprocessor receives a mute command from the viewer.

Troubleshooting Audio Problems

Before you attempt to solve a "no audio" problem in one of these sets, rule out the EEPROM as the cause. If you haven't already done so, read what I have written in this chapter about the EEPROM and audio problems. And, having ruled out the EEPROM as the culprit, use a signal generator to inject a signal at the tuner before you proceed.

It is almost always best to divide in half the circuit you are troubleshooting. It cuts down on time spent and makes the problem a little easier to find. A good midpoint in the audio circuit is pin 3 of U1001 (*Figure 3-41*), where you will be looking for the wideband audio out signal. If it is not present, you may have to go back to the IF circuit to find the problem. You can confirm an IF problem by injecting an external audio signal at the aux audio input. If you have audio from the jack pack but not the tuner, you pretty much know you are losing the audio portion of the signal either in or just before the IF block.

If wideband audio is present, you know the IF block is working and can move on to the next step, which is checking for audio out on pins 13 and 14 of U1701. If it is not present, suspect the IC, the Vcc to pin 19, or a defective coupling capacitor in the decoder circuit. If audio is present at pins 13 and 14 of U1701, check for it at pins 59 and 60 of U1001. If audio is present at pins 4 and 5 but not at pins 59 and 60, you should suspect U1001. If it is present at pins 59 and 60, check for it at pins 1 and 5 of U1901 and at pins 8 and 10 which go to the speakers. If you have it at the input and not at the output, U1901 will probably be the culprit.

Before you change the audio output IC, make a couple of checks. Do you have +26 volts at pin 9? Do you have mute voltage at pin 3? You should have about +12 volts on pin 3 when the speakers are <u>not</u> muted.

Concluding Comments

I am going to resist the temptation to talk about the picture-in-picture feature, partly because I have never been called upon to service a PIP problem, and partly because I have not even heard of a problem with it. I either live an isolated life or the feature doesn't give trouble. If you are interested, you can read about it in the last few pages of *The CTC/175/176/177 Technical Training Manual*.

I do want to end this chapter by previewing some "oddball" problems that have arisen. I have become acquainted with them either through service bulletins or conversations I have had with other techs.

(1) The set will not enter the stereo mode when a known stereo signal is present, and the stereo OSD does not appear. Check capacitor C1701 which couples the audio signal to the stereo pilot detector in U1701. If the pilot signal is not detected, the TV will not enter the stereo mode. The capacitor is located at 23C on the copper foil side of the PCB. RCA's part number is 205230.

(2) The TV will either be dead or will intermittently shut down. It is losing regulated B+ because of poor solder around T4101 and/or the switching IC. Just put these two items on your "to solder" list when you get one of these sets in for service.

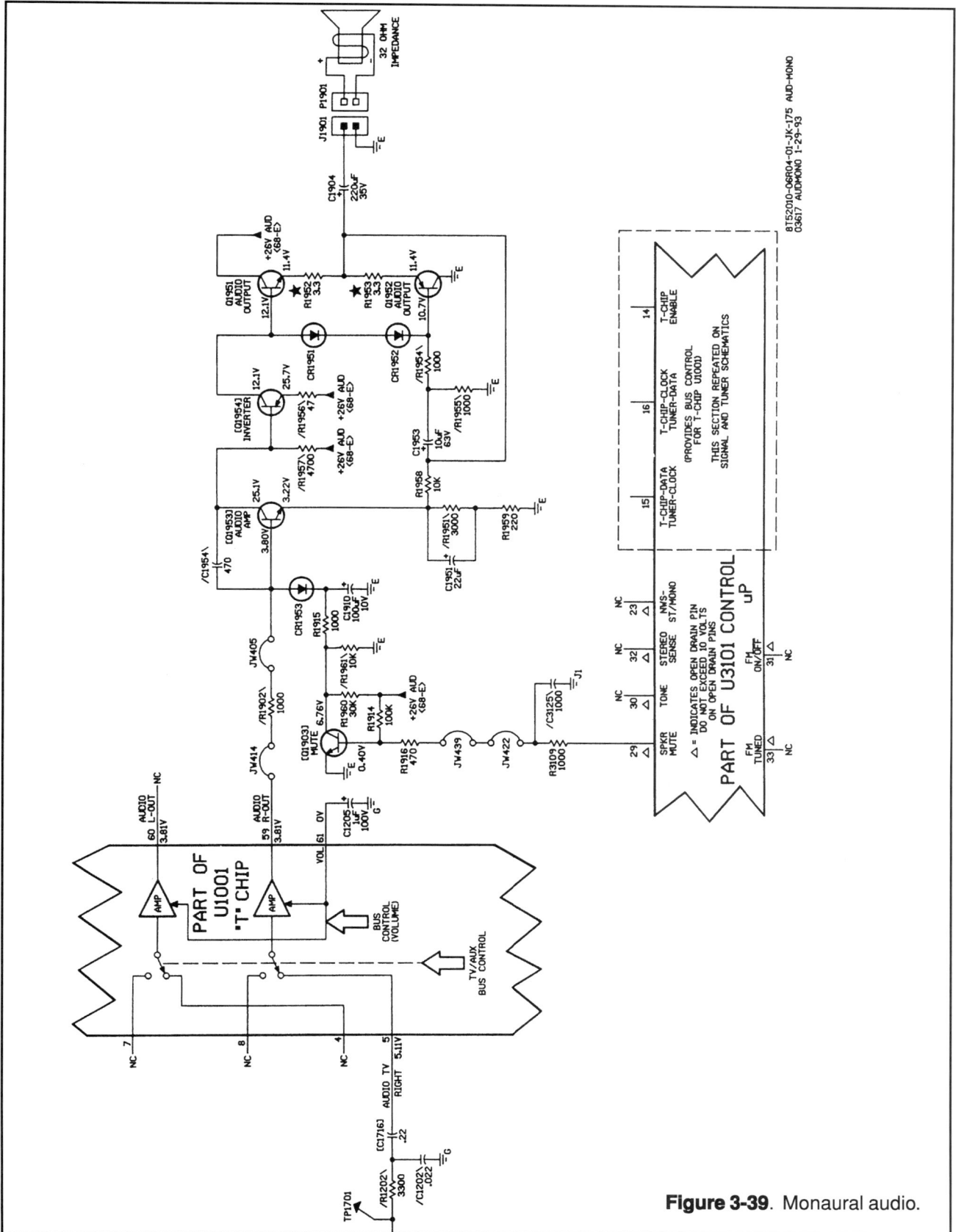

Figure 3-39. Monaural audio.

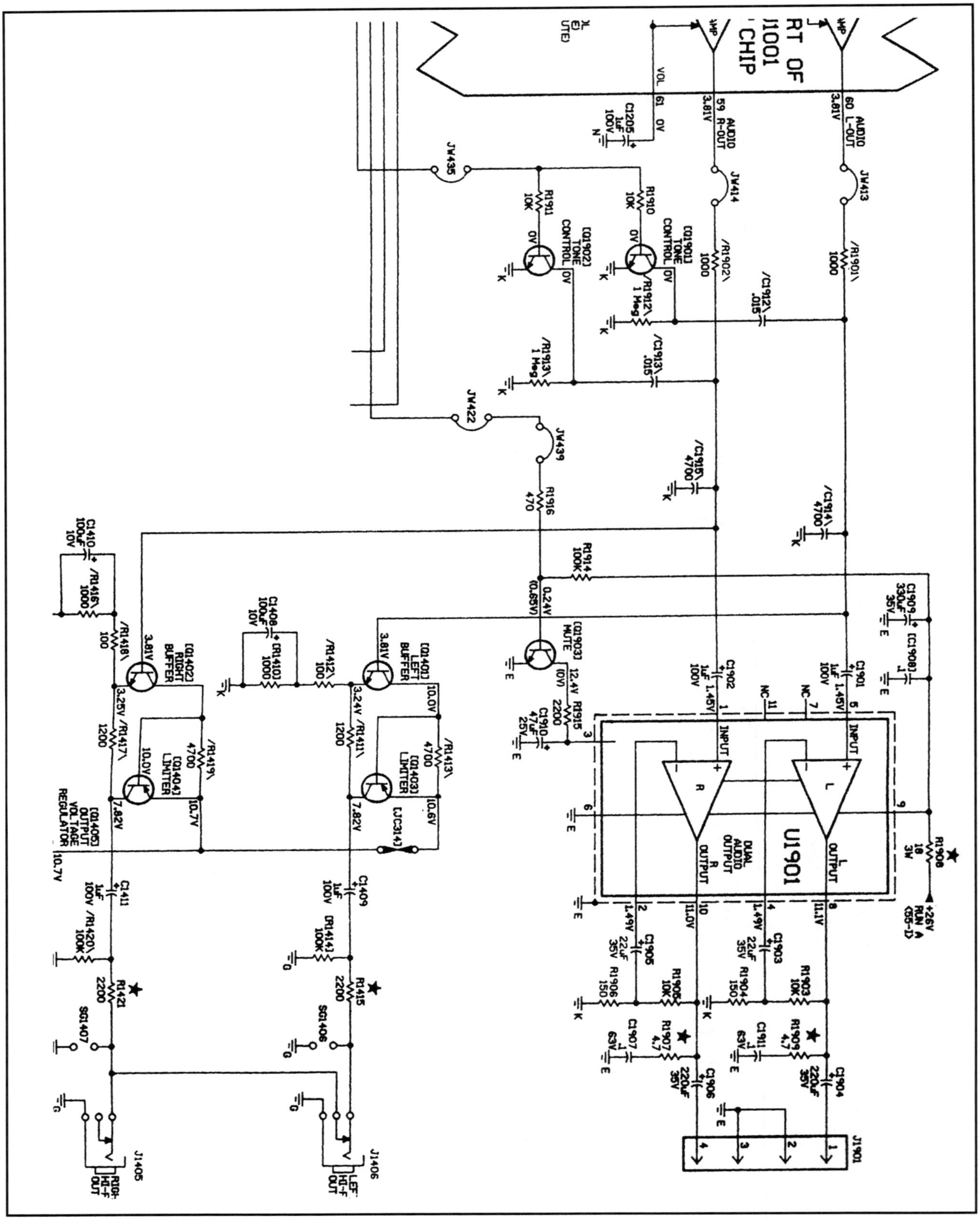

Figure 3-40. Stereo audio.

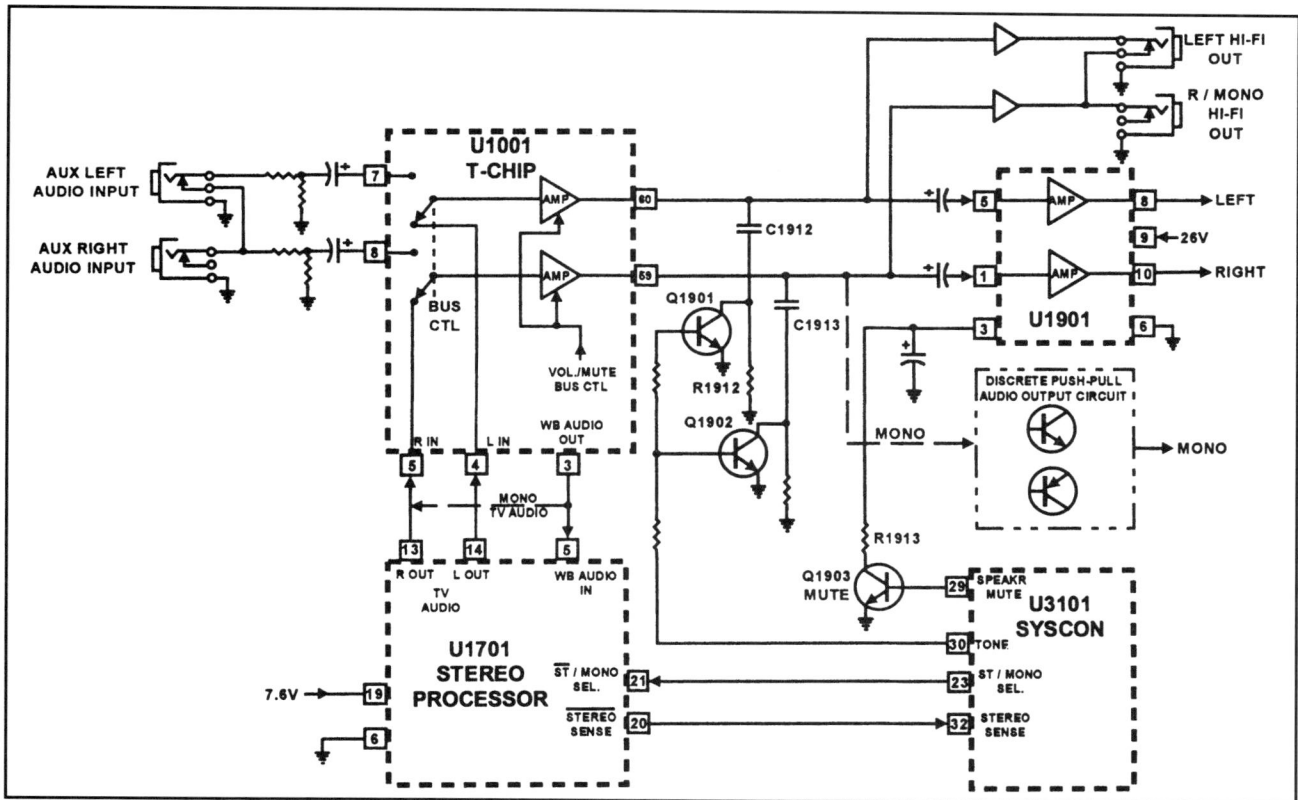

Figure 3-41. Audio block diagram.

(3) The picture will be normal when the TV is first turned on, but gets snowy as it warms up. The problem may be the 4 MHz crystal (Y7401). Remove the top tuner cover and spray the crystal with freeze spray when the TV acts up. If the picture stabilizes, use a heat source like a hot air gun to heat it up to see if you can reproduce the symptom. If you must replace the crystal, use part number 182839.

(4) The TV will blank for a couple of seconds after a channel change. In some instances, it also won't autoprogram. C3301 on the foil side of the PCB near pin 42 of the microprocessor may have cracked. It routes tuning sync from Q2705 to the base of Q3301, which couples it to pin 38 of the microprocessor. Replace it with stock number 195689.

(5) The picture blanks when the front-panel menu button is pressed. C3113 may have opened. It is located near the microprocessor at location 15J. Replace it with part number 195689.

(6) Some CTC187 chassis have an arc of fine horizontal lines in the top third of the picture. The problem has been traced to the pincushion correction circuit. Replace R4863, a 3.3 ohm resistor, with a 2.2 ohm, 2 watt resistor (part number 215211). Then position the wires that go above L4853 between L4853 and C4851. Route the wires that go from the pin board to the chassis so that they are between C4950 and C4113. Be sure to keep them away from L4001 by pressing them down between the two capacitors and away from T4101. Use cable ties to keep the wires in place if necessary.

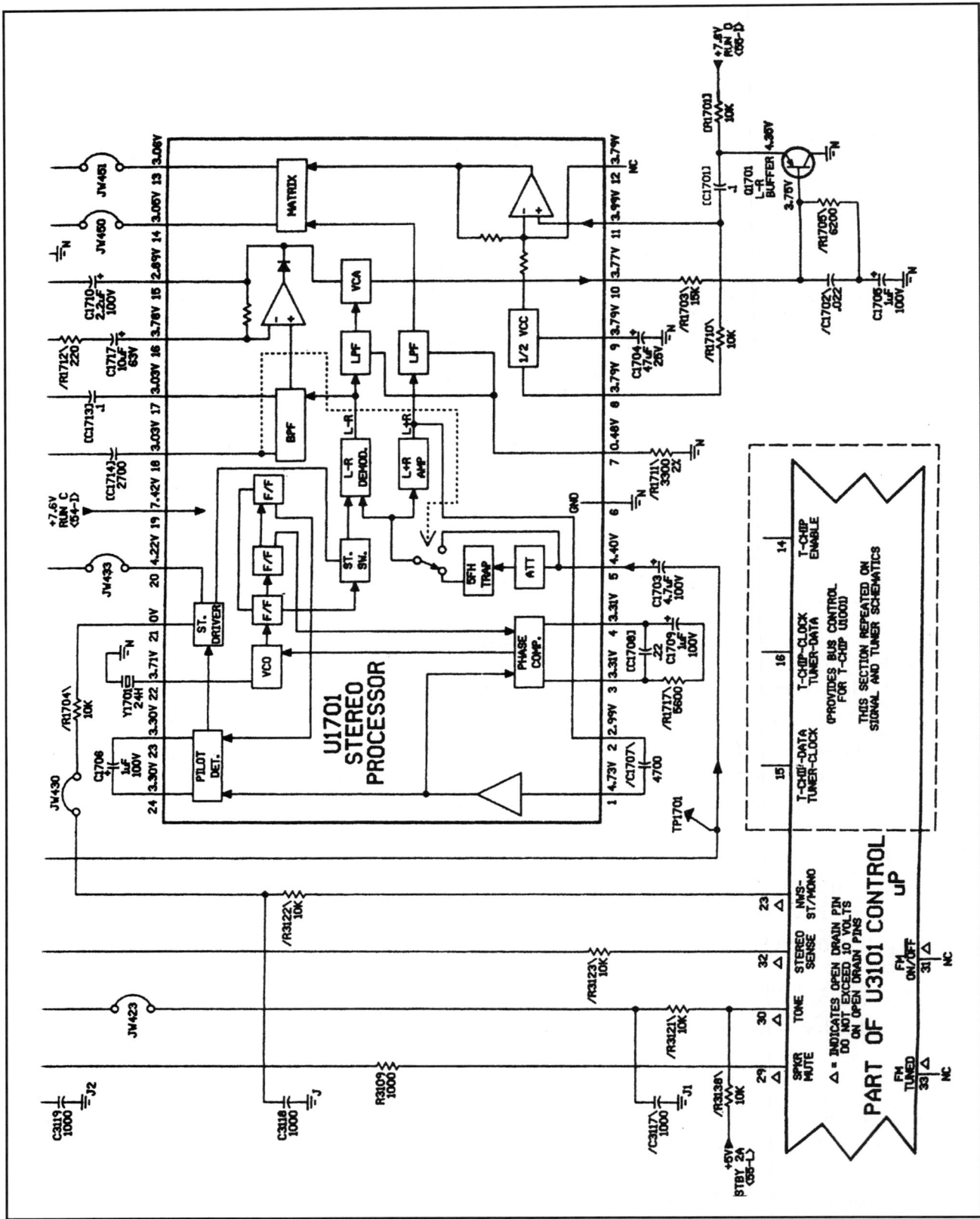

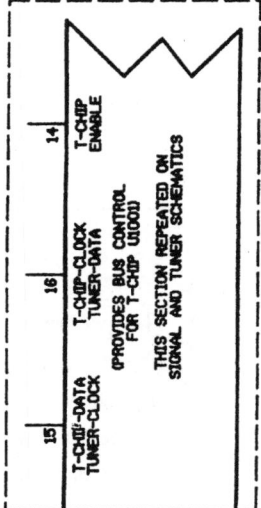

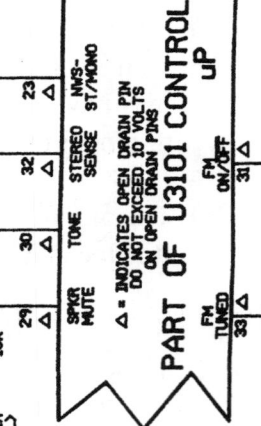

Figure 3-42. Stereo decoder U1701.

CHAPTER 4

THE CTC185 CHASSIS

The CTC185 chassis is remarkably similar to the CTC175/176/177/187 family. It is contained on a single circuit board and designed around a serial bus controlled 64-pin color television (CTV) processor, working in conjunction with a system control microprocessor. The tuner is a tuner-on-board. Virtually all service and tuner adjustments are serial bus controlled and stored in an EEPROM. Thomson follows its present-day scheme by employing surface-mount technology extensively. But there are significant differences. The integrated circuits have been redesigned, the circuit board is quite a bit smaller, the power supply is different, the service menu has been redesigned, and the customer menus have been changed. These changes are the most obvious, and there are others less noticeable.

According to the literature, the CTC185 chassis is used in fifteen models—CTC185A (three models), CTC185B (two models), CTC185AA (two models), CTC185AB (six models), and CTC185M (two models). CRT sizes range from the standard 19" to 27". The chassis will be accompanied by one of three remote transmitters, CRK10A1, CRK20A1, or CRK74B2.

Figure 4-1 will give you an overview of the circuit board and the notations Thomson uses in its technical diagrams. The illustration is taken from *Color Television Basic Service Data: CTC185* (pages 1-4).

Two factors make component location relatively easy: (1) the circuit board has been roadmapped, as you can see by studying its outer edges in *Figure 4-1*; and (2) it uses a component numbering system which relates a specific component to a general circuit board area. The following table reproduces the component numbering system:

1200	Wideband audio processing	**4100**	B+ regulated and standby voltage supplies
1700	Stereo audio processing	**4200**	The degauss circuit
1900	Audio output	**4300**	Horizontal drive
2300	IF input/output processing	**4400**	Horizontal output
2700	Luminance/chroma processing	**4500**	Vertical deflection
2800	Luminance/chroma processing	**4700**	Scan-derived voltages
3100	Instrument control (system control)	**4900**	X-ray protection
3300	On-screen display	**5000**	Picture tube socket
4000	AC input and raw B+	**7100-7400**	Tuner-on-board circuits

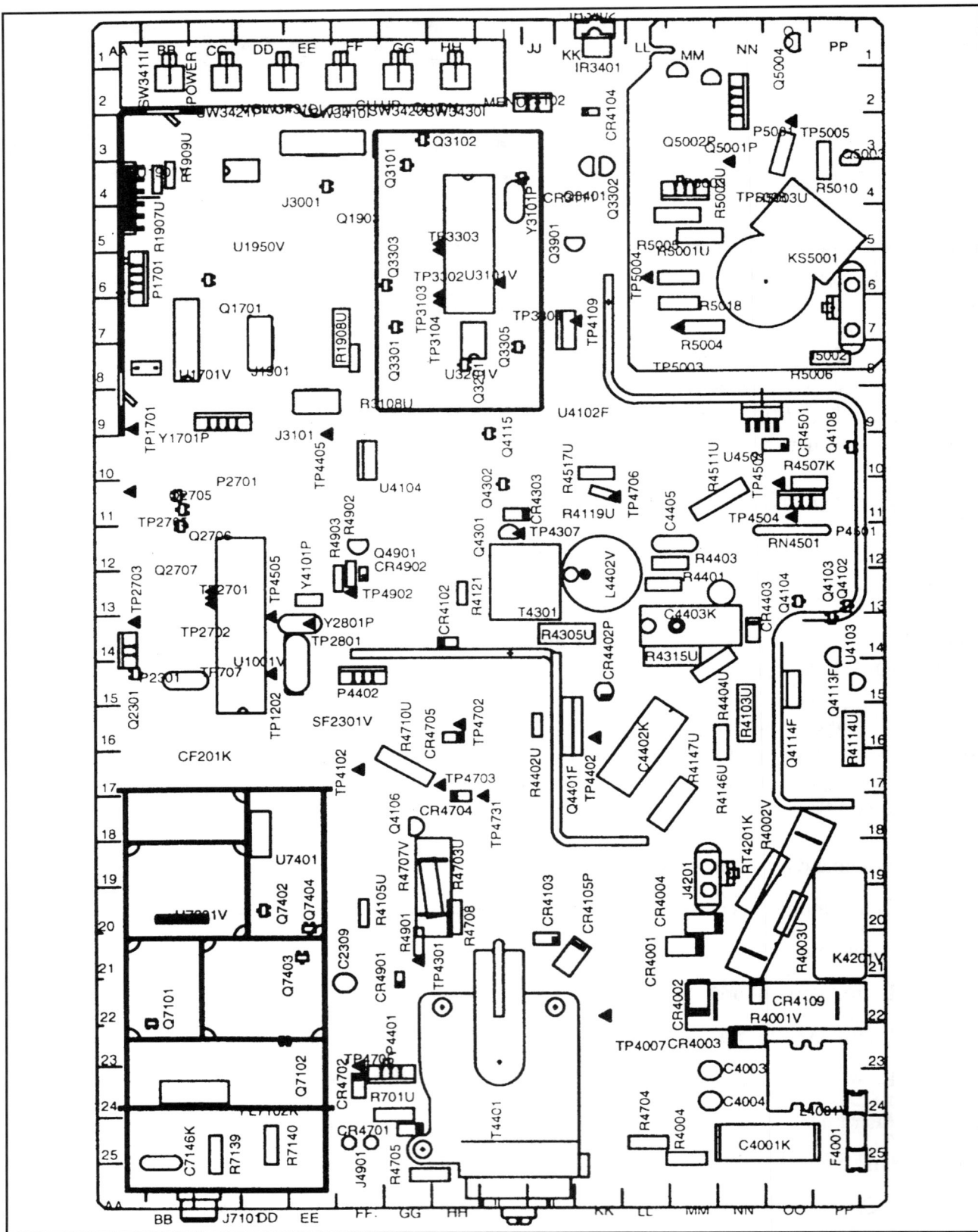

Figure 4-1. Main circuitboard.

One of the neat things about the newly designed factory service literature is its component location guide found on pages 1-34 through 1-37. I worked on a CTC185 today and needed to locate parts associated with the reset circuitry. The TV was shutting down. X-ray protection voltage on pin 24 of the microprocessor was in good order, as was the B+ to the horizontal output transistor. The only voltage off was the reset voltage on pin 2 of U3101. It should have been over +5 volts but was dropping to less than 1.5 volts. I needed to check a transistor and two capacitors tied to pin 2, all of which were surface-mount components. One of the capacitors was C3113. If you have ever tried to find one of those little devils you know what a chore it can be. Look at how easy the new literature makes it. C3113 is located on the bottom of the board at location H4.

Speaking of factory service literature, I shall conclude this brief introduction with a note about available literature. First, there is *CTC185 Technical Training Manual*. It gives a rather detailed description of the following circuits: power supply, system control, horizontal deflection, vertical deflection, video/audio IF, video processing, and audio processing. The second is *Basic Service Data: CTC185*. I rate the factory service data highly. The schematics are about the easiest to read Thomson has ever produced, and there is a wealth of troubleshooting tips, including troubleshooting flowcharts. If the genie were to pop out of the bottle and grant me a wish, I would wish Thomson had kept to this format and never even thought of going to a CD-ROM! Anyway, you can get these booklets from your local RCA distributor or from Thomson Consumer Electronics at the address I gave you in the introduction. You might want to look at a brief article of mine in the October 1998 issue of *Electronic Servicing and Technology*, "Servicing a 'Dead Set' RCA CTC185 Chassis." It's more to the point and far more brief than what I cover here. Note that current RCA literature is available only on CD-ROM.

The Service Menu

I was surprised to learn that Thomson does not discuss alignment procedure in the technical training manual. You will have to consult the service literature for those particulars. Fortunately, the chassis and tuner alignments are discussed in depth.

The service menu holds the alignment information for the chassis as it does for the chassis I discussed in Chapter 3. When a parameter value is modified, the corresponding T-Chip register, tuner register, and EEPROM register are updated as they were in the CTC175 family. The difference is the CTC185 service menu has been "re-engineered." The parameters are now in two groups instead of three:

(1) group "0" lists the instrument and chassis parameters (*Figure 4-2*);

(2) group "1" lists the tuner parameters (*Figure 4-3*).

You access the service menu using the same procedure as before: (1) turn the set on; (2) hold the menu button down; (3) press and release the power button; (4) press and release the volume+ button; and (5) release the menu button. You will get information on the screen that corresponds to *Figure 4-4*. The group number will be on the lower left side of the screen. The parameter to be adjusted will appear in the center and will be changed by channel up/channel down buttons. The value stored in the parameter will be on the right and will be changed by the volume up/volume down buttons. Remember, you can

Parameter # Chan to Change	Parameter Name	Value Range Vol to adjust	Comment:
0 00	Pass No. for Serv. adjust	Must set to 76	May not advance until value set
Service Adjustment Parameters			
0 01	Horiz. Phase	00-15	
0 02	Vertical DC	00-63	
0 03	Vertical S Correction	00-15	
0 04	Vertical size	00-127	
0 05	Red Bias	00-127	Press Menu button for setup line
0 06	Green Bias	00-127	Press Menu button for setup line
0 07	Blue Bias	00-127	Press Menu button for setup line
0 08	Red Drive	00-63	
0 09	Green Drive	00-63	
0 10	Blue Drive	00-63	
0 11	Sub-Brightness	00-127	
0 12	RF AGC	00-63	
0 13	FM Level	00-31	
0 14	VCO Tuning	00-127	
0 15	APC Detector Adjust	00-63	Defeat IF AGC
0 16	Tint Preset	00-127	
0 17	Color Preset	00-127	
0 18	Video Level	00-07	
0 19	Vertical Linearity	00-15	
0 20	Vertical Countdown Mode	00-03	

Figure 4-2. Service menu chart, group "0".

use the remote control functions to change parameters and parameter values once you have gained access to the service menu.

When you first call up the service menu, the parameter will be "0." If you attempt to change parameters with the channel up/channel down buttons while the "0" is present, the TV will exit the service mode. The service adjustments are protected by a security code, such as they were for the chassis in Chapter 3. The security code for Group 0 is 76. Use the volume UP control on the TV or the remote control to change the value to 76 to gain access to the instrument and chassis alignments. The security code for Group 1 is 77. To gain access to Group1 settings, press channel UP or channel DOWN controls to return to parameter 0 and change the value to 77.

Two concluding comments:

First, when you are setting the CRT bias and drive parameters, you can collapse the raster to obtain a service setup line by pressing the menu button once. When you want to return to the full raster, press the menu button again. Second, press the power button on the TV or the remote control to exit the service menu and save adjustments.

Electronic Tuner Alignment Parameters

Parameter # Chan to Change	Parameter Name	Value Range Volume to adjust	Parameter # Chan to Change	Parameter Name	Value Range Volume to adjust
1 01	Ch. 2 secondary	00-62	1 28	Ch. 46 secondary	00-62
1 02	Ch. 2 primary	00-62	1 29	Ch. 46 primary	00-62
1 03	Ch. 2 single	00-62	1 30	Ch. 46 single	00-62
1 04	Ch. 6 secondary	00-62	1 31	Ch. 50 secondary	00-62
1 05	Ch. 6 primary	00-62	1 32	Ch. 50 primary	00 62
1 06	Ch. 6 single	00 62	1 33	Ch. 50 single	00-62
1 07	Ch. 98 secondary	00-62	1 34	Ch. 51 secondary	00-62
1 08	Ch. 98 primary	00-62	1 35	Ch. 51 primary	00-62
1 09	Ch. 98 single	00-62	1 36	Ch. 51 single	00-62
1 10	Ch. 15 secondary	00-62	1 37	Ch. 61 secondary	00-62
1 11	Ch. 15 primary	00-62	1 38	Ch. 61 primary	00-62
1 12	Ch. 15 single	00 62	1 39	Ch. 61 single	00 62
1 13	Ch. 17 secondary	00-62	1 40	Ch. 75 secondary	00-62
1 14	Ch. 17 primary	00-62	1 41	Ch. 75primary	00-62
1 15	Ch. 17 single	00-62	1 42	Ch. 75 single	00-62
1 16	Ch. 18 secondary	00-62	1 43	Ch. 101 secondary	00-62
1 17	Ch. 18 primary	00-62	1 44	Ch. 101 primary	00-62
1 18	Ch. 18 single	00-62	1 45	Ch. 101 single	00-62
1 19	Ch. 9 secondary	00-62	1 46	Ch. 114 secondary	00-62
1 20	Ch. 9 primary	00-62	1 47	Ch. 114 primary	00-62
1 21	Ch. 9 single	00-62	1 48	Ch. 114 single	00-62
1 22	Ch. 29 secondary	00-62	1 49	Ch. 122 secondary	00-62
1 23	Ch. 29 primary	00-62	1 50	Ch. 122 primary	00-62
1 24	Ch. 29 single	00-62	1 51	Ch. 122 single	00-62
1 25	Ch. 39 secondary	00-62	1 52	Ch. 125 secondary	00-62
1 26	Ch. 39 primary	00-62	1 53	Ch. 125 primary	00 62
1 27	Ch. 39 single	00-62	1 54	Ch. 125 single	00-62

Figure 4-3. Service menu chart, group "1".

The Chipper Check

Somewhere I need to say a few words about the Chipper Check™, and I suppose here is as good a place as any. The Chipper Check has evolved to meet the challenge of servicing today's televisions, so the line goes, ". . . and to provide a more user-friendly method of diagnosing problems and performing adjustments" (*Chipper Check User Manual*, pg. 3). It is supposed to be helpful diagnosing the problems of the CTC175/176/177 and CTC187 chassis, but speaking from my experience, I confess I have found it somewhat less than helpful. It a useful but not necessary tool for working on the CTC185 chassis; it <u>is</u> a necessary tool for servicing the CTC195/197 chassis if you plan to do in-depth repair.

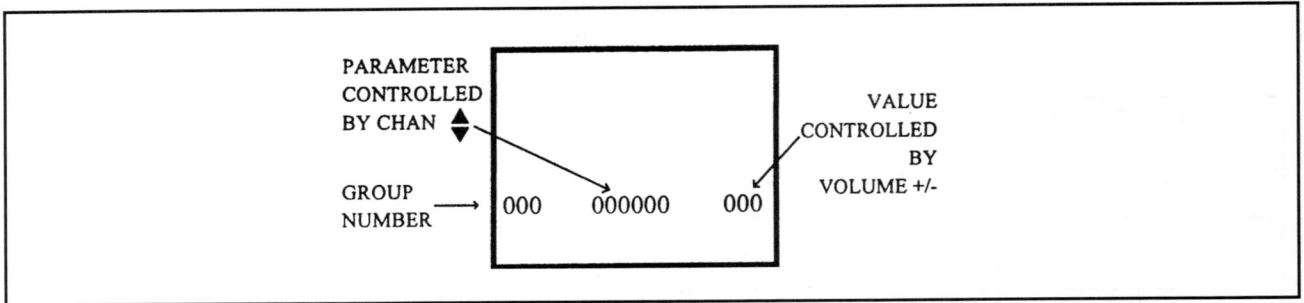

Figure 4-4. Service menu screen map.

The Chipper Check consists of an interface box, a cable, two adapter boards, a power pack, and a software package. If you want to know what the package looks like, I invite you examine *Figure 4-5*. If you would like to order the package, you will find the part numbers in the upper part of the picture.

The Chipper Check serves as an interface between your computer and the television on which you are working. The interface box attaches to your computer with a standard printer cable. The "interface cable" attaches the box to the television via one of two "adapter boards." *Figure 4-6* is a picture of a CTC185 chassis. The plastic plug inside the white box in the center of the picture shows you where the adapter board plugs into the chassis.

The Chipper Check interface package includes the following items:

1. Interface box (part number CCF002)
2. Interface cable (stock number 212118)
3. DC adapter (stock number 52325)
4. Adapter board, 7 and 3 pin (stock number 212117)
5. Adapter board, 6 pin (stock number 212121)

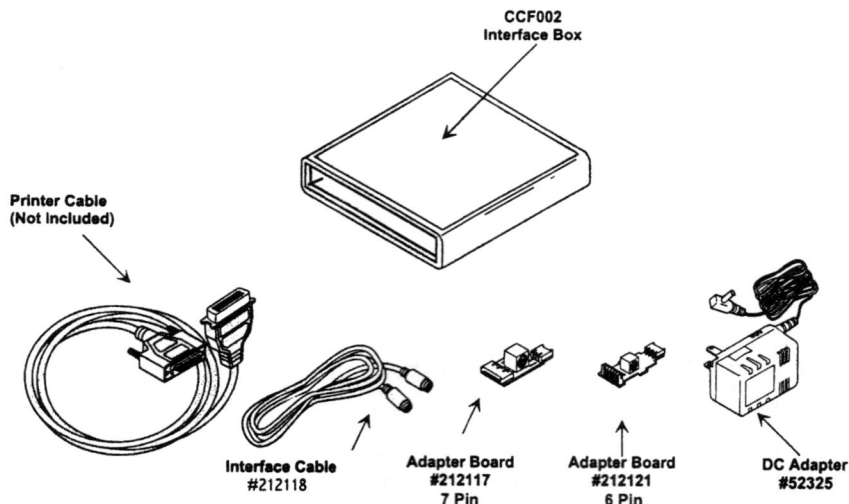

Figure 4-5. Chipper Check inventory.

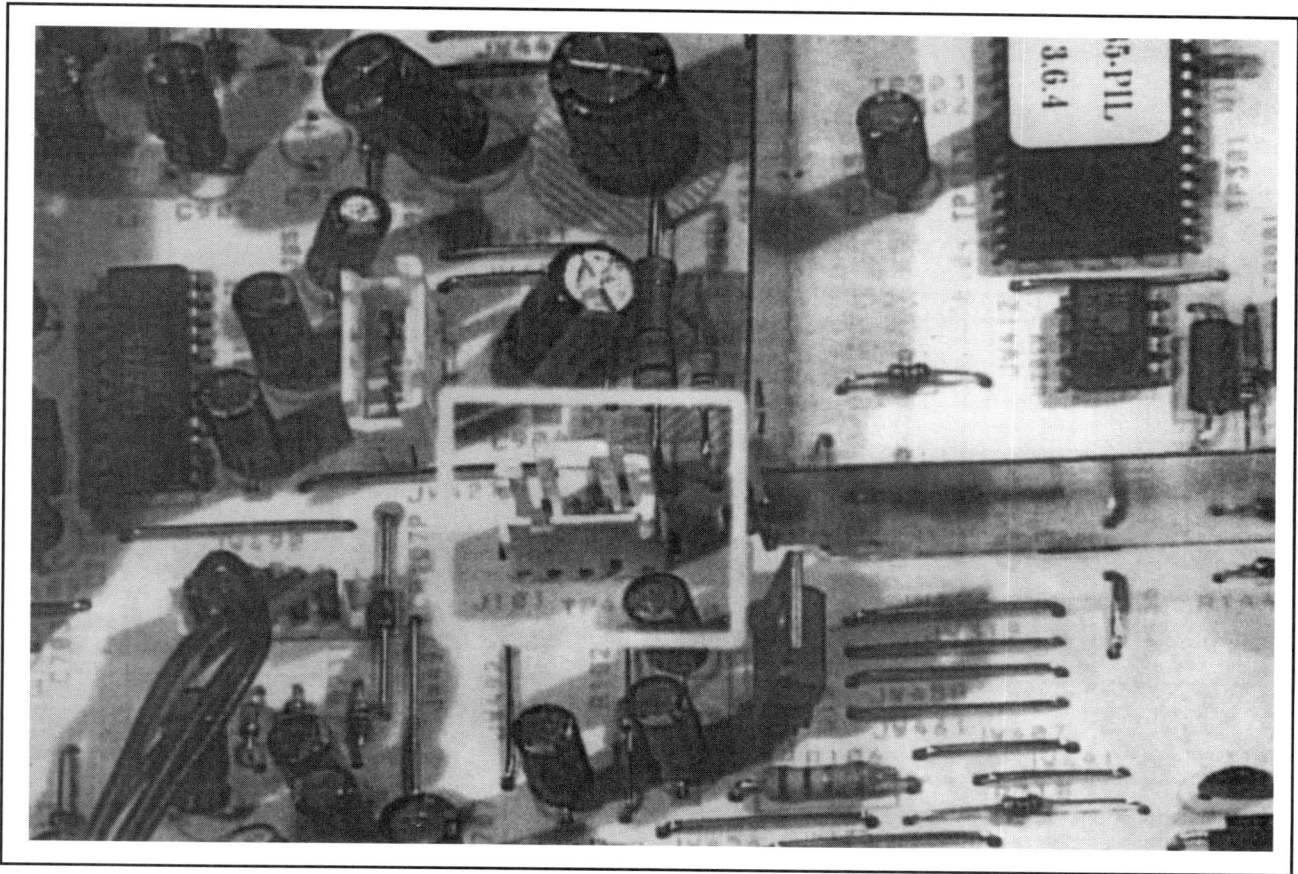

Figure 4-6. Interface box connection location.

How does the Chipper Check work?

Take a look at *Figure 4-7*, a representative system control section from the CTC185 chassis. There are four main elements in the system control block: the microprocessor, the EEPROM, the data bus, and various output devices. The microprocessor communicates with these devices to control the television. It also receives input information from the front-panel controls, the remote transmitter, the EEPROM, and other devices on the data bus. If we had the equipment to listen in on the communication, we would hear a constant chatter as information is taken in and instructions given out with hardly a pause!

The EEPROM stores information used by the microprocessor to control the television. With respect to the chassis under discussion, these values include alignment information, initializing signals, and customer values. The "customer values" information includes channel scan list, color, brightness and audio levels. The memory is nonvolatile, which means it stays put even if power is lost.

The microprocessor begins communication with the devices on the data bus whenever AC power is applied. Unlike the EEPROM, its memory is volatile—You lose power, you lose data stored there. Just here the EEPROM becomes important. When you plug the TV into the AC outlet, the microprocessor receives B+, which brings it to life. It also receives reset voltage and initializes its program at a specific

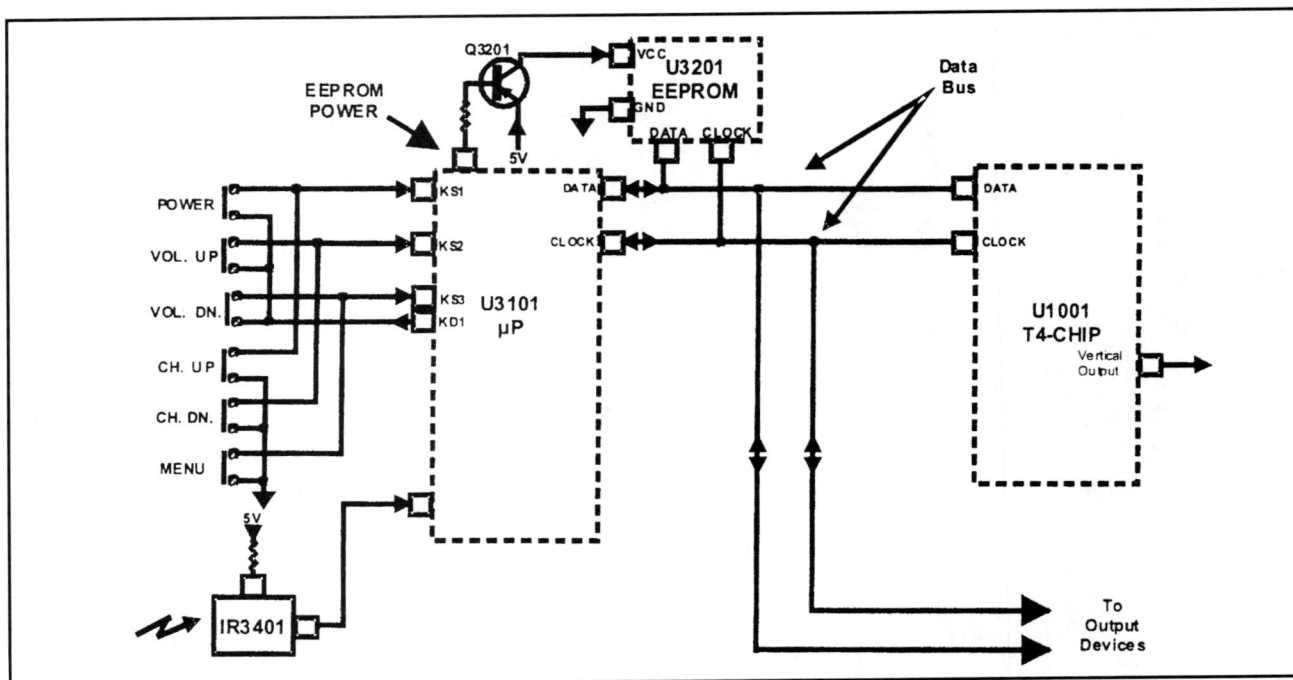

Figure 4-7. System control.

place. Once reset has occurred, the microprocessor signals the EEPROM to begin downloading its information. The microprocessor receives the download and uses it to operate the television. Moreover, the microprocessor updates itself from the EEPROM once every second to keep the data used by the various output devices current and reduce the effect of corrupted data.

The next question is, "Whence the information in the EEPROM?" Most of the values were stored there during manufacturing. The factory has automatic test equipment to perform the various alignments and write the data into the EEPROM. The customer will add information when he/she begins to use the TV. The third source of data for the EEPROM is the Chipper Check, which adds data when it is connected.

The Chipper Check merely takes control of the data bus, which is why when it is connected the front controls cease to operate. Once Chipper Check is in control of the bus, it sends data to the various output devices. It also uses the data bus to write data to the EEPROM. Using it, you can make whatever adjustment you want and write that adjustment directly into the EEPROM.

There is a great deal more to the Chipper Check than I have said. If you want the additional information, I suggest you order *Chipper Check User Manual*. It is a brand new publication, sporting a 1998 copyright date, and can be ordered from the address I have already give you.

The Power Supply

The first system I want to deal with is the power supply. There are three of them: the main power supply, the standby power supply, and the scan-derived power supply.

The Main Power Supply

Thomson describes the power supply as "a non-isolating switching power supply that uses a MOSFET as a switching device." The service literature is a bit more specific, calling it "a non-isolating modified buck converter." It is rather complicated and has a relatively high component count, but it seems to work quite well.

We will begin our review with *Figure 4-8*, a block diagram of the supply. Raw B+ is fed through a winding of the IHVT (pins 1 and 3) into the regulator circuit. The winding of the IHVT sums an inverting trace onto the raw B+ to provide a "pre-boost" which enables the power supply to regulate even when the raw B+ falls below 130 volts. The number of turns on the IHVT winding determines the lowest voltage at which the supply will regulate. The 19" and 20" sets will operate quite well with AC as low as 90 volts; the 25" and 27" TVs will continue to operate with AC as low as 95 volts.

After pre-boost, the power supply acts like a buck converter (output voltage = input x duty cycle). As Q4114 turns on, current flows through CR4105, L4102, and Q4114 to charge C4153. When horizontal retrace begins, the voltage on the anode of CR4105 begins to drop, but current continues to flow because of L4102. The voltage at the junction of the cathodes of CR4105 and CR4103 (V2 of *Figure 4-8*) continues to decrease until it reaches about -0.7, at which point CR4105 turns off while CR4103 continues to conduct until current flow drops to zero. Q4114 remains on until current flow drops to zero.

Now what about regulation? The horizontal sync-sawtooth generator block outputs a sawtooth waveform which is combined in the comparator with a reference voltage to generate gate drive for Q4114. The gate drive circuit outputs 9 volts, which floats above the regulated B+ and turns on the MOSFET. Its turn-off time is fixed, but its turn-on time varies in response to the control circuit. I will spare you the details of how this lovely little circuit works. If you insist on a complete analysis, I suggest you read the booklets I have cited.

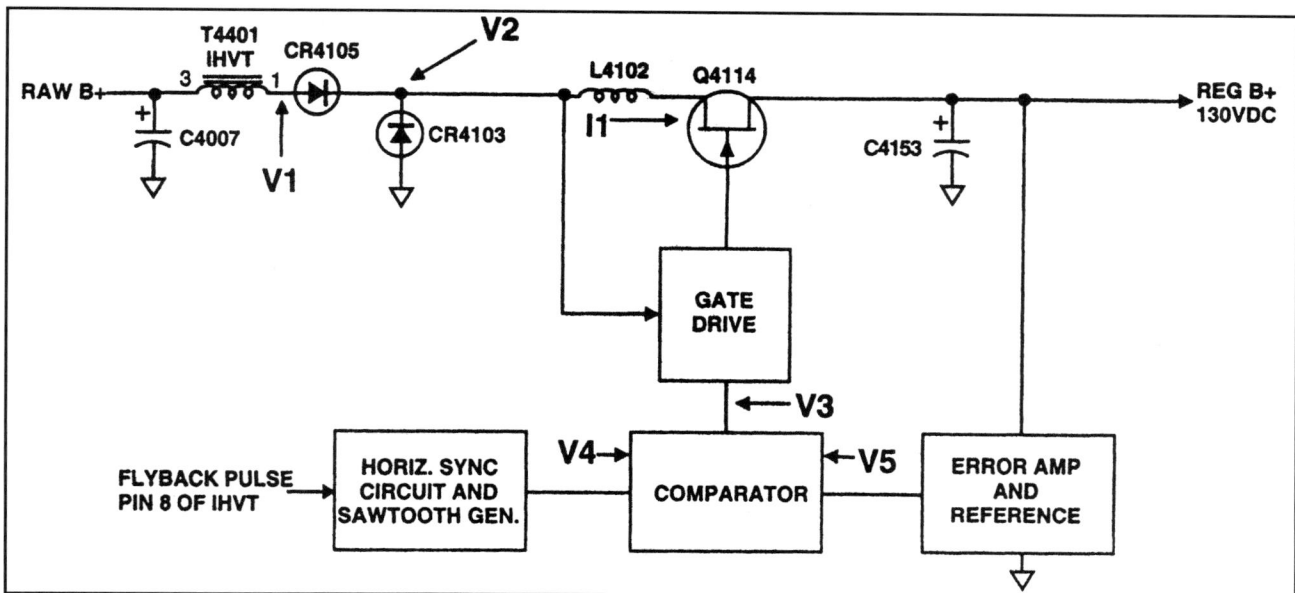

Figure 4-8. Power supply block diagram.

Even though I will forego a detailed analysis, I will place the components that make up the circuit into their blocks using *Figures 4-8* and *4-9* as references:

(1) the horizontal sync and sawtooth generator: Q4104, Q4108 and associated components;

(2) the comparator: Q4102, Q4103 and associated components;

(3) the error amp and reference: U4103 and voltage-sense resistors, R4137, R4111, and R4112 (These are precision resistors.);

(4) the gate drive circuit: C4106 and CR4111. (The gate is switched on through R4138 and turned off when Q4113 turns on and bleeds the gate charge off through R4114.)

One other component is important to the operation of the power supply, namely R4108 (*Figures 4-9* and *4-12*). It is a 220k resistor connected between the gate and drain to provide gate drive for Q4114 when the TV is in standby. Remember, in standby the output voltage equals raw B+ because the regulator circuit will not be working.

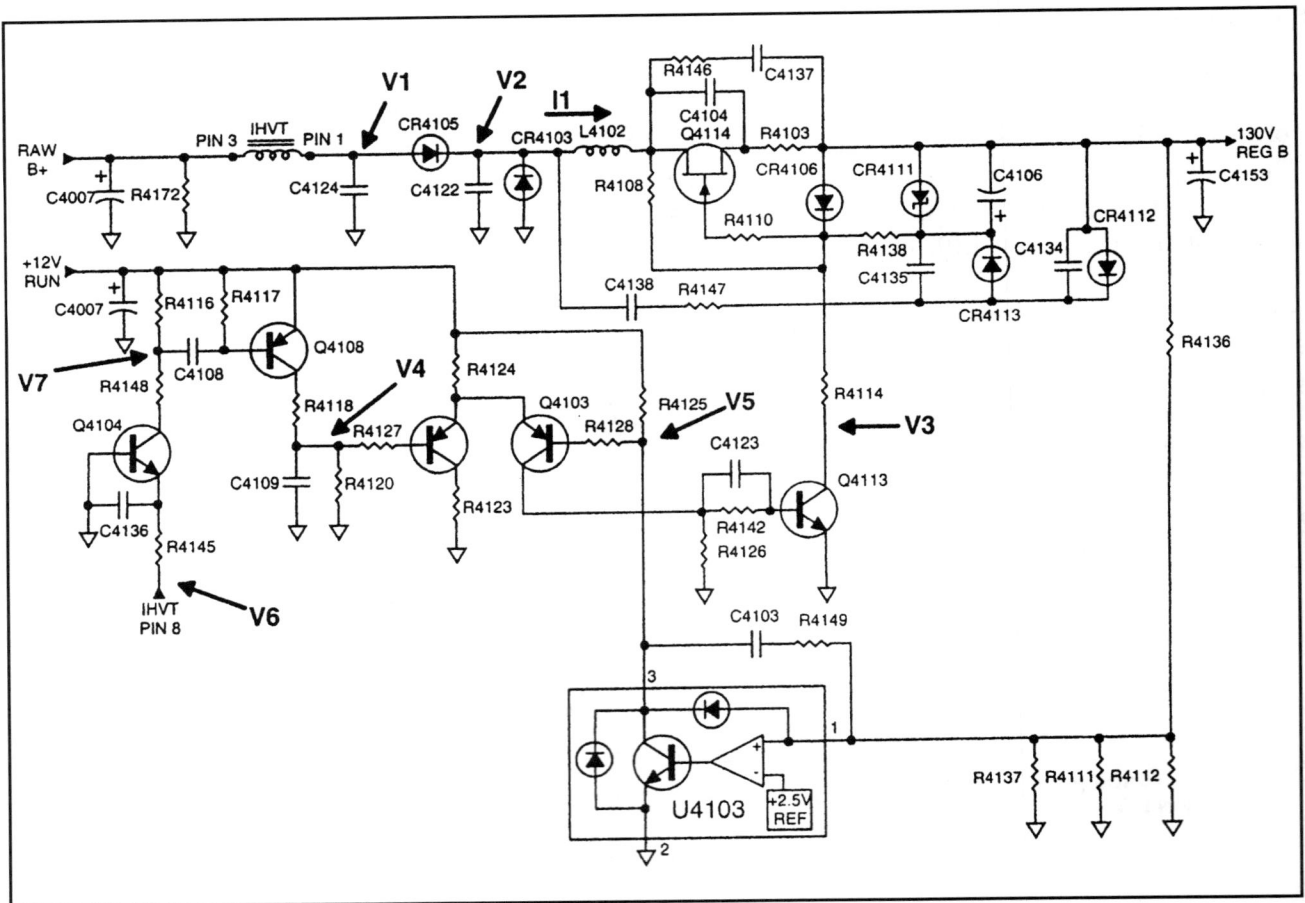

Figure 4-9. Power supply detail.

Troubleshooting the supply is difficult because it will not work unless horizontal deflection works. If you have a dead set, where do you begin to look for the problem, horizontal deflection or power supply? Is it the "chicken or egg first" dilemma? Not necessarily. You can bypass Q4114 by shorting source to drain, but you absolutely must use an isolation transformer-variac. If you do not use an isolation transformer, you risk damaging expensive test equipment or the TV or yourself, or all three. If you don't use a variac, the TV will shut down because B+ will be too high. So, to get the TV up and running, adjust the variac for an output of 90-95 VAC. When (and if) the set comes on, readjust the variac so the voltage at the collector of the horizontal output transistor is as close to 130 VDC as possible. If the TV comes on and everything works, you know the problem is in the power supply. You can use voltage and waveform analysis to locate the problem. If it doesn't come on, you can proceed with your troubleshooting, having eliminated the power supply as the problem.

I am including two troubleshooting flowcharts. *Figure 4-10* deals with the power supply, and *Figure 4-11* deals with horizontal deflection. You can file both under the heading "Dead Set: TV Will Not Come On" because both may be necessary to help you figure out why the television won't start up. These trouble trees are not cure-alls. A trouble tree never is. They, however, represent a common-sense and rather obvious approach to the problem, which is just what we need when we get into a situation where we chase our tails dog-like as we go round-and-round the problem.

Two service notes before I move on:

First, don't ever de-energize a power supply or yoke lead by shorting it directly to ground. Let me be specific. Don't use a screwdriver to bleed a filter capacitor. You will probably destroy the regulator if you do. This observation does not pertain to discharging the CRT to the ground braid around it. That's okay. If you need to discharge a capacitor, locate a large watt 1,000 to 5,000 ohm resistor, put clip leads on both ends, wrap it in tape or heat shrink tubing to keep it from shorting other components, and use the contraption to bleed off the offending capacitor. It's a safe and easy way to do it. In fact, I keep one of these devices on my work bench at all times. I learned the hard way not to use a screwdriver to discharge a fully-charged capacitor!

Second, two components in the power supply are more likely to fail that others: Q4114 and U4103. The MOSFET fails most often due to excessive loads like a shorted horizontal output transistor. U4103, on the other hand, fails most often because of excessive voltage on one or more of its pins. RCA had some problems in its early production runs because Q4114 ran too hot and eventually failed. In those early sets, specifications called for C4106 to be a 1 µF capacitor. The capacitor could and often did reduce in value, causing Q4114 to run hot. It was later replaced with a 0.47 µF polyester capacitor. The part number for the new capacitor is 226431. However, you can get a kit of parts which includes the new capacitor, the MOSFET, and a few other critical components, by ordering part number 231523. I suggest you order the kit of parts and install all of them. The kit is not very expensive, and you just might avoid a call-back. Note: The transistor in the kit will not work in all CTC185 chassis. Consult the service literature for the part number for the chassis on which you're working.

When you replace Q4114, please be certain to bend its leads so it will fit snugly against the heatsink, and use thermal compound on both sides of the insulator. Use part number 237641 to replace a defective insulator. RCA also tells you to replace the heatsink clip before you solder the new transistor in

POWER SUPPLY TROUBLESHOOTING FLOWCHART

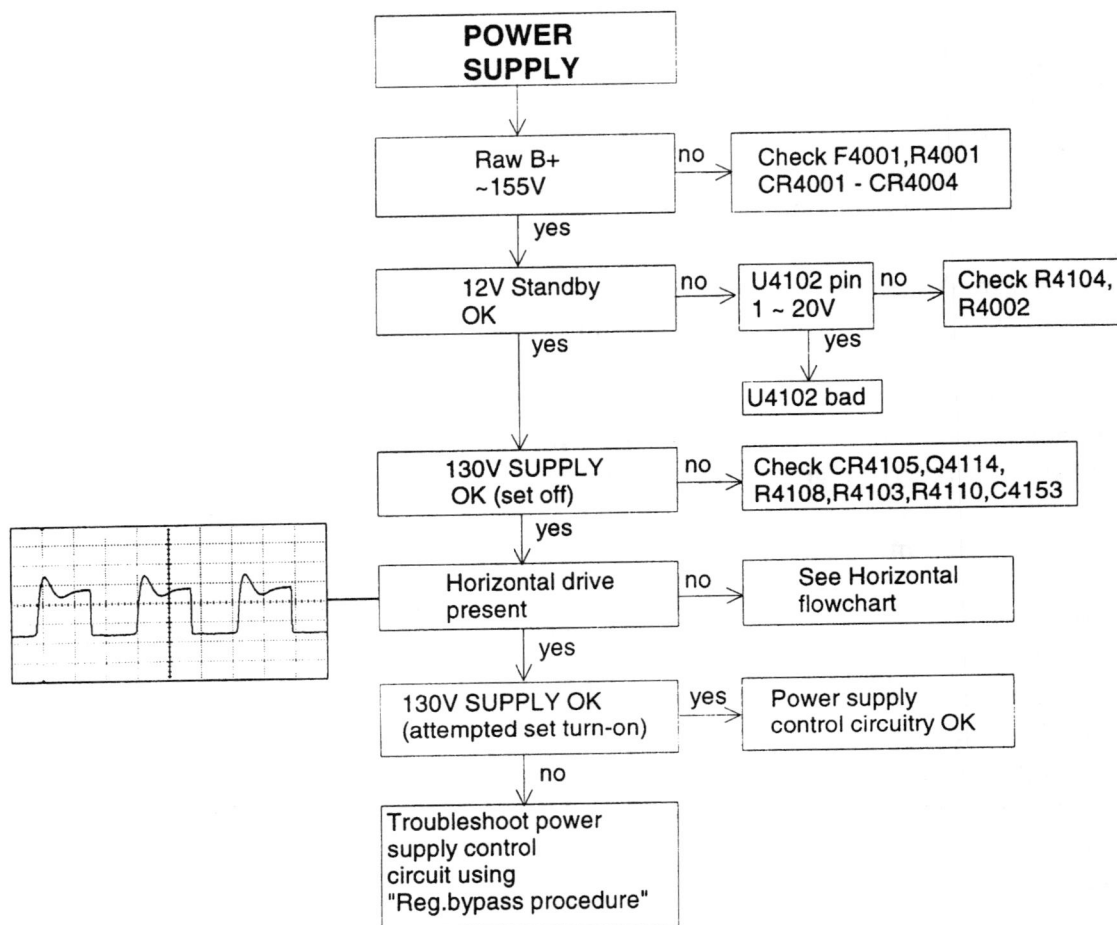

REGULATOR BYPASS PROCEDURE

Trouble shooting of the supply is difficult because the supply will not operate unless the horizontal system is operating and the horizontal will not work unless the supply is operational. The solution is to fool the horizontal section into thinking the supply is operational and then the problem can diagnosed. One method to accomplish this is to apply a short across the drain and source terminals of Q4114 and use a variac on the ac line to control the Reg B+ voltage. If the Reg B+ is allowed to get too high, The XRP circuit will shut off the horizontal oscillator. In order to get the chassis to turn on, the variac must be set to provide between 90 and 95 volts ac rms to the chassis. After the chassis is operational use the variac to set the reg B+ voltage as close as possible to 130Vdc. Now the power supply circuit can be diagnosed using the waveforms shown on the schematics.

Figure 4-10. Power supply troubleshooting flowchart.

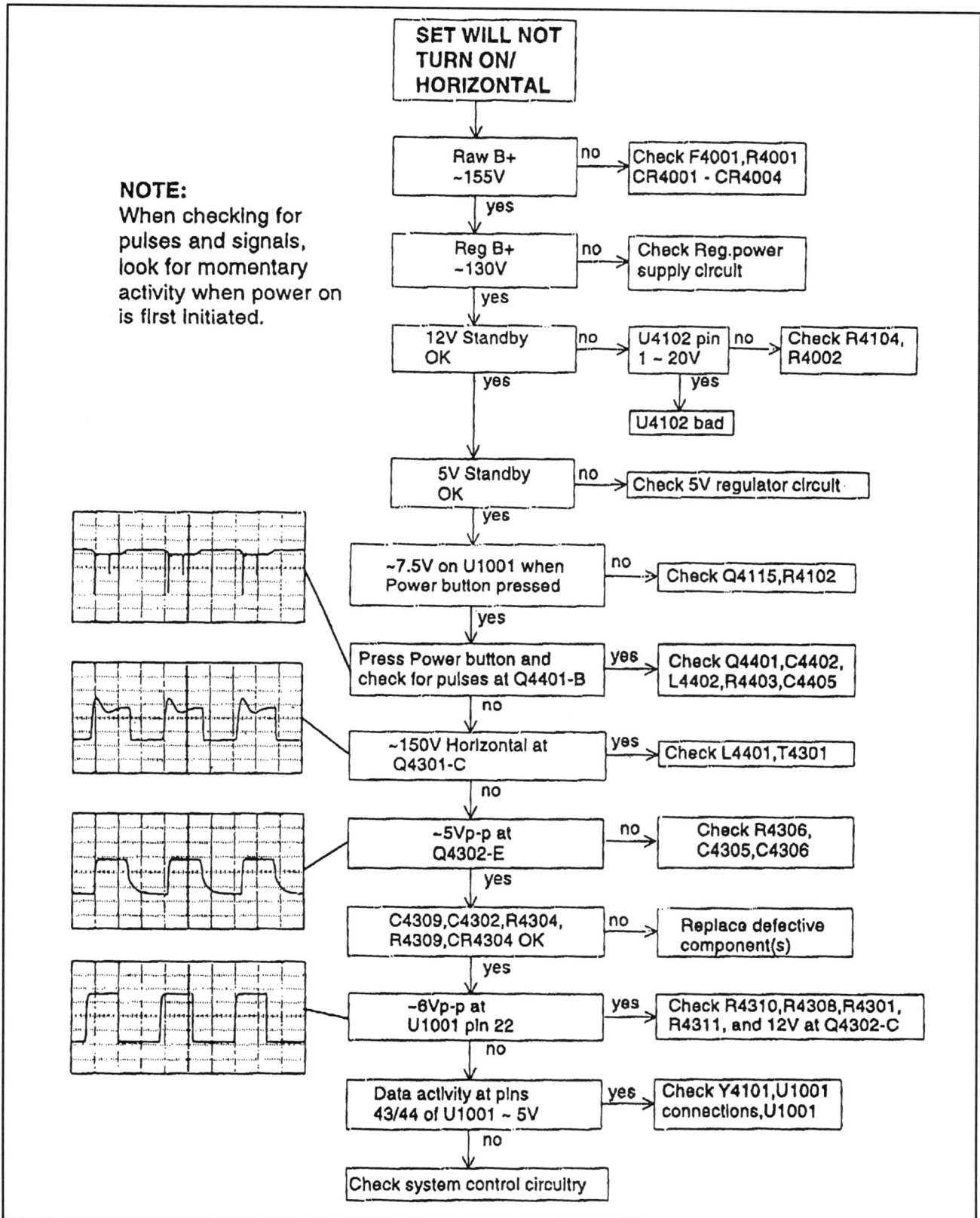

Figure 4-11. Horizontal troubleshooting flowchart.

place to ensure a snug fit between transistor and sink. The snug fit permits adequate heat transfer, which in turn will prevent premature failure of Q4114. You may be wondering why I am explaining in detail something obvious. Many of these TVs were manufactured, sold, and repaired under warranty because somebody or something failed to seat Q4114 against its heatsink. RCA responded by sending a service bulletin to its service centers, alerting us to the problem and the fix.

The Standby Power Supply

The second power supply is the standby power supply. I am including *Figure 4-12* from Sams PHOTOFACT® number 3823, for model number F25209WTTX1, a CTC185AA chassis. It shows in full detail the full power and standby power supplies and will serve as a reference for our discussion of standby supply. I also include *Figure 4-13* to give you a bare-bones sketch of the standby supply.

The standby power supply is really very simple. It is derived from a voltage dropping resistor connected to a 3-terminal 12-volt regulator (U4102). The power comes directly from a half-wave rectified AC waveform. CR4109 and CR4004 do the rectification; R4002 is the resistor that feeds the voltage to U4102; C4154 provides the filtering. CR4110, a 27-volt zener, limits the applied voltage to 27 volts. CR4703 supplements the standby supply from the +26-volt run voltage source when the TV powers up. The 12 volts from U4102 is also applied to the collector of Q3901, which is biased to make available at its emitter the +5 volts for the microprocessor and the IR receiver. CR4104 provides an additional 5.6- and 5.9-volt reference source for system control.

The TV gets by with such a limited standby power supply because it needs very little power when it is off. Unlike, say the CTC177, the microprocessor turns off the T4-Chip (no longer just the T-Chip) when it is not needed. Of course, this creates an additional problem because timing for switching the T4-Chip on is important. Why? Because the current needed to run the microprocessor and the T4-Chip together is derived from a single capacitor (C4154) until the +26-volt run supply comes up! If the timing is not correct, the television will attempt to start and shut down because it will not have sufficient power to start up and run.

I now call your attention to *Figure 4-14*, the circuit responsible for turning the T4-Chip on and off. The circuit consists of Q4115, R4143, R4144, and the microprocessor. Pin 29 of the microprocessor (U3101) is labeled "standby sw." When U3101 executes an ON command, pin 29 goes low, turning on Q4115, which applies Vcc to pin 20 of U1001, bringing the rest of the TV to life.

You are probably already a step ahead of me, because you have realized the kinds of problems this configuration can cause. For example, it adds a few details to troubleshooting a dead set. But it is not as indomitable as it might seem. Suppose you do have a dead set. Begin by checking the output of U4102 for +12 volts. If it is not there, <u>do not</u> proceed on the assumption something is wrong with the standby power supply! Remember the function of C4154? It provides just enough current to keep system control alive. It cannot supply current to U3101 and U1001 at the same time; therefore, check pin 29 of the micro first. If you measure a voltage greater than zero, the switch transistor (Q4115) is turned on, attempting to supply current to U1001. If it is turned on, U1001 and U3101 will pull the standby line

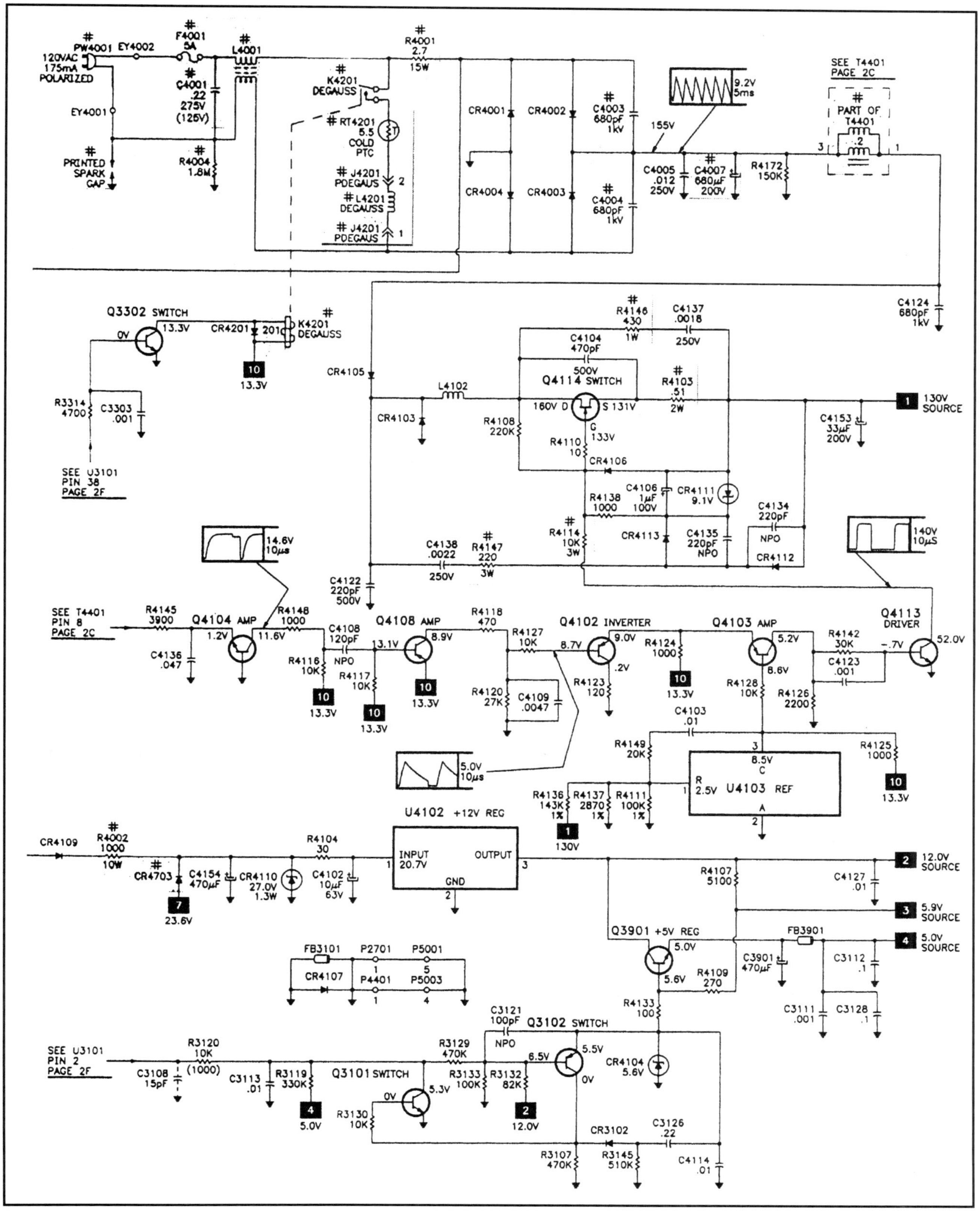

Figure 4-12. Full and standby power supplies.

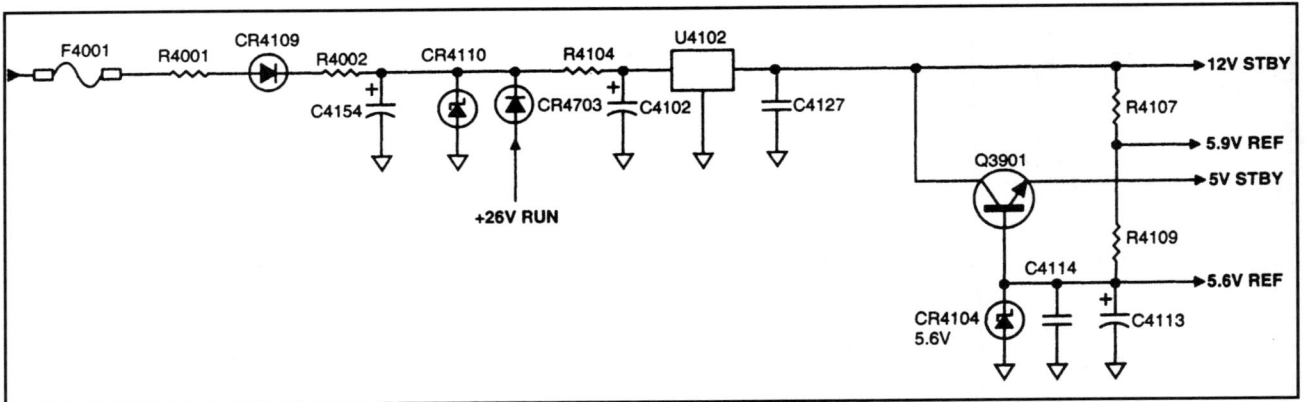

Figure 4-13. Standby supply.

low. If Q4115 is turned on, you will have to look in several areas for the problem. For example, the switch transistor, the T4-Chip, or the microprocessor itself could be defective.

Of course, the 12-volt supply could be missing. You can supplement the standby power supply by using an external +26 volts applied to the cathode of CR4703, which will provide a constant voltage permitting U1001 to run. If the set fires up and works fine, you can legitimately suspect a problem with the standby supply. I suggest you look back at *Figure 4-11*, which provides troubleshooting information about this particular problem.

The Scan-Derived Power Supply

The operating voltages for these televisions are scan-derived. You can get a glimpse of these various voltages by examining *Figure 4-15*. The configuration is vintage Thomson, and at this point should be obvious to you; therefore, I am going to resist the temptation to elaborate.

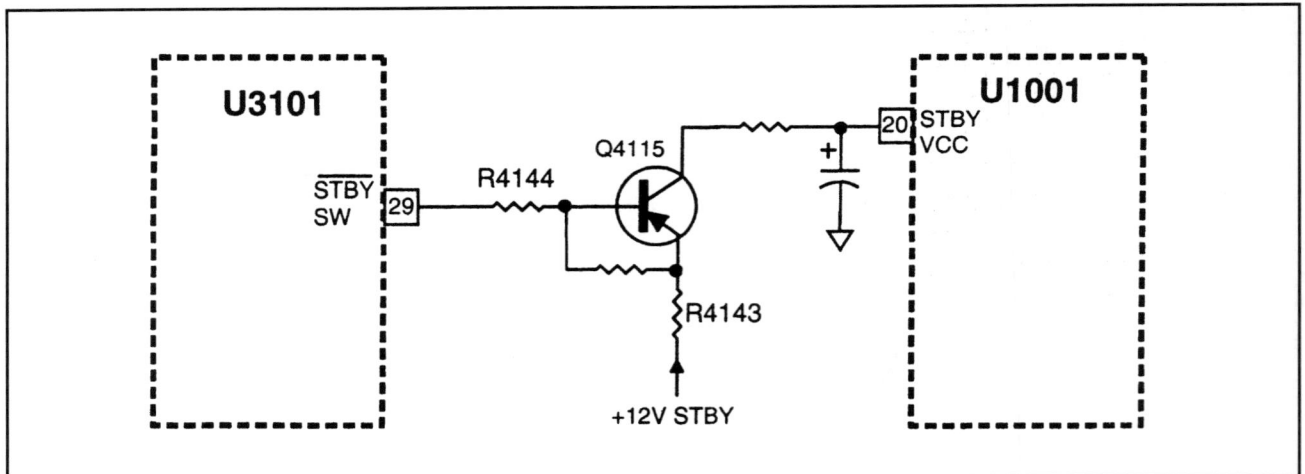

Figure 4-14. T4-Chip power control.

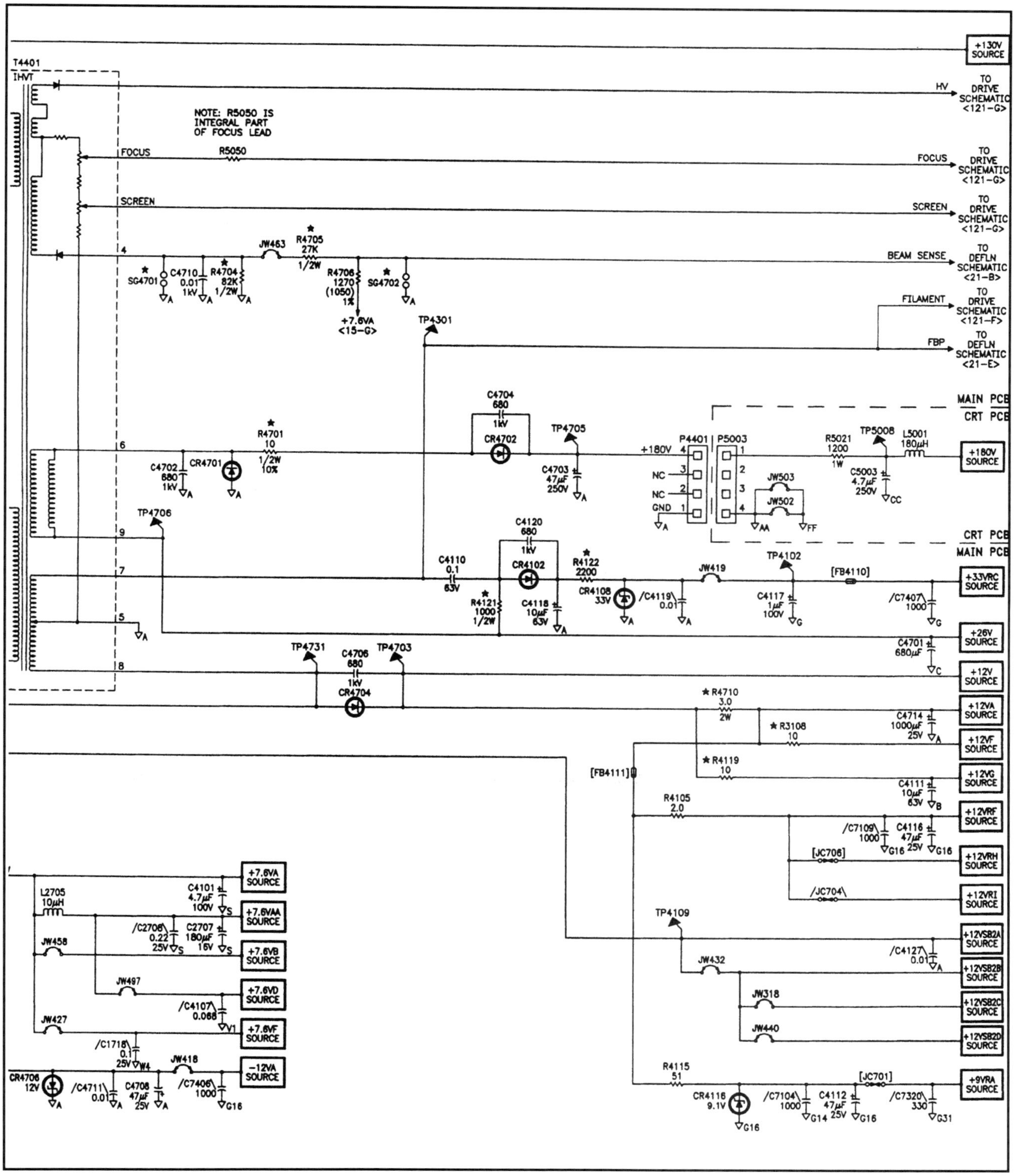

Figure 4-15. Scan-derived power.

The Tuning System

The literature describes the tuner used in the CTC185 chassis as "a single conversion, electronically aligned tuner with a hot/cold barrier" which is capable of tuning signals in the 57 MHz to 801 MHz range in three bands (*Color Television Basic Service Data: CTC185*, pg. 1-16). I am including a table (*Figure 4-16*) to give you an idea of what this tuner is capable of doing.

One of the most common problems with the new tuners is a broken or missing RF connector. I gave you the information in the last chapter, but I want to repeat it here. You can purchase an exact replacement RF connector by ordering part number 215543. MCM lists them for $1.83 (1998 price). The new connector is far superior to one you can salvage because it fits into the tuner wrap without modification. I suggest you order several because you will have to replace the RF connector regularly. The zinc tuner wrap is softer than the stainless steel or copper wrap. If someone tries to move a television while the cable is connected, he/she will almost surely rip the RF connector loose.

The TOB is remarkably similar to the TOB found in the CTC175 chassis, but there are significant differences, and you need to be aware of these differences. The troubleshooting procedure you use will not be markedly different, but other things will. I will summarize the differences like this:

(1) a new PLL (U7401) with an internal digital-to-analog converter (DAC), which eliminates the need for the interface circuitry (U7501) found in the CTC175 family of chassis;

(2) a newly-designed mixer (U7301);

(3) the elimination of the surface acoustic wave (SAW) filter preamp;

(4) a newly designed hot/cold isolation barrier (which in my opinion is not as mechanically stable as the older design), and;

(5) self-biased dual-gate MOSFETs as RF amplifiers.

Since I have spent a lot of time discussing tuner basics and the TOB in particular, I will spend very little time dealing with this one. The *CTC177/187 Troubleshooting Guide* is still one of the best places to go for information about tuner basics and tuner troubleshooting. I will, however, comment on the differences and include some charts and diagrams to aid in troubleshooting.

RCA says your best tool for troubleshooting the tuner is a good digital multimeter. By making voltage and resistance checks, the tech should be able to isolate tuner problems in a reasonable amount of time. My reply is, "Yes and no." You also need a signal generator capable of tuning many channels, a very good magnifier with an even better light, a small-watt soldering iron, steady hands, and a temper that you keep on a short leash. With a little practice and some reasonably good service aids and tools, you can make competent repairs in a reasonable amount of time.

CABLE CHANNEL		PIX FREQ.	SOUND FREQ.	LO FREQ.
BAND 1	2	55.25	59.75	101.00
	3	61.25	65.75	107.00
	4	67.25	71.75	113.00
	1	73.25	77.75	119.00
	5	77.25	81.75	123.00
	6	83.25	87.75	129.00
	98	109.25	113.75	155.00
	99	115.25	119.75	161.00
	14	121.25	125.75	167.00
	15	127.25	131.75	173.00
	16	133.25	137.75	179.00
	17	139.25	143.75	185.00
BAND 2	18	145.25	149.75	191.00
	19	151.25	155.75	197.00
	20	157.25	161.75	203.00
	21	163.25	167.75	209.00
	22	169.25	173 75	215.00
	7	175.25	179 75	221.00
	8	181.25	185.75	227.00
	9	187.25	191.75	233.00
	10	193.25	197.75	239.00
	11	199.25	203.75	245.00
	12	205.25	209.75	251.00
	13	211.25	215.75	257.00
	23	217.25	221.75	263.00
	24	223.25	227.75	269.00
	25	229.25	233.75	275.00
	26	235.25	239.75	281.00
	27	241.25	245.75	287.00
	28	247.25	251.75	293.00

CABLE CHANNEL		PIX FREQ.	SOUND FREQ.	LO FREQ.
BAND 2	29	253.25	257.75	299.00
	30	259.25	263.75	305.00
	31	265.25	269.75	311.00
	32	271.25	275.75	317.00
	33	277.25	281.75	323.00
	34	283.25	287.75	329.00
	35	289.25	293.75	335.00
	36	295.25	299.75	341.00
	37	301.25	305.75	347.00
	38	307.25	311.75	353.00
	39	313.25	317.75	359.00
	40	319.25	323.75	365.00
	41	325.25	329.75	371.00
	42	331.25	335.75	377.00
	43	337.25	341.75	383.00
	44	343.25	347.75	389.00
	45	349.25	353.75	395.00
	46	355.25	359.75	401.00
	47	361.25	365.75	407.00
	48	367.25	371.75	413.00
	49	373.25	377.75	419.00
	50	379.25	383.75	425.00
BAND 3	51	385.25	389.75	431.00
	52	391.25	395.75	437.00
	53	397.25	401.75	443.00
	54	403.25	407.75	449.00
	55	409.25	413.75	455.00
	56	415.25	419.75	461.00
	57	421.25	425.75	467.00
	58	427.25	431.75	473.00

Figure 4-16. Tuner frequencies (continued next page).

	CABLE CHANNEL	PIX FREQ.	SOUND FREQ.	LO FREQ.
	59	433.25	437.75	479.00
	60	439.25	443.75	485.00
	61	445.25	449.75	491.00
	62	451.25	455.75	497.00
	63	457.25	461.75	503.00
	64	463.25	467.75	509.00
	65	469.25	473.75	515.00
	66	475.25	479 75	521.00
	67	481.25	485.75	527.00
	68	487.25	491 75	533.00
	69	493.25	497.75	539.00
	70	499.25	503.75	545.00
	71	505.25	509.75	551.00
	72	511.25	515.75	557.00
	73	517.25	521.75	563.00
BAND 3	74	523.25	527.75	569.00
	75	529.25	533.75	575.00
	76	535.25	539 75	581.00
	77	541.25	545 75	587.00
	78	547.25	551.75	593.00
	79	553.25	557.75	599.00
	80	559.25	563.75	605.00
	81	565.25	569.75	611.00
	82	571.25	575 75	617.00
	83	577.25	581.75	623.00
	84	583.25	587.75	629.00
	85	589.25	593.75	635.00
	86	595.25	599.75	641.00
	87	601.25	605.75	647.00
	88	607.25	611.75	653.00

	CABLE CHANNEL	PIX FREQ.	SOUND FREQ.	LO FREQ.
	89	613.25	617.75	659
	90	619.25	623.75	665
	91	625.25	629.75	671
	92	631.25	635.75	677
	93	637.25	641.75	683
	94	643.25	647.75	689
	95	91.25	95.75	137
	96	97.25	101.75	143
	97	103.25	107.75	149
	98	109.25	113.75	155
	99	115.25	119.75	161
	100	649.25	653.75	695
	101	655.25	659.75	701
	102	661.25	665.75	707
	103	667.25	671.75	713
BAND 3	104	673.25	677.75	719
	105	679.25	683.75	725
	106	685.25	689.75	731
	107	691.25	695.75	737
	108	697.25	701.75	743
	109	703.25	707.75	749
	110	709.25	713.75	755
	111	715.25	719.75	761
	112	721.25	725.75	767
	113	727.25	731.75	773
	114	733.25	737.75	779
	115	739.25	743.75	785
	116	745.25	749.75	791
	117	751.25	755.75	797
	118	757.25	761.75	803

Figure 4-16. Tuner frequencies (continued).

CABLE CHANNEL		PIX FREQ.	SOUND FREQ.	LO FREQ.
	119	763.25	767.75	809
	120	769.25	773.75	815
	121	775.25	779.75	821
BAND 3	122	781.25	785.75	827
	123	787.25	791.75	833
	124	793.25	797.75	839
	125	799.25	803.75	845

AIR CHANNEL		PIX FREQ.	SOUND FREQ.	LO FREQ.
	14	471.25	475.75	517.00
	15	477.25	481.75	523.00
	16	483.25	487.75	529.00
	17	489.25	493.75	535 00
	18	495.25	499.75	541.00
	19	501.25	505.75	547 00
	20	507.25	511.75	553.00
	21	513.25	517.75	559.00
	22	519.25	523.75	565.00
	23	525.25	529.75	571.00
	24	531.25	535.75	577.00
	25	537.25	541.75	583.00
	26	543.25	547.75	589.00
	27	549.25	553.75	595.00
	28	555.25	559.75	601.00
BAND 3	29	561.25	565.75	607.00

AIR CHANNEL		PIX FREQ.	SOUND FREQ.	LO FREQ.
	44	651.25	655.75	697
	45	657.25	661.75	703
	46	663.25	667.75	709
	47	669.25	673.75	715
	48	675.25	679.75	721
	49	681.25	685.75	727
	50	687.25	691.75	733
	51	693.25	697.75	739
	52	699.25	703.75	745
	53	705.25	709.75	751
	54	711.25	715.75	757
	55	717.25	721.75	763
	56	723.25	727.75	769
BAND 3	57	729.25	733.75	775

Figure 4-16. Tuner frequencies (continued next page).

30	567.25	571.75	613.00
31	573.25	577.75	619.00
32	579.25	583.75	625.00
33	585.25	589.75	631.00
34	591.25	595.75	637.00
35	597.25	601.75	643.00
36	603.25	607.75	649.00
37	609.25	613.75	655.00
38	615.25	619.75	661.00
39	621.25	625.75	667.00
40	627.25	631.75	673.00
41	633.25	637.75	679.00
42	639.25	643.75	685.00
43	645.25	649.75	691.00

58	735.25	739.75	781
59	741.25	745.75	787
60	747.25	751.75	793
61	753.25	757.75	799
62	759.25	763.75	805
63	765.25	769.75	811
64	771.25	775.75	817
65	777.25	781.75	823
66	783.25	787.75	829
67	789.25	793.75	835
68	795.25	799.75	841
69	801.25	805.75	847

Figure 4-16. Tuner frequencies (continued).

I mentioned service aids. What am I talking about? You will need something like the information I am about to give you in *Figures 4-18* through *4-23*.

(1) *Figure 4-18* is a layout for the components underneath the shield on the top of the circuit board (tuner, top view).

(2) *Figure 4-19* is the layout for components underneath the circuit board (tuner, bottom view.)

(3) *Figure 4-20* is the bandswitching and tuner voltage chart.

(4) *Figure 4-21* is captioned, "Varactor Diode Data Channel Tuning Voltage Chart." It is intended as a guide for typical voltages on the alignment channels. Don't use them for alignment purposes!

(5) *Figures 4-22* and *4-23* are tuner troubleshooting flowcharts, which should be very helpful when you have to tackle tuner problems.

Having said that, I will now try to be specific about the differences the CTC175 and CTC185. I will reference my comments to the block diagram in *Figure 4-17* rather than to a full-blown schematic because signal flow is often easier to follow in a block diagram than a full-size schematic.

RF enters the tuner through the antenna terminal, which is also the isolation block. Remember, the CTC185 is a hot-ground chassis, and therefore requires a barrier between the world of the circuit board and the world of the customer. Those two worlds, for safety's sake, must not meet! You, therefore, have enough sense not to defeat it.

Once inside the tuner, the RF signal is split and sent to the VHF and UHF tuning networks, where the single-tuned filter selects the desired band (1, 2, or 3), and channel frequency. Dual-gate MOSFETs do the job of amplification. The RF signal is input to gate 1 while AGC voltage from the IF stage is input to gate 2. The amplified signal then scoots on to the double-tuned filter, where it is more selectively filtered and matched with respect to impedance for processing by later stages.

Both frequency bands (VHF and UHF) are input to U7301, where the local oscillator and mixer stages heterodyne the necessary frequency with the incoming RF to produce the IF. To refresh your memory, the video signal is at 45.75 MHz and the audio is at 41.25 MHz. The IF frequency is 6 MHz wide.

U7301 (the oscillator/mixer IC) is controlled by U7401, the tuner control IC. This integrated circuit receives its commands from U3101 (the system control microprocessor) via clock and data lines and performs the necessary frequency division to form a frequency synthesizer. Pins 3 and 5 output the tuning voltage that controls the local oscillator. Unlike the CTC175, this tuner control IC incorporates a DAC to control the tuning of the varactor diodes in the single-tuned, and primary and secondary double-tuned filters. These voltages exit the chip at pins 6, 7, and 8. The tuner control IC also performs the necessary bandswitching chore.

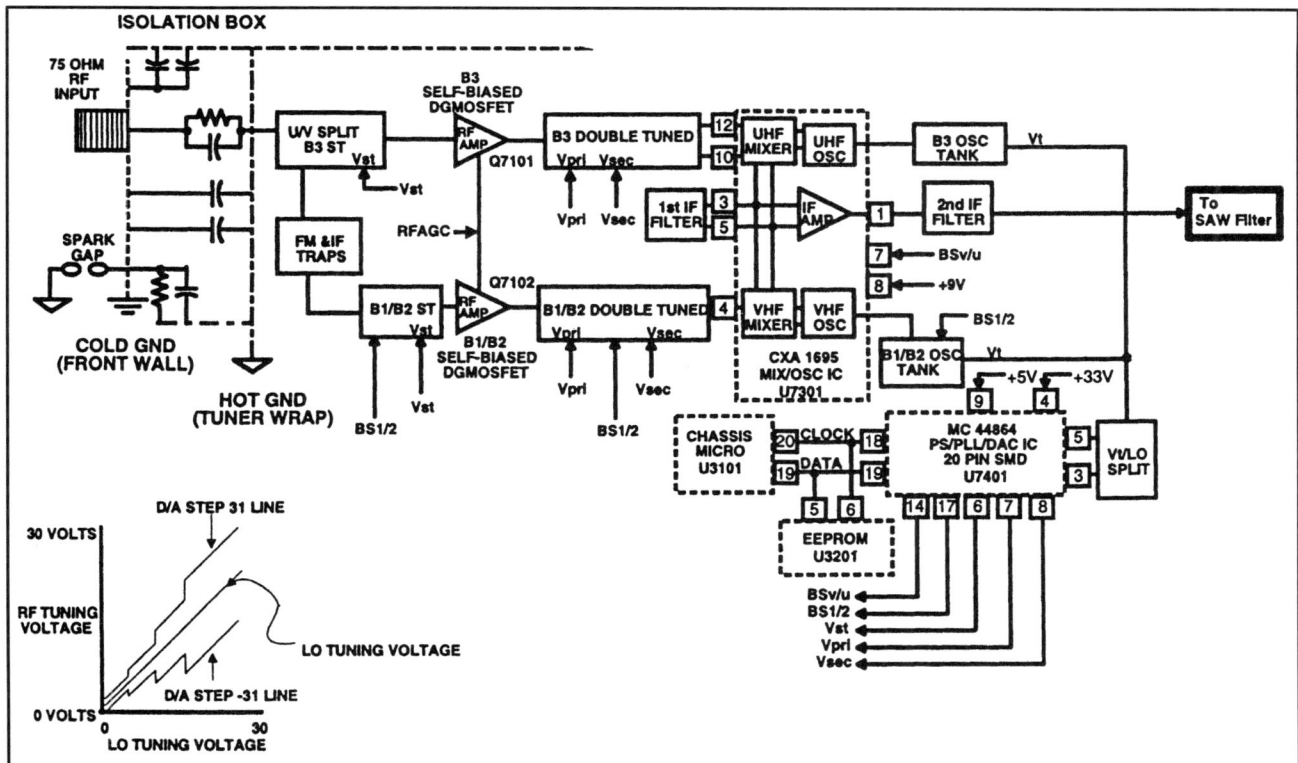

Figure 4-17. Tuner block diagram.

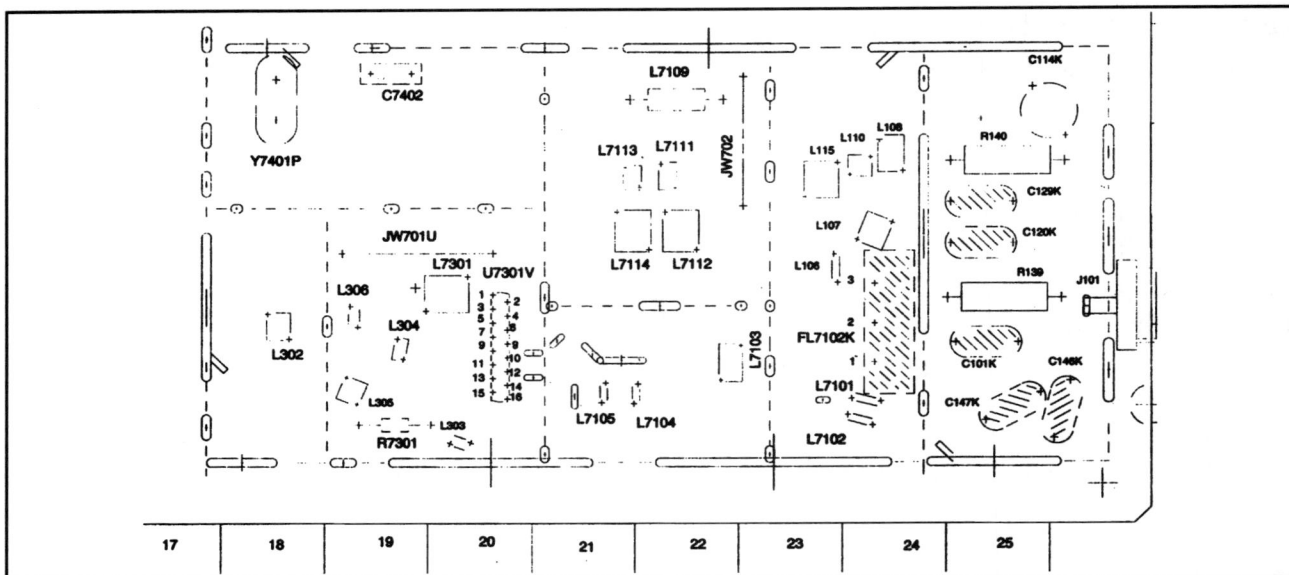

Figure 4-18. Tuner, top view.

BAND SWITCHING AND TUNING VOLTAGE CHART								
CABLE CHANNEL		**U401-14**	**U401-17**	**U401-8**	**U401-7**	**U401-6**	**U401-5**	**U401-3**
BAND 1	2	11.5	12.2	1.8	1.6	1.3	2.3	2.6
	3	11.5	12.2	2.9	2.6	2.5	3.0	2.6
	6	11.5	12.2	7.4	6.8	7.2	5.9	2.6
	98	11.5	12.2	12.4	12.3	10.6	9.9	2.6
	15	11.5	12.2	17.0	17.8	215.7	14.5	2.6
	17	11.5	12.2	23.1	28.3	21.7	20.9	2.6
BAND 2	18	11.3	.1	2.4	2.1	2.4	2.5	2.6
	9	11.3	.1	5.7	5.3	5.7	5.1	2.6
	29	11.3	.1	10.5	9.7	10.6	8.9	2.6
	39	11.3	.1	15.9	15.6	15.9	13.6	2.6
	46	11.3	.1	21.3	23.6	21.8	19.2	2.6
	50	11.3	.1	26.3	33.0	28.7	26.5	2.6
BAND 3	51	.2	.1	.4	.7	.6	.6	2.6
	61	.2	.1	2.6	3.1	2.7	2.9	2.6
	75	.2	.1	6.7	7.1	6.6	7.1	2.6
	101	.2	.1	12.5	12.8	12.6	12.8	2.6
	114	.2	.1	17.1	17.3	17.3	17.3	2.6
	122	.2	.1	21.5	21.5	21.6	21.6	2.6
	125	.2	.1	23.8	23.6	23.8	23.7	2.6

Figure 4-20. Bandswitching and tuner voltage chart.

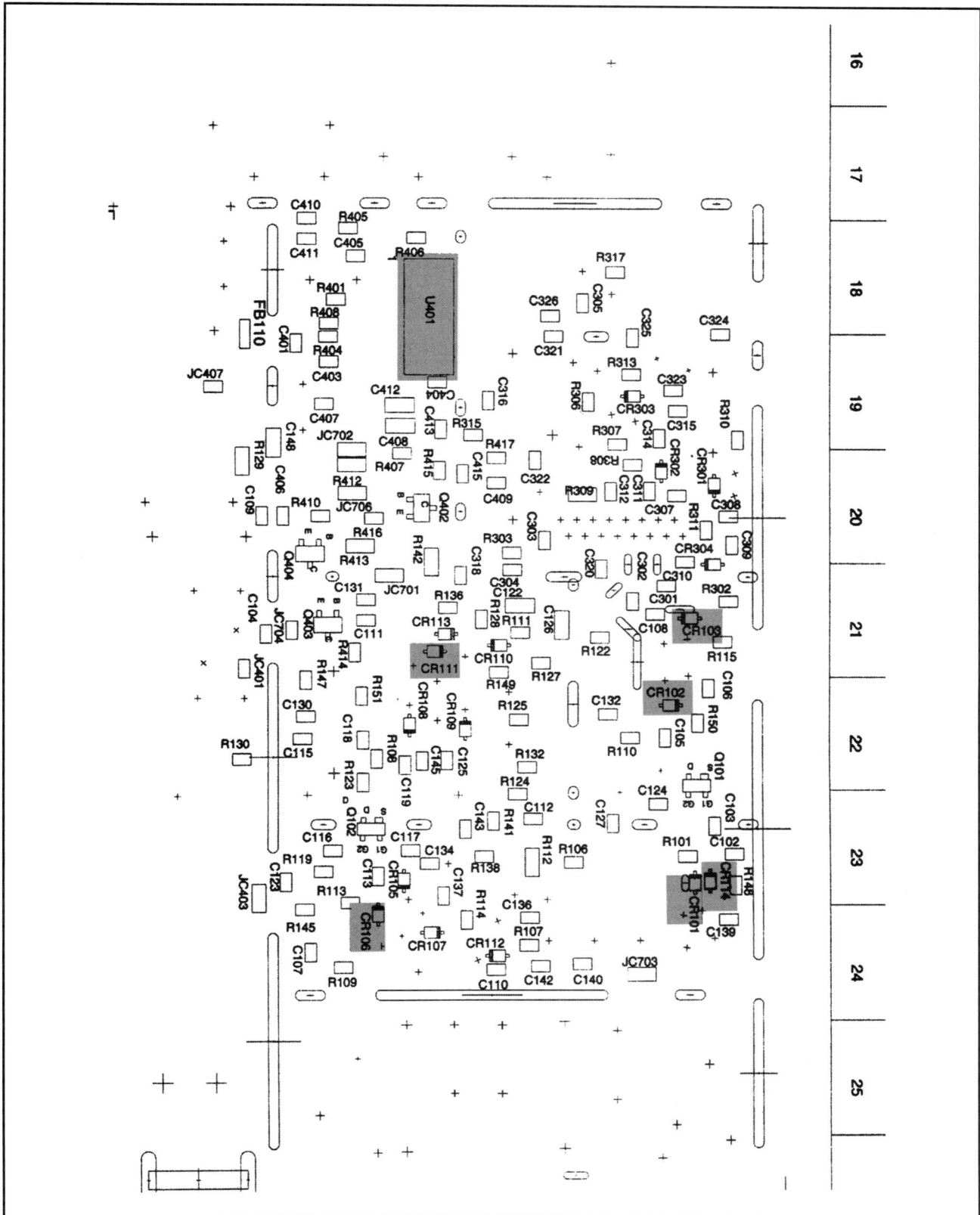

Figure 4-19. Tuner, bottom view.

Varactor Diode Data Channel Tuning Voltage Chart			
CABLE CHANNEL	**CR106 CR7107**	**CR108**	**CR111**
BAND 1 — 2	1.3	1.6	1.7
3	2.4	2.6	2.8
6	7.1	6.8	7.2
98	10.5	12.3	12.2
15	15.6	17.8	16.7
17	21.7	28.2	22.7
BAND 2 — 18	2.3	2.1	2.4
9	5.6	5.3	5.6
29	10.5	9.7	10.3
39	15.8	15.6	15.6
46	21.7	23.6	20.8
50	28.9	32.9	25.8
CABLE CHANNEL	**CR101 CR7114**	**CR102**	**CR103**
BAND 3 — 51	.6	.7	.4
61	2.6	3.0	2.5
75	6.6	6.9	6.5
101	12.4	12.5	12.2
114	17.1	16.9	16.6
122	21.3	21.0	20.9
125	23.4	23.1	23.1

Note: Voltages are aproximate cathode voltages only and will vary from set to set. This chart is supplies as a basic guide for typical voltages on the alignment channels. DO NOT USE THESE VOLTAGES AS A BASIS FOR TUNER ALIGNMENT.

Figure 4-21. Varactor diode data channel tuning voltage chart.

TUNER TROUBLESHOOTING FLOWCHART

1. Connect a service generator to the antenna input (no attenuation).

2. Check operation of all bands. If only 1 or 2 bands are not functioning limit troubleshooting checks to those bands.

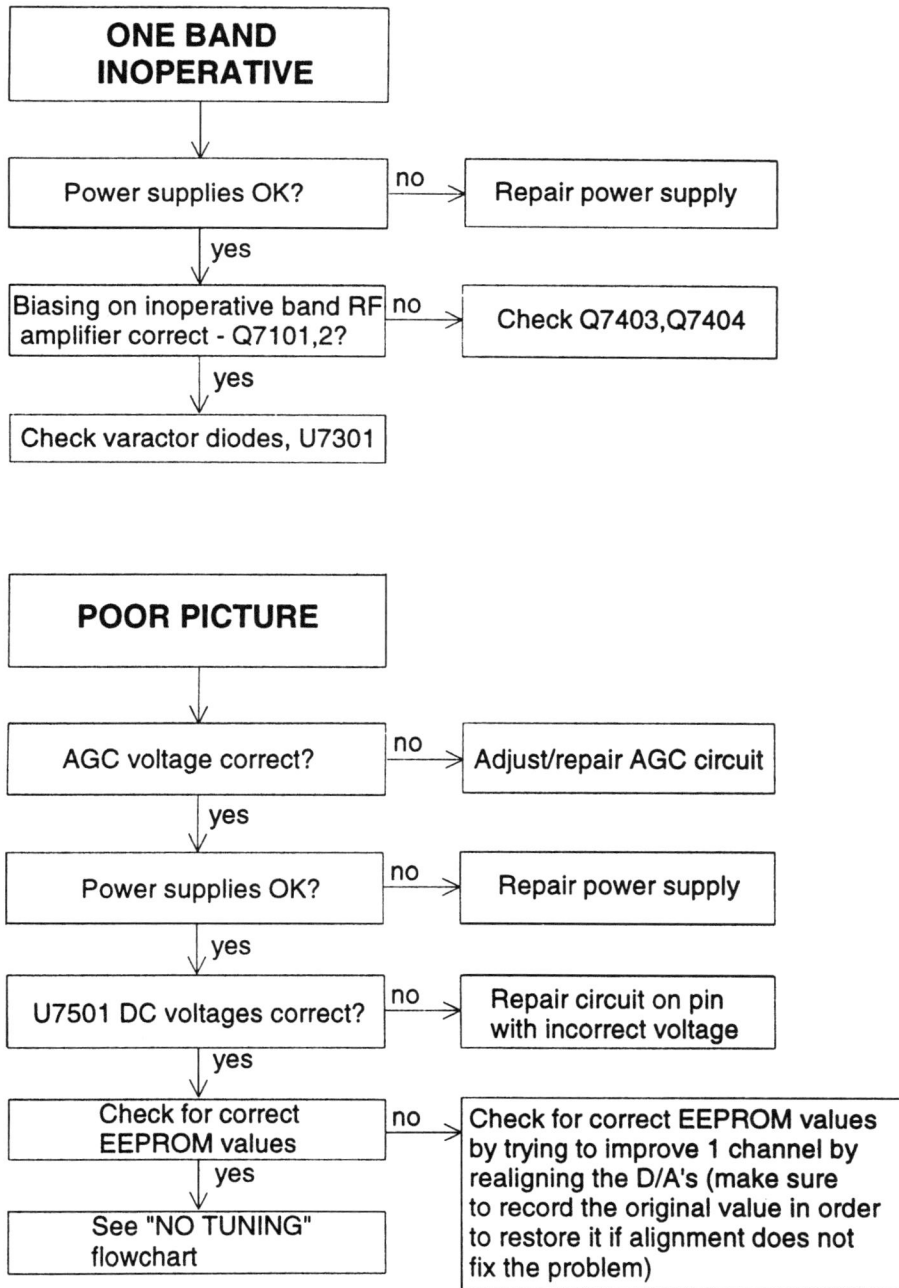

```
┌─────────────────────┐
│   ONE BAND          │
│   INOPERATIVE       │
└─────────────────────┘
          │
          ▼
┌─────────────────────┐  no   ┌─────────────────────┐
│ Power supplies OK?  │─────▶ │ Repair power supply │
└─────────────────────┘       └─────────────────────┘
          │ yes
          ▼
┌─────────────────────────┐  no  ┌─────────────────────┐
│ Biasing on inoperative  │────▶ │ Check Q7403,Q7404   │
│ band RF amplifier       │      └─────────────────────┘
│ correct - Q7101,2?      │
└─────────────────────────┘
          │ yes
          ▼
┌────────────────────────────┐
│ Check varactor diodes,     │
│ U7301                      │
└────────────────────────────┘
```

```
┌─────────────────────┐
│   POOR PICTURE      │
└─────────────────────┘
          │
          ▼
┌─────────────────────┐  no   ┌────────────────────────┐
│ AGC voltage correct?│─────▶ │ Adjust/repair AGC circuit│
└─────────────────────┘       └────────────────────────┘
          │ yes
          ▼
┌─────────────────────┐  no   ┌─────────────────────┐
│ Power supplies OK?  │─────▶ │ Repair power supply │
└─────────────────────┘       └─────────────────────┘
          │ yes
          ▼
┌───────────────────────────┐  no  ┌──────────────────────────┐
│ U7501 DC voltages correct?│────▶ │ Repair circuit on pin     │
└───────────────────────────┘      │ with incorrect voltage   │
          │ yes                     └──────────────────────────┘
          ▼
┌───────────────────────────┐  no  ┌────────────────────────────────┐
│ Check for correct         │────▶ │ Check for correct EEPROM values│
│ EEPROM values             │      │ by trying to improve 1 channel │
└───────────────────────────┘      │ by realigning the D/A's (make  │
          │ yes                     │ sure to record the original    │
          ▼                         │ value in order to restore it   │
┌───────────────────────────┐      │ if alignment does not fix the  │
│ See "NO TUNING"           │      │ problem)                       │
│ flowchart                 │      └────────────────────────────────┘
└───────────────────────────┘
```

Figure 4-22. Tuner troubleshooting.

TUNER TROUBLESHOOTING FLOWCHART

```
              ┌─────────────────────────┐
              │       NO TUNING         │
              └─────────────────────────┘
                           │
                           ▼
              ┌─────────────────────┐   no   ┌──────────────────────────┐
              │ OSD channel         │───────▶│ Repair system control    │
              │ numbers change?     │        └──────────────────────────┘
              └─────────────────────┘
                      │ yes
                      ▼
              ┌─────────────────────┐   no   ┌──────────────────────────┐
              │ Power supplies OK?  │───────▶│ Repair power supply      │
              └─────────────────────┘        └──────────────────────────┘
                      │ yes
                      ▼
              ┌──────────────────────────────┐ no  ┌──────────────────────────────┐
              │ Proper band switching voltages│────▶│ Repair band switching         │
              │ on U7301-pin 7,Q7403,Q7404    │     │ circuit,check for data/clock  │
              └──────────────────────────────┘     │ activity at U7401 - 18,19     │
                      │ yes                          └──────────────────────────────┘
                      ▼
              ┌──────────────────────────────┐ no  ┌──────────────────────────────┐
              │ Does tuning voltage (VT)      │────▶│ Q7402, U7401                 │
              │ change with channel change    │     └──────────────────────────────┘
              └──────────────────────────────┘
                      │ yes
                      ▼
              ┌──────────────────────────────┐ no  ┌──────────────────────────────┐
              │ Does RF AGC voltage rise if   │────▶│ Repair RF AGC circuit,        │
              │ input signal to TV is attenuated│   │ A7102,R7119,R7130,            │
              └──────────────────────────────┘     │ R7106,R7109                   │
                      │ yes                          └──────────────────────────────┘
                      ▼
              ┌──────────────────────────────┐ no  ┌──────────────────────────────┐
              │ Check for correct             │────▶│ Q7404,bias components         │
              │ Q7101,Q7102 bias              │     └──────────────────────────────┘
              └──────────────────────────────┘
                      │ yes
                      ▼
              ┌──────────────────────────────┐
              │ Check IF and associated       │
              │ components                    │
              └──────────────────────────────┘
```

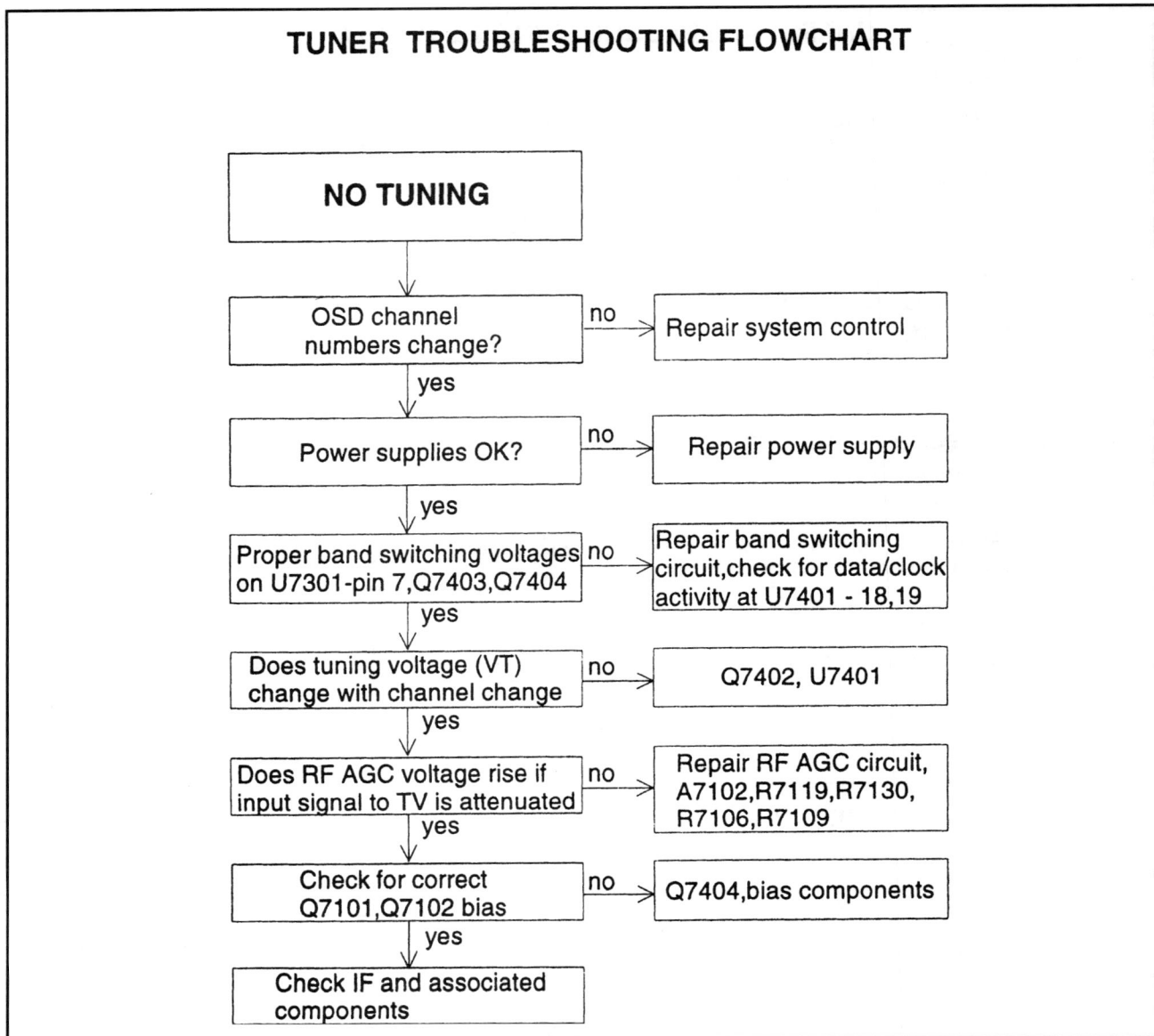

Figure 4-23. Tuner troubleshooting.

The EEPROM stores alignment values for the eighteen "data channels" (See *Figure 4-3* for a list of the data channels.). These values provide the settings for the three tuned filters. Like the CTC175 setup, the CTC185 utilizes "linear interpolation" to adjust the tuning voltages across the range of channels the tuner can receive. On the chance you might need it, I am including the alignment data for the tuner in *Figure 4-24*. The procedure is a lot like the one I discussed in the preceding chapter, except that you connect your meter to pin 11 of U1001. You can take it from there.

A word of caution before I exit the tuner discussion: Don't dare touch the coils found on the top of the circuit board. These coils have been deformed for a reason. If you try to reshape them, you are going to have to realign them, and folks, that is a chore in and of itself! So, leave them alone.

Electronic RF Alignment

Test Point:	U1001 PIN 11 (IF AGC)	Main PCB
Adjust:	Parameter #'s 1 thru 54	Range: 0 - 62

1. Connect the TAG001 Service Generator as described in the TAG001 user's manual.

2. Connect the DC voltmeter to U1001 pin 11 (IF AGC).

3. Set the service generator for Channel 2 output. Tune the instrument to receive Channel 2.

4. Adjust *Channel 2 Secondary* (parameter #1) for minimum DC voltage.

5. Adjust *Channel 2 Primary* (parameter #2) for minimum DC voltage.

6. Adjust *Channel 2 Single-Tuned* (parameter #3) for minimum DC voltage.

7. Repeat steps 4 through 6. The adjustments must be repeated to assure correct alignment.

8. Change the service generator and the instrument to Channel 6. Adjust *CH6 Sec/Pri/Sgl* (parameter #'s 4, 5 and 6) for minimum DC voltage, then repeat the adjustment.

9. Change the service generator and the instrument to Channel 98. Adjust *CH98 Sec/Pri/Sgl* (parameter #'s 7, 8 and 9) for minimum DC voltage, then repeat the adjustment.

10. Change the service generator and the instrument to Channel 15. Adjust *CH15 Sec/Pri/Sgl* (parameter #'s 10, 11 and 12) for minimum DC voltage, then repeat the adjustment.

11. Change the service generator and the instrument to Channel 17. Adjust *CH17 Sec/Pri/Sgl* (parameter #'s 13, 14 and 15) for minimum DC voltage, then repeat the adjustment.

12. Change the service generator and the instrument to Channel 18. Adjust *CH18 Sec/Pri/Sgl* (parameter #'s 16, 17 and 18) for minimum DC voltage, then repeat the adjustment.

13. Change the service generator and the instrument to Channel 9. Adjust *CH9 Sec/Pri/Sgl* (parameter #'s 19, 20 and 21) for minimum DC voltage, then repeat the adjustment.

14. Change the service generator and the instrument to Channel 29. Adjust *CH29 Sec/Pri/Sgl* (parameter #'s 22, 23 and 24) for minimum DC voltage, then repeat the adjustment.

15. Change the service generator and the instrument to Channel 39. Adjust *CH39 Sec/Pri/Sgl* (parameter #'s 25, 26 and 27) for minimum DC voltage, then repeat the adjustment.

16. Change the service generator and the instrument to Channel 46. Adjust *CH46 Sec/Pri/Sgl* (parameter #'s 28, 29 and 30) for minimum DC voltage, then repeat the adjustment.

17. Change the service generator and the instrument to Channel 50. Adjust *CH50 Sec/Pri/Sgl* (parameter #'s 31, 32 and 33) for minimum DC voltage, then repeat the adjustment.

18. Change the service generator and the instrument to Channel 51. Adjust *CH51 Sec/Pri/Sgl* (parameter #'s 34, 35 and 36) for minimum DC voltage, then repeat the adjustment.

19. Change the service generator and the instrument to Channel 61. Adjust *CH61 Sec/Pri/Sgl* (parameter #'s 37, 38 and 39) for minimum DC voltage, then repeat the adjustment.

20. Change the service generator and the instrument to Channel 75. Adjust *CH75 Sec/Pri/Sgl* (parameter #'s 40, 41 and 42) for minimum DC voltage, then repeat the adjustment.

21. Change the service generator and the instrument to Channel 101. Adjust *CH101 Sec/Pri/Sgl* (parameter #'s 43, 44 and 45) for minimum DC voltage, then repeat the adjustment.

22. Change the service generator and the instrument to Channel 114. Adjust *CH114 Sec/Pri/Sgl* (parameter #'s 46, 47 and 48) for minimum DC voltage, then repeat the adjustment.

23. Change the service generator and the instrument to Channel 122. Adjust *CH122 Sec/Pri/Sgl* (parameter #'s 49, 50 and 51) for minimum DC voltage, then repeat the adjustment.

24. Change the service generator and the instrument to Channel 125. Adjust *CH125 Sec/Pri/Sgl* (parameter #'s 52, 53 and 54) for minimum DC voltage, then repeat the adjustment.

Figure 4-24. Tuner alignment data.

System Control

There is a significant number of similarities between the CTC185 and the chassis discussed in the last chapter; therefore, I shall confine my comments, more or less, to the differences between the two systems. I will make some concessions, but not many.

System control consists of the microprocessor (U3101), the T4-Chip (U1001), the EEPROM (U3201), and the tuner PLL integrated circuit (U7401) in the configuration illustrated by *Figure 4-25*. The chips communicate over the "I-squared-C" (the inter-integrated) bus, a two-wire arrangement by which serial data and serial clock signals are transmitted and received. You will be glad to know Thomson has gotten away from the "three-bus protocol" used in the CTC175 family. One bus now serves all.

When the brains of the outfit (U3101) resets, it downloads initial configuration data from U3101. If you scope the clock and data lines, pins 5 and 6 of the EEPROM, you will see a flurry of activity as the chassis initializes. Once the EEPROM has shaken hands and downloaded data, clock and serial lines go low. If the EEPROM in the CTC175 chassis did not respond, clock and data lines would not go low, a diagnostic bit of information that pointed to a corrupted chip. The CTC185 chassis uses a different arrangement. Take a close look at *Figure 4-25*, paying attention to Q3201. This transistor is a switch which the microprocessor uses to turn Vcc to the EEPROM on and off. If the EEPROM fails to respond to initialization commands, the microprocessor proceeds to turn Q3101 off for about fifty milliseconds and then back on in an effort to reset it. If the microprocessor does not receive an acknowledgment, it will switch the power on-off and on-off in a effort to get the information.

When it receives an ON command from the remote control or front-panel control, the microprocessor initiates a flurry of activity. Like the CTC177/187 chassis, it checks on certain bits of information in U1001. The status registers report the status of power-on-reset, x-ray protection fault, horizontal lock detector, and automatic fine tuning. Power-on-reset (POR) refers to a circuit which detects when the standby voltage has dropped below about 6 volts and shuts the IC off by stopping both the PWM and horizontal outputs. X-ray protection (XRP) informs the microprocessor when an XRP or POR condition has occurred. When the XRP input goes above the reference value, the TV will be shut down. The horizontal lock detector compares the position of the flyback pulse with the sync of the video signal. It can be used to detect the presence of an active channel, but it is not used for tuning.

In addition to reading the status register, the microprocessor constantly updates the registers inside U1001 about once every second with data it pulls out of the EEPROM. The constant data update prevents electrical disturbances such as CRT arcs from corrupting the information inside U1001. The updating does not take place if the TV is in the service mode.

There are two other microprocessor inputs I want to comment on before I discuss troubleshooting procedures:

The first is the "run sense" input at pin 13. The microprocessor uses pin 13 to monitor the presence of the run 12-volt supply which, as you recall, is scan-derived. If the 12 volts is not present, U3101 will

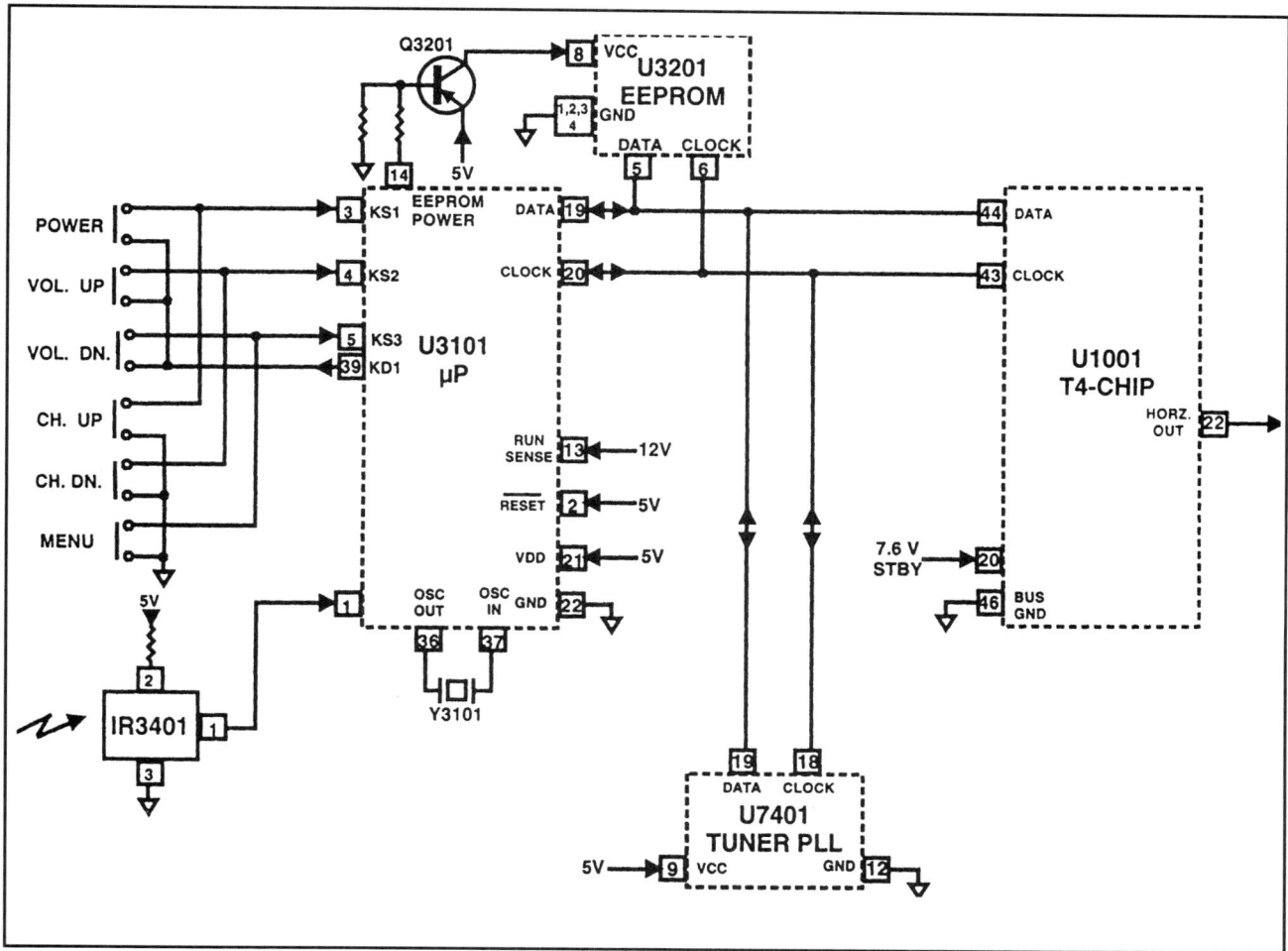

Figure 4-25. System control.

place the TV in the off mode and then try to restart it. If the TV does not start after three off-on attempts in one minute, the microprocessor will place the television in the off mode. You must then issue an ON command to start the process again.

The second is the reset circuit, which is like the reset circuit used in the CTC175 series. I covered the circuit in some detail in the last chapter and just want to make a passing comment on it now. Refer to *Figure 4-26* as you read. This circuit does not normally cause trouble, but it can. I mentioned in the last chapter at least two failures caused by a defective Q3102. Just last week, I faced a different sort of problem which the reset circuit generated. The TV would come on and play fine for about ten minutes and would then power down. If I tried to turn it back on, the TV would not respond. I had to remove AC for several minutes and reapply it to get the microprocessor to respond. The TV would then come on and go right back off. All voltages checked good, except the reset voltage. After a little bit of head-scratching, I found out that the reset voltage started at 5 volts but gradually dropped to about 1.41 volts. The culprit turned out to be C3113, a .01 μF surface-mount capacitor! When the chassis warmed up, C3113 developed a leak which shunted the reset voltage to ground.

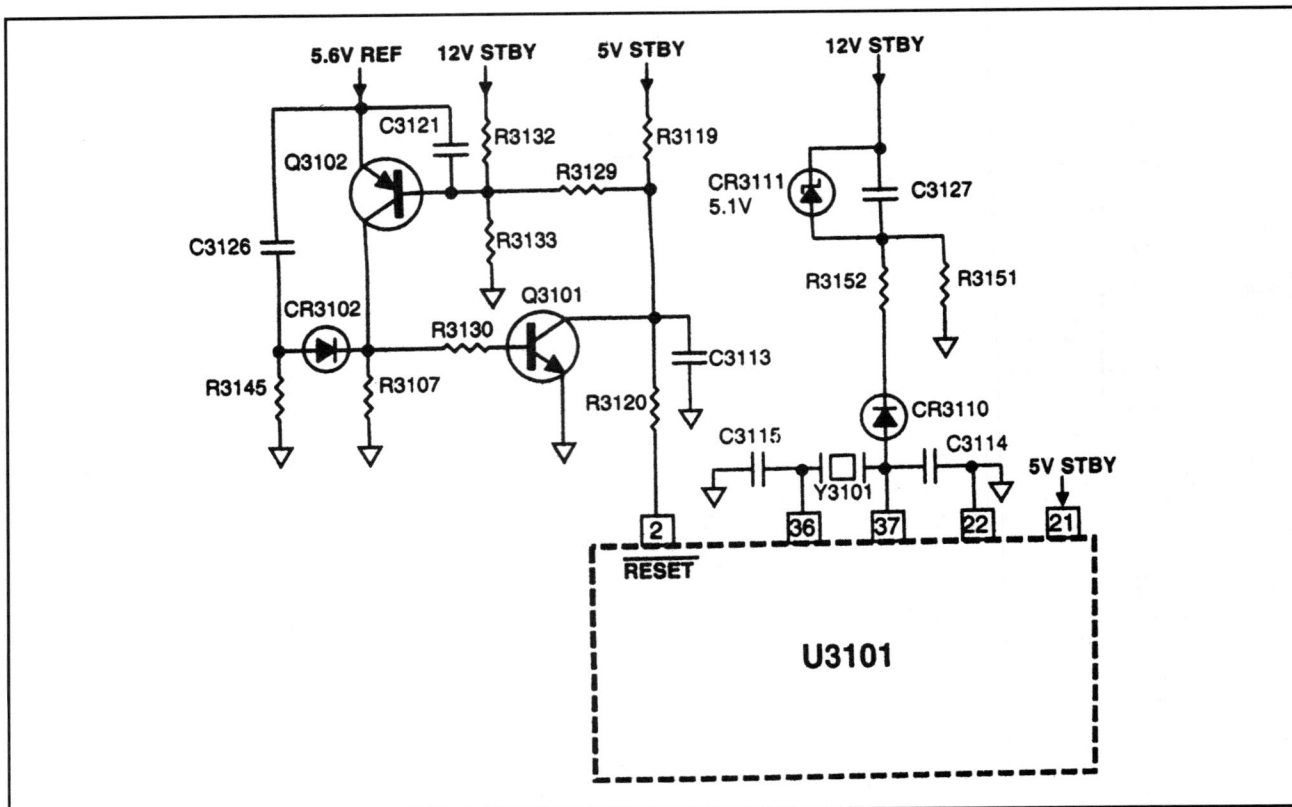

Figure 4-26. Reset circuit.

Troubleshooting System Control Problems

As you know, the system control circuit controls every function of the TV. Any failure in this circuit will cause the entire TV to malfunction. Because the four ICs that make up system control are so interrelated, the troubleshooting procedure will also have some overlapping. I will use *Figure 4-27* as foundation for the discussion, and I am including *Figure 4-28*, a system control troubleshooting flowchart, from the factory service manual. The technical training booklet suggest a slightly different approach, which I will reproduce in modified format.

(1) Make sure the standby supplies are up and working. Use an external +26 volts to support the standby supply if necessary. I have already described how to implement the procedure.

(2) Check for horizontal drive out of pin 22 of U1001 when the power button is depressed. If the pulses are present even momentarily, system control is working and the problem is in (a) the deflection circuit or (b) the 12-volt run supply.

If the pulses don't appear, check the 7.6-volt standby voltage at pin 20 of U1001. If you need to, place a +26-volt supply at the cathode of CR4703 to provide the voltage. If the supply is not present at pin 20, wick out the pin to see if the supply is present at the pad. If it is, U1001 will need to be replaced. If it isn't, trace the supply back to its source.

SYSTEM CONTROL SCHEMATIC

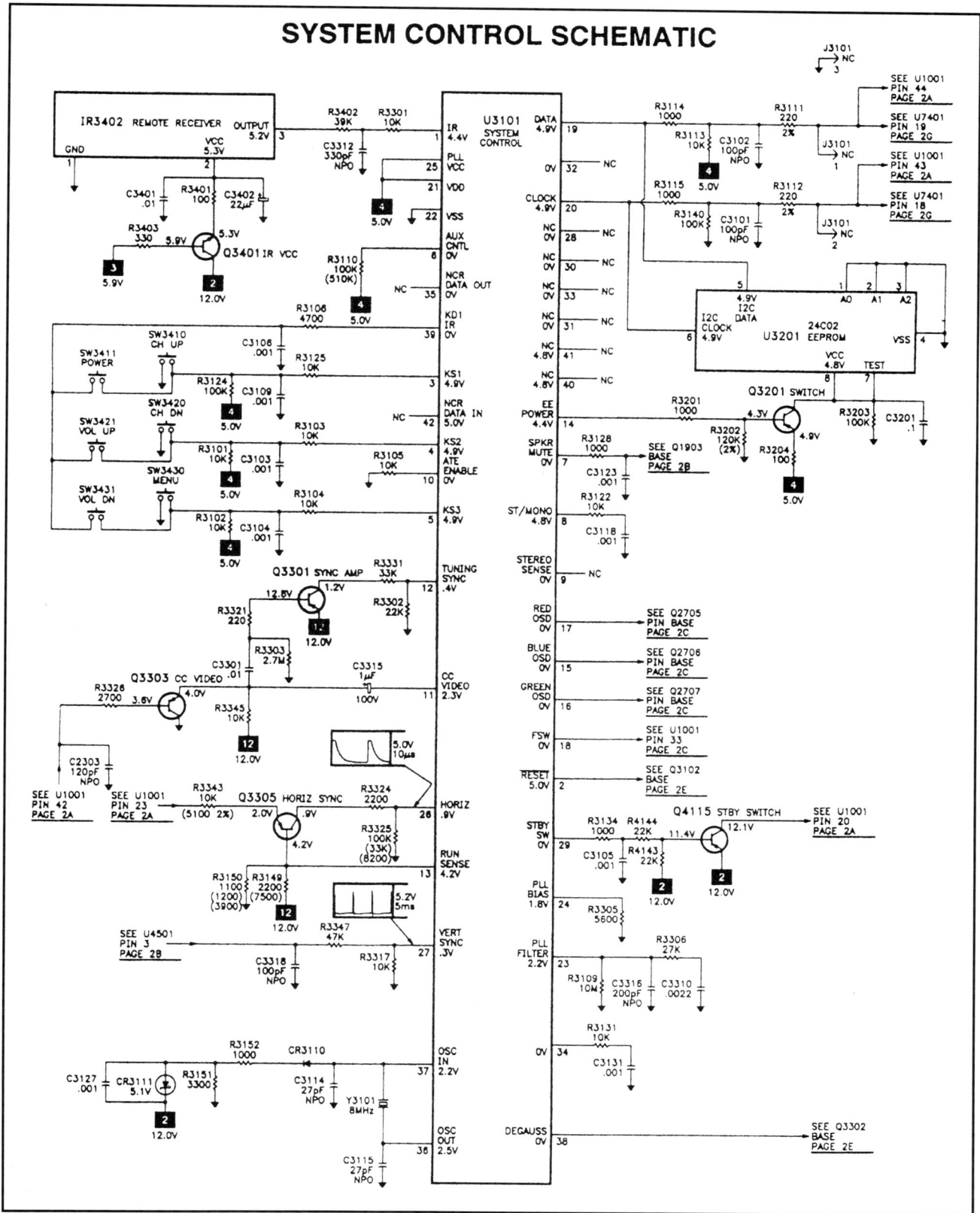

Figure 4-27. System control schematic.

| SET WILL NOT TURN ON | DEGAUSS RELAY REPEATEDLY SWITCHING ON AND OFF | yes → U7401 |

U3101 pin 21 = 5V → no → Repair power supply

↓ yes

U3201 pin 8 = 5V → no → Repair power supply, Q3201

↓ yes

U3101 pin 2 = 5V → no → Q3101, reset circuit

↓ yes

8MHz on pin 36 of U3101 → no → CR3110, Y3101

↓ yes

Monitor U3101 pin 3, does voltage go from 5V to 0V when the channel up button is pressed → no → Keypad button

↓ yes

Repeat for pin 4 (channel down button) and pin 5 (menu button) → no → Keypad button

↓ yes

Monitor U3101 pin 39, does a 20msec high pulse followed by a 20msec low pulse occur when the volume up/down or power button is pressed → no → Keypad button, U3101

↓ yes

0 to 5V signal present on pin 1 of U3101 when pressing a button on the remote control → no → IR 3401 or power supply

↓ yes

U3101 pins 19 & 20 = 5V with set plugged in and power off. After power button is pressed pin 20 should have a 31 kHz (5V) signal present. Pin 19 should have data activity present → no → U3101

↓ yes

U3101 pin 13 > 3V? → no → U1001, 12V run supply

↓ yes

System control OK

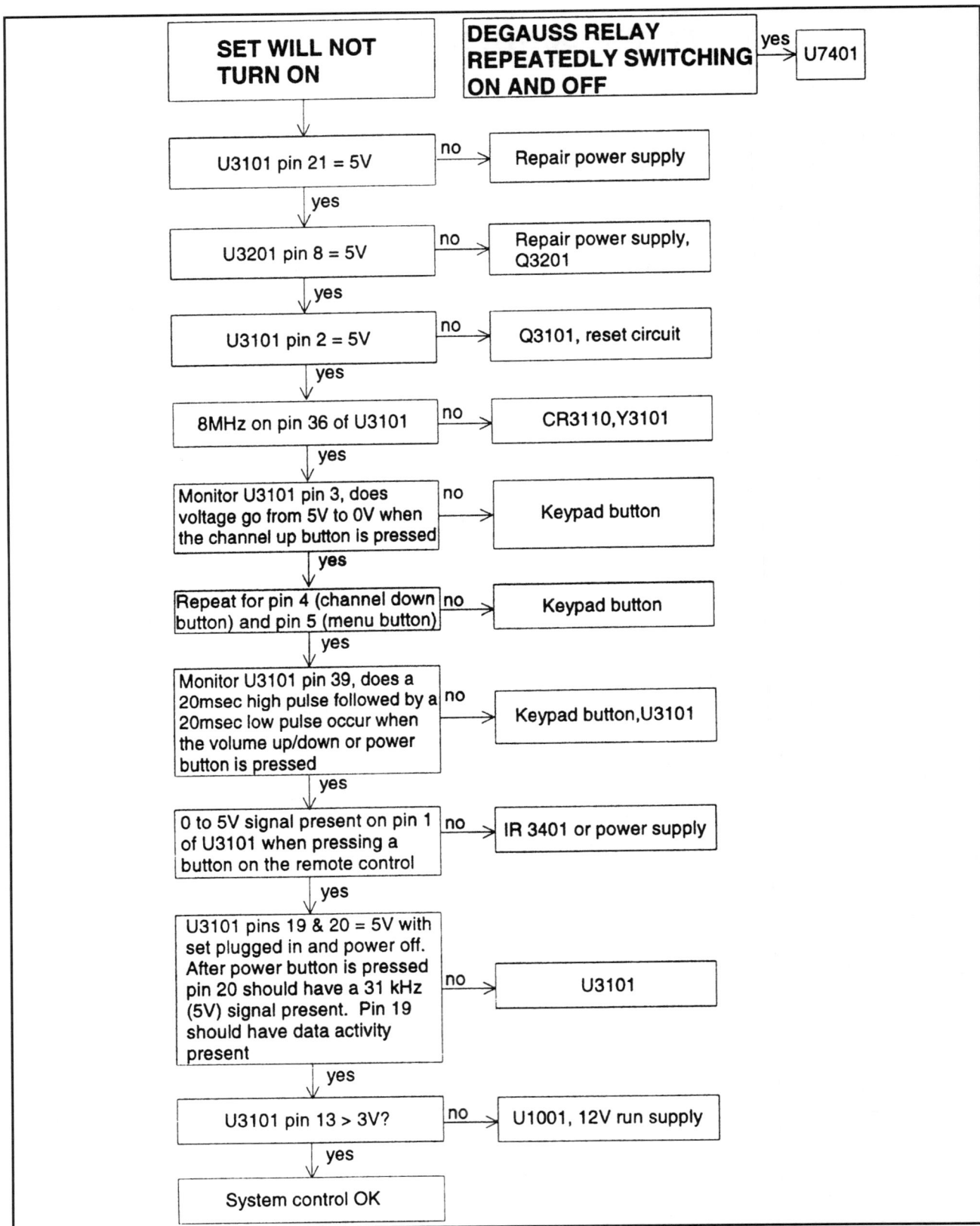

Figure 4-28. System control troubleshooting flowchart.

(3) Check for Vcc at pin 21 of U3101.

(4) Check reset voltage at pin 2 of U3101.

(5) Check for the oscillator signal at pins 36 and 37. The signal should be at the 5 Vp-p level and at a frequency of 8 MHz.

(6) Monitor pins 19 and 20. These pins should have no activity while the TV is in the standby mode and should go to 5 Vp-p when a power ON command is executed. If pulses don't appear when the power button is pressed, unsolder the pins and see if activity resumes. If no activity is seen on pins 19 and 20 after they have been unsoldered, U3101 is probably defective.

If activity appears on the pins when they have been unsoldered (wicked out), something is pulling the lines low. With the lines disconnected, U3101 will output a constant stream of data as it searches for U3101. This condition is normal and indicates the microprocessor is working. You will have to find out what is pulling them low. The best procedure is to resolder pins 19 and 20 and begin by disconnecting clock and data lines at each IC on the bus until you find out which is causing the trouble.

There is one bit of information you really should underscore. If the tuner control IC (U7401) fails to acknowledge, the microprocessor will turn the set off and then back on to try to clear the fault on the bus. The off-on cycle will continue indefinitely, causing the degauss relay to "click" about twice a second. The literature suggests that the tech first check for 5 volts at pin 9 of U7401. If it is present, suspect a defective chip. If the 5 volts isn't present, troubleshoot the 5-volt run supply.

Horizontal and Vertical Deflection Circuits

As you can see by examining *Figure 4-29*, the horizontal drive circuit is quite similar to its cousin used in the CTC175 family. The various functions are performed by U1001 and are controlled by the I-Squared-C bus. These functions include AFC, APC, horizontal drive, east-west pincushion correction, x-ray protection and horizontal Vcc standby regulation. The difference, however, is the addition of Q4115, a switch employed by the microprocessor to turn on and off B+ to pin 20 of U1001. Since I have already discussed its function in some detail, I will merely note its presence here.

The horizontal drive and output circuits are also about the same, which makes them quite conventional in design. I will therefore refer you to the previous chapter if you want a review of how the circuits work.

Like the CTC177, the CTC185 uses two horizontal output configurations. *Figure 4-30* illustrates the circuit used in 19" and 20" televisions, and *Figure 4-31* illustrates the circuit used in the 25" and larger sets. The difference, of course, will be the addition of a diode-modulated pincushion correction circuit.

If you need to replace the horizontal output transistor, remember to use the correct part for the chassis on which you are working. The CTC185A, B, and M use part number 233182; the CTC185 AA and AB use 231532. If the horizontal output transistor has shorted, be sure to check the regulator (Q4114) for damage because a shorted HOT will almost always damage the regulator.

I should also point out an error in the service data which gave an incorrect part number for the flyback in the 06-28-96 revision. Even though the error has been corrected, if you are not using the current revision you might get the wrong part and install it. If you do, the symptom will be a TV that either shuts down or won't start after you have installed the new flyback. The correct part numbers are:

CTC185AA, AA2, AB (25" and 27")	**231449**
CTC185M (20")	**231450**
CTC185A, B (19")	**231530**

Troubleshooting the horizontal circuit is pretty much straightforward. Since I have already covered the procedure in some depth, for example *Figure 4-11*, I refer you to what I have already written.

Now just a very few words about the vertical deflection circuit:

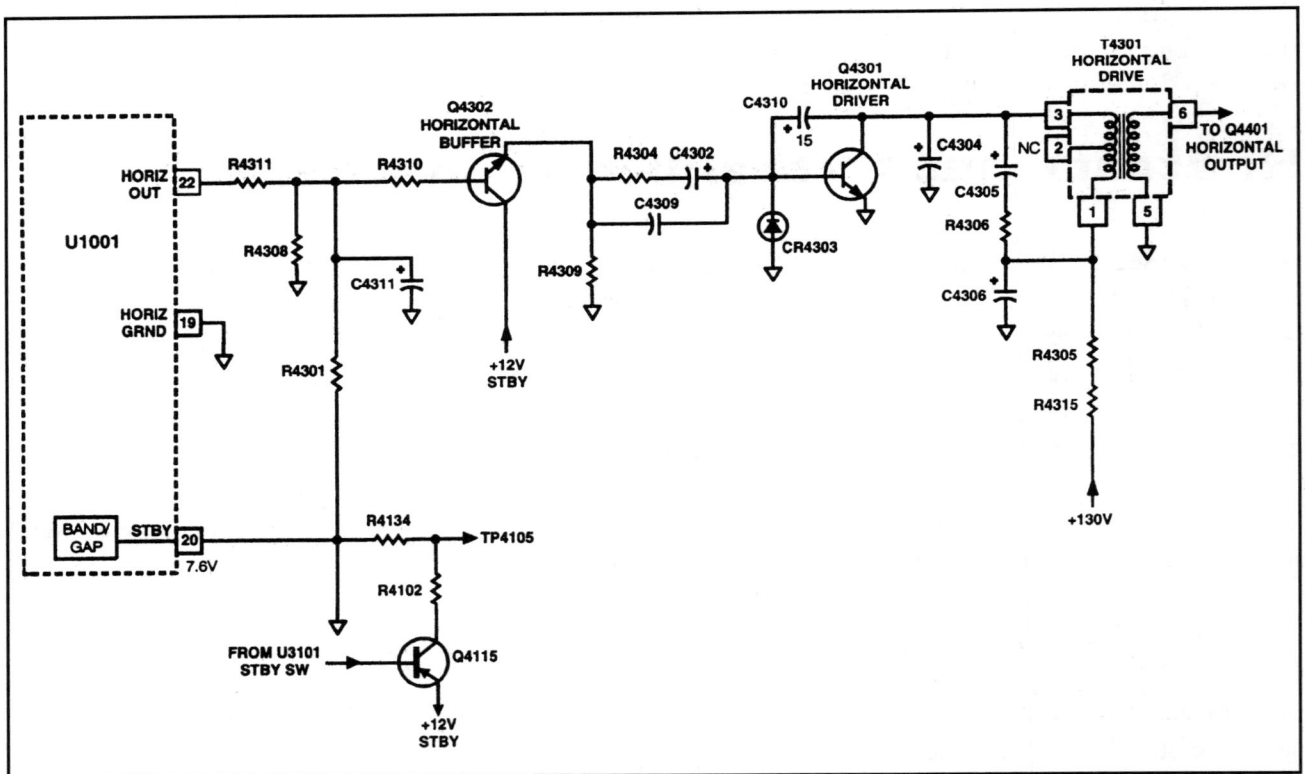

Figure 4-29. Horizontal drive.

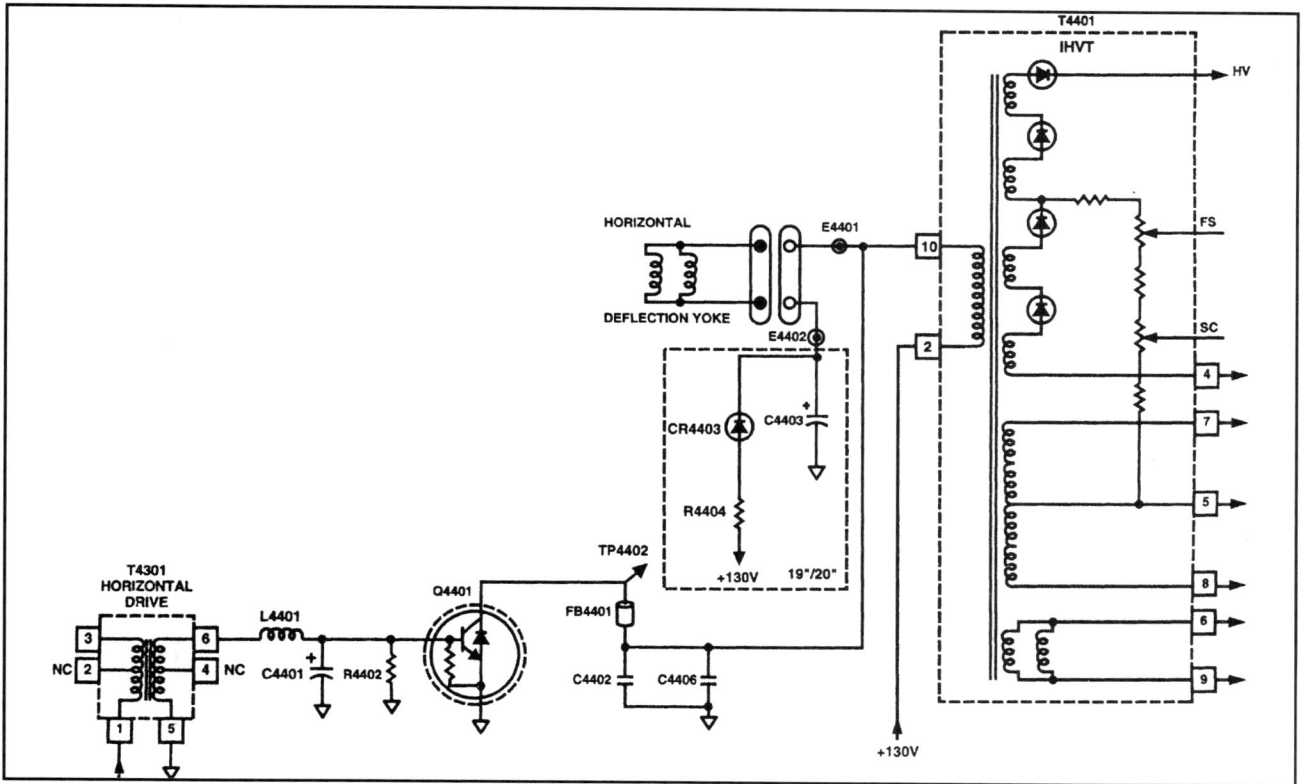

Figure 4-30. 19" and 20" circuits.

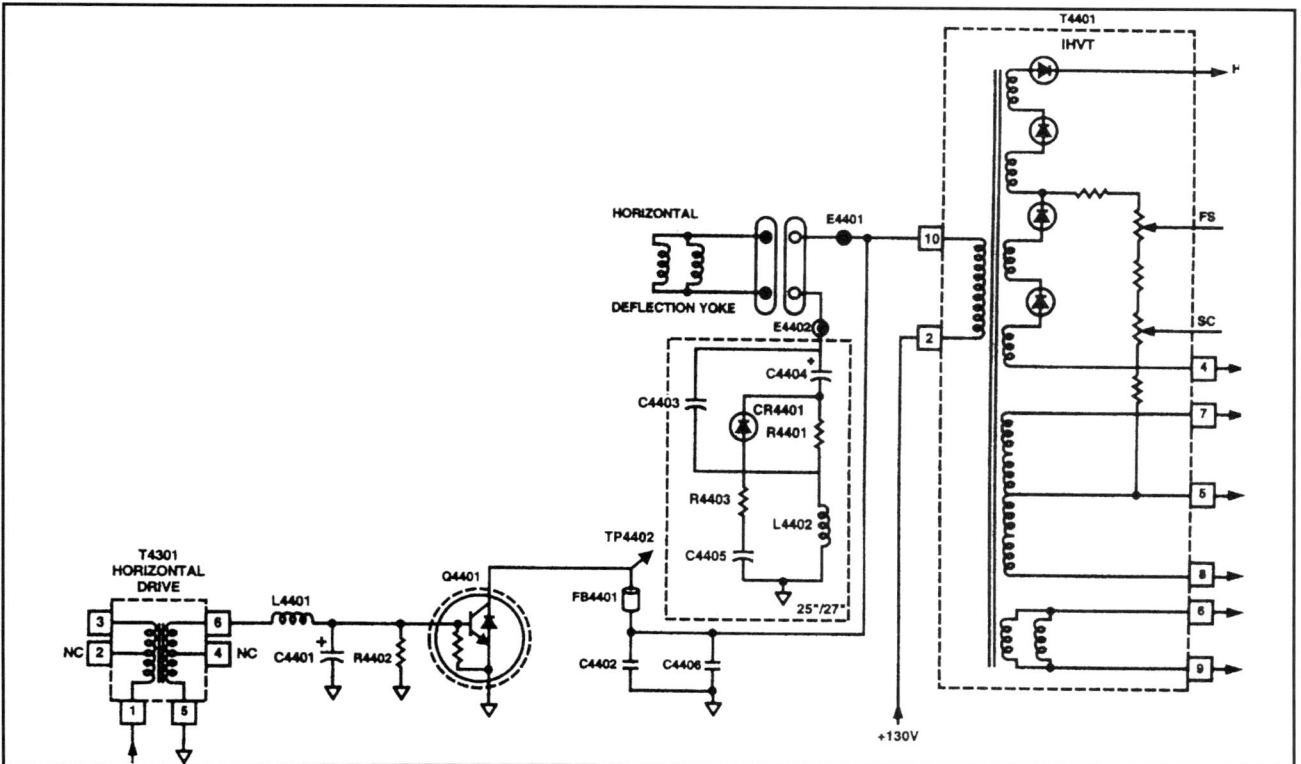

Figure 4-31. 25" and larger circuits.

A glance at *Figure 4-32* shows you the circuit is about the same as the one used in the CTC175 family. I refer you to the last chapter for a discussion of how the circuit works and how to troubleshoot it when the need arises.

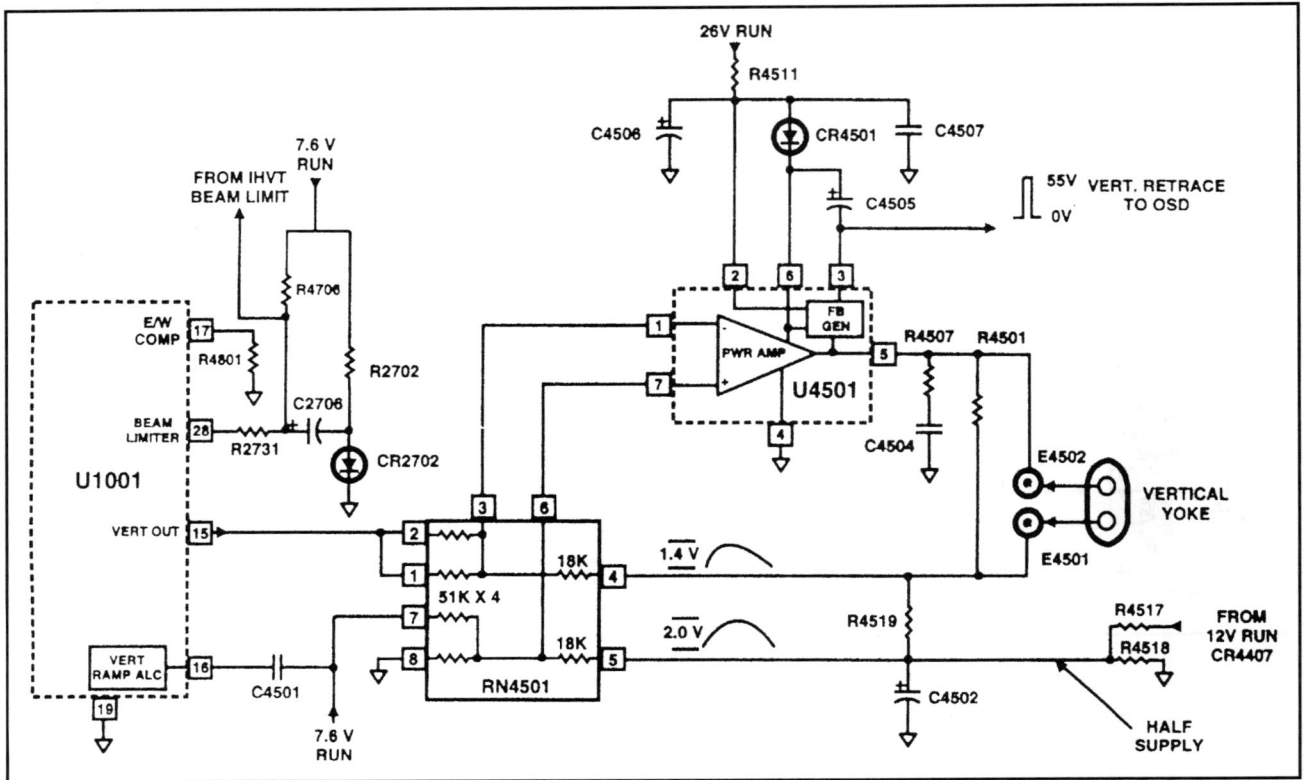

Figure 4-32. Vertical deflection circuit.

The T-4 Chip: Video, Chroma and Audio Processing

As the literature points out, most of the picture and sound IF circuits are essentially the same as those in the CTC177/87 family of chassis. But there are some new features, specifically analog AFT with status register, and auto-tuned quadrature FM detector. Since I covered the circuits in some detail in the last chapter, I will confine myself to a few passing comments and give you plenty of illustrations to keep you from having to leaf back to the previous chapter. Since U1001 has been redesigned and designated the "T4-Chip," I am including in *Figure 4-33* an overall block diagram of it which should be a handy troubleshooting reference.

For the present, I will confine my comments to *Figure 4-34*, a block diagram of the video and audio intermediate frequency stages within U1001. Because it is pretty self-explanatory, I will keep my remarks relatively brief. Tuner video enters U1001 through the SAW filter at pins 9 and 10, where it is amplified by an amplifier whose gain is controlled by the AGC loop. It then goes to the detector. The detector (video and sound) uses a voltage-controlled oscillator which is phase-locked to the picture

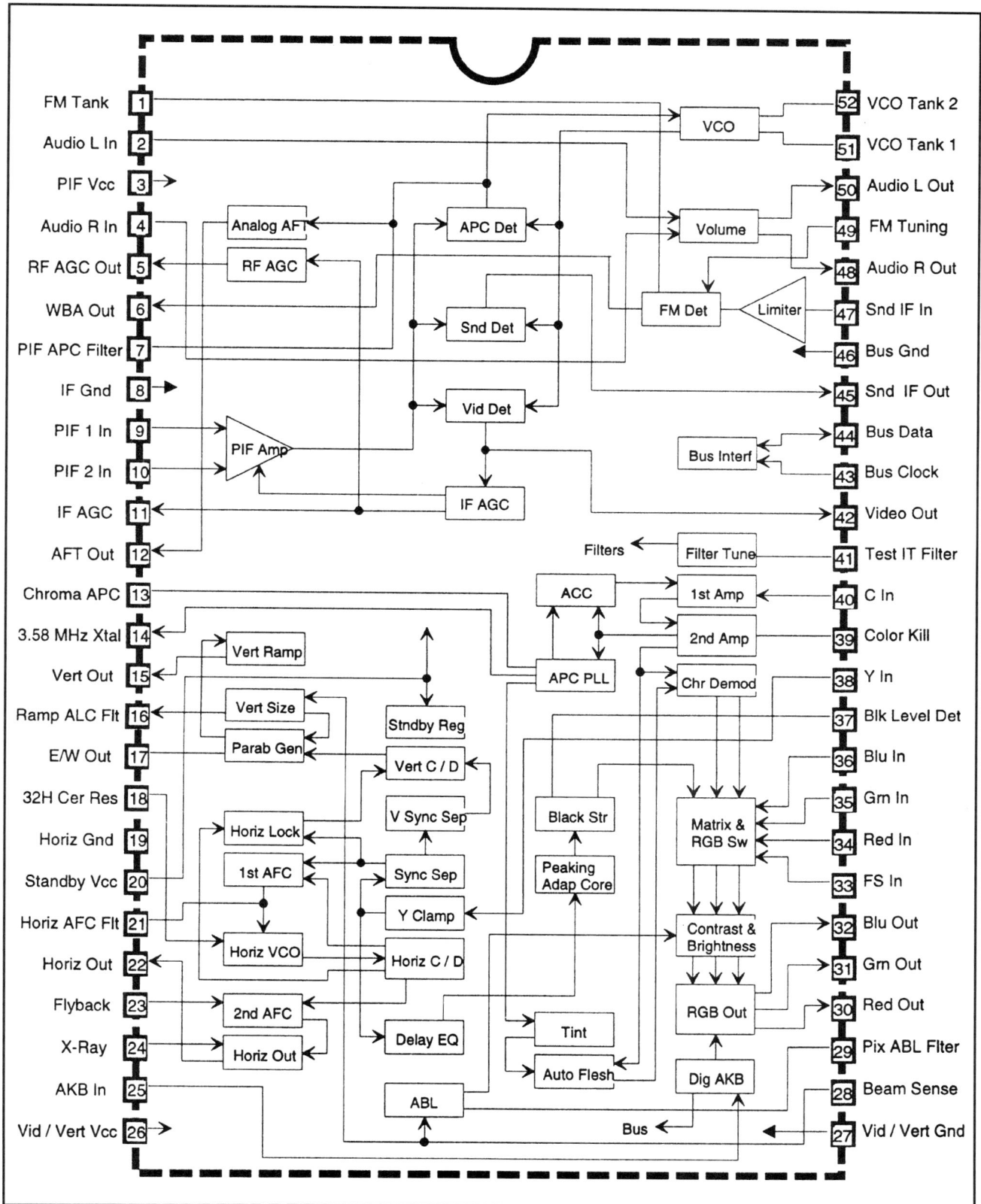

Figure 4-33. T4-Chip block diagram.

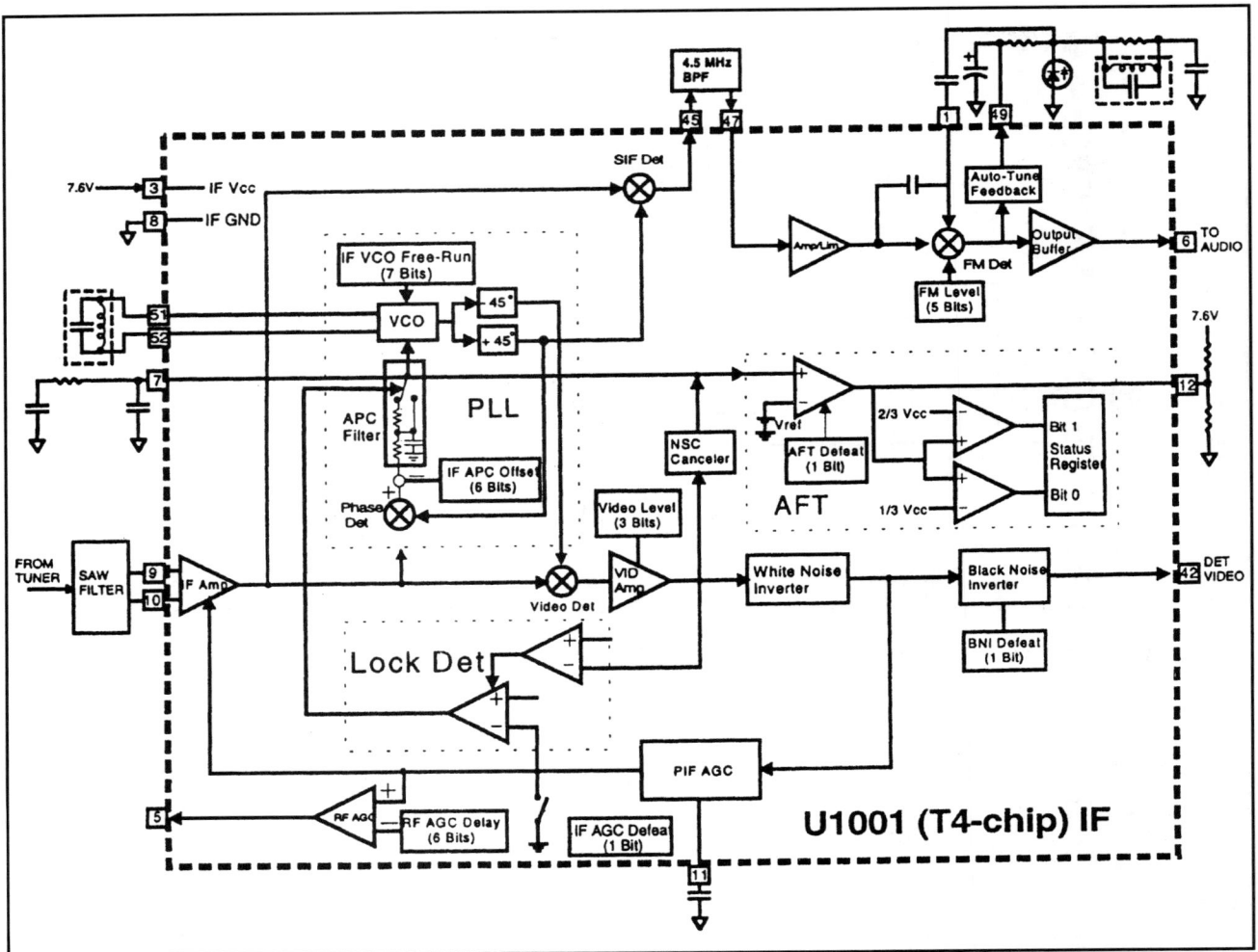

Figure 4-34. Video and audio IF block diagram.

carrier to provide it with a stable 45.75 MHz signal. The phase-locked loop APC detector output voltage is used as the IF frequency indicator for AFT. This voltage is amplified by an amplifier whose gain is set by the resistor network at pin 12, and it is applied to a "window comparator" with a 2-bit register. The microprocessor adjusts the tuner local oscillator and then reads the AFT register to determine if the IF frequency is correct.

If you follow the signal flow after the video detector, you will notice some baseband signal processing, which includes black and white noise inverters. These circuits provide some immunity against noise. After processing, the signal exits the chip at pin 42, where it is sent to external amplifiers and then back into U1001 for luminance and chrominance processing.

It reenters U1001 at pin 38. You can easily follow the signal path inside the T-4-Chip by consulting the block diagram given in *Figure 4-35*. Pin 37 can place the IC in an "RGB mode" by turning the luminance path off if the voltage is less than about 1.5 volts—In other words, luminance is enabled when the voltage is greater than 1.5 volts. On-screen display information from the microprocessor is input at

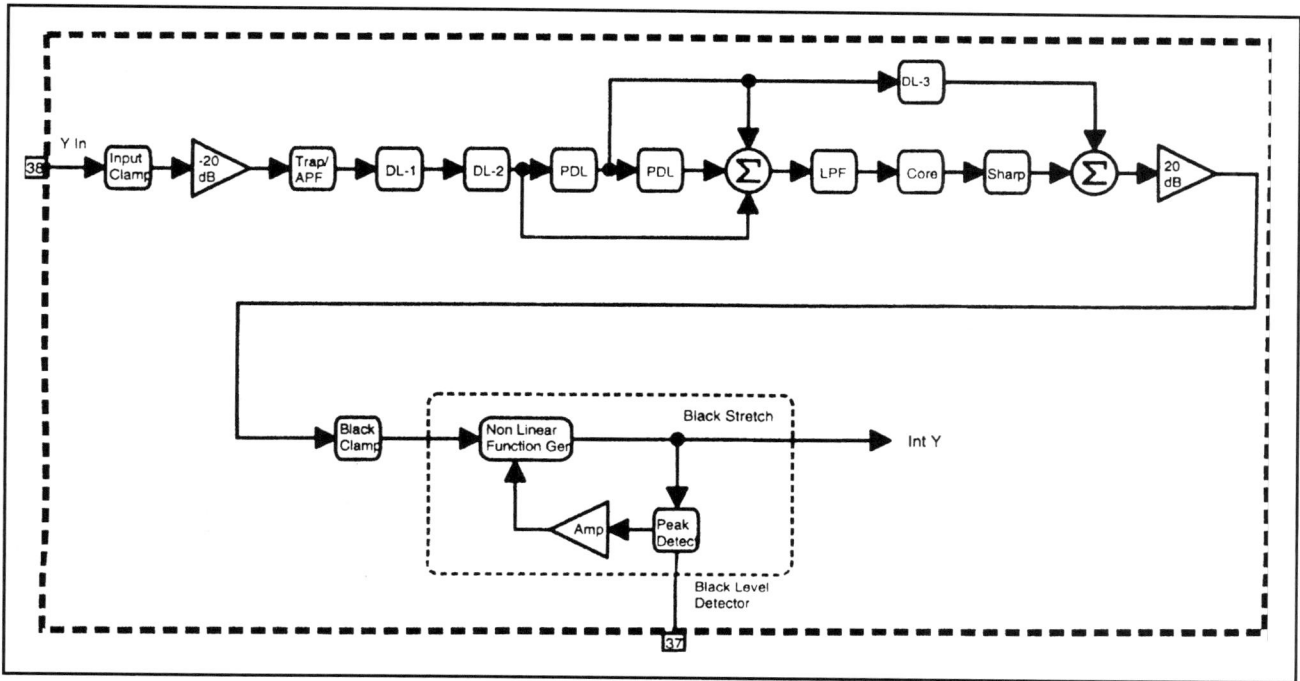

Figure 4-35. T4-Chip signal path.

pins 34, 35, and 36 (*Figure 4-36*). The fast switch input at pin 33 controls whether OSD or video is switched out. If the voltage is greater than 2.5 volts, external video (OSD) is switched. If it is less than 2.5 volts, internal video is switched.

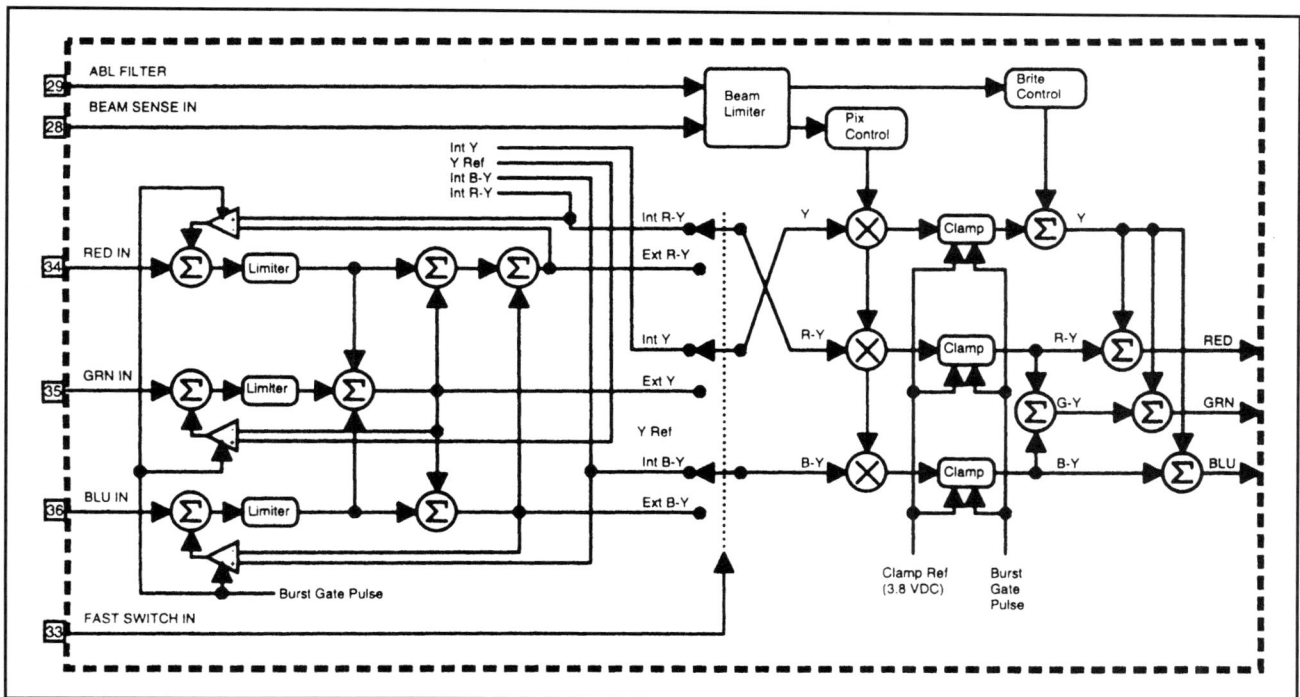

Figure 4-36. OSD input to microprocessor.

The output of the switches produces the Y, R-Y and B-Y signals. Each signal is routed to its respective circuit (contrast and clamp controls). Note that these processing circuits are serial bus controlled. The processed red, green, and blue signals exit at pins 30, 31, and 32 and go on to the kine drive circuits (*Figure 4-39*).

Pin 28 is the "beam-sense input." Like the CTC175 family, this input is used to reduce brightness and contrast during high-beam-current producing scenes to keep the picture tube from "doming" and the picture from "blooming." The circuit is active below about six volts.

Troubleshooting luminance path problems follows the same route I discussed in the last chapter. The procedure I gave you then is valid for the CTC185. First, check the brightness and contrast controls in the user menu. If you can't see the menu, press "reset" on the remote control. Second, look for a 1 Vp-p signal at pin 38. Third, check the voltages on the following pins: pin 31, which is active when it has less than about 6.2 volts on it; pin 33 for a voltage less than 2.5 volts; and pin 37 for a voltage greater than 1.5 volts. The factory service literature has taken the information about the luminance circuit and created a "luminance troubleshooting flowchart" which is good for the CTC185 and, in a modified form, for the CTC175 family. I am including it in *Figure 4-37* for your benefit.

The chrominance processing circuitry inside U1001 is given in block diagram fashion in *Figure 4-38*. There is basically no difference between the CTC185 and the CTC175 family; therefore, I will not go into detail here as I did in the previous chapter, but I will underscore two things. First, you can defeat the color killer circuit by applying 3.8 VDC to pin 39. Second, the 3.58 MHz oscillator is connected to pin 14. You cannot view the signal with a scope because when it is locked on frequency, there will be no

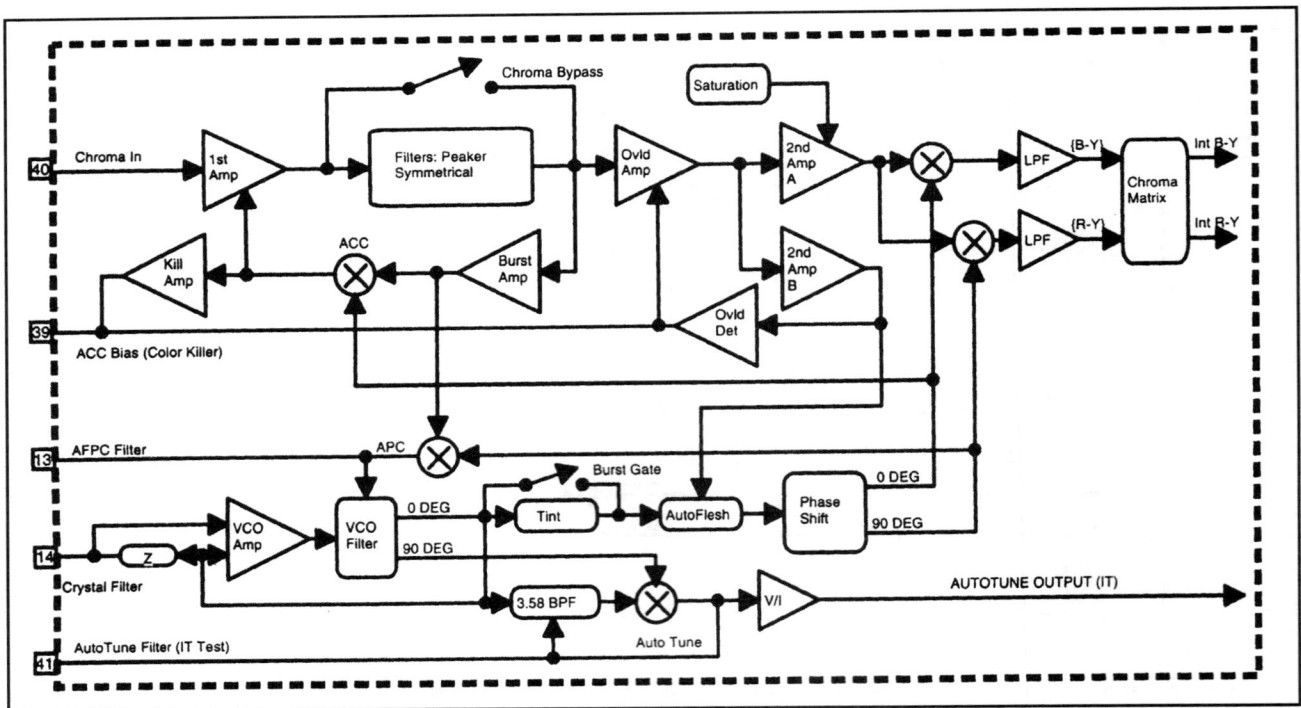

Figure 4-38. U1001 processing.

NO LUMA

Reset Pix,
Bright Controls → <u>Control System</u>

Luma Okay? → Yes → Exit

No ↓

Check Signal @ Pin 38-IC1001 → <u>Luma Input</u>

~ 1 Vpp? → No → Check Signal @ Pin 42 → ~ 2 Vpp? → No → Troubleshoot Tuner/IF

Yes ↓ ... ~ 2 Vpp? → Yes → Troubleshoot Video Takeoff

Check Signal @ Pins 30, 31, 32 → <u>RGB Outputs</u>

~ 1Vpp/3 VDC? → Yes → Troubleshoot Kine Drivers

No ↓

Check VDC @ Pin 28 → <u>Beam Limiter</u>

> 6.25 VDC? → No → Troubleshoot Pin 28

Yes ↓

Check VDC @ Pin 37 → <u>Black Level Detector</u>

> 1.5 VDC? → No → Troubleshoot Pin 37

Yes ↓

Check VDC @ Pin 33 → <u>Fast Switch</u>

< 1.2 VDC? → No → Troubleshoot Pin 33

Yes ↓

Replace U1001

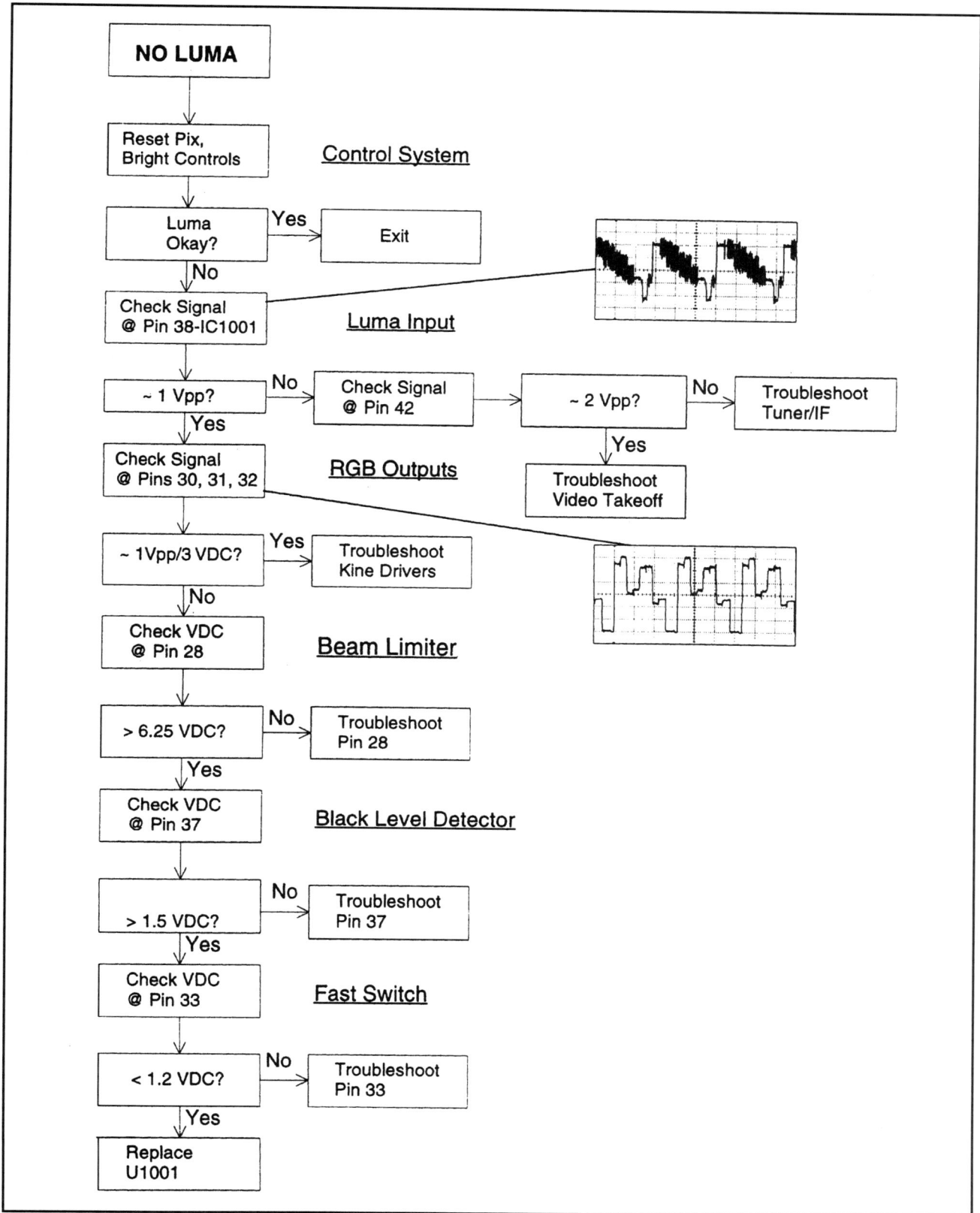

Figure 4-37. Luminance troubleshooting flowchart.

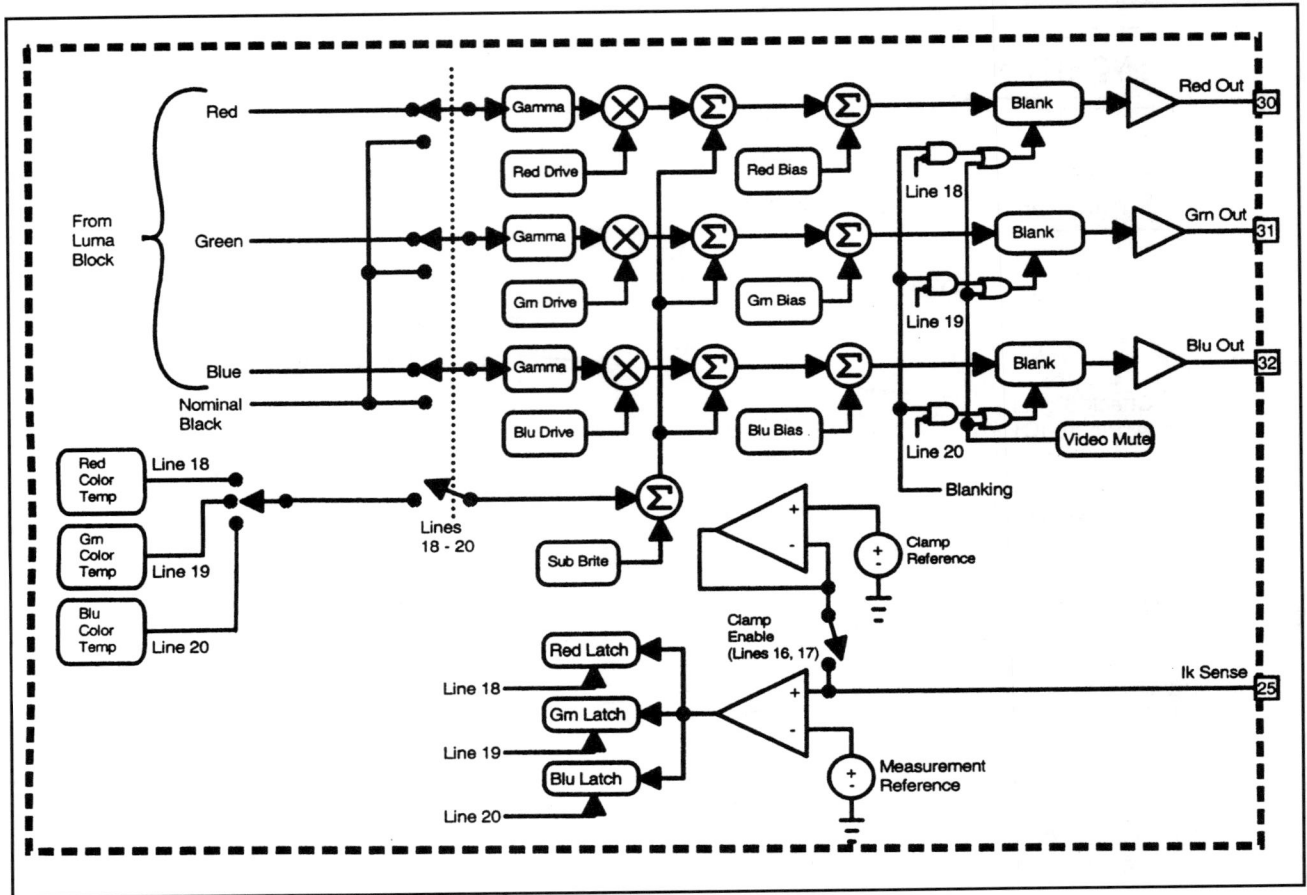

Figure 4-39. Luma processing.

voltage on pin 14! A problem with the oscillator can cause tuning and/or IF-type problems because its signal is used for chroma processing and as a reference signal for the other oscillators inside U1001. I am adding a troubleshooting tree (Figure 4-40) to help you solve a "no-color" problem.

The 4.5 MHz sound signal exits at pin 45, passes through a bandpass filter to remove any video that might be present, and reenter the chip at pin 47. It is then applied to the FM detector. Notice the varactor at pin 49. The voltage applied to the pin varies the diode's capacitance and thus the tuning of the network (a phase-shift network) at pin 1. Recovered wideband audio (L+R, L-R, and SAP) exits at pin 6.

If the TV is mono, the audio from pin 6 is looped back in at pin 4 (*Figure 4-41*) where the right channel audio is processed to serve as the mono channel. If the TV is stereo, the audio is routed from pin 6 to the stereo decoder, where it is processed and sent back to pins 4 and 2 of U1001 and then output to the audio amps. Volume control is accomplished inside U1001 via the serial bus. There are no tone adjustments.

Troubleshooting a no-audio problem generally follows the same path as a no-audio problem in the CTC175 family. If you want additional help, consult the trouble trees in *Figure 4-42* (monaural audio) and *Figure 4-43* (stereo audio).

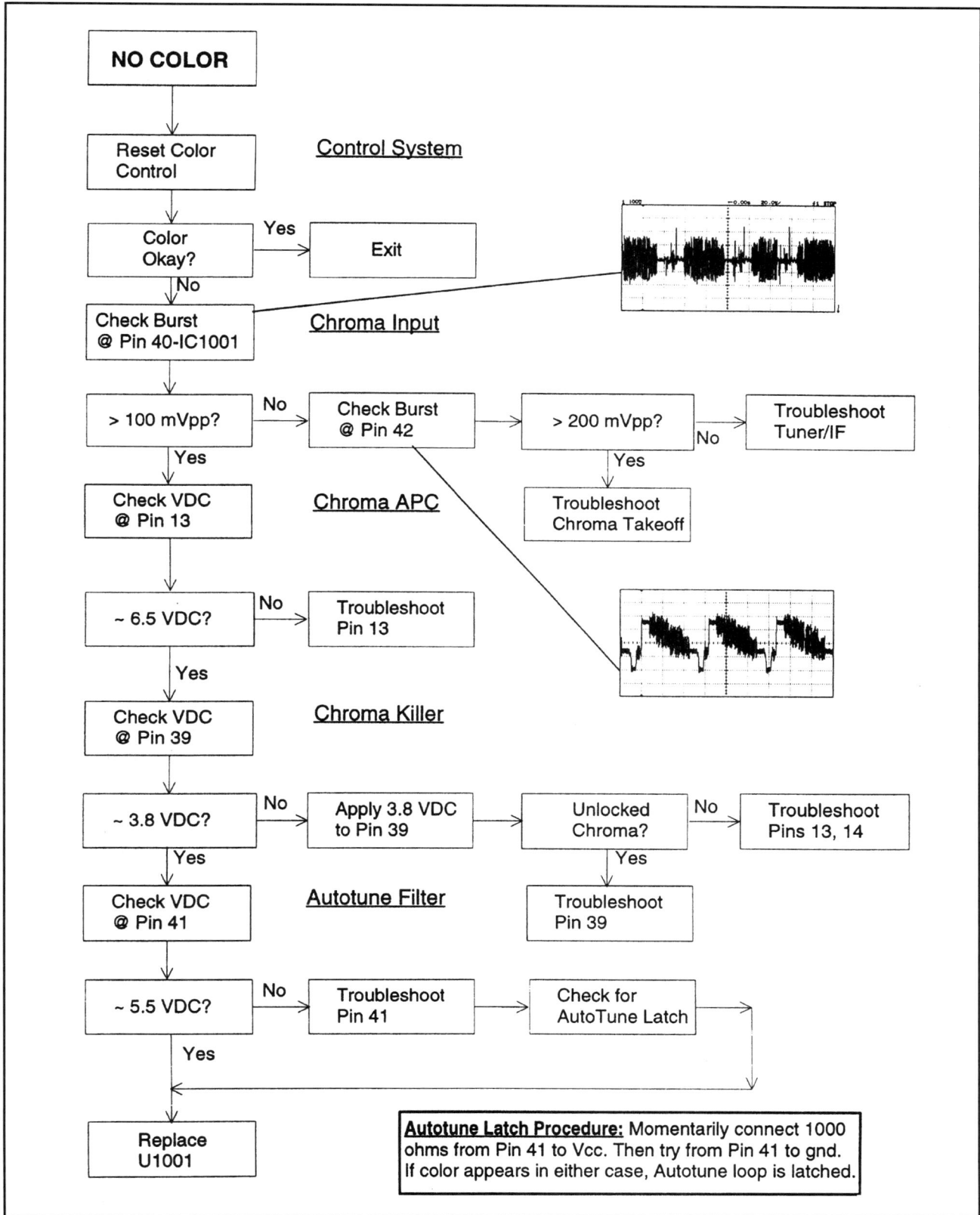

Figure 4-40. Chrominance troubleshooting flowchart.

Flowchart text content:

NO COLOR

Reset Color Control — **Control System**

Color Okay? — Yes → Exit

No ↓

Check Burst @ Pin 40-IC1001 — **Chroma Input**

> 100 mVpp? — No → Check Burst @ Pin 42 → > 200 mVpp? — No → Troubleshoot Tuner/IF

> 200 mVpp? — Yes → Troubleshoot Chroma Takeoff

> 100 mVpp? — Yes ↓

Check VDC @ Pin 13 — **Chroma APC**

~ 6.5 VDC? — No → Troubleshoot Pin 13

~ 6.5 VDC? — Yes ↓

Check VDC @ Pin 39 — **Chroma Killer**

~ 3.8 VDC? — No → Apply 3.8 VDC to Pin 39 → Unlocked Chroma? — No → Troubleshoot Pins 13, 14

Unlocked Chroma? — Yes → Troubleshoot Pin 39

~ 3.8 VDC? — Yes ↓

Check VDC @ Pin 41 — **Autotune Filter**

~ 5.5 VDC? — No → Troubleshoot Pin 41 → Check for AutoTune Latch

~ 5.5 VDC? — Yes ↓

Replace U1001

Autotune Latch Procedure: Momentarily connect 1000 ohms from Pin 41 to Vcc. Then try from Pin 41 to gnd. If color appears in either case, Autotune loop is latched.

Miscellaneous Information

(1) Tuner alignment. If the microprocessor is marked XX15053-470, you must use The Chipper Check to align the tuner. If it is marked XX15053-47A, you can use the built-in service mode. An error in the microprocessor code causes the incorrect values to be store in the EEPROM, hence the need for The Chipper Check.

(2) Parental Control. This feature reverts to off when the TV is unplugged or power is lost if the microprocessor is marked XX15053-470. To store parental control codes in the nonvolatile memory of the EEPROM where they should be stored, you will need to replace the microprocessor with part number 233180.

(3) How Parental Control Works. The front-panel buttons, except power on/off, will not work, but the TV will work fine using the remote control. To turn parental control off, enter the "channel" menu with the remote control and turn "parental control" off.

(4) If the television locks up when channel "01" is selected, replace the microprocessor with part number 233180. The original microprocessor will be marked XX15053-470.

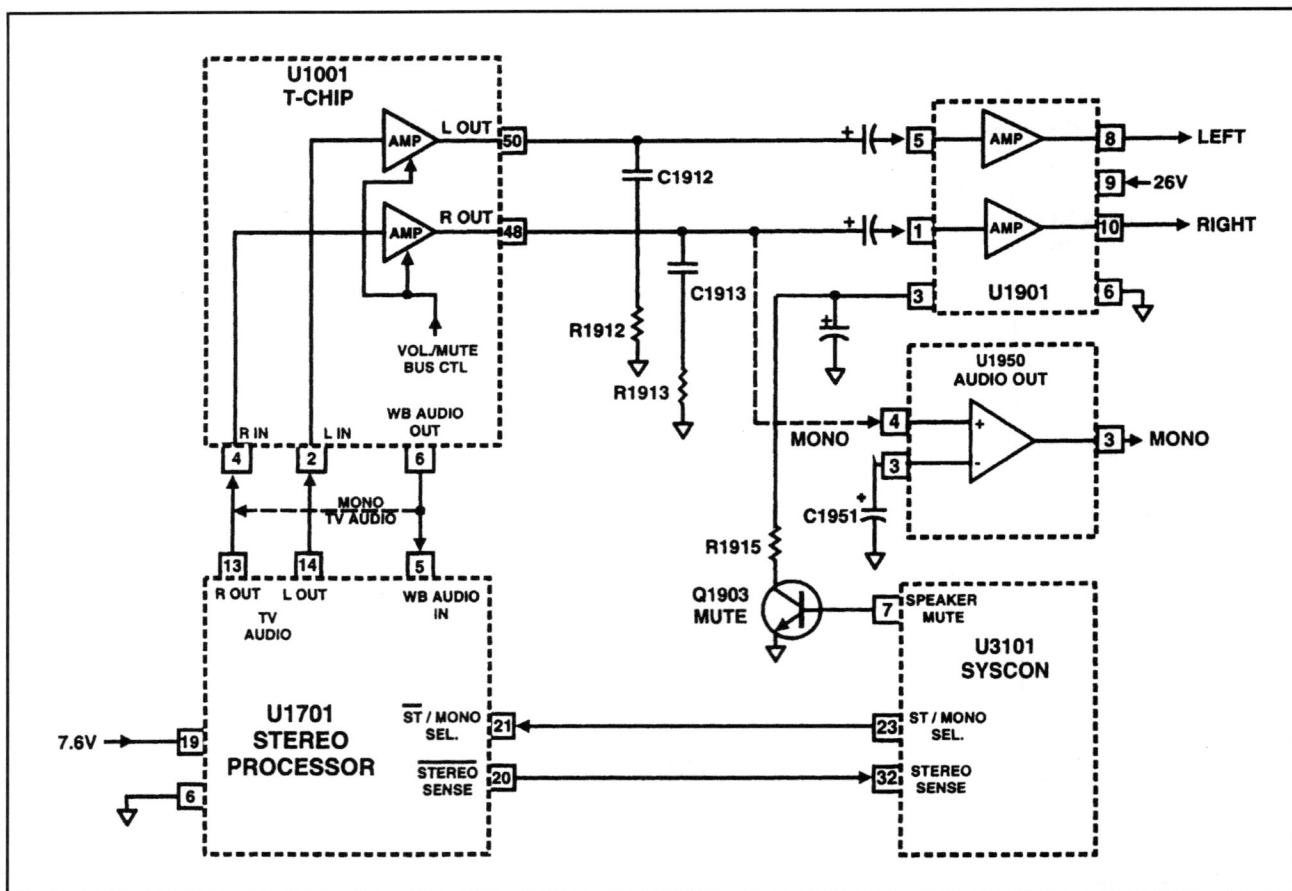

Figure 4-41. Audio circuit block diagram.

AUDIO TROUBLESHOOTING

SYMPTOM	CHECK
Power amp oscillation	R/C stability networks on power amp output pins (.1uF/4.7ohm)
Distortion or cut-out after warmup	Power amp properly clamped to heatsink
Frequency response wrong	Components in L-R signal path

MONAURAL AUDIO TROUBLESHOOTING FLOWCHART

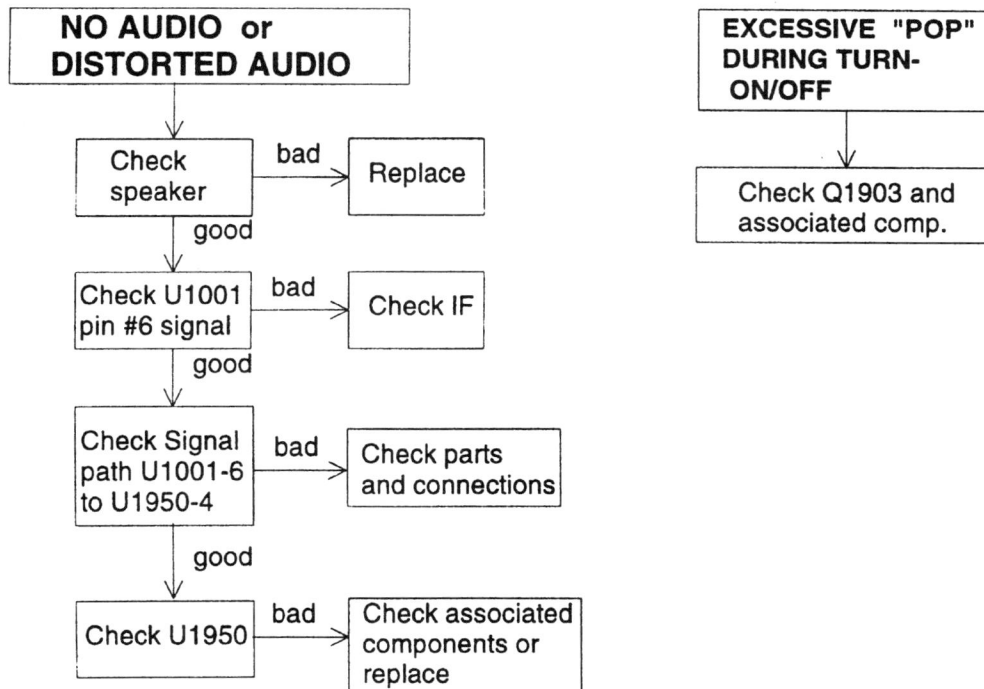

NO AUDIO or DISTORTED AUDIO
- Check speaker — bad → Replace
- good → Check U1001 pin #6 signal — bad → Check IF
- good → Check Signal path U1001-6 to U1950-4 — bad → Check parts and connections
- good → Check U1950 — bad → Check associated components or replace

EXCESSIVE "POP" DURING TURN-ON/OFF
- Check Q1903 and associated comp.

Figure 4-42. Monaural audio troubleshooting flowchart.

237

NO AUDIO or DISTORTED AUDIO

Similar to MONO for totally dead audio, except also check stereo IC U1701

NO STEREO BUT STEREO INDICATOR OK

Verify absence of stereo on more than one stereo source. Look at Left/Right channels on oscilloscope using Lissajoux pattern. L=horiz., R=vert. If mono, there will be only a straight diagonal line. If stereo OK, there will be deviation from the line.

Make sure STEREO/MONO command from micro is OK at U1701 pin #21.

bad → Troubleshoot signal path

STEREO IMAGE SOUNDS "HOLLOW" or TOO "EXPANDED"

Suspect (L+R) signal path broken inside U1701. No external fix for this.

Suspect U1701, Q1701, Y1701, or component associated with U1701.

bad → Replace as needed

good

Check pilot level coming out of U1001 pin #6. Should be at least 40 mvrms. If lower, could be bad U1001, or low broadcast pilot.

STEREO DISPLAY BAD BUT STEREO FUNCTION SEEMS OK.

Test DC voltage at pin 20. Should be less than 1v if stereo, greater than 3.5v if mono.

bad → Suspect U1701

good

Check path from U1701 pin #20 to micro.

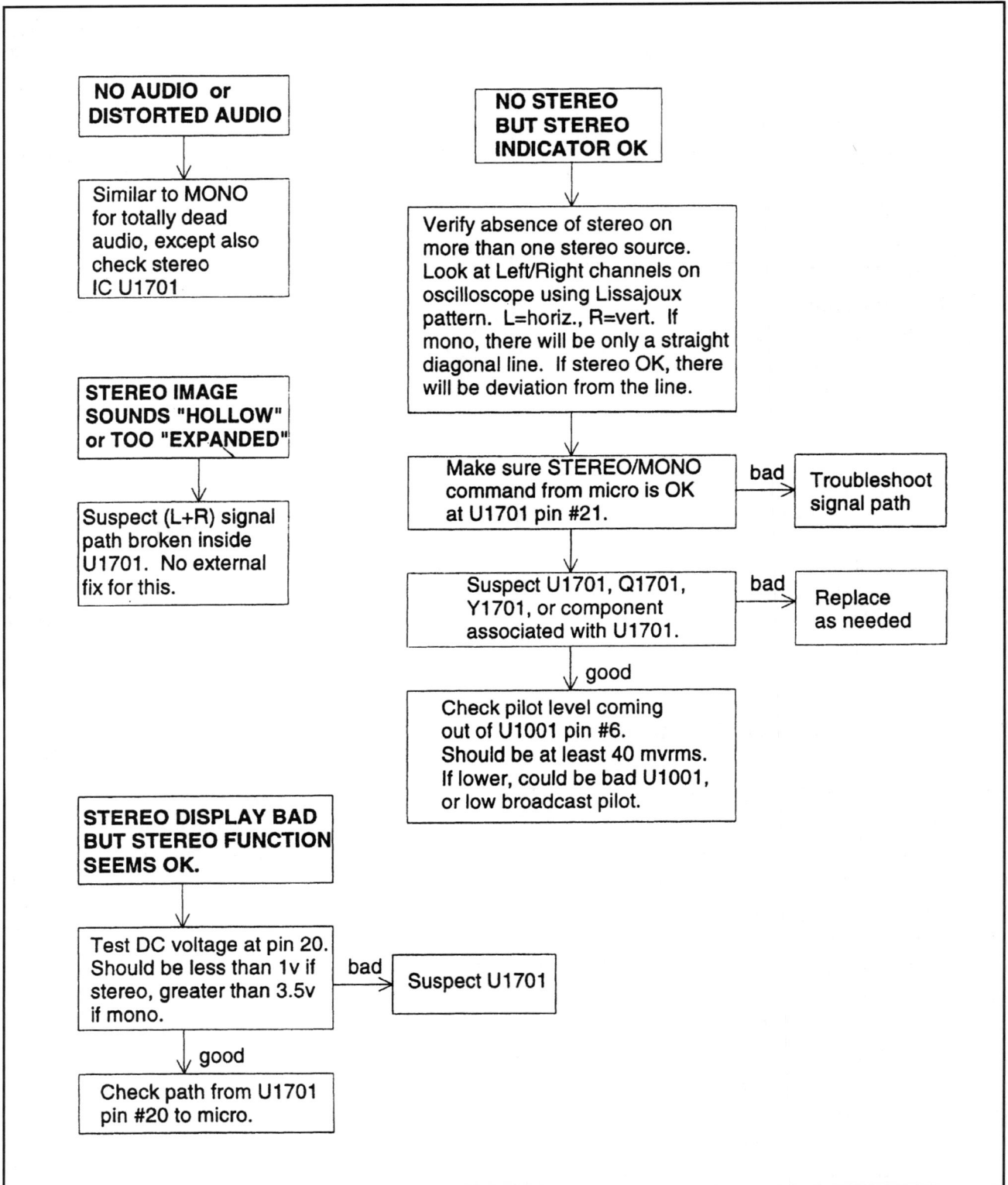

Figure 4-43. Stereo audio troubleshooting flowchart.

CHAPTER 5

The CTC 195/197 Chassis

The CTC195/197 is "the newest of the new." These chassis are built on what Thomson has already done in the CTC175 and CTC185 families, and go on to incorporate features so new most of us have scarcely even imagined them! Therefore, I begin this chapter by citing a brief bibliography which will provide you with the necessary technical descriptions and schematics. I advise you to spend some time studying these booklets because the time has passed when a tech can pop the back off a set he/she has never seen and proceed with a competent repair. Technology demands attention to the new, and attention to the new requires constant study. I remember a comment Homer Davidson made in one of his books that one should spend about an hour a day reading about what is taking place in his/her profession. That is very good advice, especially in light of our rapidly changing field.

The first book is the technical training manual, *CTC195/197 Color Television Technical Training Manual*. At 171 pages, it is one of the largest technical training manuals I have read. Would I be wrong if I said it is the largest one Thomson has published to date? The second piece of literature is *Color Television Basic Service Data*. The factory service literature for the CTC195 comes in two volumes; the literature for the CTC197 is contained in just one volume. The two volumes contain just about everything you need—circuit descriptions, alignment procedures, troubleshooting procedures, parts location guides, parts lists, voltages and waveforms, and schematics. The literature is thorough, and in my opinion among the best Thomson has ever produced.

Thomson describes the CTC195/197 venture as "the latest in the Thomson Consumer Electronics line of digitally controlled television receivers" (*CTC195/197 Color Television Technical Training Manual*, pg. 6). They are microprocessor-controlled, the control circuits being responsible for the total operation of the set and for aligning the various circuits that comprise the television. These chassis utilize the electrically erasable programmable read-only memory IC (EEPROM) and the tuner-on-board (TOB), features with which we have become very familiar. The difference between the CTC195/197 chassis and the previously discussed chassis will be newly designed circuits and integrated circuits and the various features the new models incorporate.

The CTC197 chassis will eventually replace a wide range of current TCE chassis, including the venerable CTC169 and CTC176/177 series of direct-view chassis. Video and audio features reflect a range of performance from previous core line products to midrange sets. Basic feature packages will include dBx stereo, an 8-jack panel, and on-screen program guide. The CTC195 will be used in projection televisions, from the 46" to the 61" screens. The CTC197 will be used in direct-view screen sizes from 27" to 35".

The features include, among other things, a T4-Chip, a FPIP (comb filter picture-in-picture) IC, and a new stereo IC. The television uses a dBx decoder which is serial bus-controlled. The SRS (Sound

Retrieval System) circuit was jointly developed by TCE and Hughes to provide a lower cost version of the system used in the CTC169 and CTC179. The main TOB is similar to that used in the CTC185. The PIP tuner is similar to the one used in the CTC179-2 chassis. I am including a block diagram (*Figure 5-1*) and a series of chassis layout diagrams (*Figure 5-2*) to give you a very visual idea of what you will be dealing with!

The Service Menu and Alignment Procedures

Before beginning an in-depth discussion of the service menu and alignment procedures, let's deal with some preliminary necessities.

First, the "operating conditions:"

Plug the television into a 120 VAC isolation transformer with the line voltage set at 120 ±2 volts and let it play for about 10 minutes before you attempt any adjustments. Then set the picture controls at midrange. The literature stresses that the procedures be performed in the given sequence. In other words, don't jump from one item in the list to another item further down the list.

Second, the required test equipment:

Remember the Chipper Check™ I talked about in the last chapter? It was a nice (but not necessary) piece of equipment for servicing the CTC185. Well, it is now a necessity if you plan to do in-depth work on these new televisions. You can do basic geometry and color temperature alignment from the front panel, but you will have to use the Chipper Check to go further. You will need, in addition to The Chipper Check, a computer that meets certain <u>minimum</u> standards: a 486DX processor with a minimum speed of 33 MHz and 8 meg of memory. By today's standards, that's a very minimum computer!

Based on personal experience, I encourage you to use the best computer you have. If you haven't bought one, buy the best you can afford. The so-called "minimum requirements" will as of this writing just get you by, but you will probably be dissatisfied with the way the system performs. Moreover, the time is at hand when other manufacturers will be following Thomson's example. Philips, for example, already has a program in operation by which you access a web site and download the service literature for which you have subscribed. Philips calls its program "Force." If you have been keeping up with developments, you know quite a few manufacturers are now making training material available only on CD-ROM. Your computer will gradually become the center around which your shop operates.

The Chipper Check hardware and software can be ordered from your local RCA distributor or from TCE Publications, the address for which you will find in the introduction to this book. When you order The Chipper Check, you will also receive an excellent instruction manual which is "hot off the press."

The "required equipment list" of test equipment includes: a dual-trace oscilloscope, a digital voltmeter, a frequency counter, an audio signal generator, an NTSC signal generator, an MTS signal generator, a sweep/marker generator (or standard signal generator), and the TAG001 service generator. The equipment list is not as daunting as it appears. For example, I own a Sencore SC61 scope that effectively

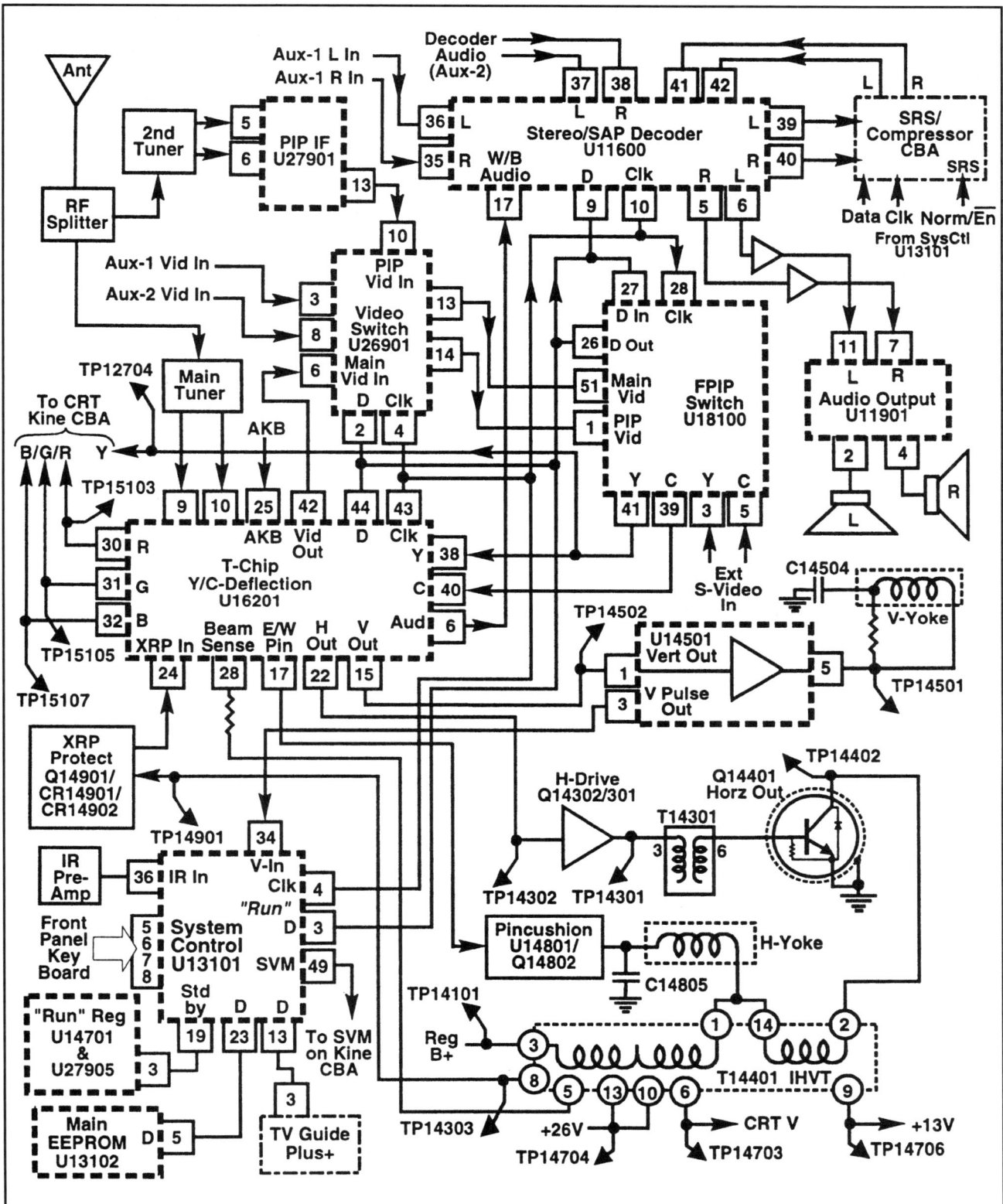

Figure 5-1. CTC195/197 block diagram.

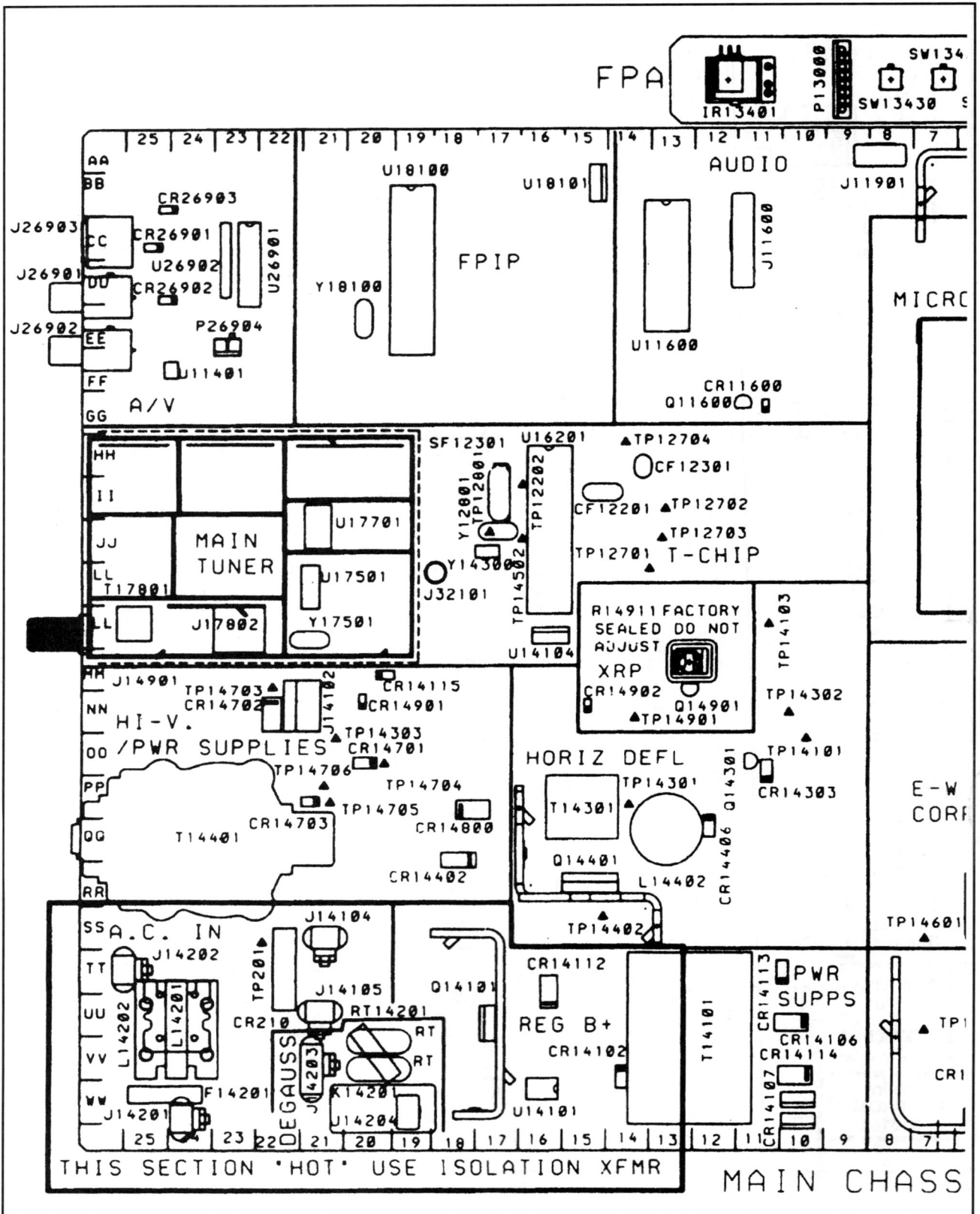

Figure 5-2. Main circuit board.

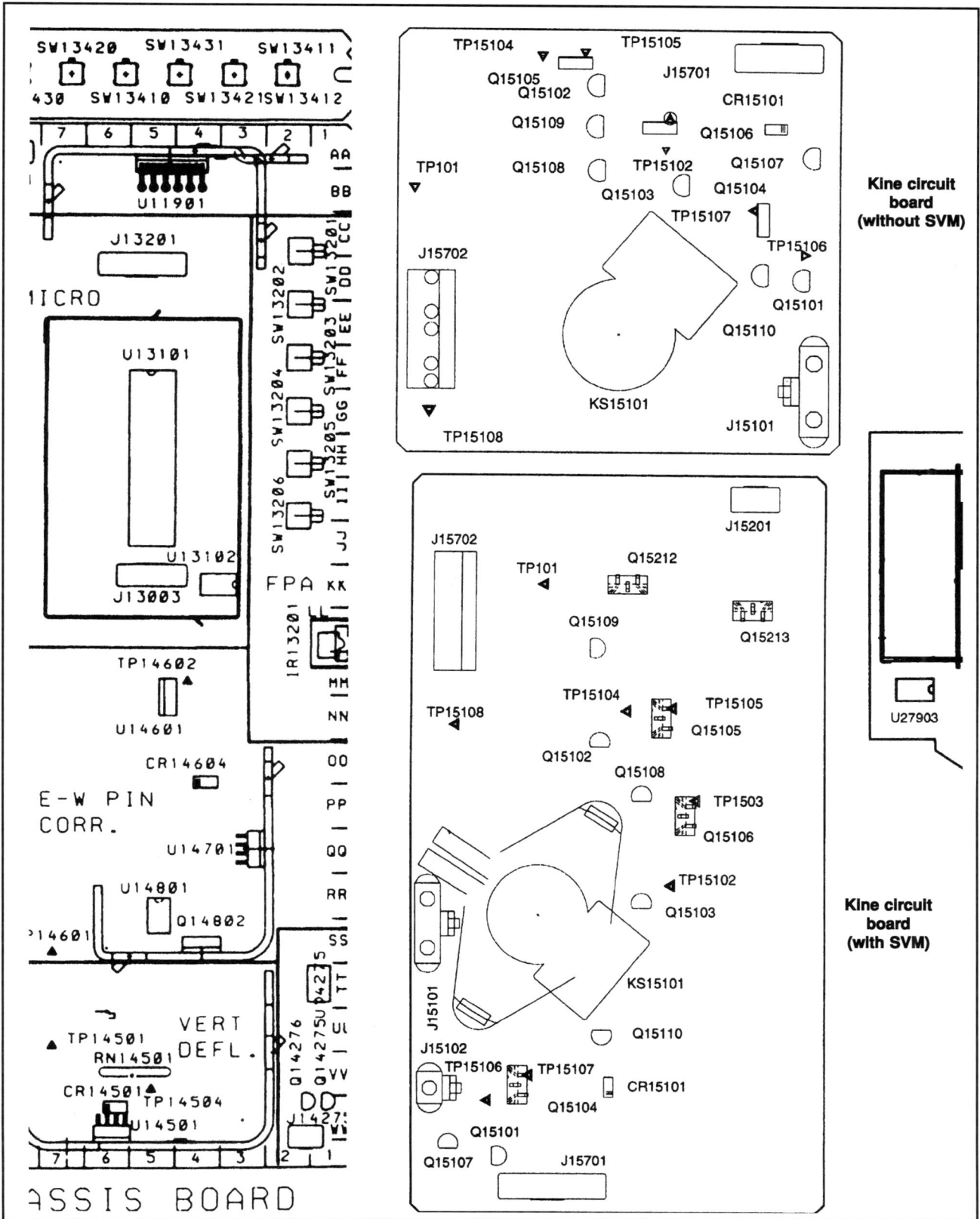

Figure 5-2. Main circuit board (continued).

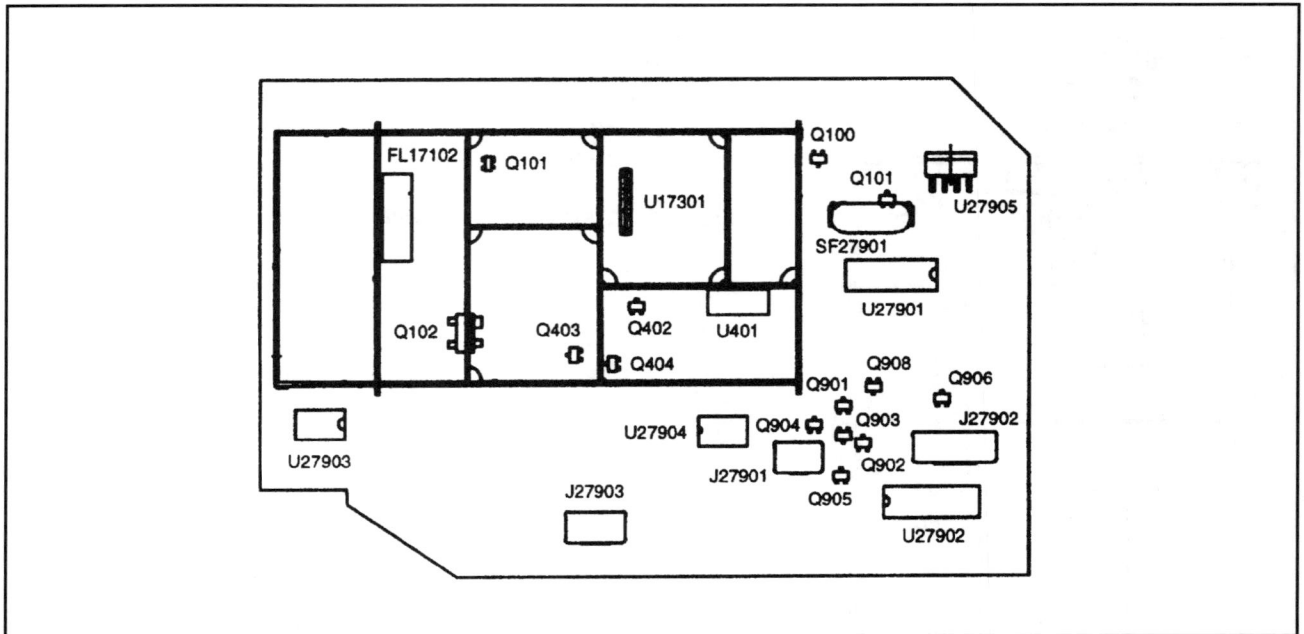

Figure 5-2A. FPIP IF circuit board.

combines into one unit the first three pieces listed. A good signal generator (B&K, Sencore, Tenma, etc.) will let you combine into one unit the audio, NTSC, MTS, and sweep/marker generators. And the TAG001 is a handy, but really not necessary, addition to your test bench because your current signal generator will probably tune the channels you need.

The Service Mode

Most of the alignments are software-driven; that is, adjustments are made by modifying parameter values in the service menu using either The Chipper Check or the front-panel controls (where possible). Modifying parameters precipitates a change in the corresponding T-Chip and tuner registers as updates are written into the EEPROM locations. There is nothing new here yet. It's basically the same as it has been since the introduction of the CTC175 family of chassis. You even use the same procedure as you have used to enter the service mode: (a) Press the power button to turn the TV on; (b) press and hold the menu button; (c) press and release the power button; (d) press and release the volume+ button, and; (e) release the menu button.

When it enters the service mode, the TV will display the parameter (P) setting on the lower left of the screen and the value (V) for the parameter on the lower right side of the screen. The channel up/down buttons change the parameter setting while the volume up/down buttons change its value. Once you have gotten into the service mode, you can use the front-panel controls or the remote control to go from one parameter to the next and to change the value setting. Nothing new here, you see.

When the service mode first comes up, the parameter will be "0." If you attempt to change it, the TV will exit the service mode, and you will have to start over. Getting into the service mode to make adjustments requires entering the necessary "security codes." Again, nothing new. The security code for front-panel access is 76. Once you have entered it via the volume UP control on the front panel or remote, you can proceed to make adjustments to the parameters listed in *Table 5-1*. The security code for access with the Chipper Check is 200. Exit the service mode simply by pressing the power button.

New Features in the Service Menu

Some features about the service menu, however, are new.

For example, parameters 1-3 are for "error codes." Certain failures in the chassis cause error codes to be written to and stored in the EEPROM. *Table 5-2* gives you the error code in hex and what the indicated error code might mean. Parameter 1 stores the code for the first failure experienced by the TV; parameter 2, the second failure; and parameter 3, the most recent failure. A "0" in the error code value indicates, of course, that there has been no failure. These error codes can and should be reset to "0" after servicing the set. Use volume up/down buttons to reset them.

Parameter	Description
0	Security Pass
1	Error Code
2	Error Code
3	Error Code
4	Horizontal Phase
5	EW DC (Horizontal size)
6	Pincushion Amplitude
7	Pincushion Tilt
8	Pin Top Corner Correction
9	Pin Bottom Corner Correction
10	Vertical Centering
11	Vertical Size
12	Vertical Countdown Mode
13	Red Bias
14	Green Bias
15	Blue Bias
16	Red Drive
17	Green Drive
18	Blue Drive

Table 5-1. Service mode codes.

If an error code for one of the integrated circuits on the I-Squared-C bus appears, system control is telling you the device did not acknowledge a handshake request. For example, suppose you find error code 128 in parameter 3. You consult *Table 5-2* and note "128" corresponds to the stereo demodulator IC (U1600). The error code informs you the IC did not respond when the set was turned on. It—and this is important—does not tell you <u>why</u> the stereo demodulator IC did not respond. You must use your troubleshooting skills to determine why it did not respond. The reason could be, for example, a failure in the supply voltage to U1600. The error codes will be useful because they will at least point you to the area where the problem is.

Neat, isn't it! But there is a slight "hitch" to this cute display of information. Any of the above failures will keep the television from turning on; therefore, you can't read the error codes via the service menu. However, you can read them by using The Chipper Check. It appears, at least on the surface, that the ability to read error codes will significantly reduce troubleshooting and repair time. If it does, the codes will be a great feature with which to work.

Error Code (HEX)	Error	Condition Indicated
00	No Error Code Thrown	
02	Detected by micro	5.1V_run has fallen below acceptable voltage level
03	Detected by micro	12V_run has fallen below acceptable voltage level
06	Pal/PIP Module	Failure to communicate with matrix switch on PAL/PIP Module
08	T4 Chip	X-Ray protection was invoked
09	T4 Chip	T4 Chip power supply problem at reset
10	PIP Module Error	FPIP power supply problem at reset
11	Stereo Decoder	Stereo Decoder power supply problem at reset
12	Detected by micro	AVR input to micro is held low
16	Detected by micro	Run IIC clock or data is clamped at 0 logic state
17	Detected by micro	Standby IIC clock or data is clamped at 0 logic state
34	StarSight	Failure to receive acknowledgement from StarSight module
44	FPIP	Failure to receive acknowledgement from FPIP
56	Digital Convergence	Failure to receive acknowledgement from Digital Convergence device
64	Octal DAC	Failure to communicate with Octal DAC
128	Stereo Decoder	Failure to receive acknowledgement from Stereo Decoder IC
130	Audio	Failure to receive acknowledgement from Audio Compressor
134	Video Matrix switch	Failure to receive acknowledgement from Video Matrix switch
138	PAL/PIP Module	PAL/PIP C1
142	PAL/PIP Module	PAL/PIP C2
160	Main or 2nd Tuner EEPROM	Failure to receive acknowledgement from EEPROM
186	T4 Chip	Failure to communicate with T4 Chip
192	2nd Tuner PLL/DAC	Failure to communicate with Tuner PLL IC
194	2nd Tuner PLL/DAC	Failure to communicate with Tuner DAC IC
196	Main Tuner PLL/DAC	Failure to communicate with Tuner PLL IC
198	Main Tuner PLL/DAC	Failure to communicate with Tuner DAC IC

Table 5-2. Error codes.

One other word about error codes. The literature mentions "error code 4," which is not on the list in *Table 5-2*. The microprocessor causes it to be logged when the power supply turns off and back on for any reason. If the power supply turns off, the microprocessor will check 500 milliseconds later to see if it has turned back on. If it has turned back on, the microprocessor will instruct the EEPROM to log the "error 4" data. An "error 4," then, simply means the television has momentarily lost power. It is really nothing to get excited about as long as you know what it means.

Parameter 12, "vertical countdown mode," is another new item. If the TV is going to work properly, parameter 12 has to be set correctly. There are four settings:

0	The Standard Setting
1	The Nonstandard Setting
2	The 50 Hz Setting
3	The 48 Hz Setting

Naturally, we (in the USA) will choose "0" for the setting.

The Chassis Alignment Procedure

Remember, the adjustments should be performed in proper sequence. The suggested sequence for the adjustments are:

Horizontal Phase	Parameter 4
Vertical Raster	Parameters 10-12
Horizontal Raster	Parameters 5-9
Focus Adjustment	Focus Control on IFT
Color Temperature	Parameters 13-18

Due to new electronics and screen size, most of these adjustments should be done "by the book." For instance, we can't just can't roll a mirror in front of the TV and proceed to make grey-scale adjustments as we have done in the past. The reason will become obvious in a few moments.

Correct adjustment of horizontal phase requires that you use a dual-channel oscilloscope, connecting channel 1 to TP12704 (pin 38 of U16201, the luma signal) and channel 2 to TP14303 (pin 8 of T14401, the filament). Both test points are on the main PCB. Set channel 1 for 100mV/2microsec/div. and channel 2 to 100V/div. and adjust horizontal phase (parameter 4) so that the time delay between the leading edge of horizontal sync (channel 1) and the midpoint of the filament pulse (channel 2) is 4.68 microseconds (as shown in *Figure 5-3*).

Vertical raster adjustments require the use of a crosshatch pattern. First, center the raster vertically. Then adjust vertical size for about 7.5% overscan at top and bottom. If the crosshatch display has 14

Vertical Raster Adjustments

Test Point:	Observe Display	
Adjust:	Parameter #10	Range: 0 - 63
	Parameter #11	Range: 0 - 127
	Parameter #12	Range: 0 - 3

1. Tune the instrument to receive the crosshatch signal.

2. Preadjust *Vertical Size* (parameter #11) so that the top and bottom edges of the display are just visible.

3. Viewing the top and bottom edges of the crosshatch, adjust *Vertical Centering* (parameter #10) to center the display vertically.

4. Adjust *Vertical Size* (parameter #11) so that ~3.75% of the crosshatch pattern is hidden at both the top and bottom of the display (7.5% overscan). Tolerance is ±5%.

Note: Vertical countdown (Parameter 12) must be set properly:
0 = standard
1 = non-standard
2 = 50Hz
3 = 48Hz

EXAMPLE: If the crosshatch display has fourteen blocks vertically, adjust so that ~ 1/2 block is hidden at both the top and bottom of the display.

Figure 5-3. Vertical raster adjustments.

blocks vertically, adjust parameter 11 so that about one-half of a block is hidden at the top and bottom of the raster.

Horizontal raster adjustment is a bit more complicated than we are accustomed to because of the screen sizes. Begin by preadjusting horizontal size (parameter 5) for about a 7.5% overscan. If the crosshatch display has 14 blocks, adjust the parameter so that about half a block is hidden on the right and left sides of the screen. Adjust parameter 6 for straight vertical crosshatch lines at the left and right edges of the raster. Look at the middle section of these lines for the time being. Then adjust parameter 9 for straight vertical lines at the bottom left and right corners of the raster. Adjust parameter 8 for straight lines at the top left and right corners. Finally, adjust parameter 7 so that the vertical crosshatch lines at the left and right edges of the screen are parallel to the sides of the picture tube mask.

Adjust the focus control for best overall focus.

We can now proceed with the color temperature adjustment. Follow the steps in the order given. Do not even think of deviating from the given order!

(1) Access the customer control menu and reset the customer controls with "picture reset."

(2) Disconnect the cable, or a signal generator if you are using one at this point. In other words, pull the plug on any signal source.

(3) Set the color bias parameters (parameters 13-15) to 63.

(4) Set the drive parameters (parameters 16-18) according to the following formula. If the CRT is a VHP tube 25" or 27", set the parameter to 32. If it is a non-VHP 27" tube or larger, set it for 40. VHP tubes carry the designation XXXAEGXXXXX.

(5) Proceed to the bias adjustments. Press the menu button on the front of the TV to get a service line. Press it again to get the full raster. Using this procedure, you can toggle between the two as necessary.

(6) Adjust the G2 control on the IFT so that you have a just visible service line. It doesn't matter at this point what its color is. Do not adjust the bias parameters of the color displayed in this step for the remainder of the alignment procedures. In other words, don't touch it!

(7) Adjust the other two bias controls so that you get a thin, <u>white</u> service line.

(8) Exit the service line.

(9) Hook up a signal generator and set it to output a grey scale signal.

(10) Adjust the screen control (G2) to produce even steps of the grey scale, light grey, grey, dark grey, etc. The last bar of the grey scale has to be black.

(11) It is now time to adjust the drive controls (parameters 16-18). You want a warm white raster. If you are a purist, shoot for a 6500-degree kelvin color temperature raster.

(12) Check the grey scale tracking, the transitions from low light to high light. If you detect any color other than grey or white in low to high light areas, the color temperature settings are not correct and should be repeated. In case you have forgotten, the bias adjustments affect the low light (or dark) areas, while the drive adjustments affect the high light (white areas).

(13) The next step is to initiate the AKB (automatic kine bias) setup and exit the service mode. When you do, the screen will, just for a moment, flash either red or green. If it is green, the AKB setup is okay. If the color is red, the AKB could not be properly set up. Don't get too frustrated because in many instances the G2 control has not been correctly set (either too

high or too low). Repeat the G2 setup step using slightly different G2 settings, and you will probably be okay. By the way, exiting the service mode automatically initiates AKB.

(14) There is just one more recommended chore—namely, to check the high voltage. Maximum high voltage for the 25" to 27" CRT is 29.7 kV and 34.0 kV for the 32" and 35" sets. Your meter should have an impedance of at least a 1000 megohms and be accurate within 5%. Note that the high voltage is not adjustable.

Component Number System

Table 5-3 gives you the component numbering system used in the CTC195/197 chassis. You also might want to read the "schematic notes" accompanying the table, especially if you are not familiar with the way Thomson does things. Using the component numbering system and the chassis layout given in *Figure 5-2*, you should be able to locate a particular component in a relatively short period of time.

The Main Power Supply

Figure 5-4 is a block diagram of the main power supply. I have included a copy of the waveforms you can expect to see at the noted test points. *Figure 5-5* is the complete schematic for the primary of the power supply.

11400 Series - Audio (Input Switching)	14600 Series - Power Supply (Standby Power)
11500 Series - Audio (Tone/Volume/Balance)	14700 Series - Deflection (HV Secondary)
11600 Series - Audio (Stereo Decoder)	14800 Series - Deflection (Pincushion)
11700 Series - Audio (SRS)	14900 Series - Deflection (Shutdown)
12200 Series - Audio (IF, Baseband)	15100 Series - CRT Drivers
12300 Series - Video (IF, Baseband)	15200 Series - Scan Velocity Modulation
12500 Series - Luma Processing (Edge, SVM)	16200 Series - Signal Processing
12600 Series - Video (Comb Filter)	17100 Series - Tuner (VHF)
12700 Series - Luma Processing (Control)	17200 Series - Tuner (Power Supply)
12800 Series - Chroma Processing (Demod)	17300 Series - Tuner (UHF)
12900 Series - Chroma Processing (RGB)	17400 Series - Tuner (Band 4)
13100 Series - System Control (Control)	17500 Series - Tuner (PLL)
13200 Series - System Control (User Interface)	17600 Series - Tuner (IF)
13400 Series - User Controls	17700 Series - Tuner (Mixer/Osc)
14100 Series - Power Supply (Regulator)	17800 Series - Tuner (RF Switch)
14200 Series - Power Supply (Degauss, AC Input)	18100 Series - Pix-In-Pix
14300 Series - Deflection (Horizontal Osc)	26900 Series - Video (Input Switching)
14400 Series - Deflection (Horizontal Out)	27900 Series - Power Supply/Pix-In-Pix
14500 Series - Deflection (Vertical)	

Table 5-3. Component numbering system.

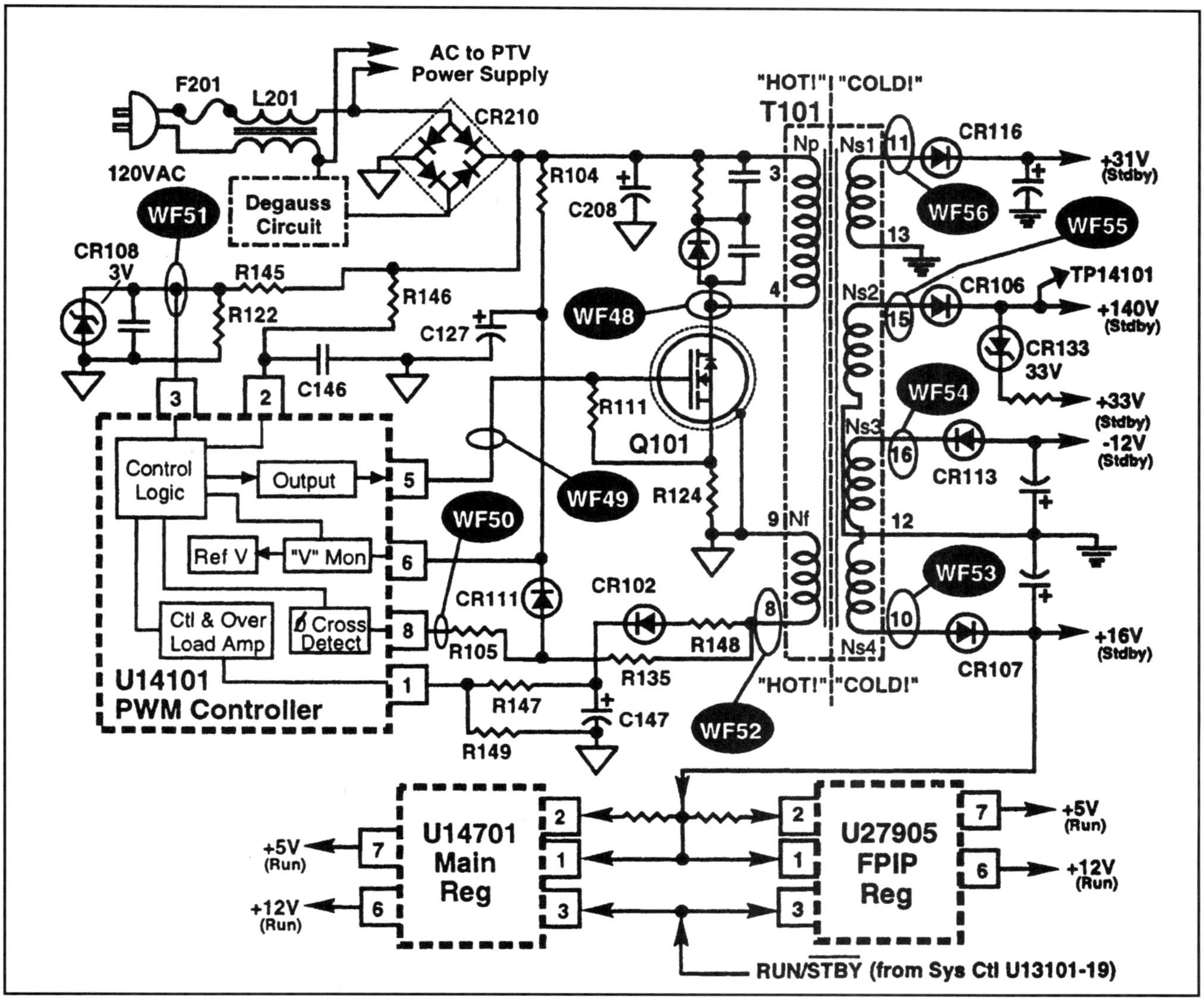

Figure 5-4. Main power supply block diagram (waveforms next page).

How The Power Supply Works

The literature describes the power supply as "a variable frequency/variable pulse-width switch-mode power supply" (*Basic Service Data*, pg. 1-17) which uses a controller IC (U14101) to drive a MOSFET (Q101). As you can tell from looking at the schematic, these TVs have a "cold" chassis which is isolated from the power supply by a ferrite core transformer, T101. Since the supply uses "hot side" regulation, there is no need for an additional transformer (as in the CTC169) or an optoisolator (as in certain Philips chassis).

The power supply is self-oscillating and begins to operate whenever it is connected to AC. It will supply current on demand up to its limit, which is a maximum input power of about 180 watts. Its operating frequency depends on the load and can vary between approximately 25 and 90 kHZ. The

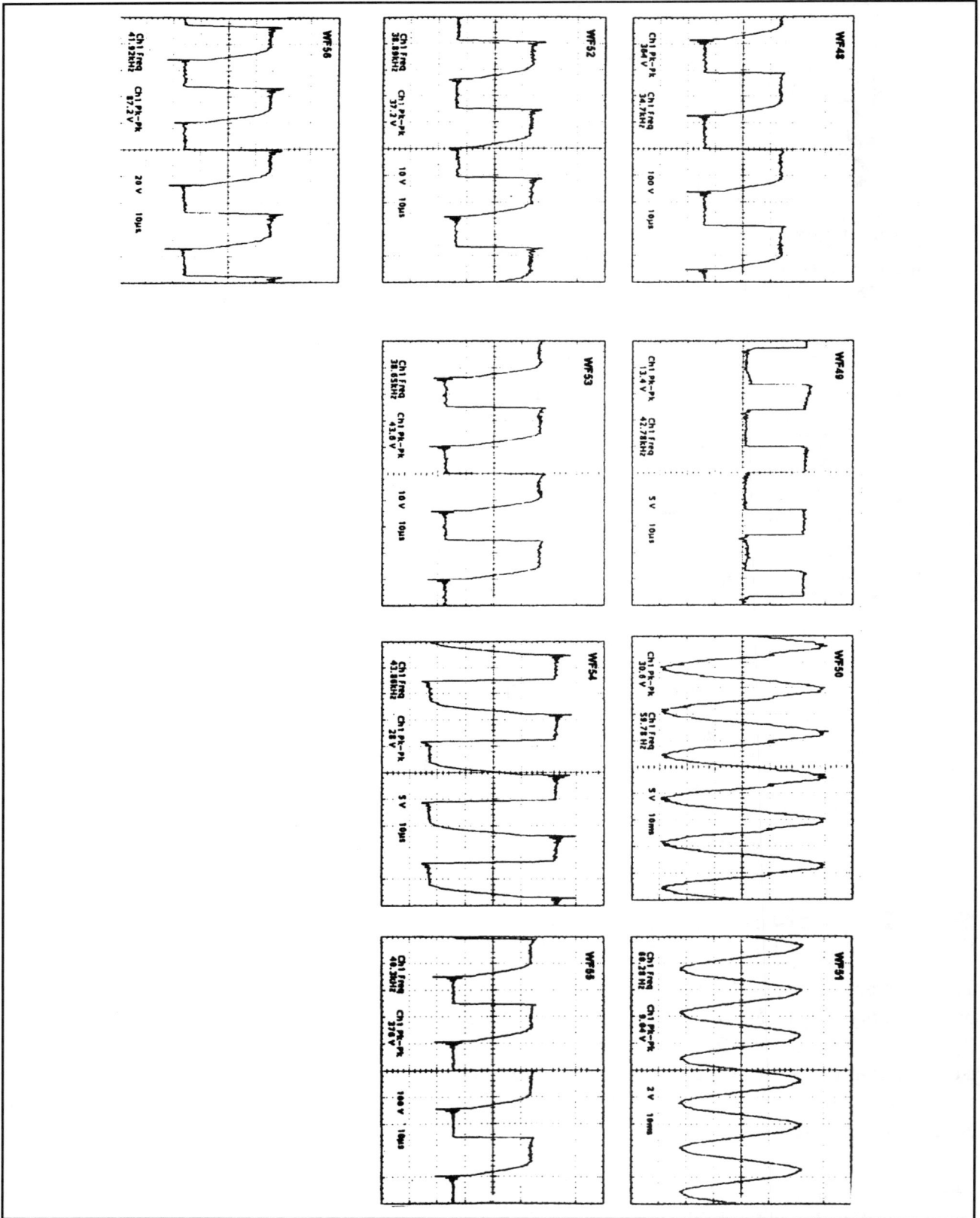

Figure 5-4. Power supply waveforms (continued).

feedback winding between pins 8 and 9 is tightly coupled to the regulated B+ windings on the secondary. Therefore, voltage variations in the regulated B+ windings are reflected in it and fed back into U14101, which has its own internal reference voltage and makes the necessary corrections.

The start sequence goes like this:

When the line cord is plugged into an AC source, the AC input circuit, bridge rectifiers, and filter capacitor (C14208) deliver about 150 volts of raw DC (raw B+) to T14101, pin 3, and out of pin 4 to the drain of the MOSFET. The MOSFET's source is taken to ground through R14124, a 0.22 ohm/2 watt resistor. The power supply is therefore ready to start, but it will not start until the regulator IC (U14101) is connected to a power source. That startup power is routed from the 150 raw B+ to pin 6 of the IC via R14104, a 100k/0.5 watt resistor.

With B+ applied to pin 6, the regulator outputs a voltage at pin 5, turning on the MOSFET and energizing the primary of the transformer. The IC senses the current flow by means of an RC network, consisting of R14146 and C14146 tied to pin 2, which is the "primary current sensing input." The capacitor will not charge until the gate of the MOSFET turns on, at which point it will charge. With the MOSFET turned on, current through it increases, and the voltage on pin 2 also starts to increase. When the voltage on pin 2 reaches about 3 volts, the IC terminates its gate drive, shutting off the MOSFET. Energy stored in the primary of the transformer transfers to the secondary. The same energy is coupled back into the feedback winding, developing a voltage at pin 8 which is rectified by CR14111 and filtered by C14127. This is the run voltage for U14101. R14104, you see, provides initial startup B+ only.

Now let's complete the cycle. After the energy just transferred to the secondary of T14101 begins to decay, the voltage at pin 8 also begins to decay. This decreasing voltage is applied to pin 8 of the IC through R14105. If you look at *Figure 5-5*, you will see that pin 8 is labeled the "zero detector." When the waveform coupled to it goes through zero, the zero detector signals the IC to begin another cycle, and it responds by turning gate drive to the MOSFET back on. And the cycle begins again.

The next question is, "What about regulation?"

Feedback voltage from the power supply is routed to pin 1 of the controller IC. The transformer has been constructed so that the feedback winding is tightly coupled to the regulated B+ winding on the secondary; therefore, its voltage tracks the regulated B+ voltage. If one rises, the other rises; if one goes low, the other goes low. The voltage at pin 8 is rectified (CR14102), filtered (C13147), and applied to a precision voltage divider network (R14147 and R14149) before it is routed to pin 1. The voltage divider has been engineered so that it corresponds exactly to the required B+ output of 140 volts DC. If the circuit delivers more than 400 mV, the controller will terminate drive to the MOSFET.

There are, then, two ways to turn off drive to Q14101. The first is when the voltage on pin 2 exceeds 3 volts; the second is when the voltage on pin 1 exceeds 400 millivolts. Let me put it differently. Pin 1 senses the output voltage of the power supply while pin 2 senses the output current. If the load on the power supply increases, more energy has to be stored in the transformer's primary. The MOSFET responds by staying on longer. If it stays on too long, C14146 charges above 3 volts and shuts down the

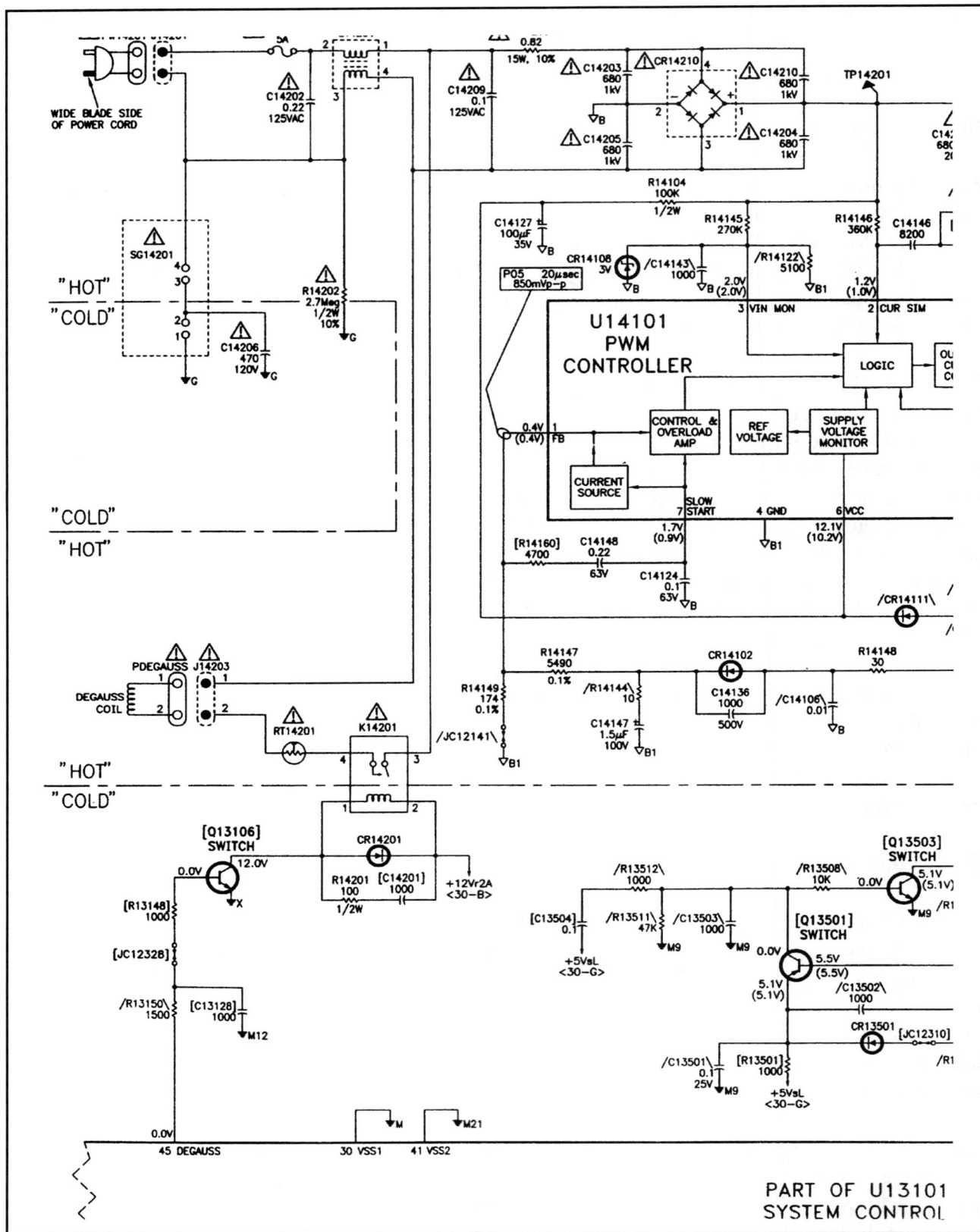

Figure 5-5. Power supply circuits.

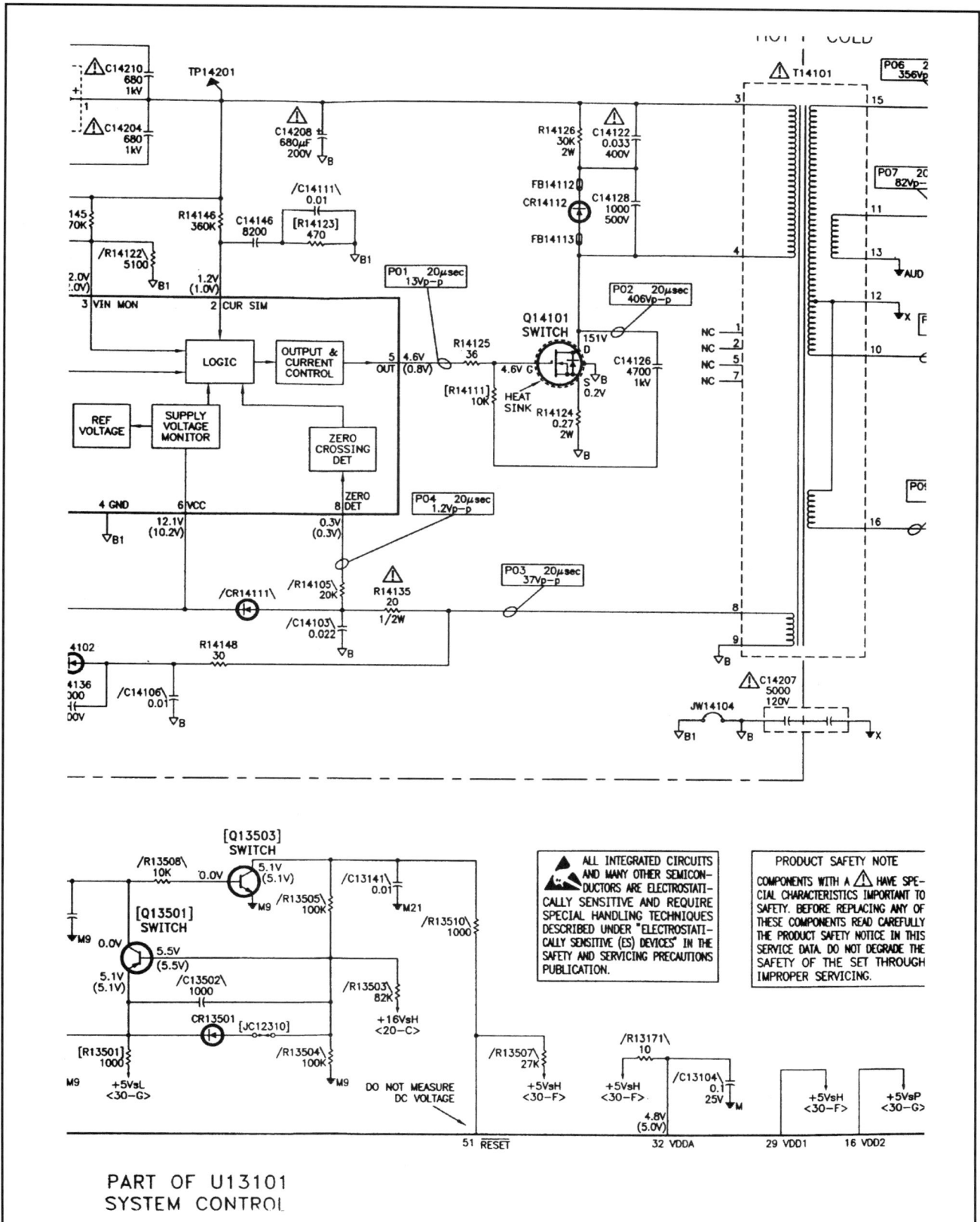

Figure 5-5. Power supply circuits (continued).

IC. This provides overcurrent protection. If the regulated B+ gets too high, the increased voltage is felt at pin 8 of the transformer and fed to pin 1 of the IC, shutting down the IC. This is overvoltage protection.

There are a few other components we need to look at.

R14145 and R14122 form a voltage divider network from raw B+ to pin 3 of the controller IC to monitor the input voltage. The circuit is designed to protect the power supply from low line voltage. If the voltage at pin 3 falls below about 1 volt, the supply will shut down. The resistor/capacitor/diode network across pins 3 and 4 of T14101 make up a snubber network to dampen the ringing when Q14101 turns on and off.

The Output Voltages

The main power supply outputs four voltages (*Figure 5-4*): +140 volts, +16 volts, -12 volts, and the audio supply, which varies depending on which audio system comes with the TV. Each voltage is available as long as the unit is plugged into an AC outlet.

The +33-volt supply for the tuner is derived from the +140-volt line via a 33-volt zener (CR14133) and its companion capacitor. You will notice (*Figure 5-4*) that the 16-volt line is also the source for two 12- volt and 5-volt switchable supplies, which are turned on or off by the system microprocessor. U14701 provides run voltages for the chassis while U27905 provides voltage for the picture-in-picture module. The +16-volt line also supplies voltage to U14601, the +5-volt standby regulator for the chassis (*Figure 5-6*).

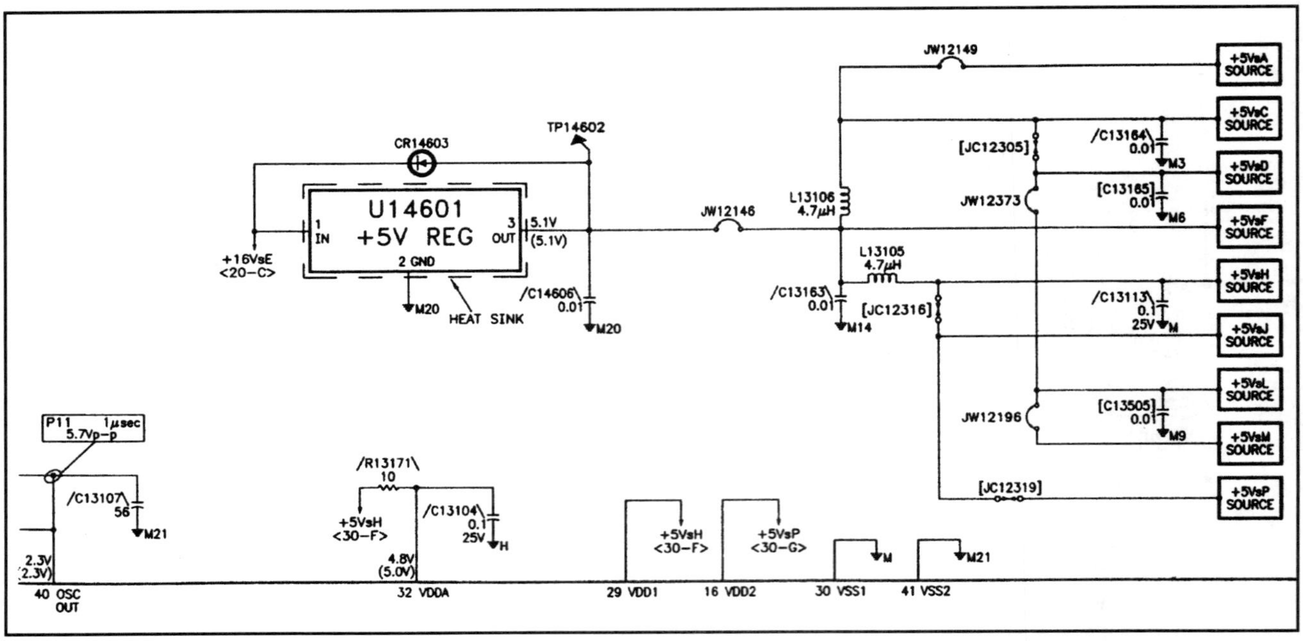

Figure 5-6. Standby regulator U14601.

Troubleshooting the Main Power Supply

Troubleshooting this power supply is relatively straightforward. You should be able to fix any problem with it by making a few voltage and resistance checks. It isn't nearly as complicated as, for example, the switching supply used in the CTC169 chassis. But you should proceed in a logical manner. You will need an isolation transformer, a digital multimeter, and an oscilloscope.

If the power supply is not running, do the obvious and check for raw B+. The main filter capacitor is a good place to make the check, as it is in the RCA chassis we have looked at in this book. The voltage at C14208 should be in the 150-155 VDC range, and the same voltage should be at the drain of the MOSFET.

If raw B+ is not present, check for an open fuse or open R14203 (*Figure 5-5*). An open fuse or open resistor could mean shorted diodes in the bridge rectifier, a shorted MOSFET, or a shorted controller IC. If raw B+ is present, use your scope to check for a varying signal on pin 6 of the controller IC. If you detect a varying signal of about 4.5 to 12 Vp-p, the IC is not getting enough voltage to start.

The problem, then, will be insufficient start-up voltage, a problem which can have one or two causes. First, the problem could be caused by the circuit consisting of R14104 and C14127. As the capacitor charges through the resistor, the voltage on pin 6 will rise and then fall as the regulator tries to turn on. Second, the problem could be caused by a defective component in the B+ line from pin 8 of T14101 to pin 6 of the regulator. The most likely causes will be an open CR14111 or R14135. If CR14111 were shorted, the voltage at pin 6 would be quite low, and there would be no oscillating signal. This is a good diagnostic fact to remember.

There is another scenario we need to examine. Suppose the power supply "chirps" when you apply AC. When it is under a very heavy load, like when a component tied to one of the secondary sources shorts, the voltage ramp at pin 2 of U14101 will exceed 3 volts. Under these circumstances, the regulator IC proceeds to shut down the power supply in the current-limit mode. It will then try to restart. If the load is still present, it will shut down again. The shut down/start up cycle will repeat itself about twice a second. When you hear the chip, you should suspect a short on one of the secondaries, like a shorted horizontal output transistor or shorted audio output IC.

The "Other" Power Supplies

You will be pleased to know there are two other power supplies in the direct-view televisions and a third in the projection sets.

The first is, as you expected, the scan-derived power supply which comes up when horizontal deflection becomes active. *Figure 5-7* will give you the details. Note that the voltage source "CrtVr" provides B+ for the video output transistors.

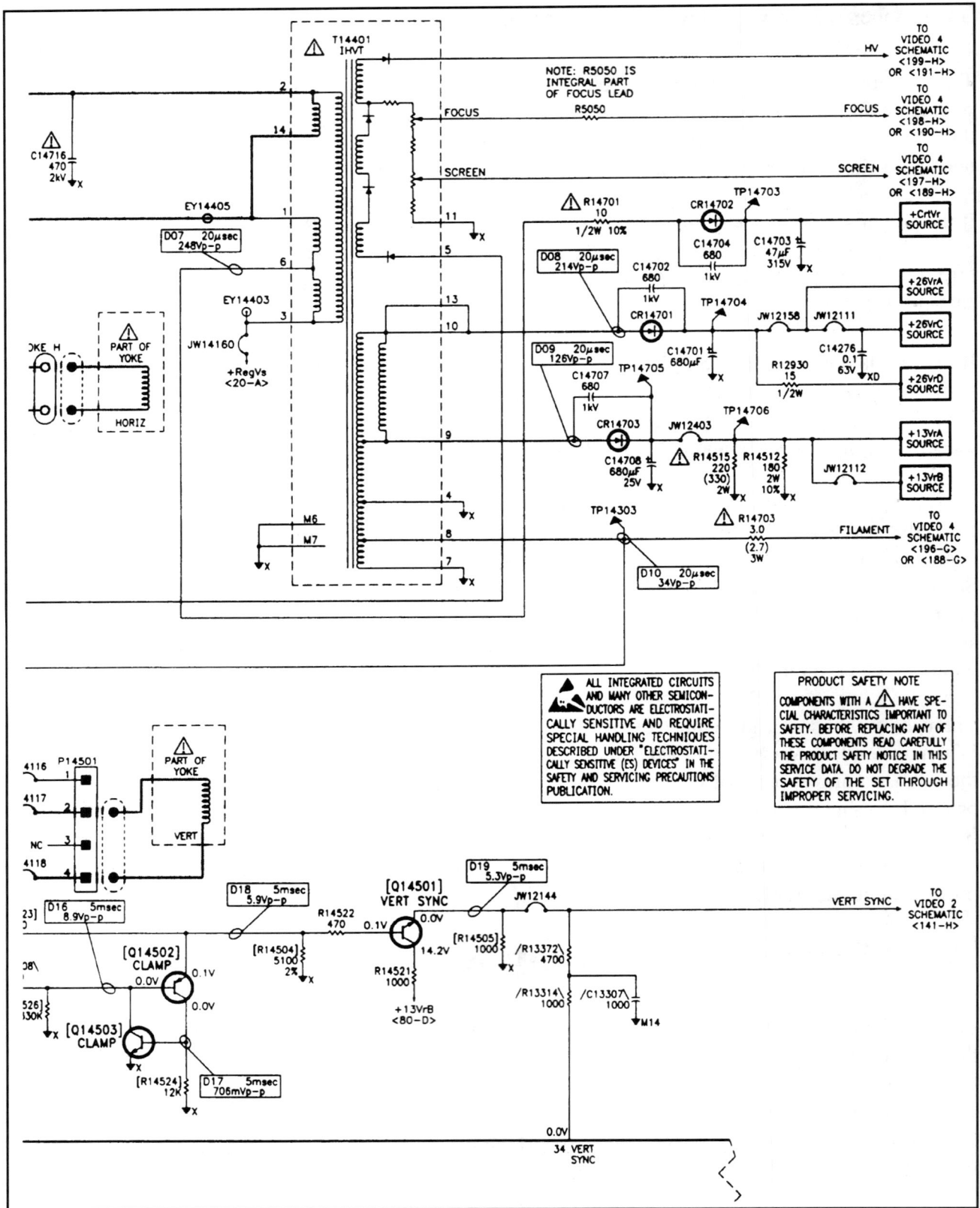

Figure 5-7. Scan-derived supply.

The second is called "the auxiliary power supply" (*Figure 5-8*). It is located on the main PCB and consists of three regulator ICs: U14104 (+7.5 volts), U18101 (3.3 volts), and U14601 (+5 volts). Q11600 is a series-pass transistor supplying +9.5 volts. These power supplies are active only when U14701 turns on. As you recall, U14701 gets its voltage from the +16-volt line and is turned on and off by the system control microprocessor. U14601 also uses the +16-volt supply and supplies +5 volts for the microprocessor, EEPROM, and remote receiver. It is on and active as long as the television is plugged into an AC outlet.

The third power supply is the supply for the digital convergence module. Since I am dealing with direct-view televisions, I will not discuss the convergence power supply here, but I will provide you with its block diagram (*Figure 5-9*).

System Control

If you don't believe televisions are getting more complicated as times goes on, take a look at *Figure 5-10*, the block diagram of system control for the CTC195/197 chassis. I am also providing the pinouts for the system control microprocessor, U13101, in *Figure 5-11*. You will find a fairly complete description of each pinout in the training manual on pages 60-65. If I covered every aspect of system control, I would repeat much of what I have written in chapters 3 and 4. Therefore, I will cover the differences.

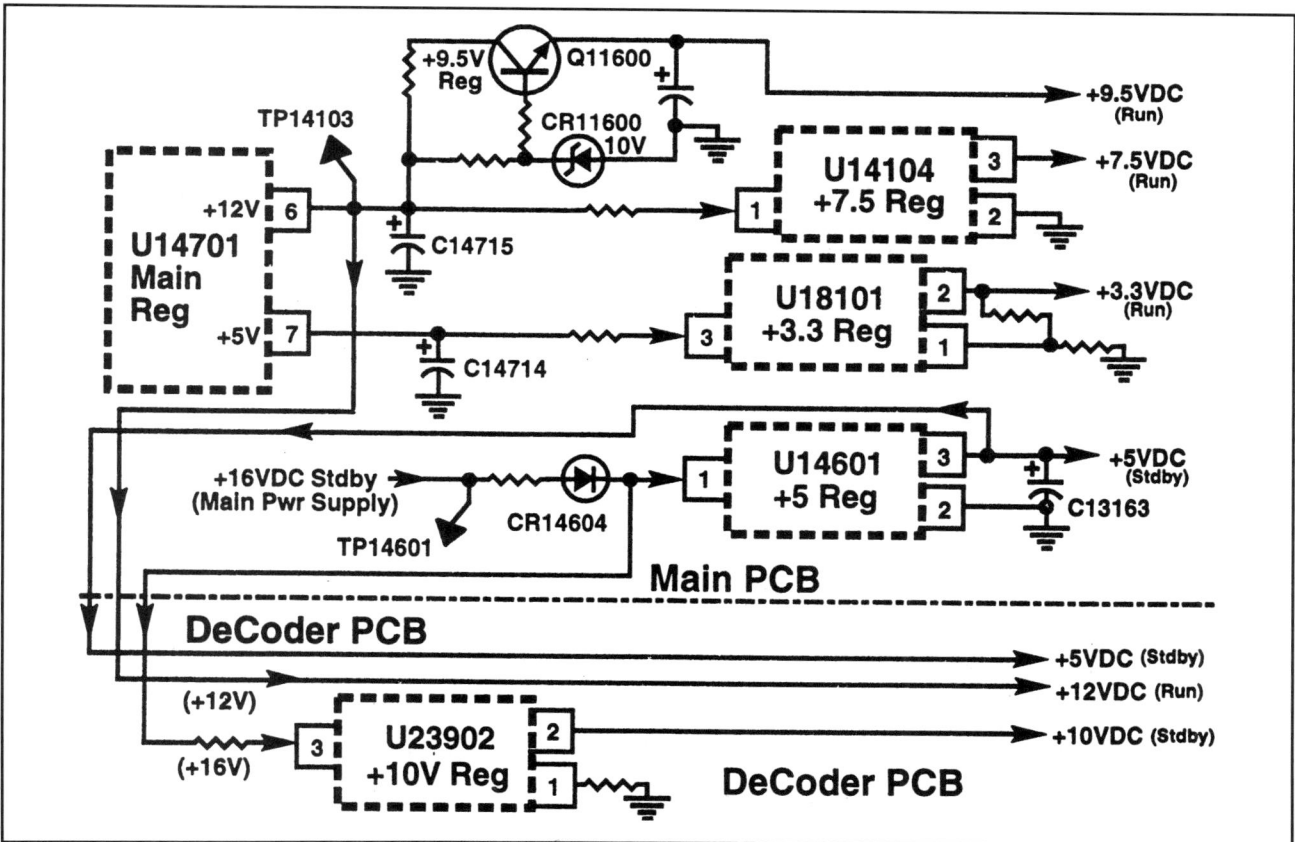

Figure 5-8. Auxiliary power supply.

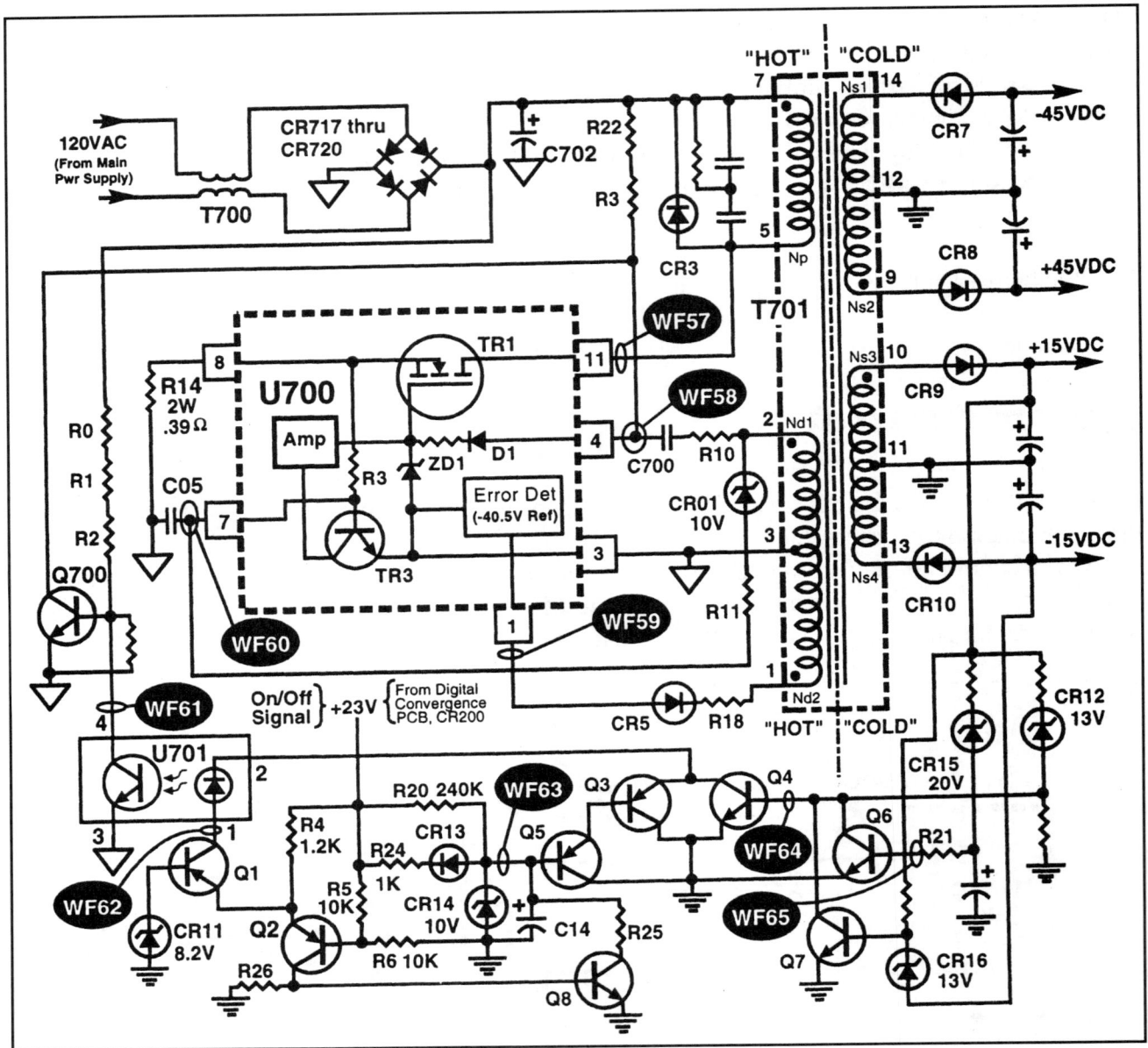

Figure 5-9. Digital convergence power supply.

The Microcomputer and Busses

System control is based on a single 8-bit ST9296 microcomputer which has several new features over those used in the past. The new features include an IR preprocessor, sync presence detector, frequency multiplier for the CPU clock, a UART (universal asynchronous receive transmit), closed-caption decoder, three A/D inputs, and an on-screen display that supports 3-bit D/A outputs.

Three I-Squared-C busses communicate with the majority of digital devices in these televisions. These busses are called the standby bus, the run bus, and the GemStar™ bus (for the TV Guide Plus+™). The standby bus connects the main EEPROM and decoder interface microcomputer with the microprocessor. The run bus connects the rest of the I-Squared-C devices. The remaining bus controls just the GemStar module when it is present. The standby and run busses operate at 50 kHz, while the GemStar bus operates at 100 kHz. The standby bus is always active, while the run bus comes up only when the TV is on. The GemStar bus can be activated via software control without powering up the chassis. If you are familiar with TV Guide Plus+, you will know it updates itself at any time through downloads from the source station. The CTC195/197 is so configured that the chassis does not need to be on for the update to take place. The block diagram in *Figure 5-10* will tell you which integrated circuits are tied together by means of the I-Squared-C bus.

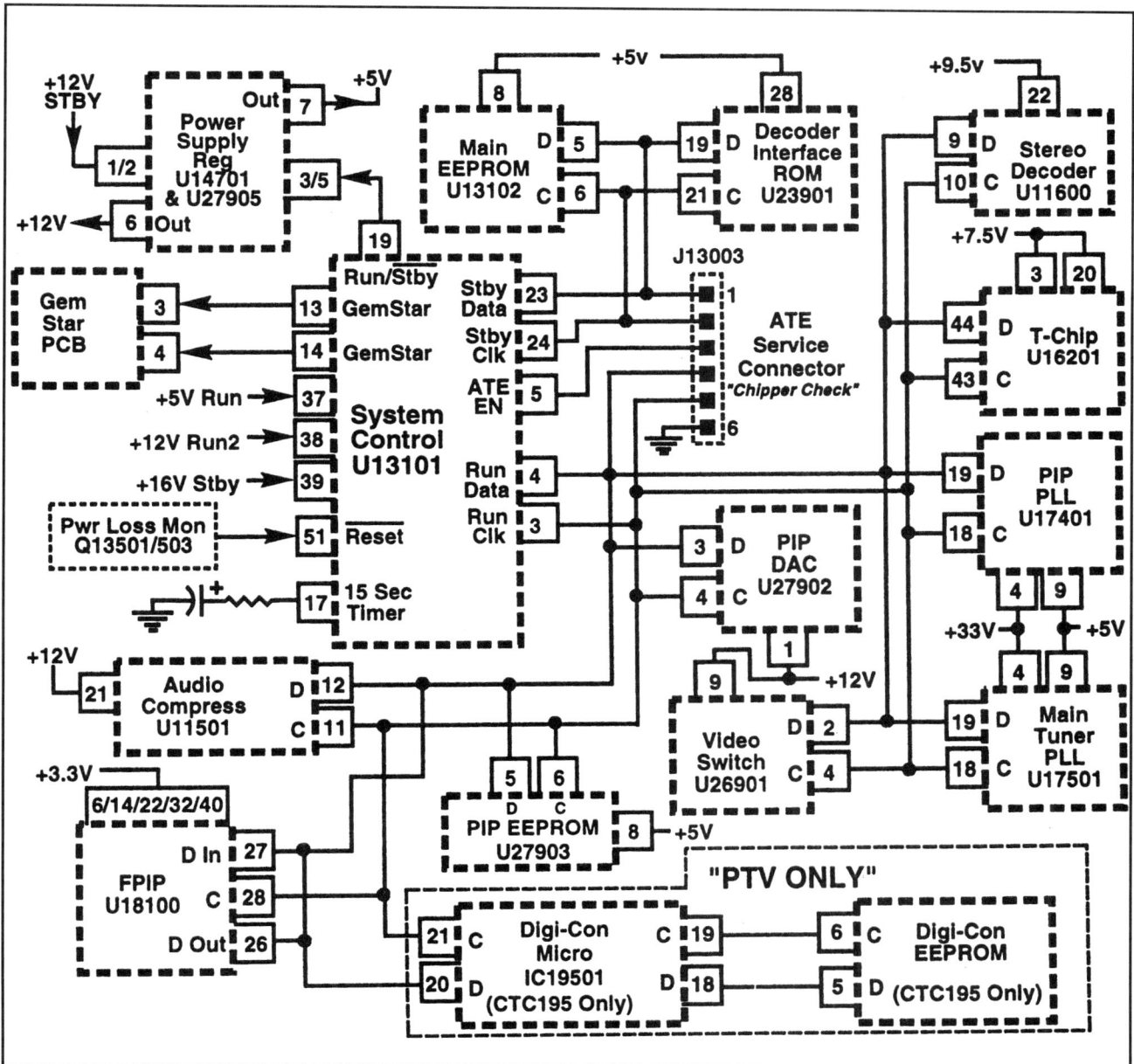

Figure 5-10. System control block diagram.

(1)	O	IF VC1	FM STEREO	I	(56)
(2)	O	IF VC2	FM TUNED	I	(55)
(3)	O	RUN I2C CLOCK	DATA OUT	I/O	(54)
(4)	I/O	RUN I2C DATA	DATA IN	I/O	(53)
(5)	I/O	KD1/ATE ENABLE	RESET I	I	(52)
(6)	I	KS1	RESET	I	(51)
(7)	I	KS2	SPEAKER MUTE	O	(50)
(8)	I	KS3	SVM	O	(49)
(9)	O	PAL 50/60 HZ	AVR	I	(48)
(10)	I	2ND TUNER AFT	TILT D/A	O	(47)
(11)	O	GEM LOW POWER	SRS NORM/EN	I/O	(46)
(12)	O	DI RESET	DEGAUSS	O	(45)
(13)	I/O	GEM I2C DATA	CONTROL #2	O	(44)
(14)	O	GEM I2C CLOCK	CONTROL #1	O	(43)
(15)	I	CC VIDEO	OSC IN	I	(42)
(16)	I	VDD2	VSS2	I	(41)
(17)	I/O	15 SECOND TIMER	OSC OUT	O	(40)
(18)	I	2ND TUNER SYNC	+16 VOLT STANDBY A/D	I	(39)
(19)	O	RUN/STANDBY	+12 VOLT RUN A/D	I	(38)
(20)	O	EEPROM ENABLE	+5 VOLT RUN A/D	I	(37)
(21)	I	MAIN TUNER H	IR	I	(36)
(22)	I/O	DI BUS ENABLE	HORIZ SYNC	I	(35)
(23)	I/O	STANDBY I2C DATA	VERT SYNC	I	(34)
(24)	O	STANDBY I2C CLOCK	FILTER OSD	I	(33)
(25)	O	FAST SWITCH	VDDA	I	(32)
(26)	O	BLUE OSD	FILTER CPU	I	(31)
(27)	O	GREEN OSD	VSS1	I	(30)
(28)	O	RED OSD	VDD1	I	(29)

U3101

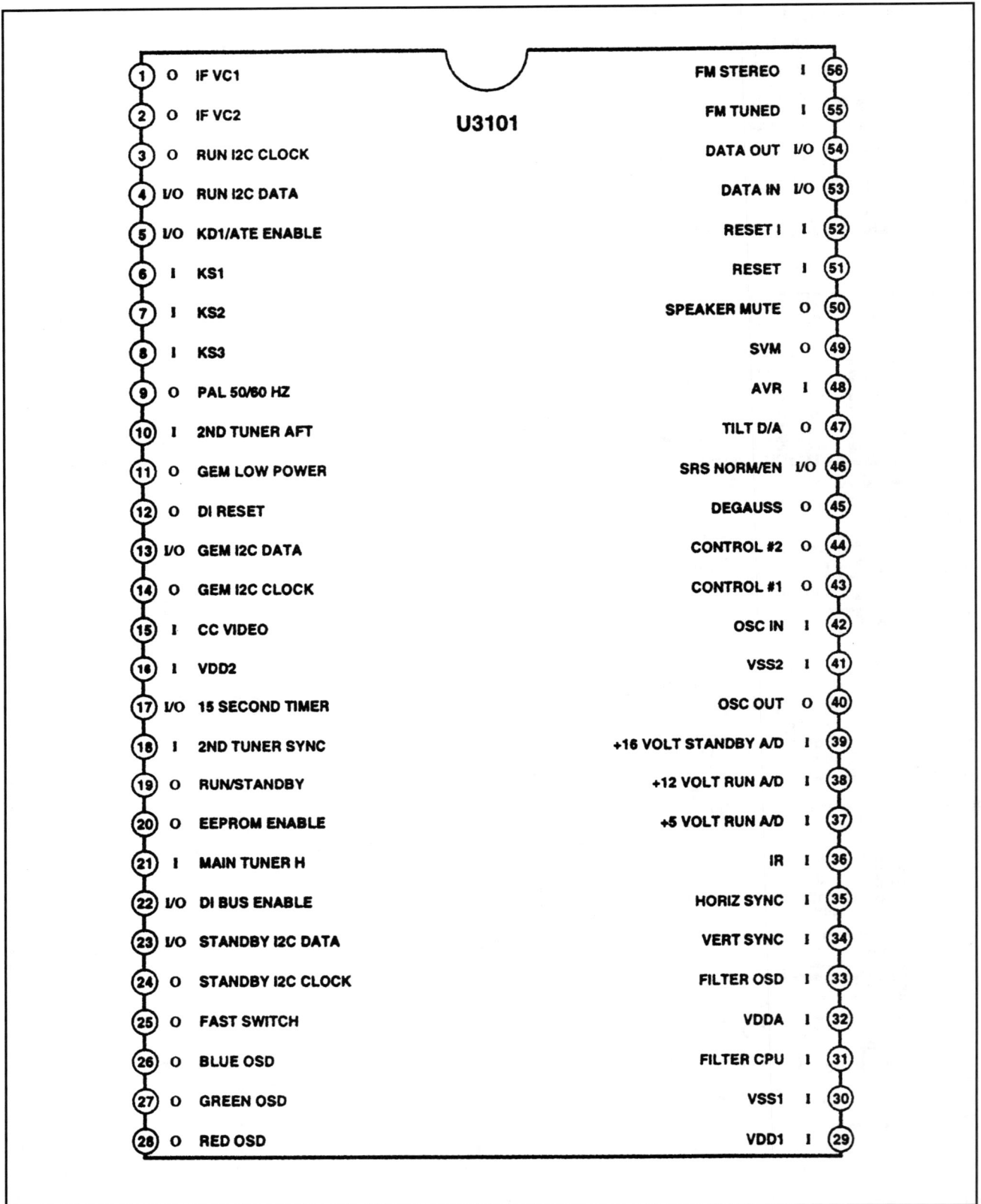

Figure 5-11. U13101 pinouts.

Power Control of the EEPROM and T4-Chip

As we saw in the CTC185 chassis, the microprocessor controls B+ to the EEPROM and T4-Chip. It uses Q13109 to turn the EEPROM on and off and U14104 to turn the T4-Chip on and off. The micro goes through a power-up sequence every time it turns on, a part of which is turning the power off to these devices, and back on to reset them

AC Line Dropout Detector and Reset

The reset circuitry monitors the standby +5 and +16-volt lines and warns the microprocessor when a power failure may be occurring. Pin 51 of U13101 is normally held high by a simple resistor network as long as the 5-volt standby source is up (*Figure 5-12*). If the +5-volt supply fails, pin 51 will go low. Q13505 is held off by the presence of the 16-volt source and will not turn on unless the +16 volts drops to about +7.5 volts. At that point, Q13505 will turn on and pull pin 51 low. In other words, if something happens to the 5-volt or the 16-volt supply, the voltage at pin 51 will go low. The microprocessor responds by disconnecting the busses internally and going into a backup routine. The backup routine also involves disabling the front-panel controls and the IR input and turning off the run/standby lines, which removes B+ to the T4-Chip and the run regulators.

"Batten Down the Hatches"

The literature refers to a system control routine called "batten down the hatches" and calls it one of the most important sequences for the technician to understand. It is invoked to save settings and alignments during any problem the microprocessor senses and causes the microprocessor to write an error code to cue the technician with respect to what has happened.

The batten-down sequence will occur when the standby 16-volt supply drops to about +9.5 volts during power-up, or to about 2 volts below the reading of the standby D/A on pin 39 about 1.5 seconds after power-up or 1.5 seconds after power-down. The 1.5-second delay is to keep the microprocessor from shutting down due to a normal power supply dip at start-up or shut-down. The delay, in other words, gives the power supply time to stabilize.

After the 1.5 second delay, if the 16-volt standby source drops below 9.5 volts, the batten-down-the-hatches sequence begins. The first actions are to reduce demands on the power left in the main power supply by turning off power to the audio amplifiers, the run supplies, OSD, and any other circuit not necessary to saving information to the EEPROM. *Figure 5-13* shows you the sequence in graph form. All instrument information is written to the EEPROM during the next 10 milliseconds, and then the EEPROM itself is disabled by switching off its B+ supply.

After the sequence of events take place, two other things may occur, depending on the state of the main power supply. First, the 15-second timer at pin 17 will tell the microcomputer how long power has been disconnected. If the power loss lasts less than 15 seconds, the microprocessor will power the TV back

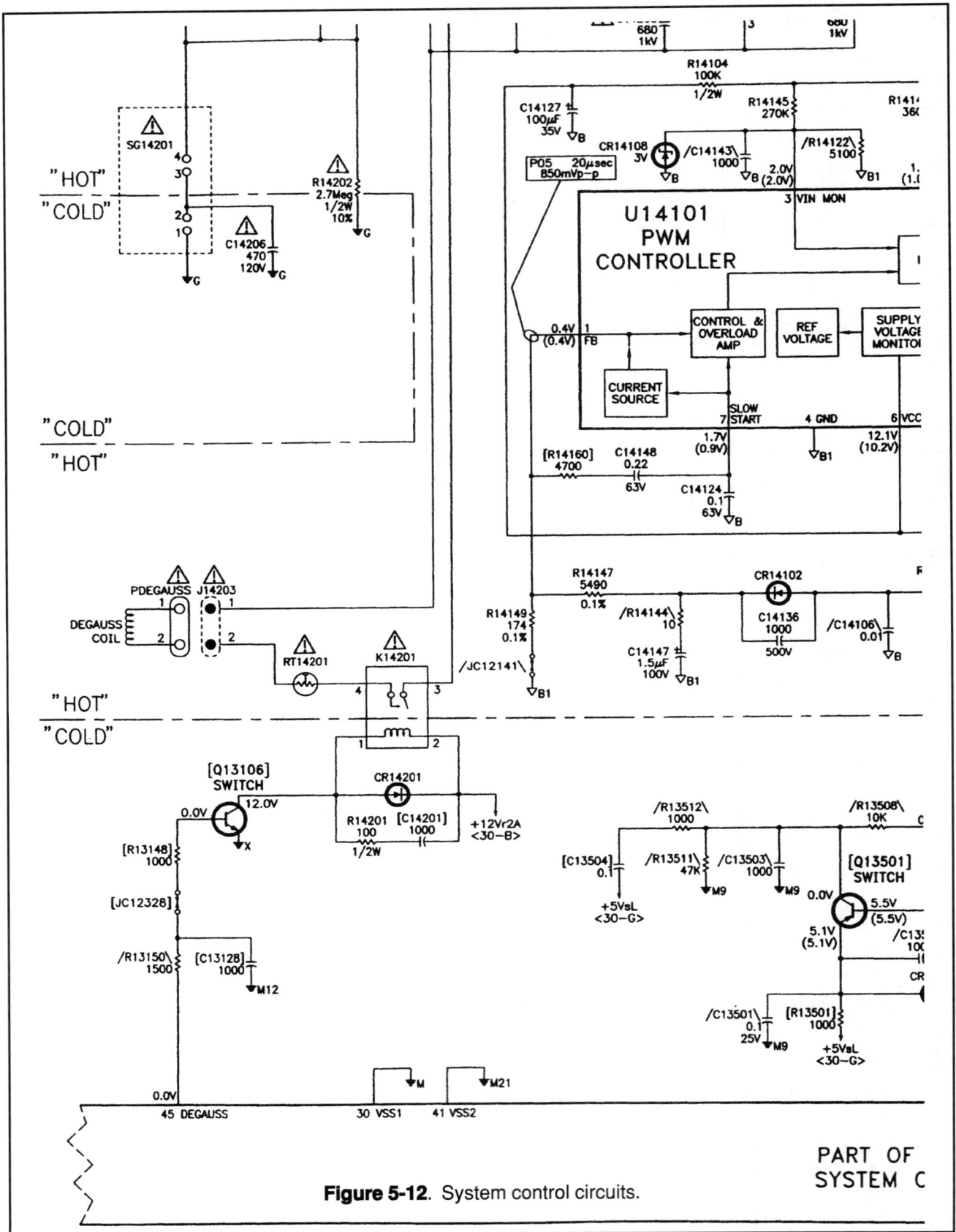

Figure 5-12. System control circuits.

PART OF
SYSTEM C

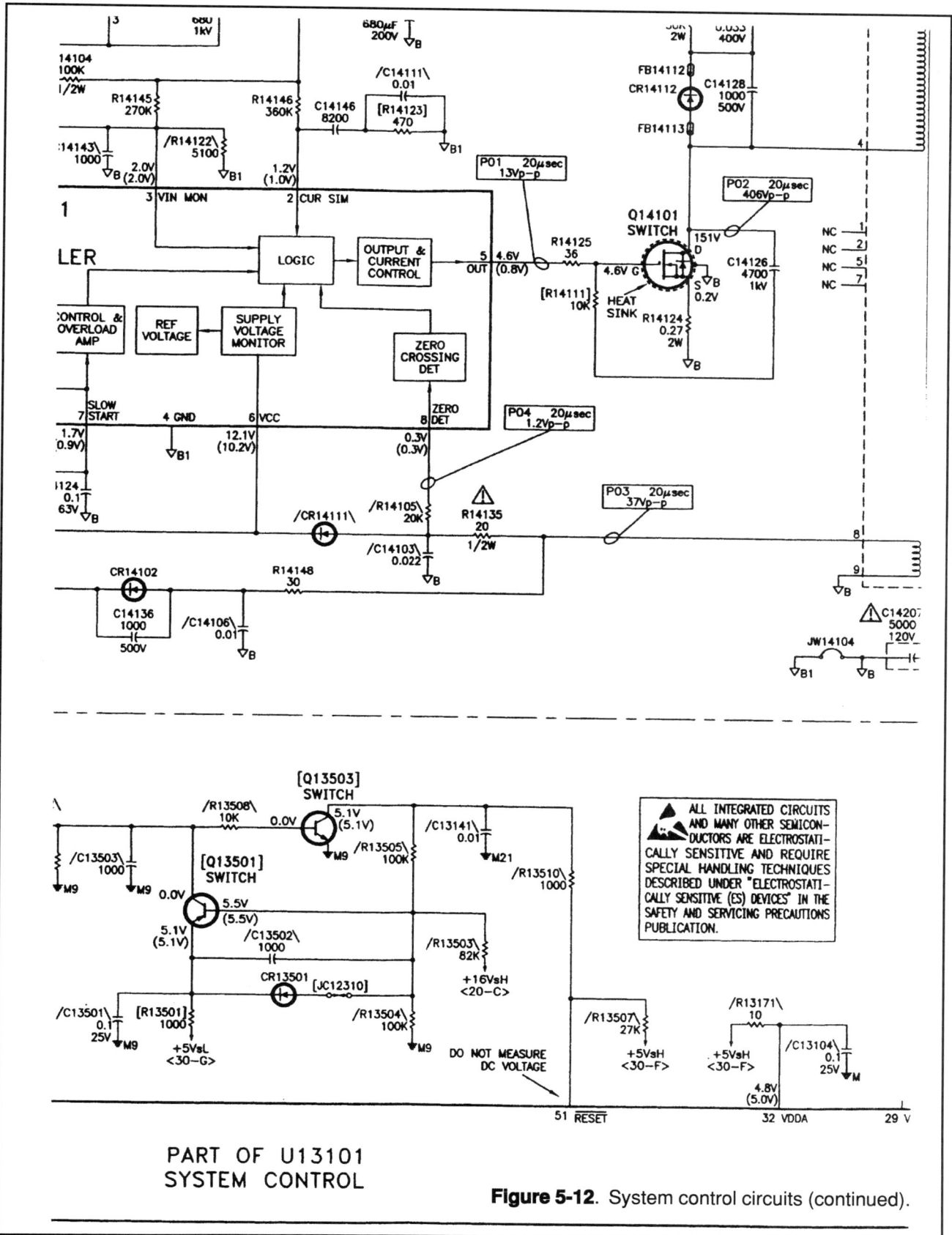

PART OF U13101
SYSTEM CONTROL

Figure 5-12. System control circuits (continued).

up with no loss of data (which also includes the OSD clock!). Second, if the power loss has been longer than 15 seconds, the clock time will be lost.

I just mentioned a "15-second timer," which Thomson uses for the very first time in the CTC195/197 chassis. Look at pin 17 of U13101 in *Figure 5-13A*, and you will see the circuit in detail. According to the description given of the function of pin 17 (see appendix), this circuit determines if the time-of-day clock information should be discarded after a power loss. If the power loss exceeds 15 seconds, the time-of-day clock information will be discarded.

Power-Off-Reset (POR)

A part of the problem I have with this chapter is knowing when to discuss what, and what should be left out. POR is a good example. Much of it is redundant, but part of it isn't. So, I will mention it here. The T4-Chip detects when the standby power source drops too low for reliable operation, and terminates deflection which turns the TV off and latches the POR. The microprocessor reads the latch via the bus. The latch has to be reset to restore operation by issuing an OFF command followed by an ON command. You must follow this sequence because the POR detector latches in the off-to-on transition of the on/off control bit in the T4-Chip. In other words, it latches while the television is on; therefore, turn the TV off and then turn it back on. If the voltage is still too low to operate, you will have to repeat the process.

Run Supply Detector

The microprocessor, as I have already indicated, monitors the +16-volt line at pin 39 via a 6-bit A/D to verify that the supply is active and within regulation. If the +16-volt line fails, the microprocessor will initiate the sequence I just discussed, namely, batten-down-the-hatches. This line is monitored to con-

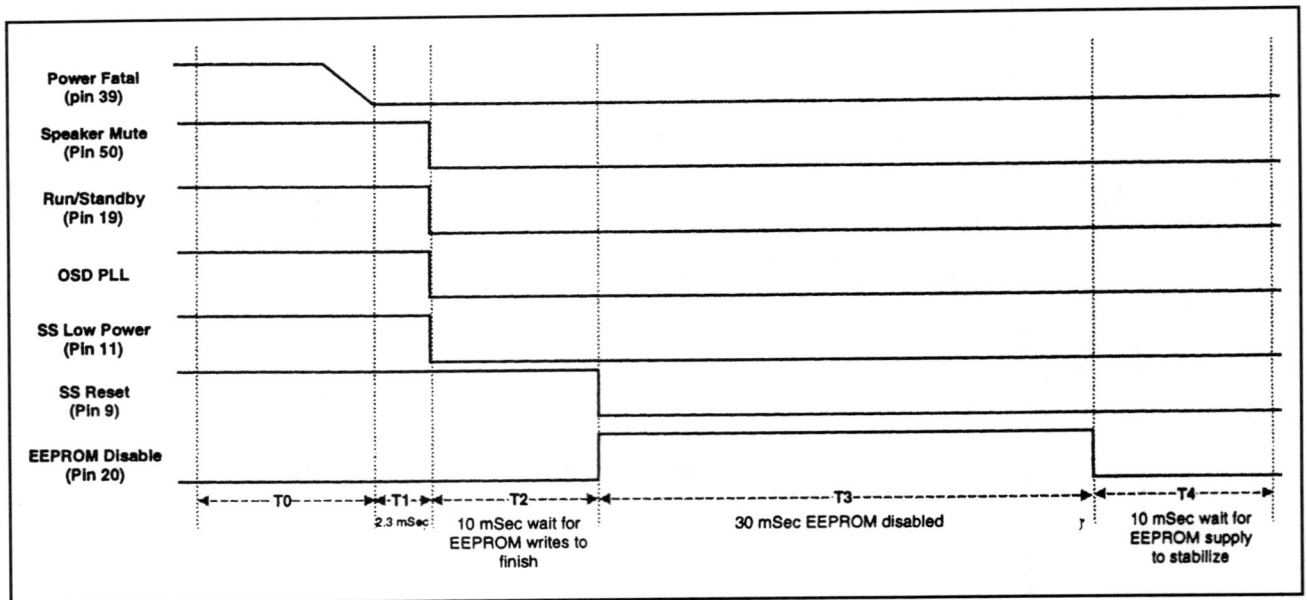

Figure 5-13. "Batten down the hatches" sequence.

trol reset of the main micro and to provide a sense level for the TV Guide Plus+ module low power monitor.

It also monitors the +5 and +12-volt run supplies from inputs to pins 37 and 38 once the TV has been turned on. If the run supply is not present when the TV is first turned on, the microprocessor will abort the start sequence and try to start the set again. It will go through the power off-on sequence three times, and if the TV does not start up, it will power the TV down. This is known as the "three strikes and you are out" sequence. You can begin the start sequence again by pressing the power button on the TV. Since there are just three error code locations, each attempt to restart the TV will cause an error code to be written. If you restart the set, the microprocessor will overwrite the error codes generated by the previous cycle with new ones.

You might also note that loss of horizontal deflection may cause the run supply detector to trip. If horizontal deflection fails, the 140-volt B+ supply will begin to climb. The power supply error amplifier, which monitors the +140-volt line, will respond by shortening the duty cycle of the MOSFET in order to reduce B+ output. But the +12-volt line will already be loaded and will respond to the shortened duty cycle by slumping. If the slump is greater than the minimum requirements the microprocessor expects, the run detector will trip, and the microprocessor will write a run-supply error. The moral is, when you see a run-supply error code, verify that it has not been caused by a failure in the horizontal deflection circuit!

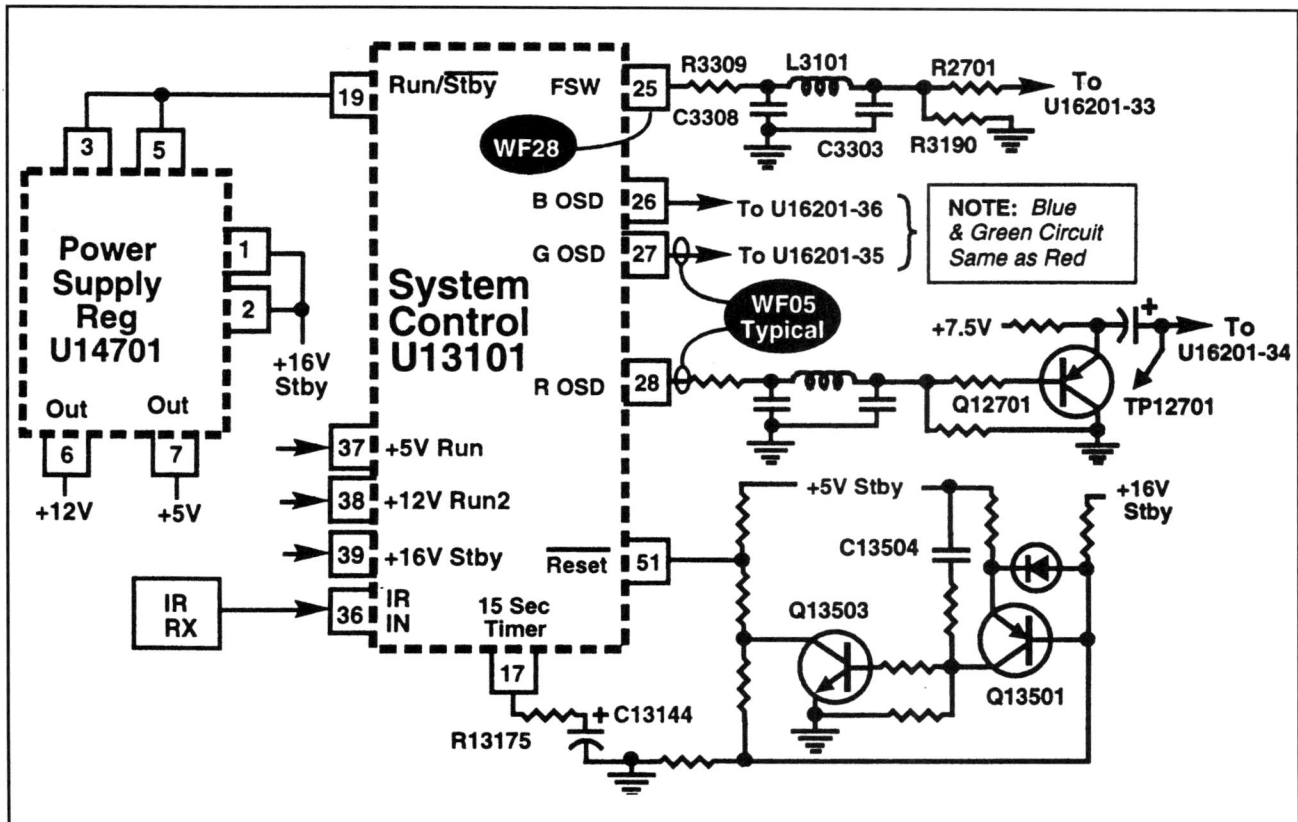

Figure 5-13A. 15-second timer circuit.

Main Power Supply On/Off Control

The TV is turned on and off by controlling the main power supply. When it is plugged into an AC outlet, the power supply comes alive, providing U13101 with the required +5 volts for B+ and reset. After it resets, it sends a HIGH to pin 20, which turns on B+ to the EEPROM. The microprocessor proceeds to check the EEPROM address for a handshake (acknowledgment). If acknowledgment is not forthcoming, the micro continues its request. You can see what is happening by scoping the clock and data lines to the EEPROM. We are going to assume the handshake has been offered and accepted.

The power supply is up and running, and system control has everything it needs to turn on the TV. To turn it on, issue an ON command via the front controls or the remote control. The microprocessor responds by pulling pin 20 momentarily low to reset the EEPROM. The +16-volt supply will momentarily dip as the set comes alive. During this brief time, the video and audio mute lines are held low so that no picture or sound will be processed by circuits having some residual voltage supply remaining.

When the circuit has stabilized, the OSD and tuner come alive. As soon as a channel has been captured, video blanking is turned off, permitting video to pass. When the high voltage reaches normal level, the CRT will display a picture. If you want a visual representation of the power-up sequence, take a look at *Figure 5-14*.

Power-down sequence follows the reverse order: (1) Volume level goes low and speakers mute; (2) T4-Chip stops the deflection circuits, and; (3) the run/standby pin (pin 19) of the microprocessor turns off the 5 and 12-volt run supplies. The TV is now in the OFF mode.

A Miscellany

The microprocessor has the capability of detecting the hardware attached to it (like the CTC179/189). If it detects the appropriate hardware, it supports it. If it does not detect the hardware, it assumes the feature is not supported in the chassis and runs without it. An example would be the convergence board in the projection sets. If it is disconnected or malfunctions, the microprocessor will not receive an acknowledgment and will revert to a direct-TV mode. The auto-detect feature extends to TV Guide Plus+, MCR (a commercial chassis), the Digital Convergence module, and the two-tuner PIP.

I will conclude this description of system control by pointing to a few of the signals input to the microprocessor.

(1) Selected video out is buffered by Q13306 and applied to pin 13 for decoding of closed-caption information.

(2) Video out of the T4-Chip is buffered by Q13101 and applied to pin 38 for tuning sync.

(3) Horizontal and vertical pulses are applied to pins 24 and 25 to provide a reference for OSD information.

Troubleshooting System Control Problems

System control is responsible for operating every function of these televisions. If it fails, the TV simply will not work. If U13101, U13102 and U16201 are in working order, you can force the TV to turn on in the service mode. Once in the service mode, you can read the error codes, which will probably point you to the defective circuit. There will be cases when you cannot force it on even in the service mode. When you encounter such a problem, you will need to use the Chipper Check to read the error codes.

Perhaps this is a good place to make an observation about the microprocessor's input and output voltages. Many of them are digital. This means the voltage will either be a logic high (in this instance 2.5 to 5 volts) or a logic low (in this instance less than 2.5 volts). This is what you will find if you use a standard multimeter to check them. You will need an oscilloscope to check clock and data activity. You won't be able to see the details of the data stream, but that is not important. You are looking for activity. If the clock line is flat, you will more than likely be facing a microprocessor problem. If the data line is flat, you will first of all have to understand what communication should be taking place before you assume the microprocessor is at fault. It is therefore very important that you learn as much as you can about this aspect of the television. I am including in an appendix a complete description of each pin of the microprocessor. The information is taken from the *Technical Training Manual*, pg. 60-64.

We will begin by assuming the television has tried to start three times and stops—"three strikes and you are out." How do you know it tried to start three times? Listen for the click of the degauss relay. If you hear the clicks, you know the EEPROM, T4-Chip, and U16201 are working; therefore, the problem is most likely in the power supply and/or the deflection circuits. I am reproducing RCA's troubleshooting chart in *Figure 5-15*, and I will give it in this text in abbreviated form.

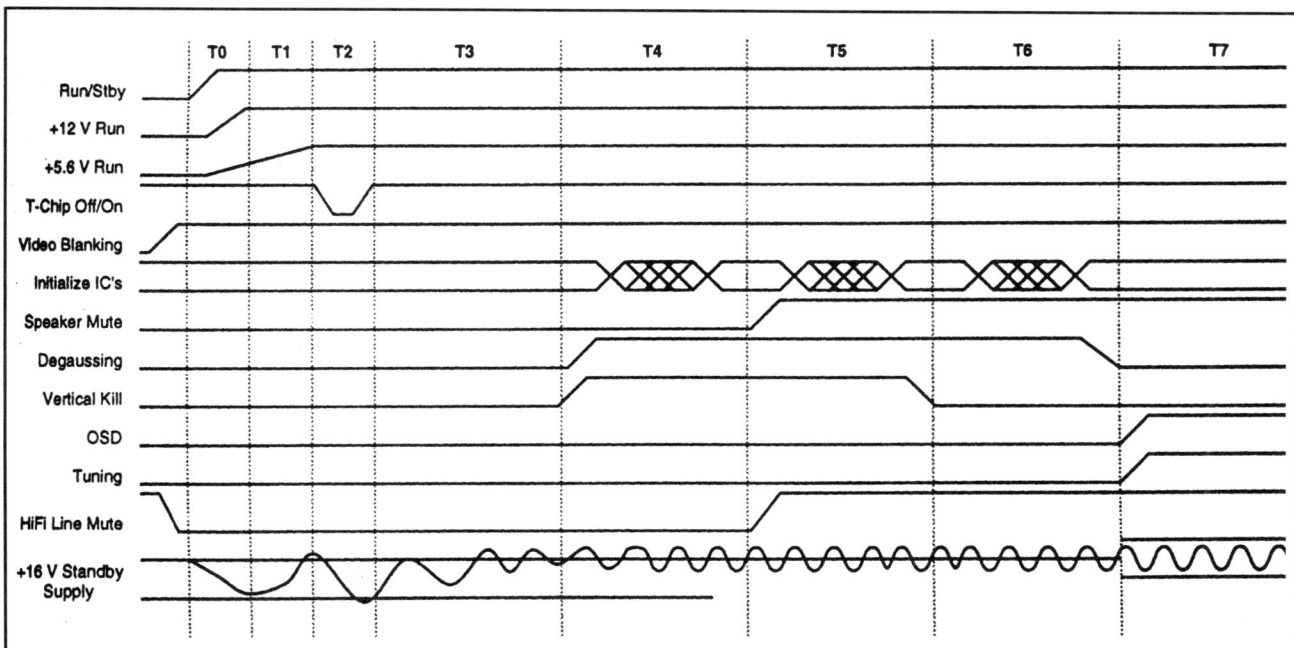

Figure 5-14. Power-up sequence representation.

(1) Check the +16-volt standby supply input at pin 39 of the microprocessor. Also check for a high at pin 19 which turns on the run/standby supply.

(2) Check pins 37, 38 and 51. Pins 37 and 38 monitor the +5-volt and +12-volt supplies, and pin 51 is reset.

(3) Apply AC and check for horizontal pulses at pin 22 of the T4-Chip when you give the TV an ON command. If pulses aren't present, think in terms of a defective T4-Chip or corrupted EEPROM.

(4) If horizontal pulses are present, unsolder the collector of the horizontal output transistor. Press the power button so the TV will attempt to start. Before the third attempt, apply an external +16 volts to the cathode of CR14107 to hold up the +12-volt supply. When the main power supply comes up, remove the external supply and confirm horizontal drive pulses. This procedure will confirm a horizontal deflection problem. I will have more to say about horizontal deflection problems later on in this chapter.

Of course the next question is, "What if I issue an ON command and the degauss relay doesn't click?" You will find the following information in the troubleshooting guide I included as a part of *Figure 5-15*.

(1) Check pin 19 of the microprocessor to confirm that it goes to about +5 volts (i.e., it goes high). If it goes high, check U14701 and associated circuitry in the power supply on/off control. If these checks don't uncover a problem, suspect a power supply problem and begin checking voltages and waveforms there.

(2) If pin 19 does not go high, check pins 16, 29 and 32 of the microprocessor for +5 volts. Remember to check pin 8 of the EEPROM for +5 volts also. Check for +7.6 volts at pins 3 and 20 of the T4-Chip. If any one of the supplies are missing, check the respective +5-volt standby source and the EEPROM-T4-Chip power control circuitry.

(3) If the voltages are normal and present, check for a normal reset signal at pin 51 of the micro. It should be at a 5-volt level. If it isn't, troubleshoot the reset circuit.

(4) If reset is normal, check pins 40 and 42 of the micro for its oscillator signal. The signal should be at about a 4.5 Vp-p level.

(5) Use your oscilloscope (10 msec/div) to monitor pin 23 of the microprocessor while you apply AC to the TV to look for momentary clock and data pulses after the data line rises to 5 volts. The data line should go high and pulses should appear.

If the data line does not go high, wick out pin 23 to see if the high appears. If the high appears, something on the bus besides the microprocessor is pulling it low. If the high still does not appear, check the 5-volt pull-up supply from R13327.

Set Will Not Turn On

If the television tries to start three times and then stops (you can hear the clicking sound of the degauss relay), the EEPROM (U13102) and T-chip (U16201) are working - hardware OK. Problem is most likely power supply or deflection related. Use the following flowchart to isolate the fault.

(A)

Degauss relay does not click when pressing power button

Press power button and monitor U3101 pin 19

Does pin 19 go Hi - (5 Volts)
— Yes → Power supply regulator/on off control circuit or associated components, U4701
— No →

U13101 pin 16, 29,32,51 U3102 pin 8=5 volts
— No → Repair defective power supply source
— Yes →

U16201 pin 3& 20=7.6 volts
— No → Repair defective power supply source
— Yes →

U3101 pin 40&42– 4.5Vp-p,.4MHz sine wave
— No → Y3101,C3106 C3107,R3107 U3101
— Yes → (B)

(B)

Scope probe (10msec/div) to U3101 pin 23. Apply 120V-AC power & check for presence of momentary clock and data pulses after data line rises to +5V

Data line goes Hi and pulses appear
— Yes →
— No → Unsolder pin 23 of U13101

Clock and data pulses on pins 43 & 44 of U16201
— Yes → U16201 U13102
— No → Buss problem

Pin 23 pad goes to +5V
— Yes → U13101, front panel/ keyboard scan or drive lines
— No → +5V pull-up supply through R13142, shorted device pulling down data line

Degauss relay clicks three times then set shuts down
— No → Go to chart "A"
— Yes →

U13101 pin 39=16 volts, pin 19=Hi
— No → Repair power supply
— Yes →

U13101 pin 37,51=5 volts, pin 51=12 volts
— No → Repair power supply
— Yes →

Place scope probe on U6201 pin 22

Horiz. pulses momentarily output when power button is pressed
— No → U6201 or corrupted EEPROM (U3102)
— Yes →

Unsolder the collector of the HOT - Q4401, press the power so the set attempts to start. Before the third attempt, apply external 16 volt power supply to the cathode of CR4107. Once the main power supply comes up, the external supply can be removed. Confirm horiz drive pulses at U6201 pin 22. If pulses are present there is a problem in the horizontal deflection circuit. If not present the problem is most likely in the system control or power supply circuitry.

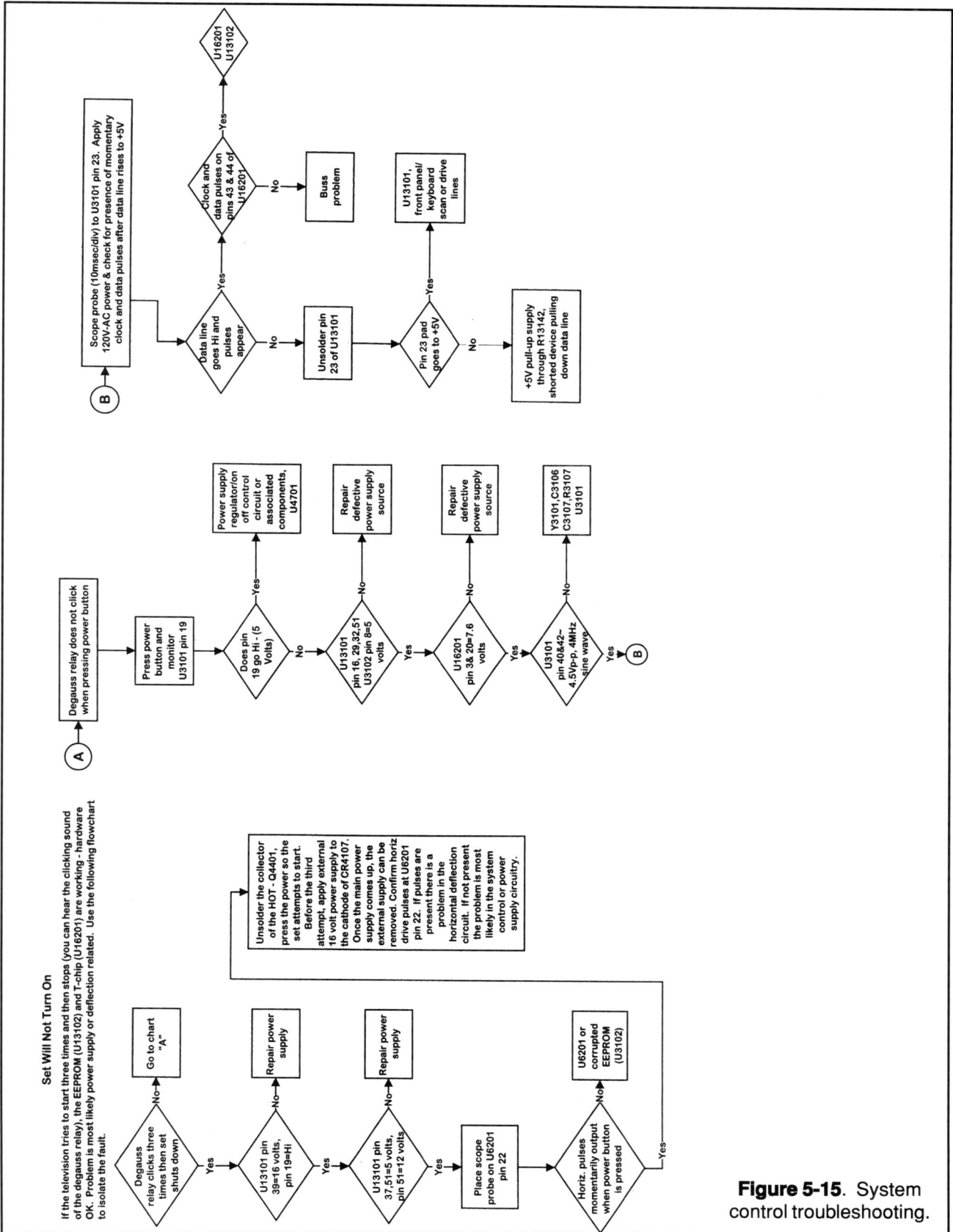

Figure 5-15. System control troubleshooting.

If it goes high but the negative-going pulses do not appear, unsolder the clock line (pin 24) and check for continuous pulses. If the pulses don't appear, think in terms of a defective microprocessor.

(6) If the negative-going pulses appear when power is first applied or when the clock line is disconnected, reconnect the clock line and press the power button as you look for data activity on pin 23. If data activity appears as the power button is pressed, check pins 43 and 44 of the T4-Chip to see if clock and data pulses are present. If they are present, suspect a problem with U16201 or U13102.

If pulses still don't appear, you should suspect a problem with the front panel or the key board drive and scan lines.

Remote Control Problems

There may be occasions when the remote control won't work because of problems with the IR receiver. Troubleshooting it is really straightforward, but I do need to comment on the receiver. Some of these IR devices may be sensitive to fluorescent lighting. If pin 36 of the microprocessor shows a 5 Vp-p signal of constant noise, remove the lighting by turning it off or by covering the remote receiver and rechecking the signal.

Also, RCA's protocol gives the keyboard priority over the remote control. If a keyboard button is stuck, the microprocessor will ignore any input from the remote receiver.

Run Bus Latch

The run bus will latch when either line is clamped to ground. A circuit path short or a power supply failure to any device on the bus line can cause the problem. The error code will tell you which device you should troubleshoot.

NOTE: Any IC connected to the I-squared-C bus has to be fully powered to prevent the ESD protection diodes on the bus line from clamping the bus.

Horizontal Deflection

The horizontal deflection circuit does two things: (1) It supplies the current for the yoke coils, which generate the magnetic field that moves the electron beam horizontally along the face of the picture tube; (2) and it provides the power to generate a number of voltages used to operate the TV. The yoke current is derived from a switch (the HOT), the primary inductance of the flyback, a retrace capacitor, the trace capacitor (sometimes called the "s-shaping capacitor"), and the horizontal yoke coils. The voltage supplies are derived from the various windings on the flyback.

The so-called "low level" horizontal signal processing is done in the T4-Chip (see *Figure 5-16*) and consists of the sync separator, automatic frequency control, automatic phase control, horizontal drive, east-west pincushion correction, x-ray protection, and the horizontal Vcc standby regulator. As you can tell from the way this discussion is going and by studying the diagrams, this T4-Chip does about the same thing the T-Chip and the T4-Chip in the CTC185 did. Because there is a great deal of overlapping, I will try to discuss what is new here and leave you to reread what I have said about horizontal signal processing in previous chapters.

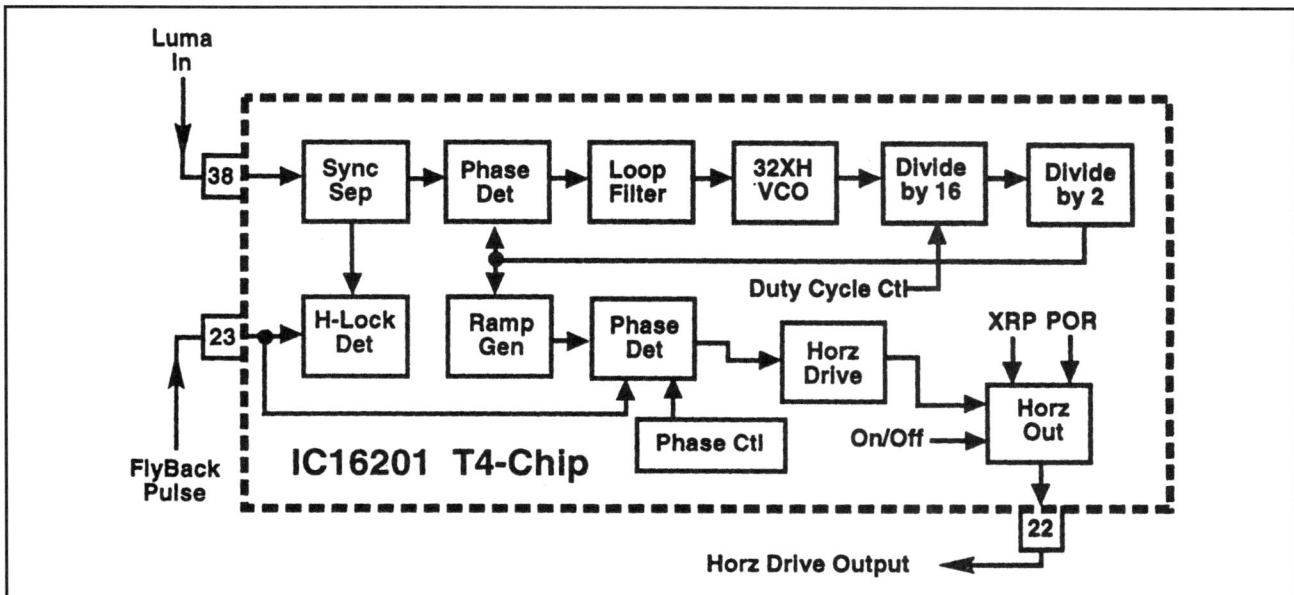

Figure 5-16. T4-Chip signal flow.

East-West Pin Correction

Low-level horizontal signal processing inside the T4-Chip includes the ability to make geometry corrections (vertical linearity, vertical S-shaping, and east/west pin correction). Look at *Figure 5-17* for the circuit particulars as this discussion progresses. L14402 makes horizontal linearity correction possible. C14405and R14403 are included to reduce ringing in this coil at the beginning of scan. C14404 helps the T4-Chip to make the necessary S-correction. C14401, R14401, and C14403 assist in reducing raster deformation.

East-west pin correction is accomplished by a diode modulator circuit. Note that this type of correction is not needed in the 25" and 27" sets because they use pin-corrected horizontal yokes. The circuit works by controlling the voltage at the junction of L14801 and C14805. Remember, the amount of horizontal scan is proportional to the voltage across the S-capacitor (C14404). The voltage at the top is at the B+ level (140 volts). The diode modulator produces a vertical rate parabolic waveform and applies it to the bottom of the capacitor to produce the desired modulation of horizontal scan. The result is east-west pincushion correction.

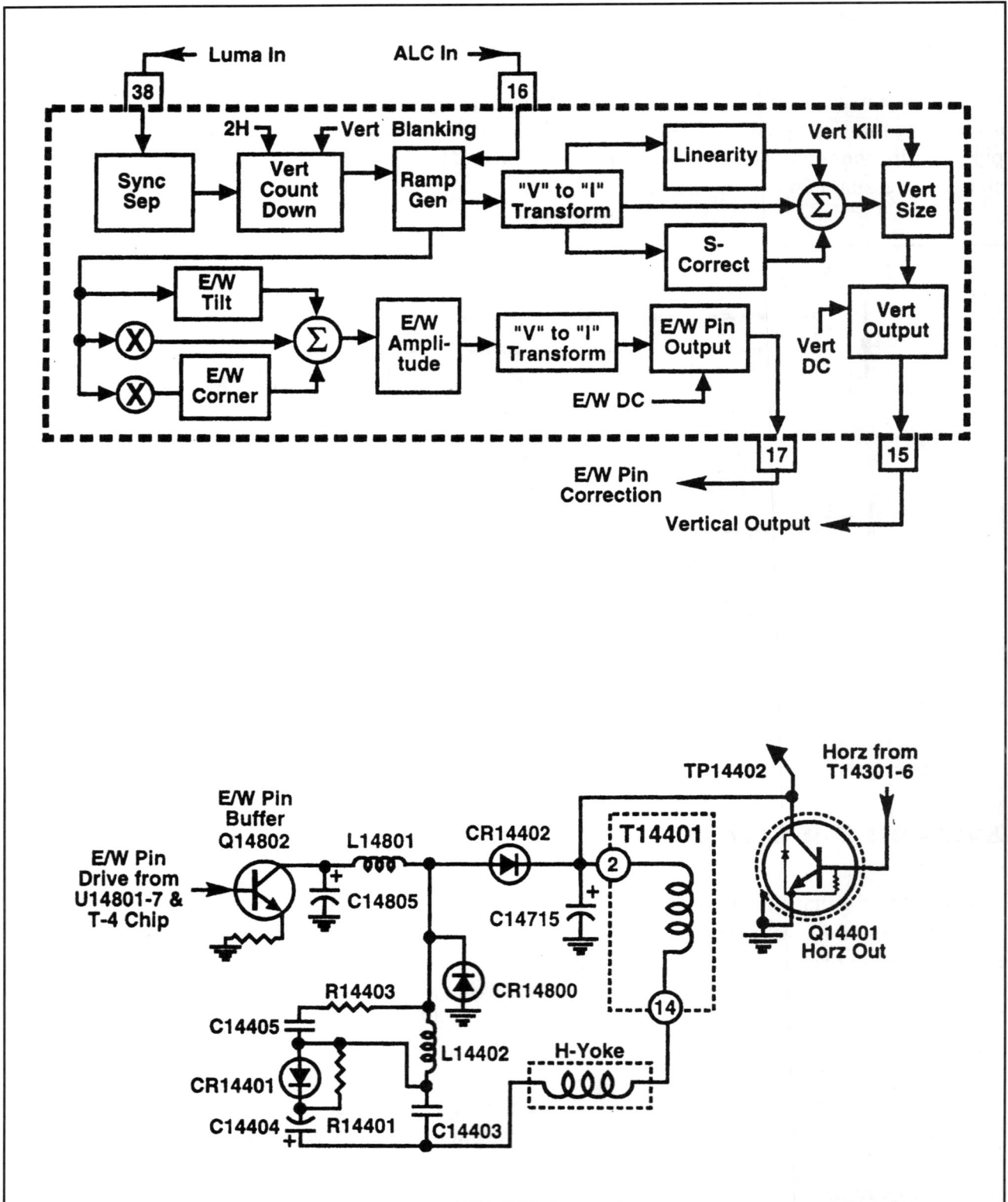

Figure 5-17. East-west pin correction.

Z-Axis Correction

Z-axis correction is necessary to counteract raster rotation when the CRT is oriented in a north to south direction. It is accomplished by adding a DC magnetic field to counteract the pull of the Earth's magnetic field. The circuit is shown in *Figure 5-18*.

Keep in mind the fact that the 32" and larger TVs use microprocessor-controlled z-axis correction. It frees space, allows adjustments to be made while the viewer is watching the screen, and makes a finer resolution possible.

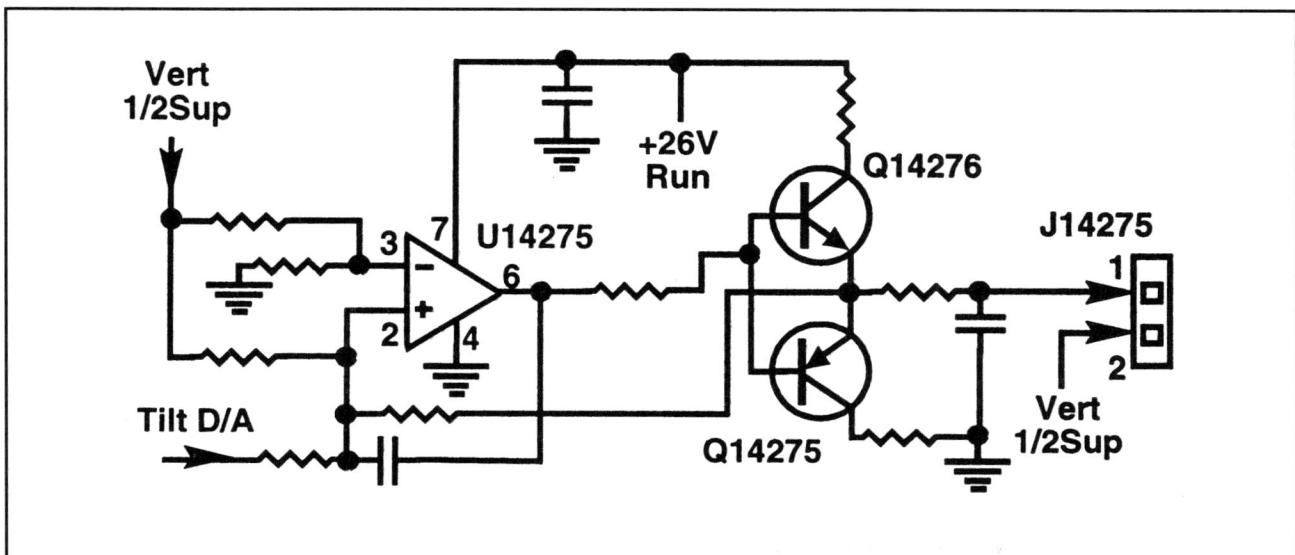

Figure 5-18. Z-axis correction.

X-Ray Protection

The input to the x-ray protection circuit is pin 24 of the T4-Chip (*Figure 5-19*). Q14902 is the switch that shuts off horizontal drive if the high voltage becomes too high. The circuit works by rectifying a filament pulse and applying the resultant voltage to the cathode of CR14902 (a 10-volt zener). If the voltage at the cathode exceeds the zener breakdown voltage, the resulting voltage will turn on Q14902, which applies a positive voltage to pin 28, turning off horizontal drive.

When an XRP condition occurs, system control will try to restart the set. If three XRP conditions occur within sixty seconds, system control will place the TV in the OFF mode and write an error code. You will have to press the power button to turn the set back on, which will start the XRP cycle over again if the high voltage is still too high.

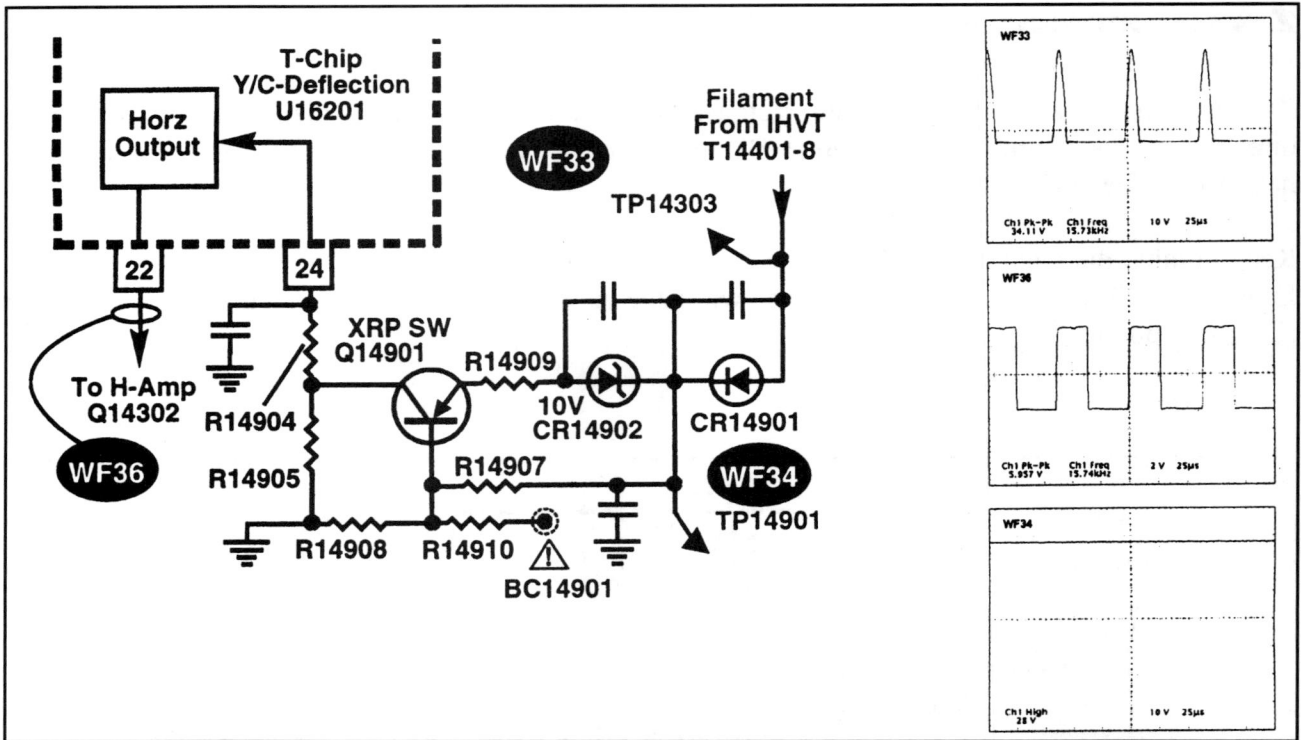

Figure 5-19. X-ray protection.

Troubleshooting Horizontal Problems

Figure 5-20 is a fairly detailed block diagram, including waveforms, of the horizontal section. You will note the addition of a buffer amplifier, Q14302, to the circuit. This transistor has been included to reduce current drain at pin 22 of the T4-Chip.

If you encounter horizontal troubles, you will probably be confronting a dead-set situation. The literature suggests this approach to resolve the problem.

   (1) Check the collector of Q14401 for +140 volts.

   (2) Check pin 20 of U16201 for +7.6 volts. If it is not there, check the 12-volt supply.

   (3) Check pin 22 of U16201 for horizontal pulses as you issue an ON command. If you don't see pulses, go to the dead-set troubleshooting section I gave you in our discussion of system control.

   (4) If pulses are present at pin 22, follow them to the stage where you lose them and troubleshoot it.

There may be occasions when you have to troubleshoot a "no horizontal sync" problem. If you do, begin by comparing the horizontal drive signal to the incoming video sync signal. One complete horizontal drive cycle begins and ends with one horizontal sync in video. Then check for AFC feedback at pin 23 of the T4-Chip. If it is missing, find out why. Finally, check the AFC filter at pin 21. If it is incorrect, suspect the components tied to pin 21.

Vertical Deflection

The vertical circuit is very similar to those we have already discussed, particularly the early versions of the CTC177. *Figure 5-21* is a schematic taken from the service manual, while *Figure 5-21A* is a rather full block diagram with the waveforms you can expect to see at the indicated points.

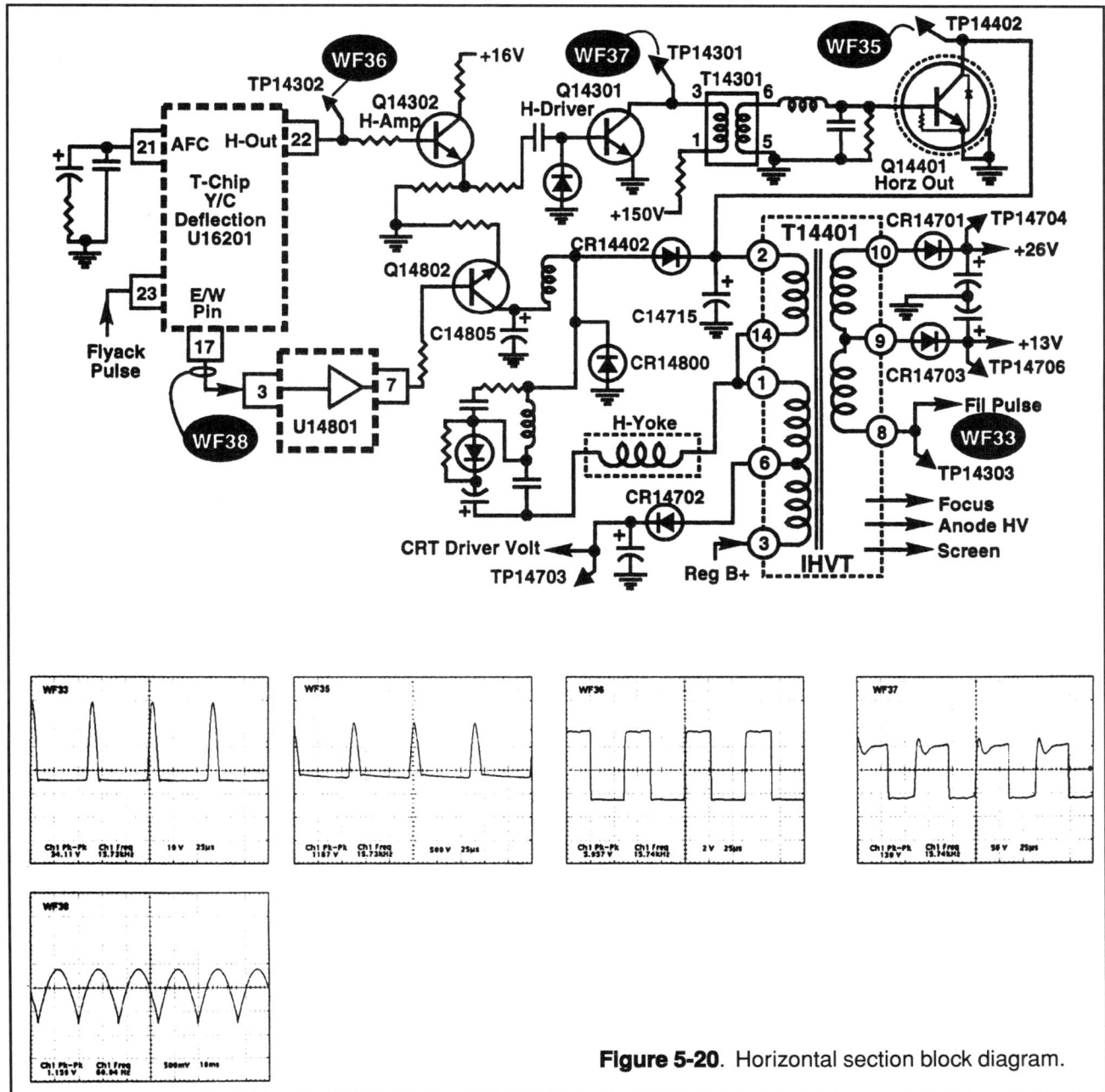

Figure 5-20. Horizontal section block diagram.

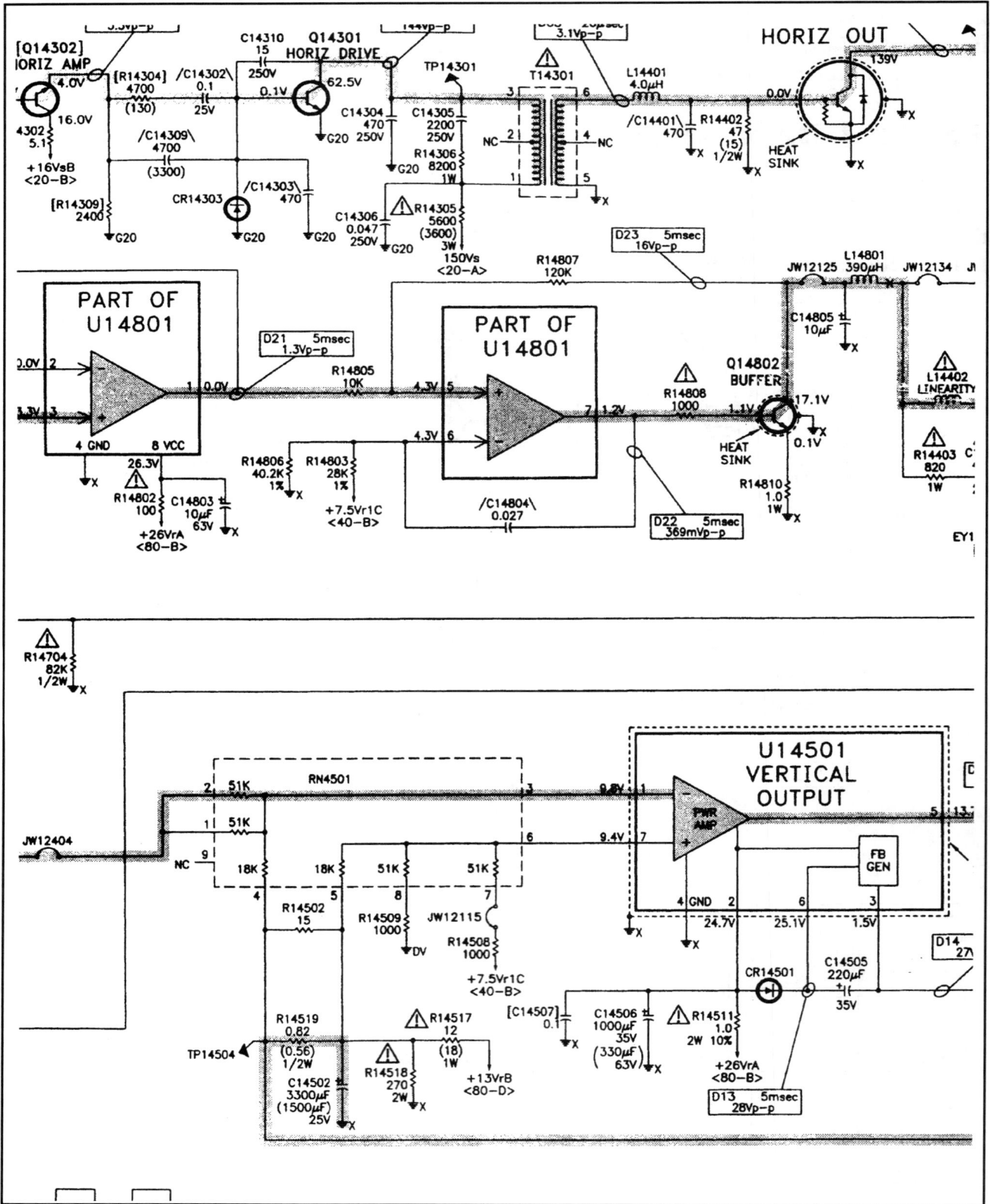

Figure 5-21. Vertical section.

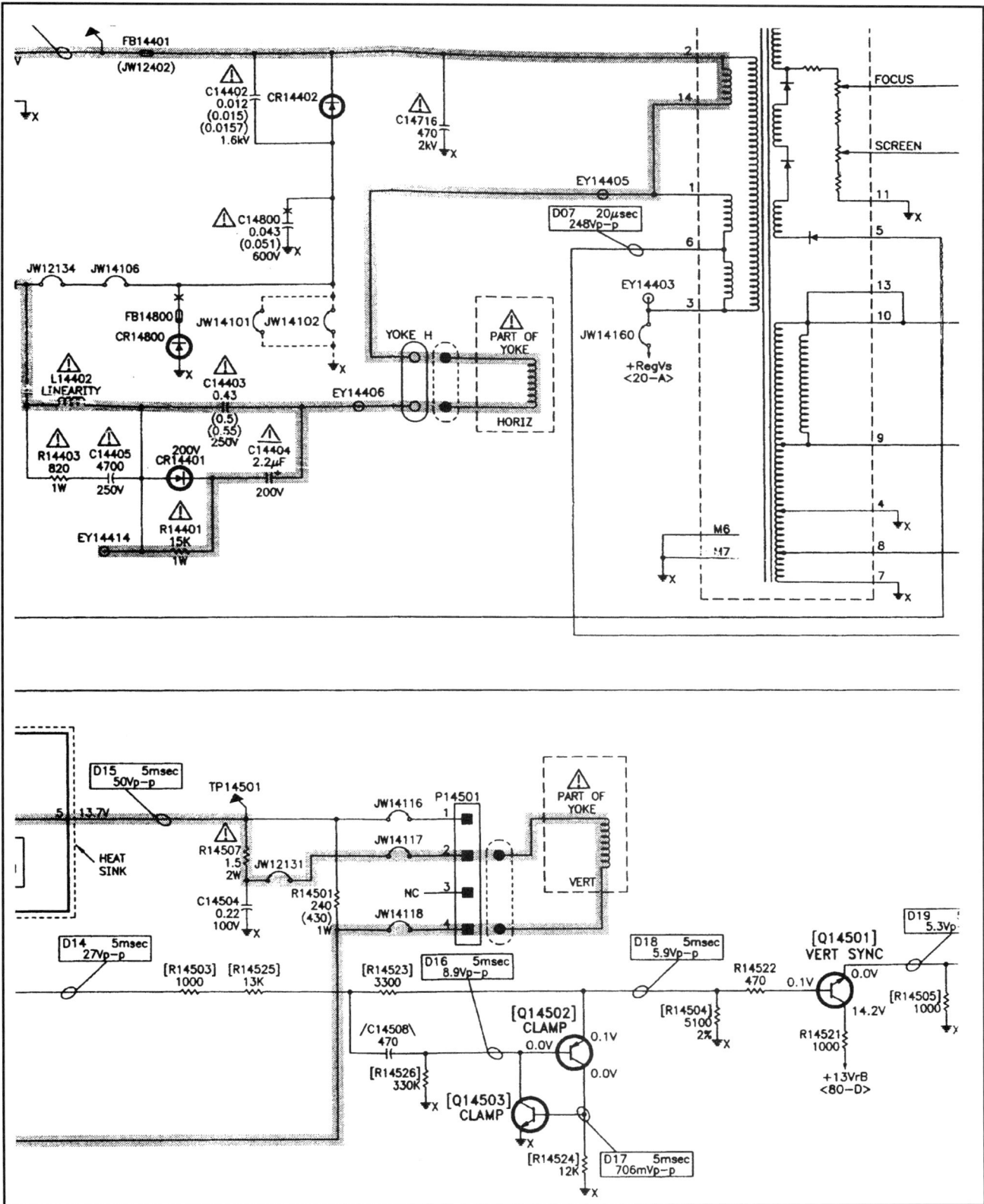

Figure 5-21. Vertical section (continued).

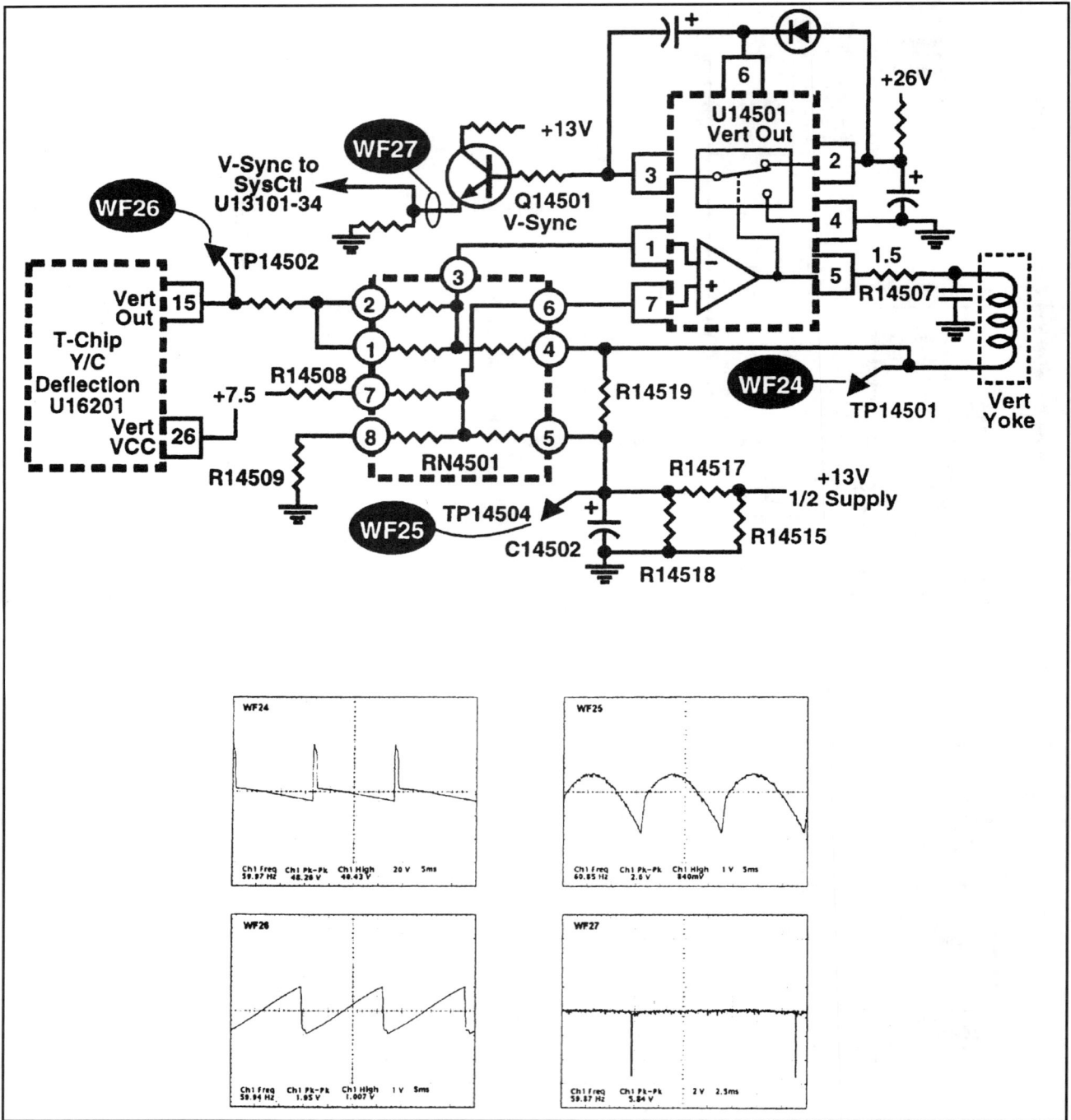

Figure 5-21A. Vertical section block diagram.

How It Works

As in other chassis, U14501 acts as a voltage-to-current converter because it changes the vertical rate DC ramp to a current ramp. It is an "inverting amplifier" that sinks current at pin 5 when pin 1 is high, and sources current from pin 5 when pin 1 is low.

Two voltage sources are critical to the operation of this circuit. The first is the +26-volt supply developed by the horizontal deflection circuit at pin 10 of the flyback (actually from the junction of pins 10 and 13) and applied to pin 2 of the vertical output IC. The second is the so-called "half-supply" taken from a rectified pulse at pin 9 of the flyback and input to pin 4 of the resistor pack, RN4501, and the low side of the yoke. It is called the half-supply because it is 1/2 the value of the B+ voltage. Since both come from a portion of the same winding on the flyback, the voltages will track each other. The purpose of the half-supply is to provide a reference voltage to the vertical circuit, around which yoke current is generated. As you can see, there really is nothing new here.

The reference voltage is important. You will remember that current through the vertical yoke must travel in two directions (*Figure 5-22*). During "active scan," current flows in the direction that causes the beam to travel from the top to the bottom of the screen. During retrace, current must reverse to deflect the beam to the top of the screen. Travel time from the top to the bottom has to occur in 1/60th of a second, but it has to return to the top much, much quicker.

Each component in the vertical circuit contributes to the job. For example, R14508 and R14509 (at pins 7 and 8 of the resistor pack) keep the beam from deflecting off the screen if the vertical output IC should short to ground or to the +26-volt source. C14502 along with R14518 reduce vertical ripple current on the +13-volt source. The parallel resistors R14519 and R14502 (Figure 5-20B) are current-sense resistors which develop a voltage drop proportional to current flowing in the yoke. RN4501 receives +13 volts at pin 5 and distributes it to various points in the circuit. Of course, you know the function of U14501.

I hesitate to go further with this description because it will duplicate what I said in Chapter 3. As far as I can tell, the circuit works pretty much the same in the CTC197 as it does in the CTC177 and CTC185.

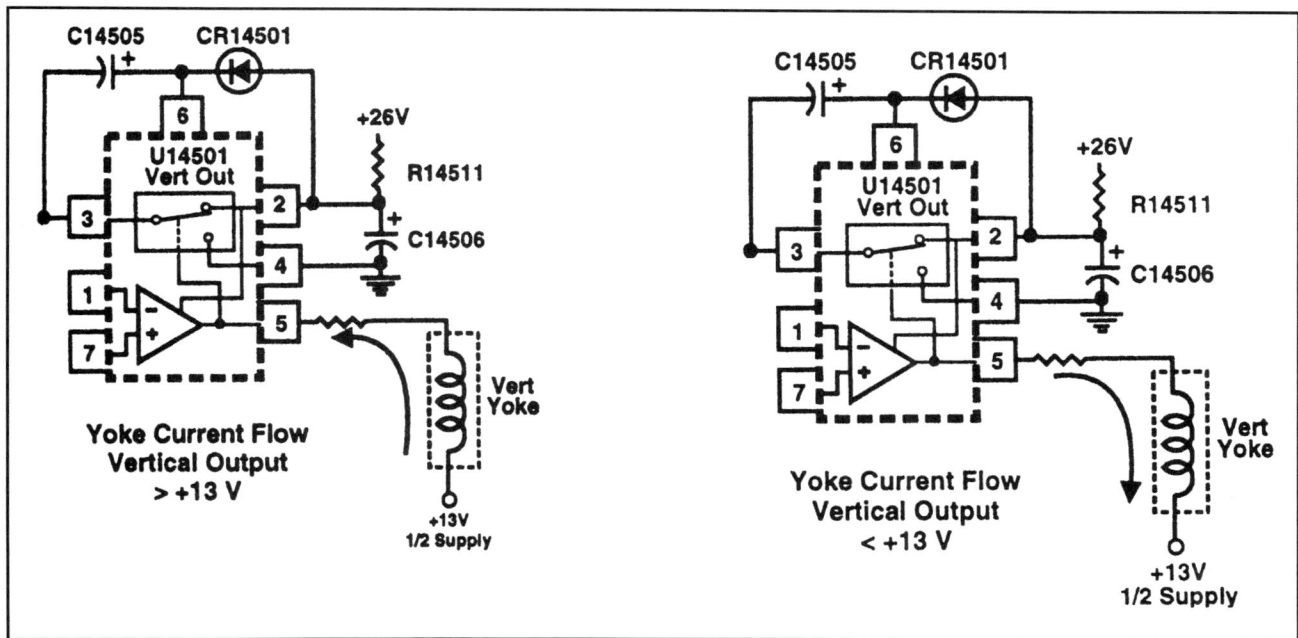

Figure 5-22. Yoke vertical current flow.

Troubleshooting Vertical Output Problems

Troubleshooting vertical problems follows much the same procedure as it did in Chapter 3. I've also included *Figure 5-23* for additional vertical troubleshooting information.

(1) Check for B+ on pin 6 of U14501. If R14511 is open, suspect a shorted vertical output chip.

(2) Check for the half-supply at TP14501, the vertical yoke connector.

(3) Check for a 2 Vp-p vertical parabola at pin 1 of U14501. If it is not there, check pin 15 of U16201.

(4) If the vertical parabola isn't present at the T4-Chip, check for 7.6 volts at pin 26 of the T4-Chip. If you find correct B+, look at pin 16 for about 3.5 volts. If this voltage is wrong, suspect C14501 and C14503 (ramp generating capacitors) and, as a last resort, the chip itself.

Scan Loss Detect

I have chosen not to deal with the CTC195 because it is used in projection TVs, which is "a whole other ballgame." However, you might like to know that the CTC195 chassis has a scan-loss detect circuit which becomes active if the chassis loses either horizontal or vertical scan. When it becomes active, the circuit will defeat video, blanking the raster. Its purpose is to protect the three picture tubes which would more than likely be damaged.

The Tuner

The CTC197 continues to use the tuner-on-board with the zinc tuner wrap. It is a single-conversion, electronically aligned tuner which the literature says is based on the CTC179/189 chassis family. These chassis will have two variations of the basic tuner, a single-input tuner and a single-input tuner with PIP RF output. The PIP tuner will be a "cold" version of the CTC185 tuner. Chassis that have the single tuner will have a VHF/UHF splitter to route the incoming signal to its proper RF amplifier. PIP chassis will also have an input splitter that will send a reduced (7 to 9 dB) RF signal to the PIP tuner and an RF signal to the main tuner, where another splitter routes the signal to its respective RF amplifier. The new splitter improves the main tuner performance at the expense of some PIP features, a fact which you might need if some customers complain about the sharpness of the PIP picture.

The Basic Sections of the Tuner

The tuner has three basic sections: the RF stage, the mixer/oscillator, and the controller IC (*Figure 5-24*). The RF stage "captures" (i.e., tunes) the incoming signal, amplifies it, and sends it on for additional processing. The mixer/oscillator takes the high-frequency signal and downconverts it to the IF frequency (45.75 MHz), which other stages in the TV use to recover video, sync and audio information.

Vertical - Collapsed

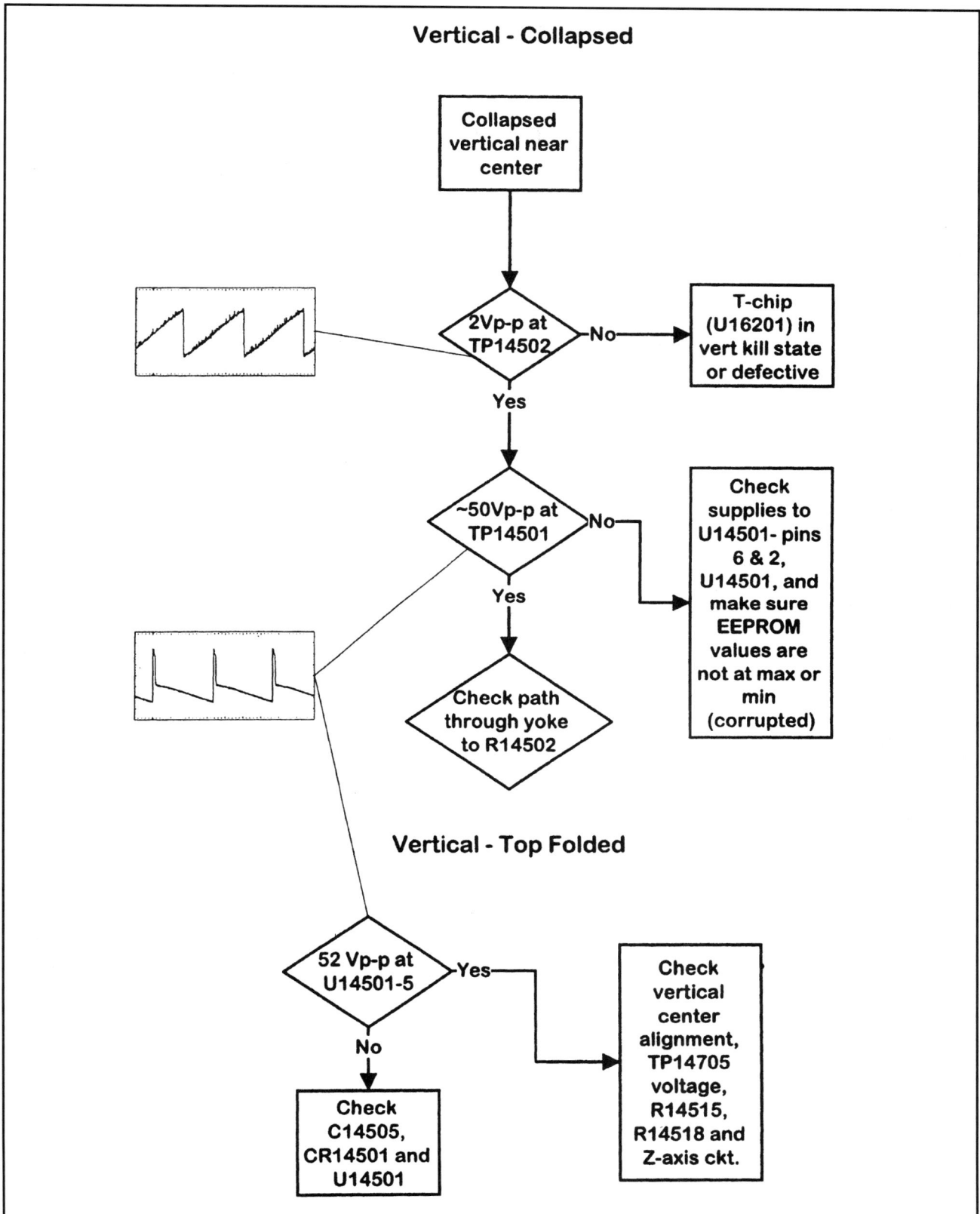

Vertical - Top Folded

Figure 5-23. Vertical troubleshooting.

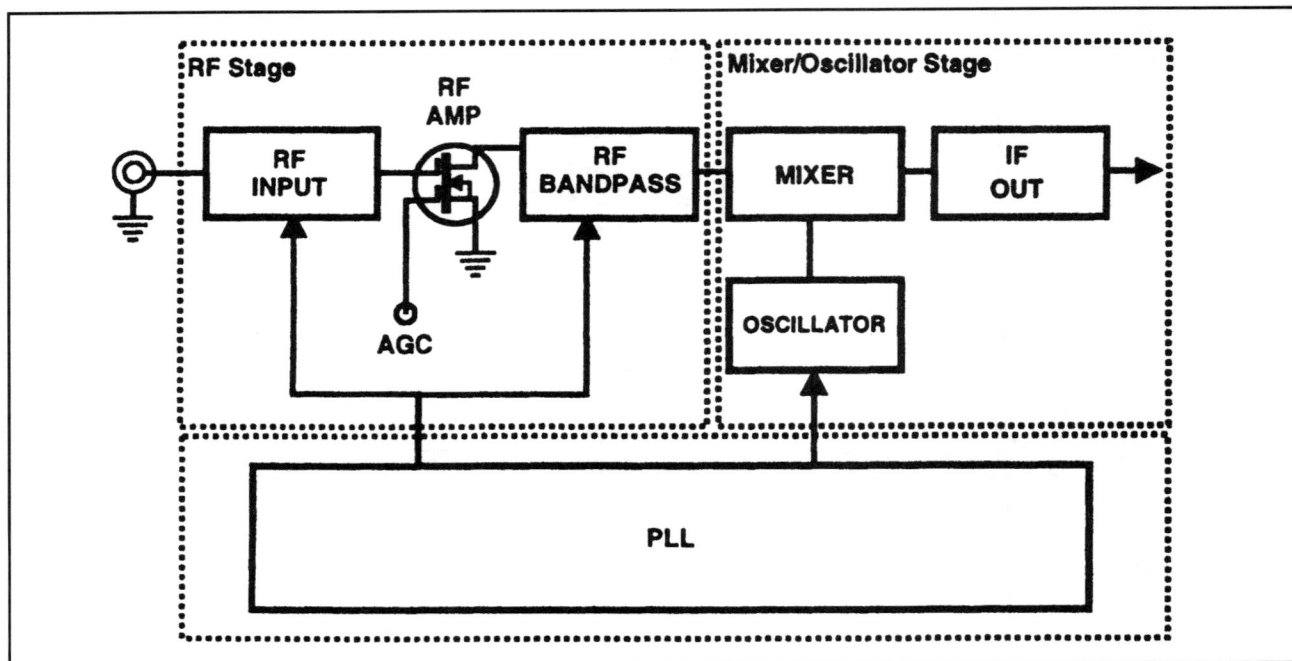

Figure 5-24. Tuner sections.

The controller IC communicates with the main microprocessor to obtain channel information. The channel information is then converted from digital to analog format and used to tune the RF and mixer/oscillator stages to obtain the proper frequency.

From this point on, I will follow the stages of the tuner as they are laid out in the block diagram presented in *Figure 5-25*.

The RF Splitter

Figure 5-26 gives you an idea about the type of RF splitter the CTC197 chassis uses. The picture-in-picture output will, of course, be used only in chassis that support that feature.

The Single-Tuned Filter

The TOB utilizes "single-tuned filtering" as the CTC175/176/177/187 and the CTC185 chassis do. The purpose is to select one particular frequency from the myriad of frequencies present at the RF input. How well the tuner does this job is a measure of its selectivity. Good selectivity in a tuner's "front end" reduces adjacent signal interference, improves signal-to-noise ratio, and improves the gain of the signal. It also improves the AGC response of the stage.

The TOB design permits the controller IC to shape the frequency response of the single-tuned filter by using voltages from pins 6 and 14 to change the frequency response characteristics of the tank circuit and RF amplifier (*Figure 5-25*). Pin 6 outputs a variable DAC (digital-to-analog conversion) output voltage, while pin 14 remains high or low, depending on which band the user has selected. *Table 5-4* lists the voltage selection for different RF bands and channels. As I pointed out in the last chapter, this arrangement permits better tuner response over the range of frequencies the tuner has to deal with than you can get from a traditional tuner.

The RF Amplifier

This chassis uses a single-stage dual-gate depletion field-effect transistor as an RF amplifier. Engineers usually choose a FET because it has a higher gain and lower noise figure than other transistors. FETs are also voltage-controlled devices that operate much like a vacuum tube. If you want to reduce their gain, apply a negative voltage to the gate with respect to the source. If the negative voltage is high enough, it will pinch off current flow altogether. The converse is also true. A positive voltage with respect to the source will increase gain.

The FET in the CTC197 is a dual-gate MOSFET. The RF signal is routed to gate 1 while AGC voltage is routed to gate 2. If the AGC voltage swings in a positive direction, the MOSFET increases gain. If the AGC voltage swings negative, the MOSFET responds by reducing gain. As you know, the T4-Chip generates AGC voltage by monitoring the signal level of the IF signal from the tuner. If the IF signal level increases, the T4-Chip responds by reducing AGC voltage. The opposite is also true.

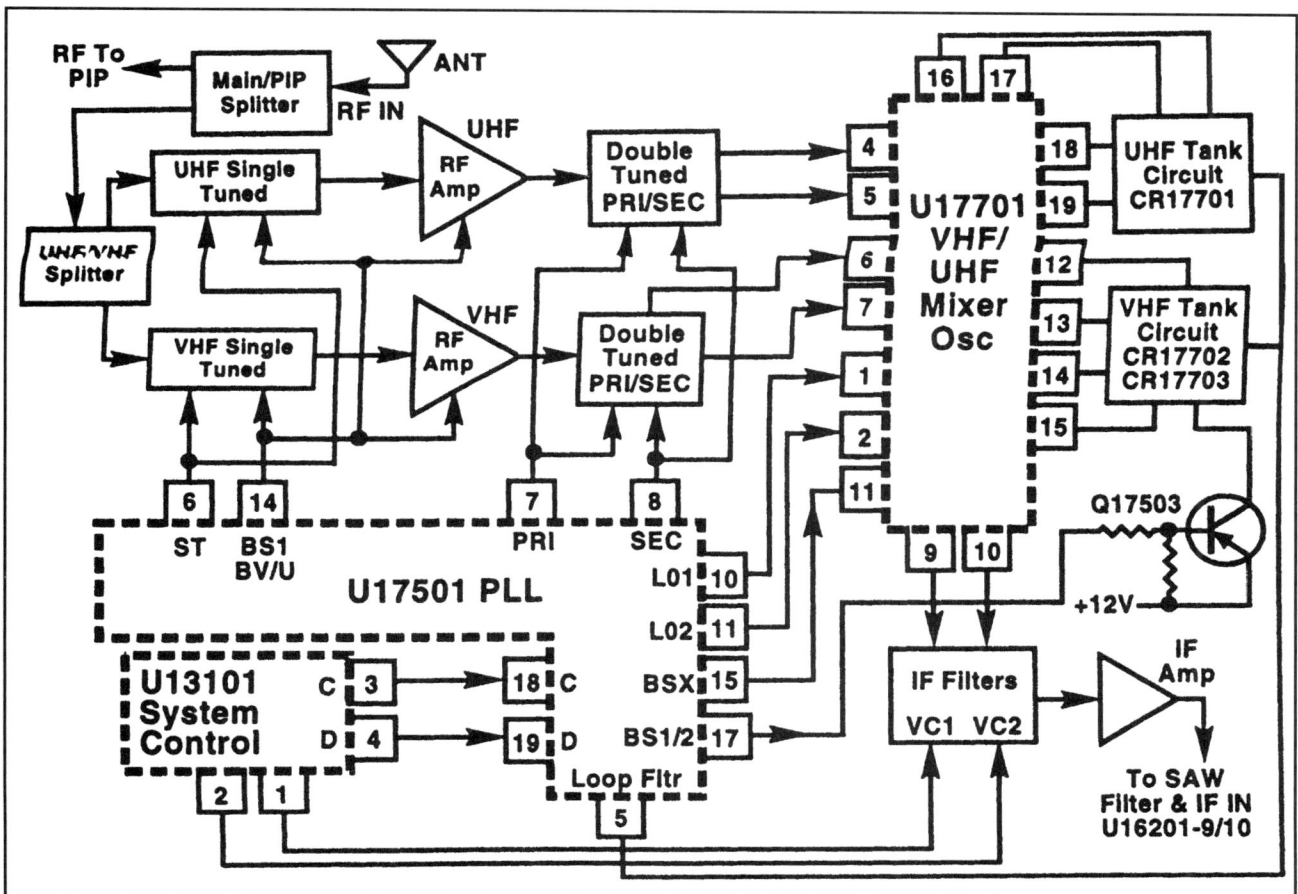

Figure 5-25. Tuner block diagram.

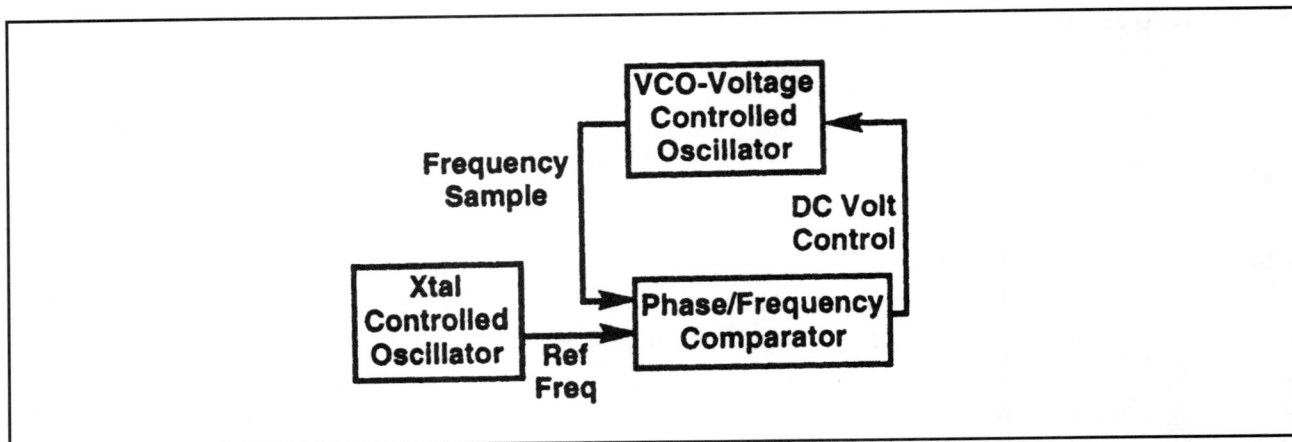

Figure 5-26. RF splitter.

The RF Bandpass Filters

As do other TOBs, the CTC197 tuner employs a double-tuned bandpass filter after the RF stage. The purpose of this filter is to increase signal selectivity and noise rejection. As the name implies, a bandpass filter possesses the ability to allow a certain "band" of frequencies to pass while rejecting signals outside the band to which it is tuned. The circuit consists of varactor and PIN diodes and inductors. The PIN diodes do the bandswitching while a tuning voltage applied to the varactor diodes vary their capacitance. Since varactor diodes are used in series/parallel with inductors, the varying capacitance changes the characteristics of the tuned circuit, which changes the bandpass parameters, permitting the filter to be tuned to meet a variety of requirements. Transformers in this part of the tuner provide impedance matching for the remainder of the RF stage.

By the way, a PIN diode is a silicon-junction diode having a lightly doped intrinsic layer serving as a dielectric barrier between the P and N layers. It is commonly used as a kind of electronic switch, just as it is here.

Channel	U17501 Pin 6 (Single-tuned Filter)	U17501 Pin 14 (Band Switching)
2	1.2 V	HIGH
6	7.8 V	HIGH
7	4.5 V	HIGH
13	6.9 V	HIGH
14	5.1 V	LOW
69	25.4 V	LOW

Table 5-4. RF bands and voltages.

The Mixer/Oscillator

U17701 is the mixer/oscillator network. You will note that it really has very few external components. It contains the circuits that do the traditional work of a mixer and oscillator to produce the IF signal which is sent on the T4-Chip.

The IF Bandpass

An IF filter and buffer transistor network stand between the mixer/oscillator and the rest of the chassis. These circuits are used to fine tune the IF bandpass by removing any remnants of the frequencies created by the mixer's heterodyne activity. The IF bandpass network uses varactor diodes, which are controlled by the main microprocessor through its own DAC ports. The microprocessor uses feedback from the T4-Chip and the original IF alignment to adjust this network. You might note that these values are the same for all channels because the IF frequency does not change as the viewer changes from one channel to another.

The Controller Integrated Circuit

The controller IC uses a Motorola PLL/DAC IC which is also used in the CTC185. As I stated there, this chip contains three DACs which the chassis uses to electrically align the single-tuned, double-tuned primary, and double-tuned secondary filters.

The purpose of this IC is to process channel select information. It receives its instructions from the microprocessor, receives feedback from the mixer/oscillator, adjusts the incoming RF filters, and tunes the local oscillator and associated tank circuits to provide the proper tuner IF output. It really is "the brains of the outfit," but it is a slave to the main microprocessor!

Of course, the controller IC has at its heart a PLL circuit. All PLLs have certain elements in common (*Figure 5-26*). For example, it has a voltage-controlled oscillator (VCO), the output of which is sampled by a phase-frequency comparator that compares the sampled frequency to a reference frequency generated by a crystal-controlled oscillator. If the sampled frequency does not agree with the reference frequency, the comparator generates an error voltage which is fed back to the VCO to correct its output. The VCO is therefore able to remain locked to the reference crystal oscillator.

This PLL has three DACs and three bandswitching outputs which can vary between 0 and +33 volts. It also has several inputs to monitor the output of the local oscillator.

The controller IC is also responsible for bandswitching. The bandswitch outputs are at pins 14, 15 and 17, and they are either high (+12 volts or +5 volts) or low (0 volts). Pin 14 selects the VHF or UHF RF amplifier. Pin 15 selects the VHF or UHF mixer which is inside U17701. Pin 17 selects the VHF or UHF local oscillator. The controller accomplishes bandswitching by turning on/off PIN diodes which switch in and out of the various circuits' inductors and capacitors to vary the tuning range. The response time of the tuner is such that it can select one channel as opposed to others in something less than 150 milliseconds.

Table 5-5 gives the bandswitching particulars. If you study these tables, you will note that the bands do not follow the standard, off-the-air or cable band designations. They are based on "the linear progression of the individual channels within the broadcast frequency range" (*Technical Training Manual*, pg. 78).

The ability to tune just one particular channel is dependent on the PLL open-collector output voltages from pins 6, 7 and 8 (*Figure 5-25*) which control the RF filter voltages used to center the RF frequency response curve over the desired channel. Pin 6 controls the response of the single-tuned filter. Pin 7 controls the response of the primary coil of the double-tuned filter, and pin 8 controls its secondary. Each of the lines provide different outputs depending on the frequency being tuned.

If you want to know the exact routine the tuner uses to capture a channel, I suggest you consult *The Technical Training Manual*, pg. 79-80. My purpose is to give you sufficient information to make a repair in the event you are called upon to do so.

Software Control

As I have stated, the tuner controller IC receives its instructions from the main microprocessor via the I-squared-C bus. Data is sent and received in packets of 2 to 5 bytes. The first byte is the address byte. Each address byte has at its beginning a start condition, a stop condition at its end, with an acknowledge condition at the end of each byte.

U17501 Pin	Callout	Band 1	Band 2	Band 3
14	BV/U	Low	Low	High
15	BSX	Low	Low	High
17	BS1/2	High	Low	High

Band	Channels		Frequency Range
	Cable	Off-Air	
1	1-6, 95-99, 14-17	2-6	54-144 MHz
2	7-13, 18-50	7-13	144-384 MHz
3	51-125	14-69	384-804 MHz

Table 5-5. Bandswitching particulars.

Data used to align the tuner electrically to receive a particular channel is stored in the EEPROM. Each alignment channel (*Table 5-6*) needs 3 bytes of data. Since there are 29 alignment channels, the EEPROM must reserve space for 87 bytes of memory for the tuner alone.

Channel	Band	Midrange (MHz)	Pix Carrier (MHz)	Local Oscillator (MHz)
2	1	57	55.25	101
3	1	63	61.25	107
6	1	85	83.25	129
98	1	111	109.25	155
14	1	123	121.25	167
17	1	141	139.25	185
18	2	147	145.25	191
13	2	213	211.25	257
29	2	255	253.25	299
35	2	291	289.25	335
41	2	327	325.25	371
45	2	351	349.25	395
48	2	369	367.25	413
50	2	381	379.25	425
51	3	387	385.25	431
57	3	423	421.25	467
60	3	441	439.25	485
64	3	465	463.25	509
68	3	489	487.25	533
76	3	537	535.25	581
83	3	579	577.25	623
88	3	609	601.25	653
93	3	639	637.25	683
105	3	681	679.25	125
110	3	711	709.25	755
115	3	741	739.25	785
120	3	771	769.25	815
123	3	789	787.25	833
125	3	801	799.25	845

Table 5-6. Tuner alignment channels.

Troubleshooting the Tuner

RCA says you need several things to troubleshoot this tuner effectively: (1) experience dealing with TOB technology; (2) basic knowledge of tuner theory; (3) a good DMM; and (4) Chipper Check. I sometimes feel like adding a fifth—namely, to know the address of a good repair depot because things are not as simple as some want us to believe! Nevertheless, how you proceed should depend on what kind of problem you are called upon to repair.

Are you facing a situation where the TV will not tune a particular band of channels? If that is the case, use the information in *Table 5-7* and the block diagram in *Figure 5-27* to narrow the problem down to a stage within the tuner and a component or components within the stage. Voltage measurements are, of course, critical. Remember, you are dealing with two power supply voltages, +5 and +12 volts.

Are you facing a situation where the TV will not select the proper channels? PIN diodes and varactors are used in the channel switching circuitry. If any of these diodes fail, the circuit experiencing the failure will not be able to change frequencies, resulting in locked or no tuning. Begin by checking the voltage at pins 6, 7 and 8 of U17501, the controller IC. If these readings are correct, follow the circuit path to the diodes. If the DC voltage disappears at any time along the path, you will probably have found the cause of the failure. Remember to replace the varactors <u>as a set</u>. If you do not follow this advice, the tuner will not work as it should.

	Channel Frequencies 54-144 MHz	Channel Frequencies 144-384 MHz	Channel Frequencies 384-804 MHz
U17501 Pin 14	+11.7V	+11.7V	+0.3V
U17501 Pin 15	+0.1V	+0.1V	+4.8V
U17501 Pin 17	+11.7V	+0.2V	+11.7V
Q17504 B	+11.7V	+11.7V	+11.0V
Q17504 C	+0.4V	+0.4V	+11.7V
Q17505 C	+0.1V	+0.1V	+11.7V
Q17503 B	+11.7V	+11.0V	+11.7V
Q17503 C	-11.1V	+11.6V	-11.1V

Table 5-7. Bandswitching voltage chart.

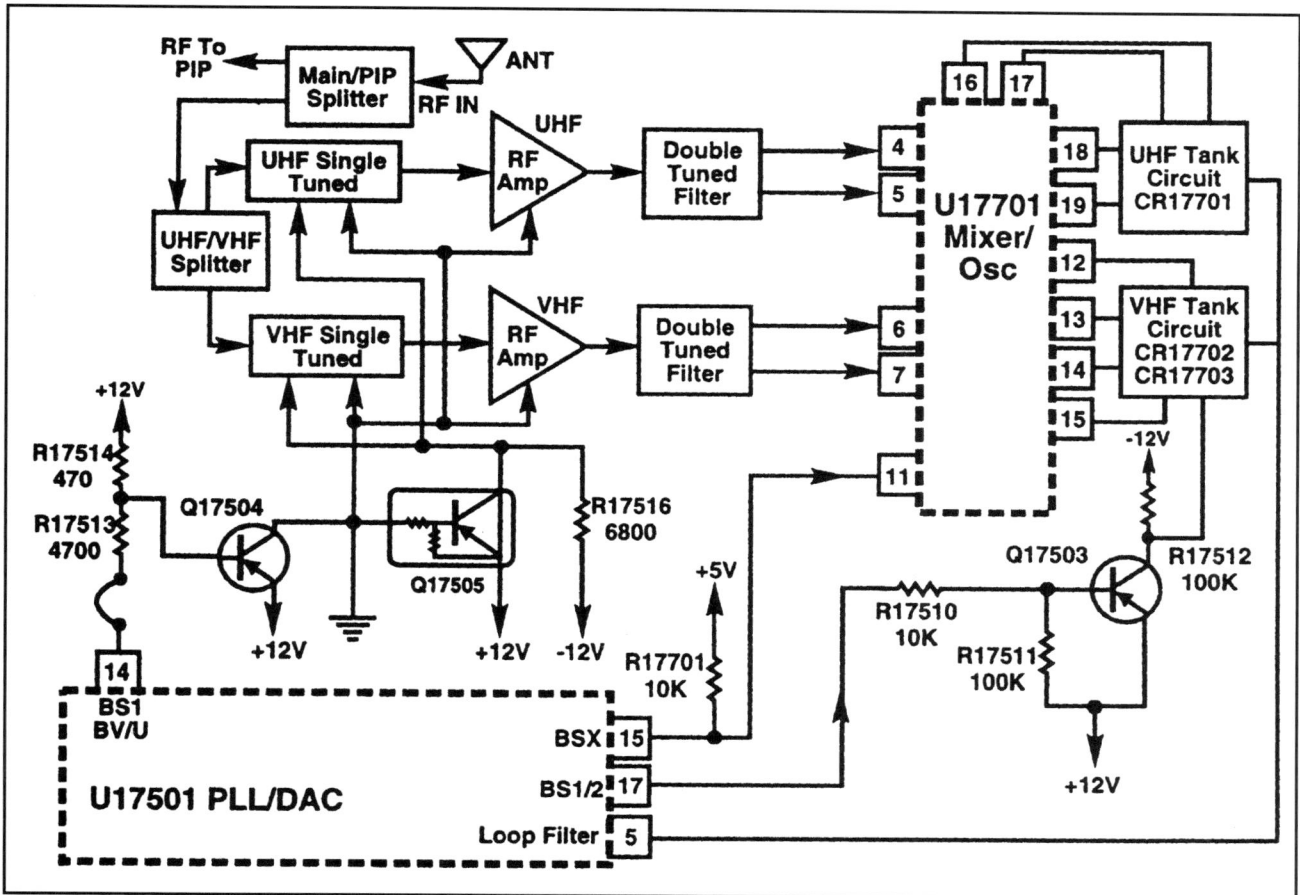

Figure 5-27. Band switch circuits.

The most common problem will more than likely be no tuning at all. RCA recommends the following procedure to deal with the "no-tuning" problem.

(1) Does the OSD show channel change? If it doesn't, the problem will be in system control and the tuner.

(2) Check power supply voltages: +5, +12, +33, and -12 volts.

(3) Check the bandswitching voltages at pins 14, 15 and 17 of the controller IC.

(4) Check the output voltages of the bandswitching transistors.

(5) Check the tuning voltage on pins 6, 7 and 8 of the controller IC. If a tuning voltage is stuck high or low, you will more than likely have a problem in the PLL loop, not the IC. Check the 4 MHz crystal (17501) for a peak-to-peak signal of about 250 millivolts. Make the check on the capacitor side of the crystal.

(6) Monitor the local oscillator voltage on pins 10 and 11 of the controller IC. The voltage will rise as channel selection goes up and will lower as it goes down.

(7) Check the single-tuned, double-tuned primary, and double-tuned secondary filter voltages at the varactors.

(8) Monitor the AGC voltage on the collector of the AGC amplifier, Q32102. If no signal is present, the reading should be about +7.5 volts.

(9) Check the B+ to the RF amplifiers. The reading should be between 10 and 12 volts.

(10) Check the IF supply voltages on the IF filter varactors. They should be between 1.4 and 3.0 volts.

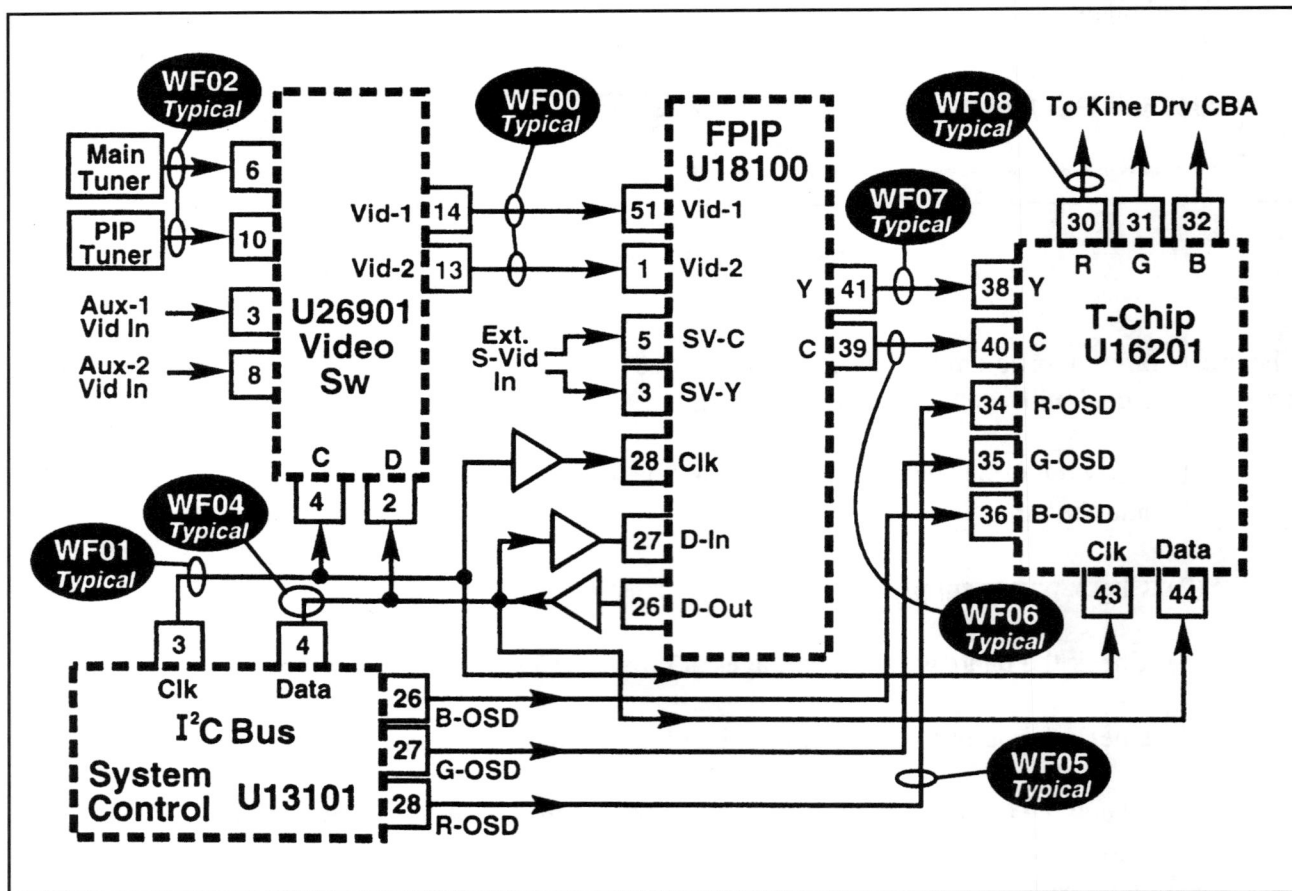

Figure 5-28. FPIP system block diagram.

FPIP and Second Tuner

FPIP stands for "comb filter plus picture-in-picture" integrated circuit. It is a CMOS IC designed to be a one-chip solution for the single moving picture-in-picture function of a television receiver. It also serves as a digital comb filter for main picture Y/C separation, and it contains analog switches to select between two composite or two component (S-Video) sources from either the main picture or the small picture. This wonder chip contains A/D converters, D/A converters, a burst-locked clock, analog video switching, and RAM needed to perform the PIP and comb filter functions. Moreover, it has been designed to keep external components to a minimum and to accept the industry standard 1 Vp-p video input and to provide the industry standard 1 V p-p output.

Figure 5-28 is a FPIP system block diagram with the waveforms you can expect at each designated point. I have put in the appendix a complete pinout diagram of the FPIP IC (U18100) and a description of the function of each pin.

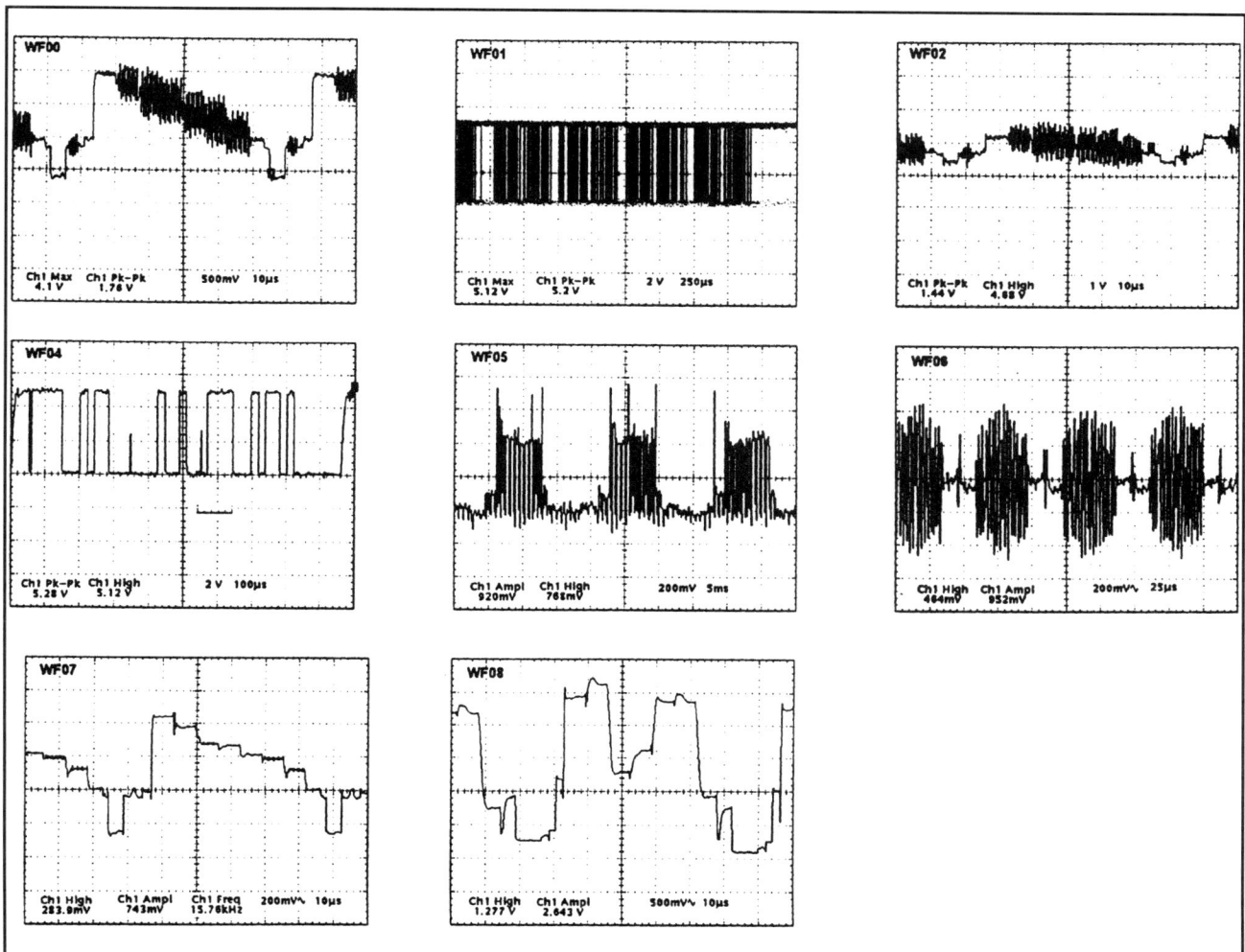

Figure 5-28. FPIP waveforms.

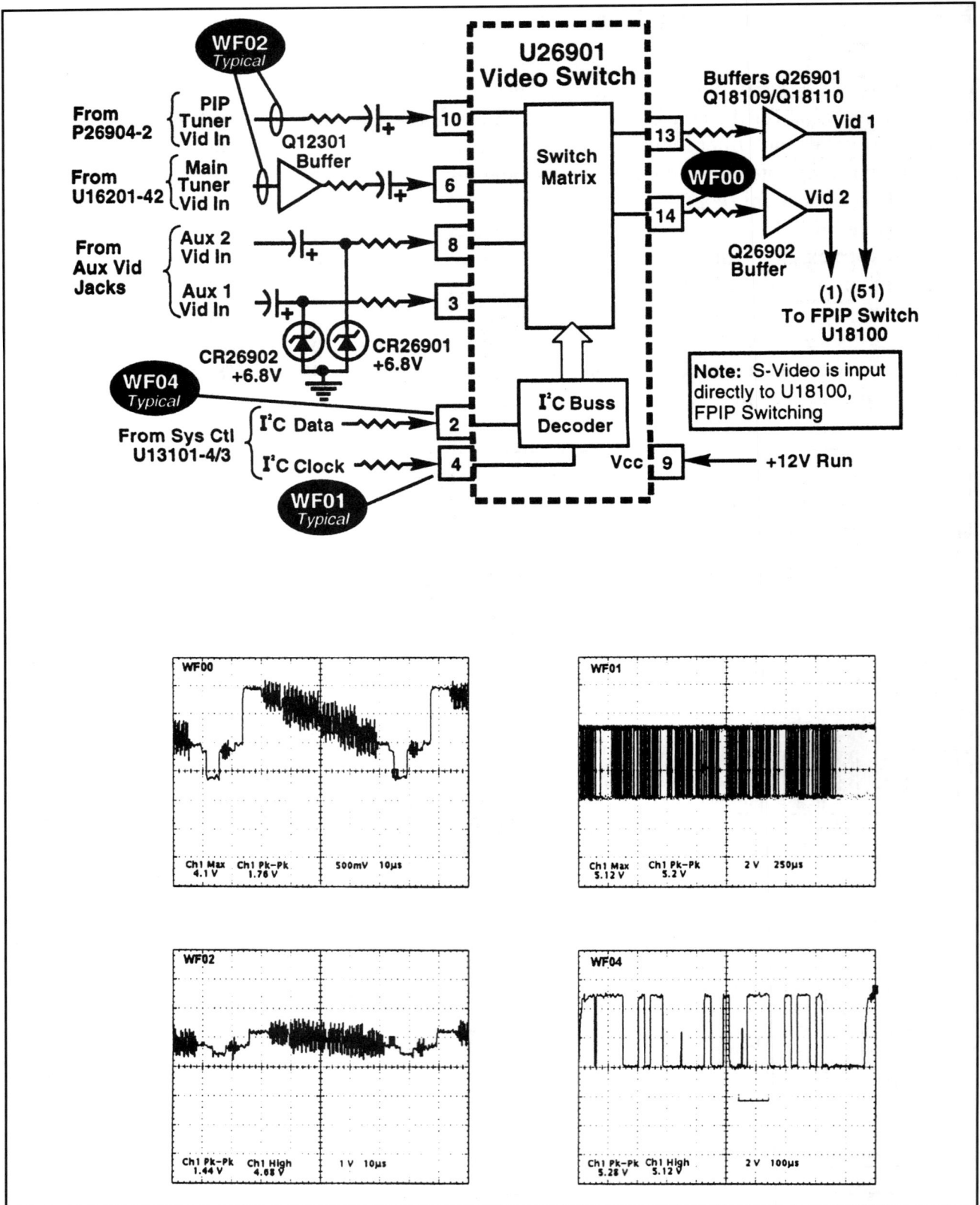

Figure 5-29. Video input switching.

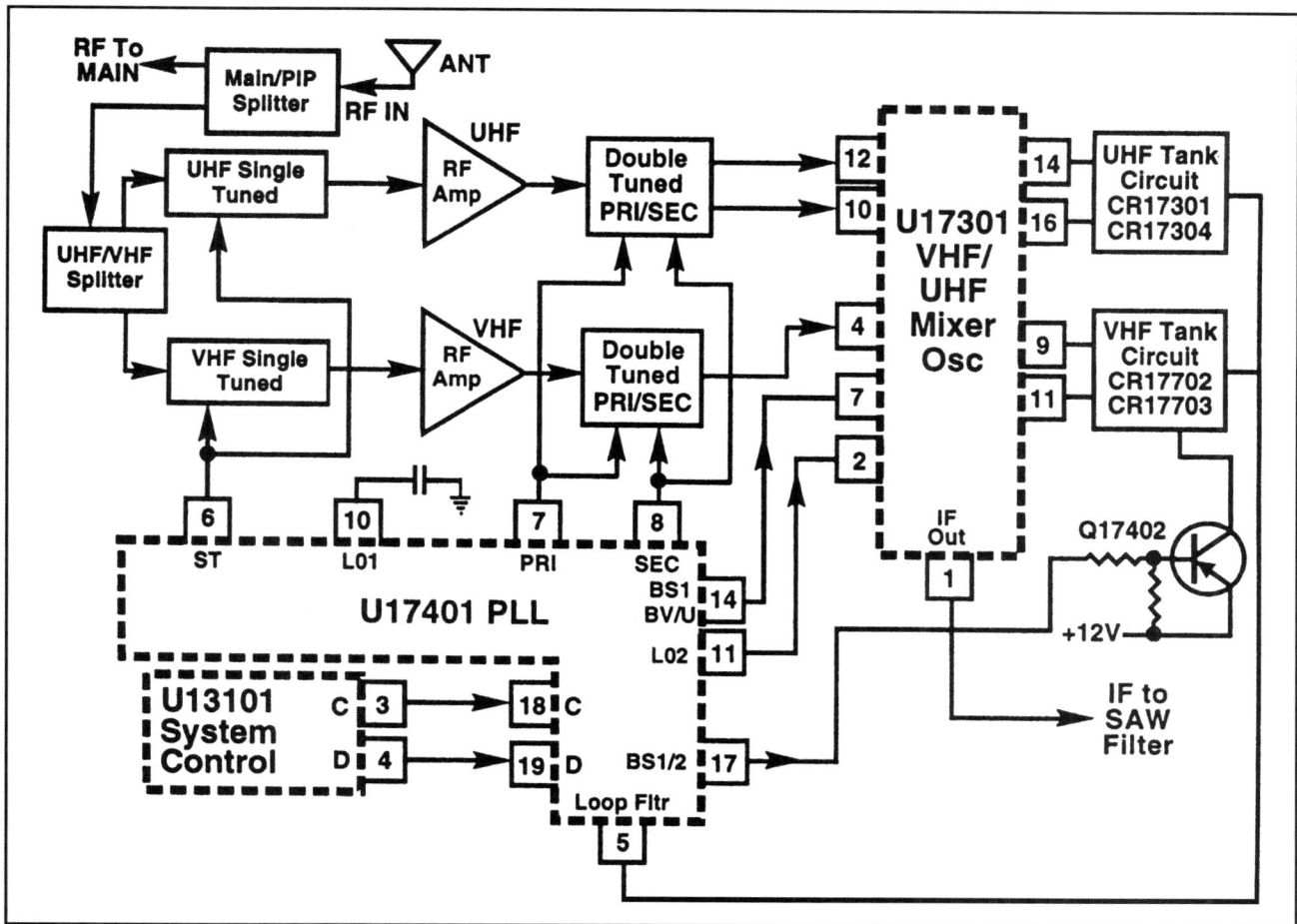

Figure 5-30. Second tuner (PIP).

Two integrated circuits, U18100 and U26901, do the video switching (*Figure 5-29*). The switch can have a maximum of eight inputs and six outputs, but only four inputs are used: PIP tuner at pin 10, main tuner video at pin 6, and the aux inputs (pins 1, 2, 3 and 8 respectively). The current scheme calls for the use of just two outputs. Switching is accomplished by the microprocessor over the I-squared-C bus.

I have debated how much information to include about the FPIP and decided to leave the bulk of the "how it works" reading to you. You will find the circuit description in the factory service literature (pg. 1-19) and in the technical training manual (pg. 92-?).

The final item under this heading is the second tuner, known also as the PIP tuner (*Figure 5-30*). It works just like the main tuner and even has a separate EEPROM to store the values it needs to operate. But there are differences.

The PIP tuner has a single input into the RF stage, whereas the main tuner uses a balanced-line input. This means that one side of the signal into the PIP tuner is at ground potential. The rationale is that it makes signal transfer between the PIP module and the main chassis less prone to interference.

Channel	Band	Midrange (MHz)	Pix Carrier (MHz)	Local Oscillator (MHz)
2	1	57	55.25	101
6	1	85	83.25	129
98	1	111	109.25	155
15	1	129	127.25	173
17	1	141	139.25	185
18	2	147	145.25	191
9	2	189	187.25	233
29	2	255	253.25	299
39	2	315	313.25	359
46	2	357	355.25	401
50	2	381	379.25	425
51	3	387	385.25	431
61	3	447	445.25	491
75	3	531	529.25	575
101	3	657	655.25	701
114	3	735	733.25	779
122	3	783	781.25	827
125	3	801	799.25	845

Table 5-8. Second tuner alignment channels.

A second difference is the tank circuits are not as selective as the main tuner, which means the PIP picture will not have the quality the big picture has. But the picture quality of the PIP window is not as important as that needed in the main picture. You should know the main picture (i.e., the big picture) is controlled by the main tuner. If the PIP and big picture are swapped, the two tuners will retune so that the main tuner remains in control of the big picture.

A third difference is the alignment channels are not the same (*Table 5-8*).

The T4-Chip (U16201)

Figure 5-31 is a complete pinout diagram of the "latest in a series of television-specific 'one chip' ICs designed to perform most low-level signal processing in a television chassis" (*Technical Training Manual*, pg. 101). As the discussion proceeds, you will see that it is very similar to the T4-Chip used in the CTC185 chassis, and in like fashion does video and audio processing, deflection processing, and CRT management under the control of the microprocessor, whose commands are transmitted over the I-squared-C bus. The T4-Chip is in a "slave configuration" because it does not initiate data and only responds to commands it is issued.

Since I have discussed these chips in some detail, I will keep my remarks here at a minimum. There is no need to repeat what I have already said.

Power-On-Reset (POR)

"Power-on-reset" is the name of a standby power supply monitor inside the T4-Chip. It detects when the standby voltage has dropped below the normal range (7.6 volts) and shuts off the IC by stopping horizontal drive. If the television is on and the standby voltage drops to 6.3 volts, the POR circuit will latch. Because the circuit latches in the ON state, the user must issue an OFF command and then an ON command to get the television started again. If the voltage is still too low (6.3 volts or lower), the IC will stay in the OFF mode, requiring the user to repeat the process.

The Bus Transceiver

The power supply monitor inside the T4-Chip latches when the standby voltage drops to 6.3 volts or lower, but the bus transceiver will remain active even if the voltage falls as low as 2.5 volts. You might like to know that each register is powered by the same Vcc as the circuit it serves. For example, the registers which control the vertical ramp are powered by the video-vertical Vcc.

IF Processing

The T4-Chip provides all the processing required to separate the IF signal into the luma and chroma signals, and then outputs a standard NTSC baseband video signal. It also provides an AGC voltage for tuner control.

Audio Processing

The T4-Chip provides the processing necessary to recover the audio signal from the IF frequency information and output a wideband audio signal, which is sent to the stereo decoder chip for final processing.

Pin	Label		Pin	Label
1	FM Tank		52	VCO Tank 2
2	Audio L In		51	VCO Tank 1
3	PIF Vcc		50	Audio L Out
4	Audio R In		49	FM Tuning
5	RF AGC Out		48	Audio R Out
6	WBA Out		47	Snd IF In
7	PIF APC Filter		46	Bus Gnd
8	IF Gnd		45	Snd IF Out
9	PIF 1 In		44	Bus Data
10	PIF 2 In		43	Bus Clock
11	IF AGC		42	Video Out
12	AFT Out		41	Test IT Filter
13	Chroma APC		40	C In
14	3.58 MHz Xtal		39	Color Kill
15	Vert Out		38	Y In
16	Ramp ALC Flt		37	Blk Level Det
17	E/W Pin Out		36	Blue In
18	32H Cer Res		35	Green In
19	Horiz Gnd		34	Red In
20	Standby Vcc		33	FS In
21	Horiz AFC Flt		32	Blu Out
22	Horiz Out		31	Green Out
23	Flyback		30	Red Out
24	X-Ray In		29	Pix ABL Filter
25	AKB In		28	Beam Sense
26	Vid/Vert Vcc		27	Vid / Vert Gnd

Internal blocks: VCO, Volume Control, Analog AFT, APC Det, FM Det, Limiter, RF AGC, Sound Detector, Video Detector, Bus Interface, IF AGC, PIF Amp, Filter Tune, ACC, 1st Amp, 2nd Amp, APC PLL, Chroma Demod, Vert Ramp, Vert Size, Parabola Generator, Standby Regulator, Vertical Count Down, Vertical Sync, Black Stretch, Matrix & RGB Switch, Horiz Lock, 1st AFC, Sync Separator, Peaking Adap Core, Horiz VCO, Y Clamp, Horizontal Count Down, Contrast & Brightness, 2nd AFC, Delay EQ, Tint, RGB Out, Horiz Out, Auto Brightness Limiter, Auto Flesh, AKB, Bus

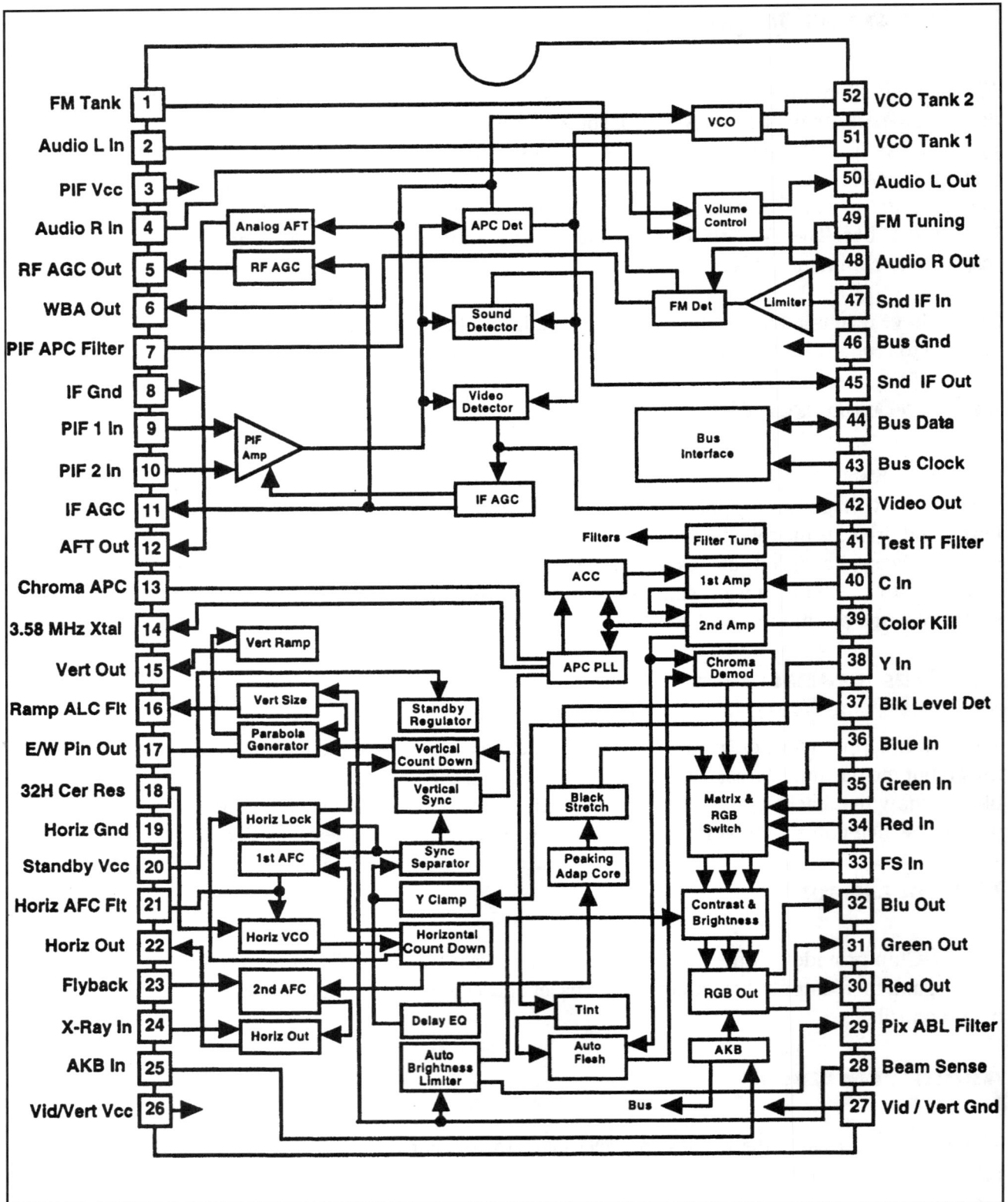

Figure 5-31. T4-Chip block diagram.

CRT Management

CRT management includes AKB, x-ray protection, E-W correction and beam-current limiting. I am reserving an independent section to discuss the AKB circuit because this circuit is relatively new. I will not discuss E-W correction and XRP circuits because that material has already been covered, but I will briefly cover beam limiting.

Beam Limiting

Beam limiting affects two circuits inside the T4-Chip, the picture and brightness controls via pin 28 (*Figure 5-35*) which monitors the secondary winding of the flyback that supplies beam current. Beam limiting is normally accomplished by reducing the amplitude (i.e., contrast) of the Y, R-Y, and B-Y signals. If the customer sets the black level too high, the brightness control will reduce the Y level first. If the circuit cannot sufficiently reduce beam current, the brightness limiter will go into emergency brightness reduction and sometimes reduce the picture to an almost black condition. The microprocessor determines when an emergency situation exists and will cause the circuit to activate. The RC network at pin 29 sets the response time, which is normally within 30 to 40 horizontal lines.

Deflection Processing

The T4-Chip performs all low-level vertical and horizontal deflection processing and control, and all sync signal processing.

Video Processing

The video processing circuitry receives the video signals from the FPIP/switch and allows the user to control the shape of the waveforms exiting to other processing circuits or the picture tube. Video processing includes brightness, color, tint, contrast, and sharpness and are controlled by commands from the microprocessor.

Video processing is made up of four areas: luma processing, chroma processing, external RGB input, and RGB outputs. Most of these functions are contained inside the T4-Chip. *Figure 5-32* will give you an idea of what is involved, including the waveforms you should expect to see at the noted points.

Luma Processing

You can follow the path of the luma signal by studying the diagram in *Figure 5-33*. Luma enters the chip at pin 38. The input voltage should be around 1 Vp-p. Too much signal will cause the internal filter to operate in a nonlinear manner, while too little will make its output unreliable. The shape of the filter depends on the signal received. It has one set of characteristics for a signal that has not been processed by an external Y/C separator, and another where Y/C separation has been performed. Remember, Y/C separation is performed in this chassis by the FPIP circuit. The new circuitry provides a sharper picture for the viewer.

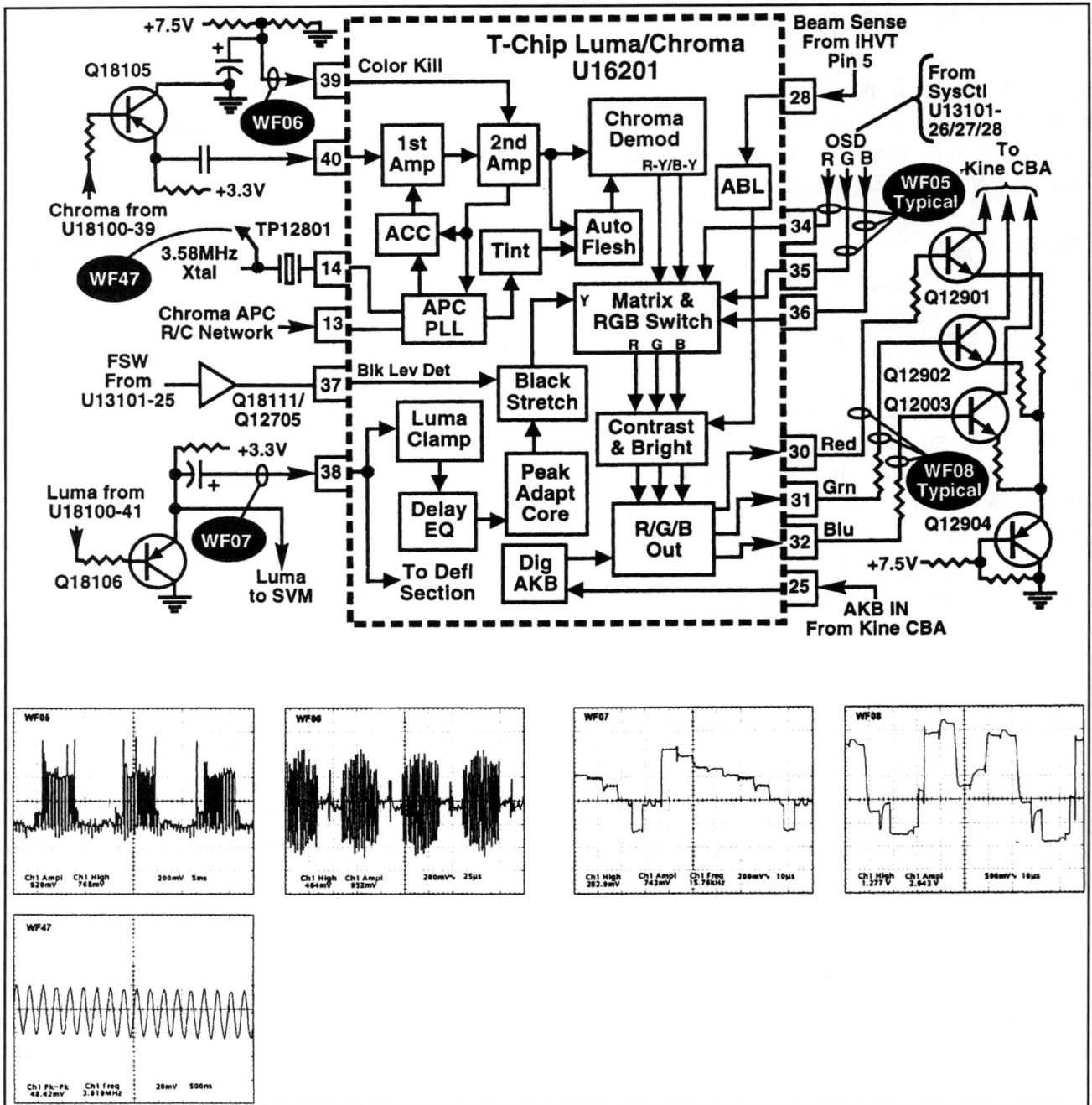

Figure 5-32. Video processing.

Chroma Processing

Chroma processing (*Figure 5-34*) is identical to the CTC185 except that the chroma has already been separated from the luma. The FPIP separates the two using digital comb filter technology. Chroma enters the T4-Chip at pin 40 and needs to be at about a 290 millivolt burst level.

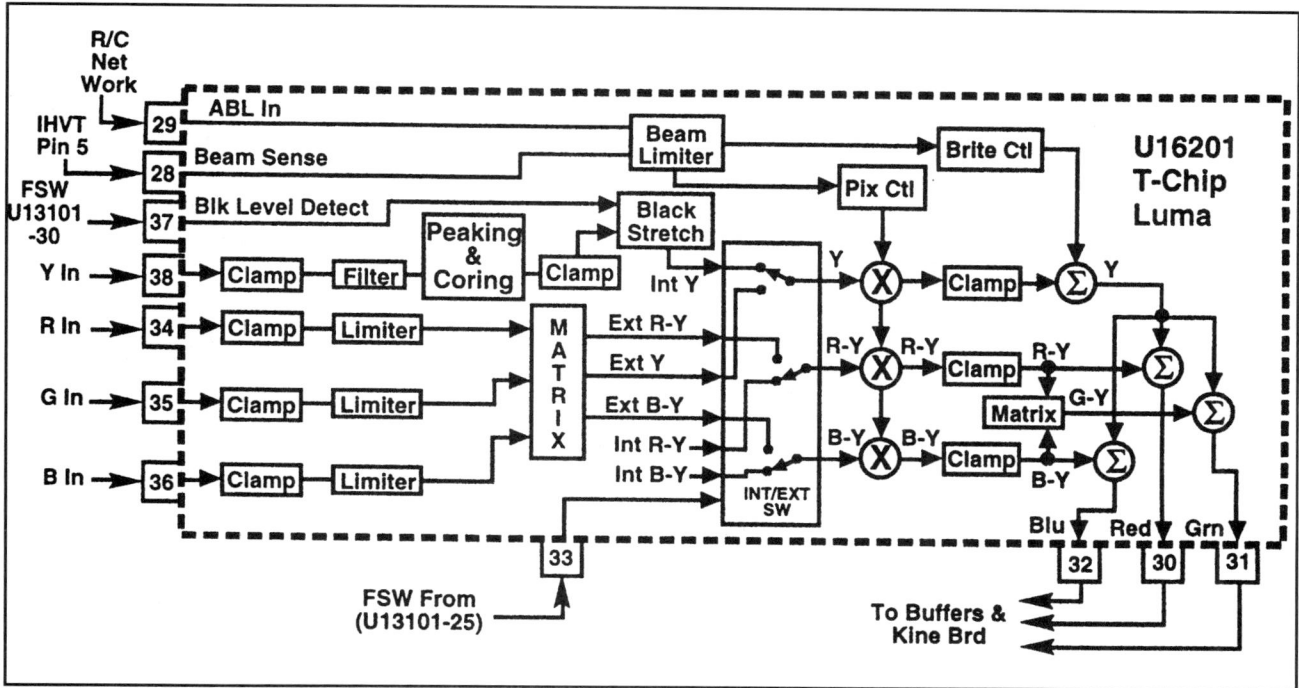

Figure 5-33. Luma processing.

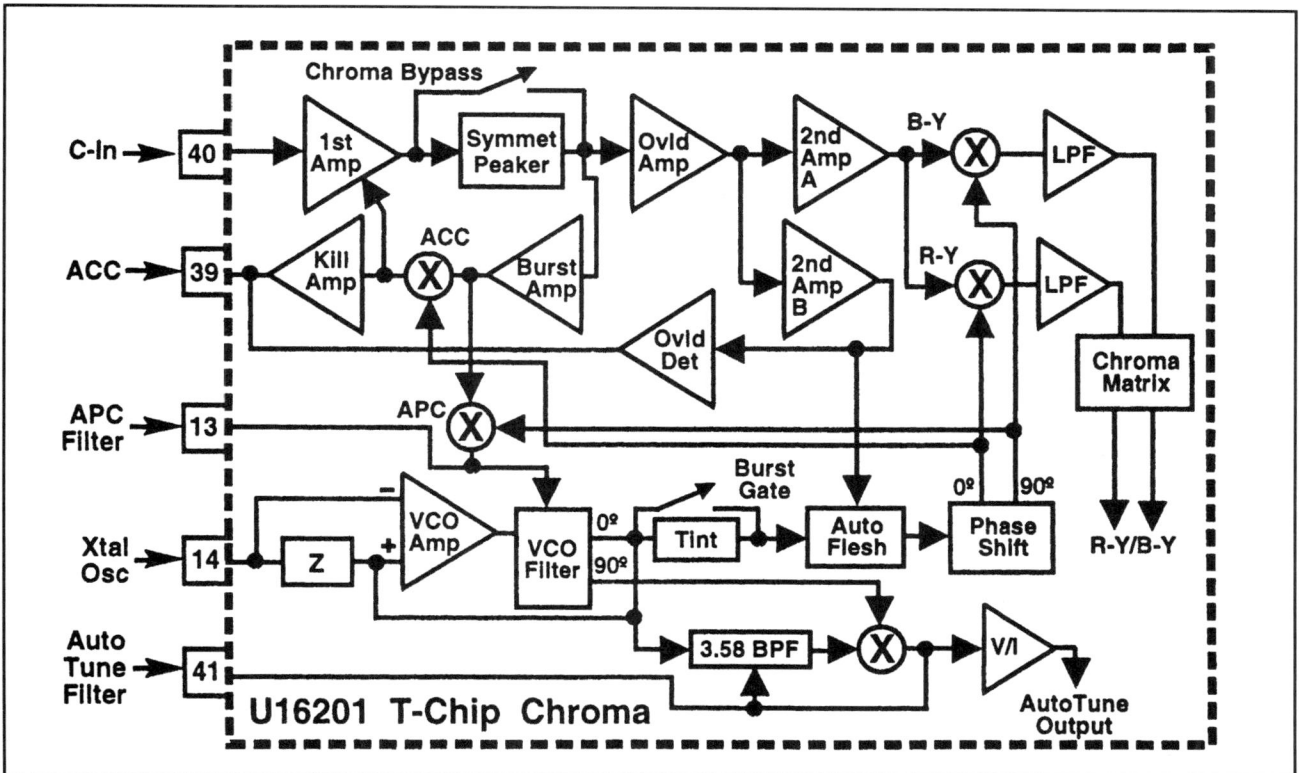

Figure 5-34. Chroma processing.

External RGB Inputs

External RGB inputs (*Figure 5-36*) are used for OSD processing. The OSD signals from the microprocessor are applied to the external RGB inputs (pins 34, 35, 36) and are then matrixed to form external Y, external R-Y, and external B-Y signals.

RGB Outputs

Several signals come together in the internal/external switch—namely, internal Y from the luma section and internal R-Y and B-Y from the chroma section (*Figure 5-35*). Pin 33, the fast switch line from the microprocessor, controls which set of signals go on to drive the CRT. A low-level signal selects the internal signals while a high level selects the external signals.

The CRT Circuits

In order to make my discussion as organized as possible, I will include under this heading the following topics: CRT drivers, scan velocity modulation, and automatic kine bias (AKB). Obviously, I could have tucked these topics under different headings, which you might want to do, but I am lumping them together for the present.

The CRT Driver Circuits

The CTC197 uses as CRT drivers a cascode circuit with PNP followers to provide a current reference for AKB (*Figure 5-36A* and *5-36B*). In case you are not familiar with the term or the circuit, a cascode

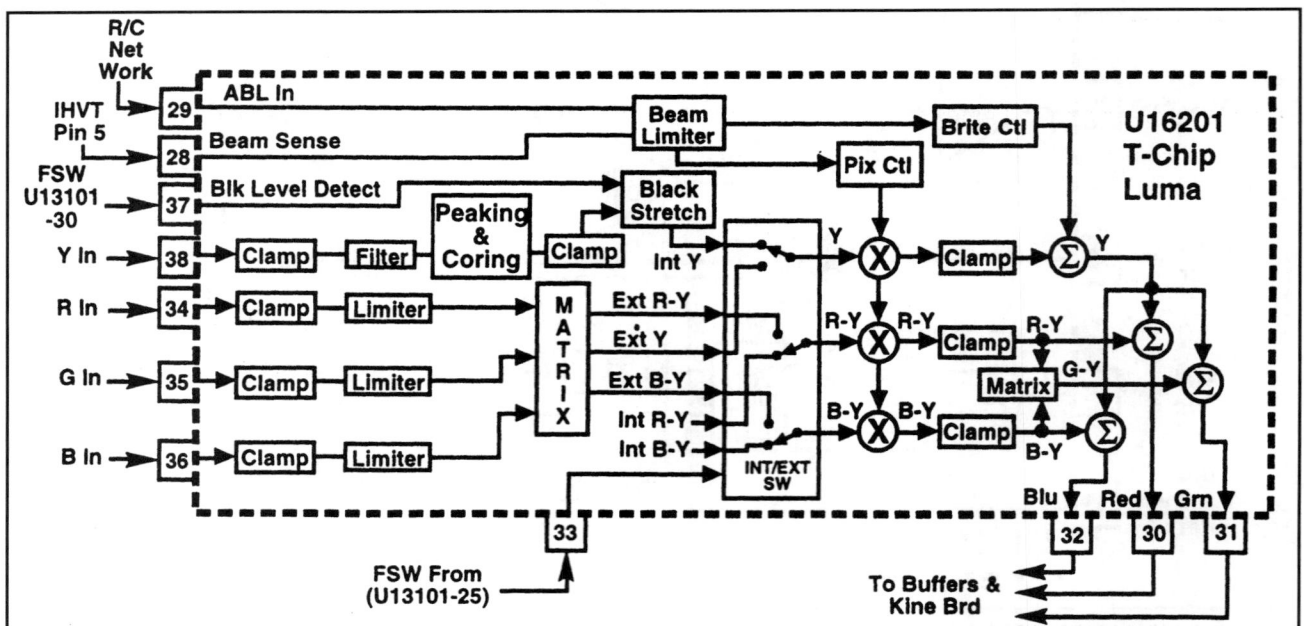

Figure 5-35. Beam limiting.

configuration is a high-gain, low-noise, high-input impedance amplifier configuration which is directly coupled. In this instance, the T4-Chip outputs are amplified from about 2 to 125 Vp-p at the CRT cathodes.

The literatures uses the green circuit as an illustration, describing the workings like this: Green drive exits U16201 at pin 31 and is amplified by Q15105 and applied equally to the bases of Q15102 and Q15109. As it increases, the signal moves toward the blanking pedestal (i.e., toward the "black"). In other words, it moves from 100 IRE (white) to 7.5 IRE (black). *Figure 5-36A* might make this a little clearer. As the signal increases, the emitter voltage of Q15102 begins to increase toward the collector supply of about +200 volts through Q15109. Remember, beam current in a picture tube is proportional to the voltage between the cathode and screen grid (bias voltage). When the bias voltage increases (the difference of potential between the cathode and screen grid), beam current increases. When it decreases, beam current decreases. The screen grid is held at a constant value of between +300 to +400 volts. As the emitter voltage tracks toward the +200 volts, bias voltage decreases, lowering the beam current. As the drive signal decreases going toward the white level, the emitter of Q15102 begins to move away from the +200-volt level, which increases the bias voltage, permitting beam current to rise.

Q15109 also serves to limit the signal voltage. Part of the drive signal from Q15105 is rectified and placed on the base of Q15109 (*Figures 5-36B* and *5-37*). If the signal increases, bias voltage to the base increases (i.e., bias voltage tracks the incoming signal). The configuration permits the emitter to follow the drive voltage until it reaches a voltage determined by R15909, the base bias resistor (*Figure 5-37*), the value of which is determined by CRT size. When the voltage reaches the predetermined limit, Q15109 turns off, effectively shutting off beam current.

The AKB circuit monitors the three CRT cathodes through Q15107, which is protected from excessive voltage by CR15101, a zener diode which limits base voltage to about 8.5 volts. NOTE: if the diode fails, the base-emitter junction of Q15107 will also likely fail and as well as the input pin 25 of the T4-Chip!

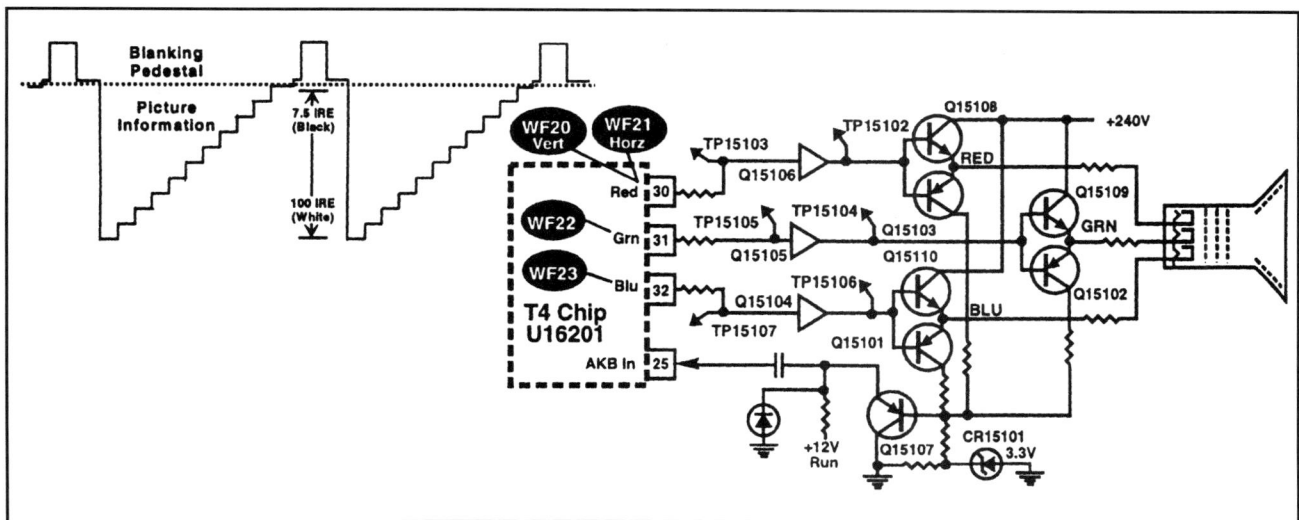

Figure 5-36A. RGB inputs.

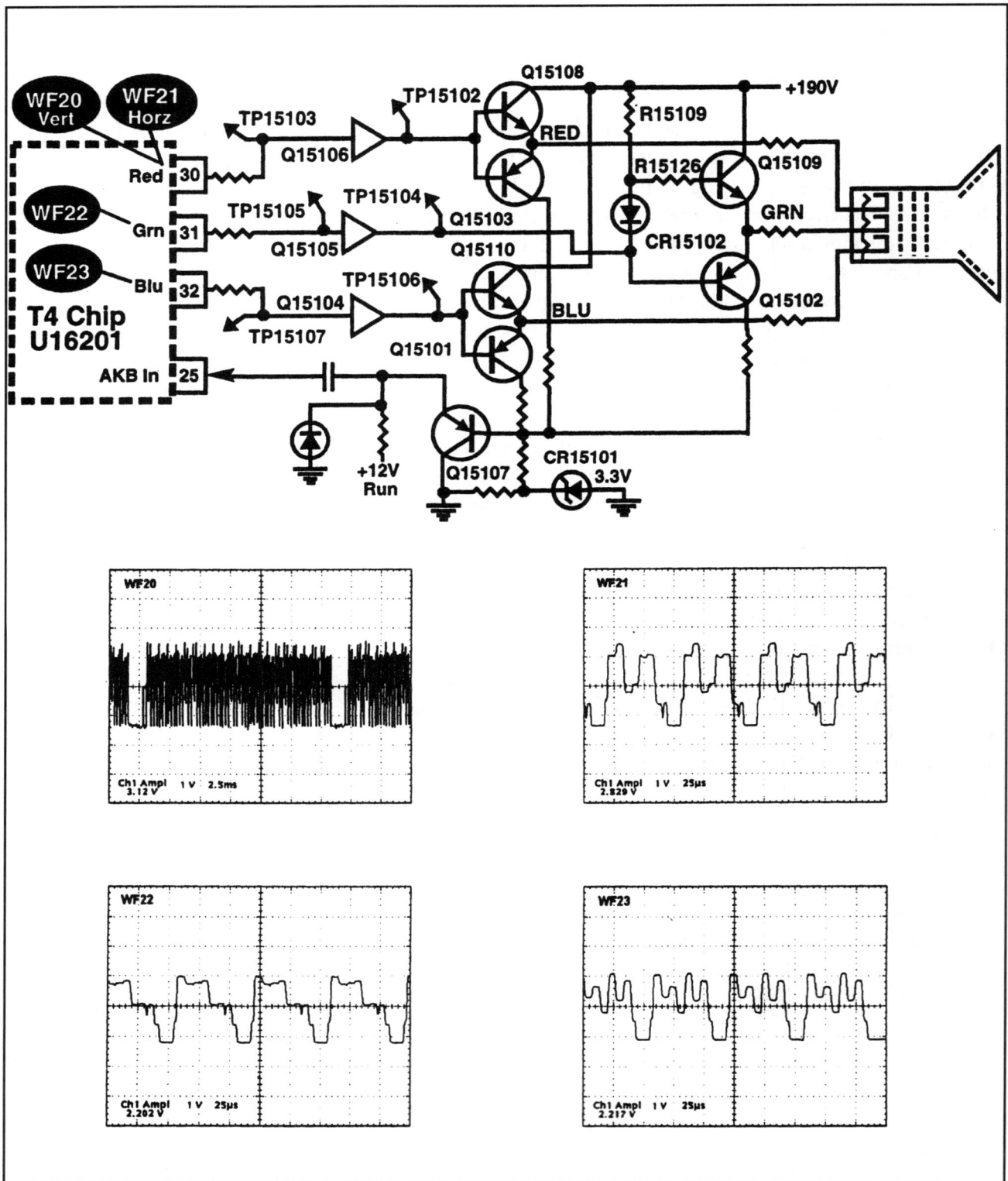

Figure 5-36B. AKB and CRT driver.

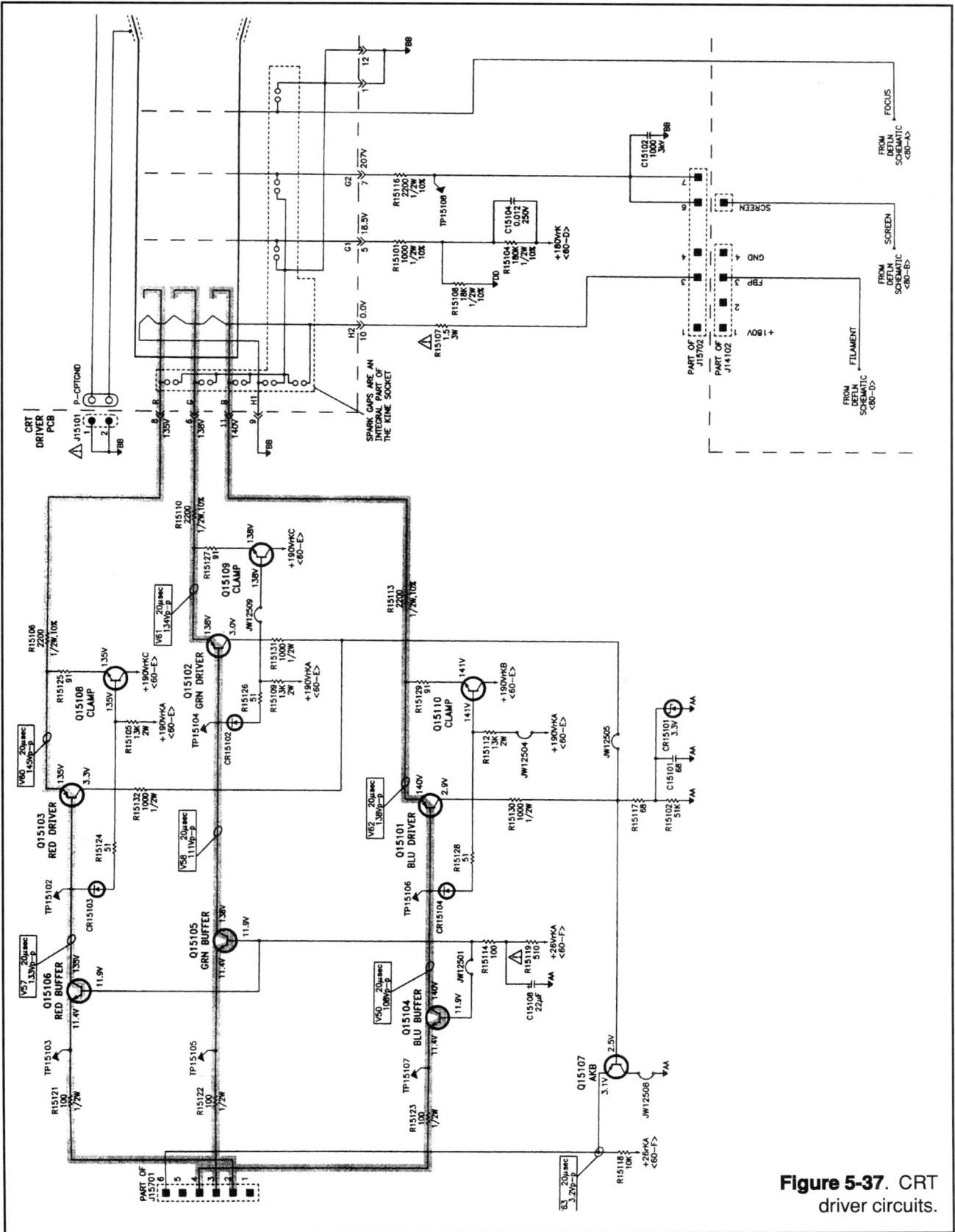

Figure 5-37. CRT driver circuits.

Scan Velocity Modulation

Scan velocity modulation is not new, and the circuit has not been changed with the exception of the addition of Q15202 (*Figure 5-38*), a switch which turns SVM off during the on-screen portion of scan. Projection televisions use the same circuit, except in triplicate.

Automatic Kine Bias

An AKB circuit tracks and compensates for the normal drift in beam current cutoff bias of a CRT. If the incoming video black level is not matched to the actual cutoff level of the CRT, the color temperature of the CRT will change. If the match is maintained, the original color temperature setup will retain over the life of the CRT.

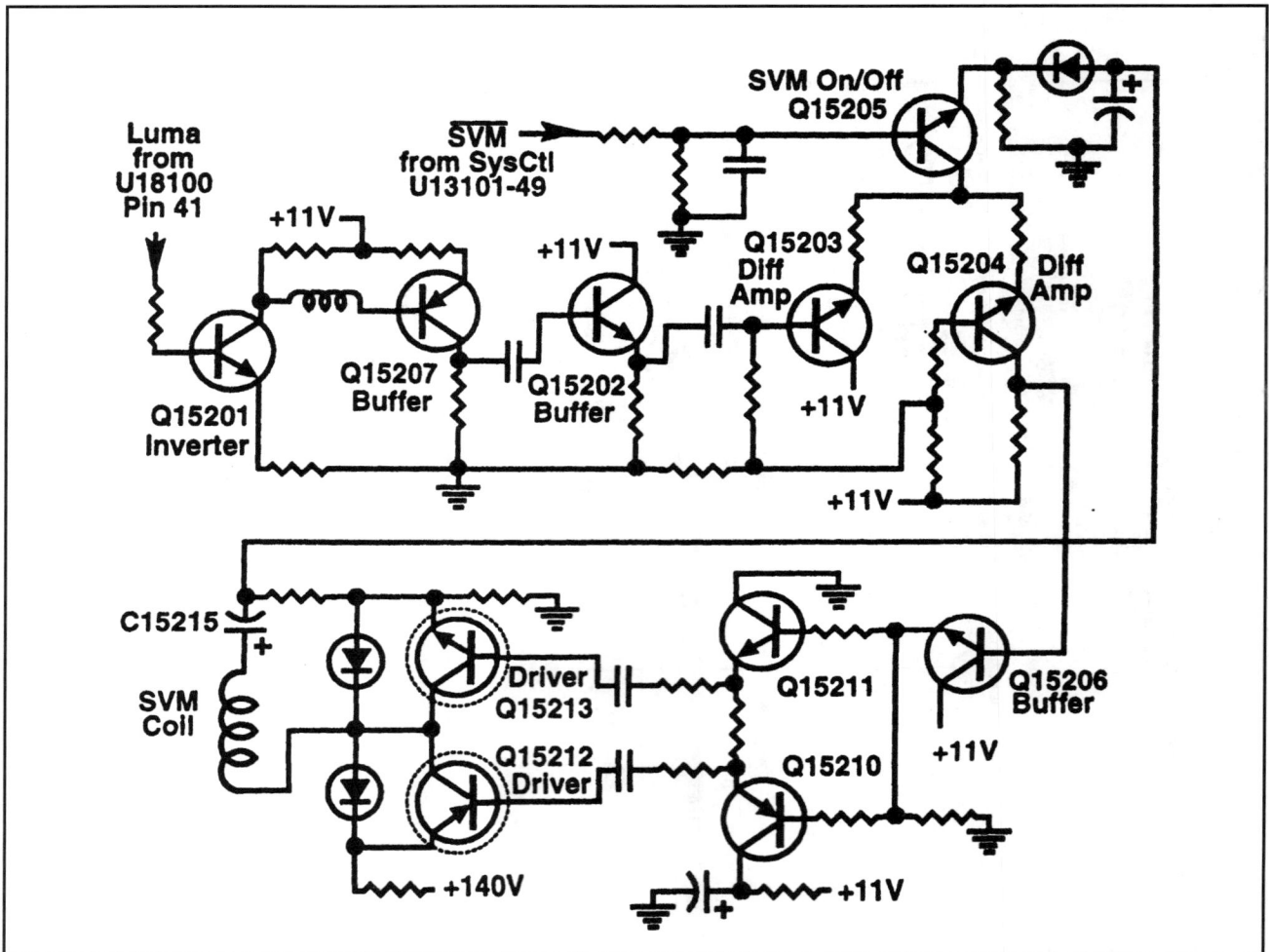

Figure 5-38. Scan velocity modulation.

The AKB circuit used in the CTC195/197 chassis differs from those used in prior chassis. Previous circuits used a sample-and-hold circuit that attempted bias correction on every field. Their workings were pretty much consigned to the hardware of the circuit. The new AKB circuit does a sample-and-compare about once every 10 seconds, which is quite sufficient to prevent rapid changes in the CRT bias and to preserve color temperature settings. The new system uses a digital comparator for cathode feedback and is therefore referred to as "digital AKB." The microprocessor does the actual measurement and correction calculations and stores the results in the EEPROM. The workings of the new circuit are based on software rather than hardware.

The digital AKB function is contained within the T4-Chip and is a one-bit system which samples the RGB cathode currents on lines 18, 19 and 20 of the raster. AKB reference pulses are added to the video black level on these lines and sampled by the T4-Chip at pin 25 (*Figure 5-36B*). The T4-Chip compares this signal to a fixed level derived from power supply voltage samplings. The output of the comparator is latched into the status register where it is read and processed by the microprocessor. The output pulse is derived from an average of the RGB AKB latch states. Said differently:

> AKB operates on the principle that if a voltage pulse imposed on the RGB pedestal is alternated in peak voltage such that the voltage resulting from the current feedback is above and below a reference voltage, the long-term average of the feedback signal should be the same as the reference voltage. If it is not, correction is required. (*The Technical Training Manual*, pg. 115)

Confusing, isn't it? Take a look at *Figure 5-39*, a graphic presentation of three NTSC waveforms which might appear at the cathodes of the CRT. The waveform labeled "cutoff high" would cause the higher IRE portion of the signal to max out beam current before the whitest portion of the picture information made it into the picture. The next waveform is labeled "cutoff low." If it found its way to the CRT, the dark or low IRE portions of the signal would occur after the cutoff bias had stopped beam current, the white portions of the signal would not have enough amplitude to reach full brightness. The last waveform is labeled "normal cutoff." If this one made its way to the CRT, beam cutoff would occur precisely at 7.5 IRE, and maximum white would occur precisely at 100 IRE. The result would be the kind of picture the CRT ought to display. The graph makes the previous verbiage easy to understand. A picture is worth a thousand words, isn't it?

The literature points out that it is not important what the cutoff voltage is. It is important, however, that AKB recognizes its value and compensates over time to ensure that the incoming signals sync with it to match the 7.5 IRE blanking level with the CRT cutoff voltage. And that is what AKB does.

I will not go into detail about how the circuit works because you can read that for yourself in the books I have already mentioned. This brief discussion should be sufficient to acquaint you with the circuit and what it is supposed to do. Because the CTC197 has AKB circuitry, you must exercise special care in setting up the color temperature. As you know, I gave you this procedure at the beginning of the chapter. The training manual goes into the setup procedure in detail, but I have given you enough information to enable you to perform it satisfactorily. Let me, however, repeat. It is <u>critical</u> to the operation of the television that the color temperature adjustments be made correctly.

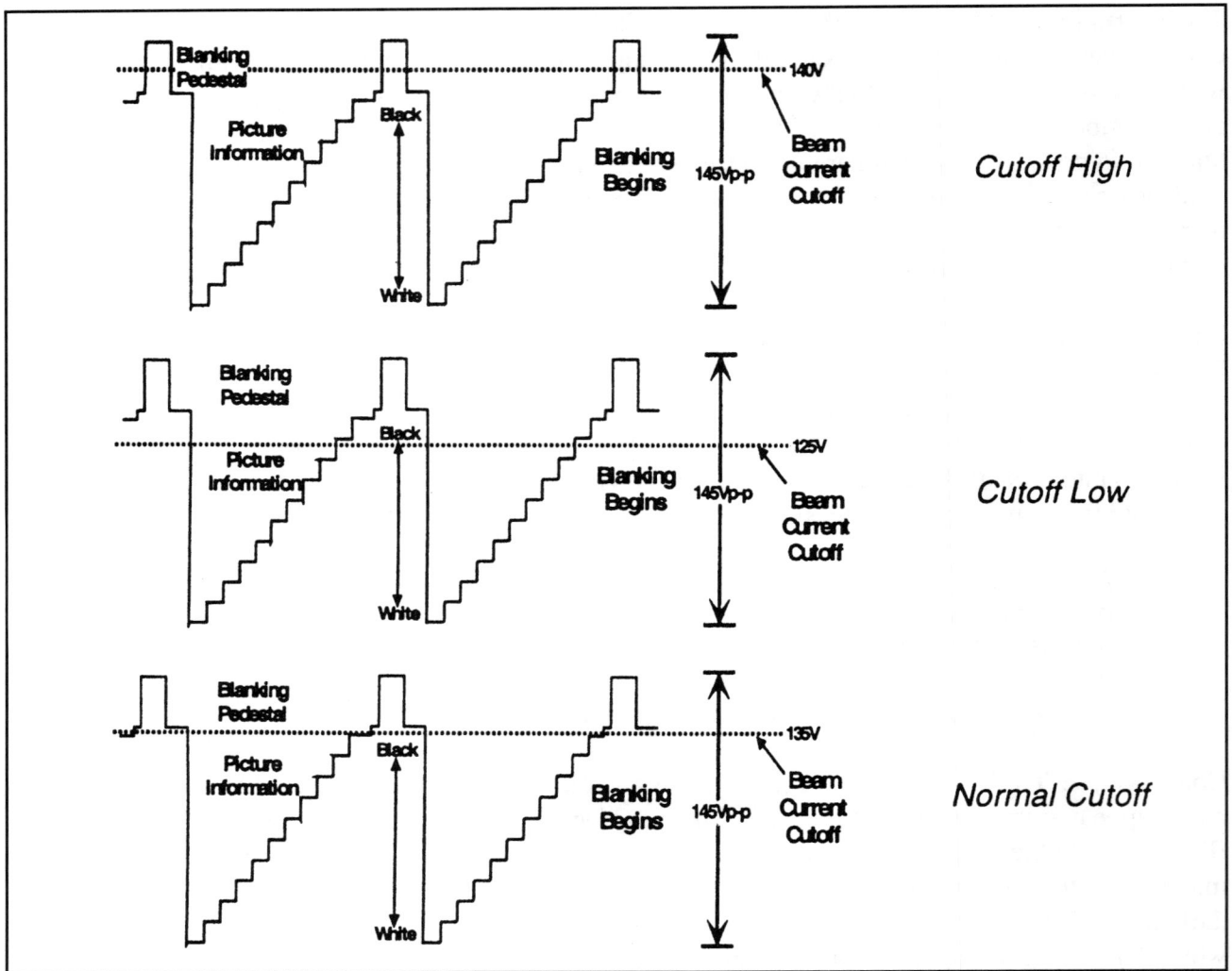

Figure 5-39. CRT cathode waveform examples.

The Audio Processing Circuits

Figure 5-40 will give you an overview of the audio processing circuits in the CTC197. In addition to the usual stereo/SAP circuits, it employs an SRS ("sound retrieval system") module and AVR ("automatic volume reduction") technology. In summary, the circuitry is rather complicated.

U11600 is the heart of the audio system. *Figure 5-41* is a block diagram of the IC, complete with the waveforms you can expect to find at the designated points. The IC contains the stereo/SAP decoders, switches, tone-volume-balance control, and the bus interface. Like the T4-Chip, U11600 is a slave with respect to the system control microprocessor that "calls the shots" via the I-squared-C bus. Alignment values are stored in the EEPROM and are rewritten to the audio processing IC any time the chassis loses power.

If you look at *Figure 5-40*, you will see inside U11600 two switches, SW1 and SW2. SW1 permits the user to select baseband audio from the T4-Chip, aux 1 or aux 2 (decoder CBA) audio. In other words, he/she can select among three audio sources. SW2 determines if the audio is routed through the SRS/compression module or sent directly to the audio output jacks. You will also notice that the AVR technology is available only in the PTV models.

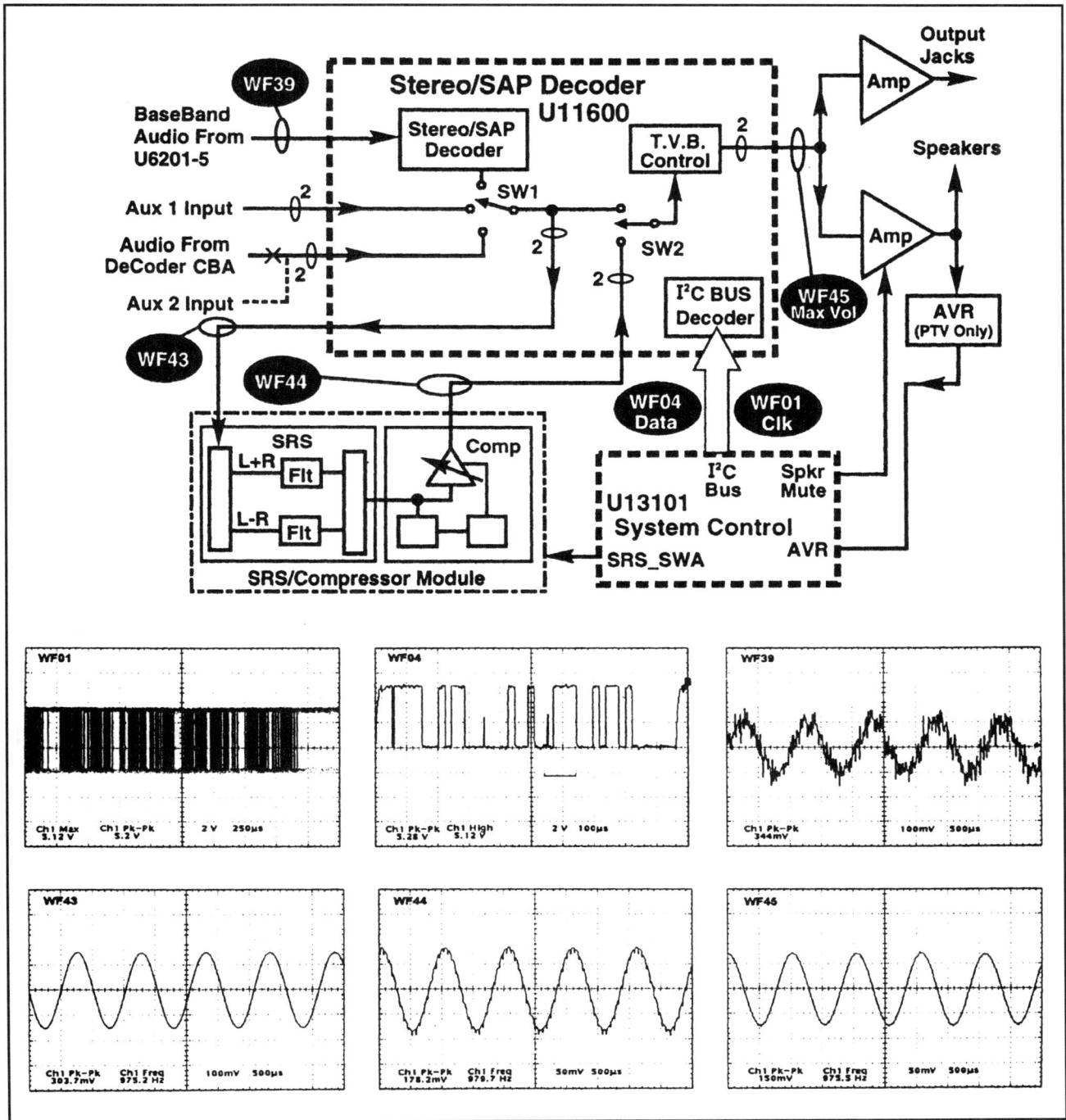

Figure 5-40. Audio processing.

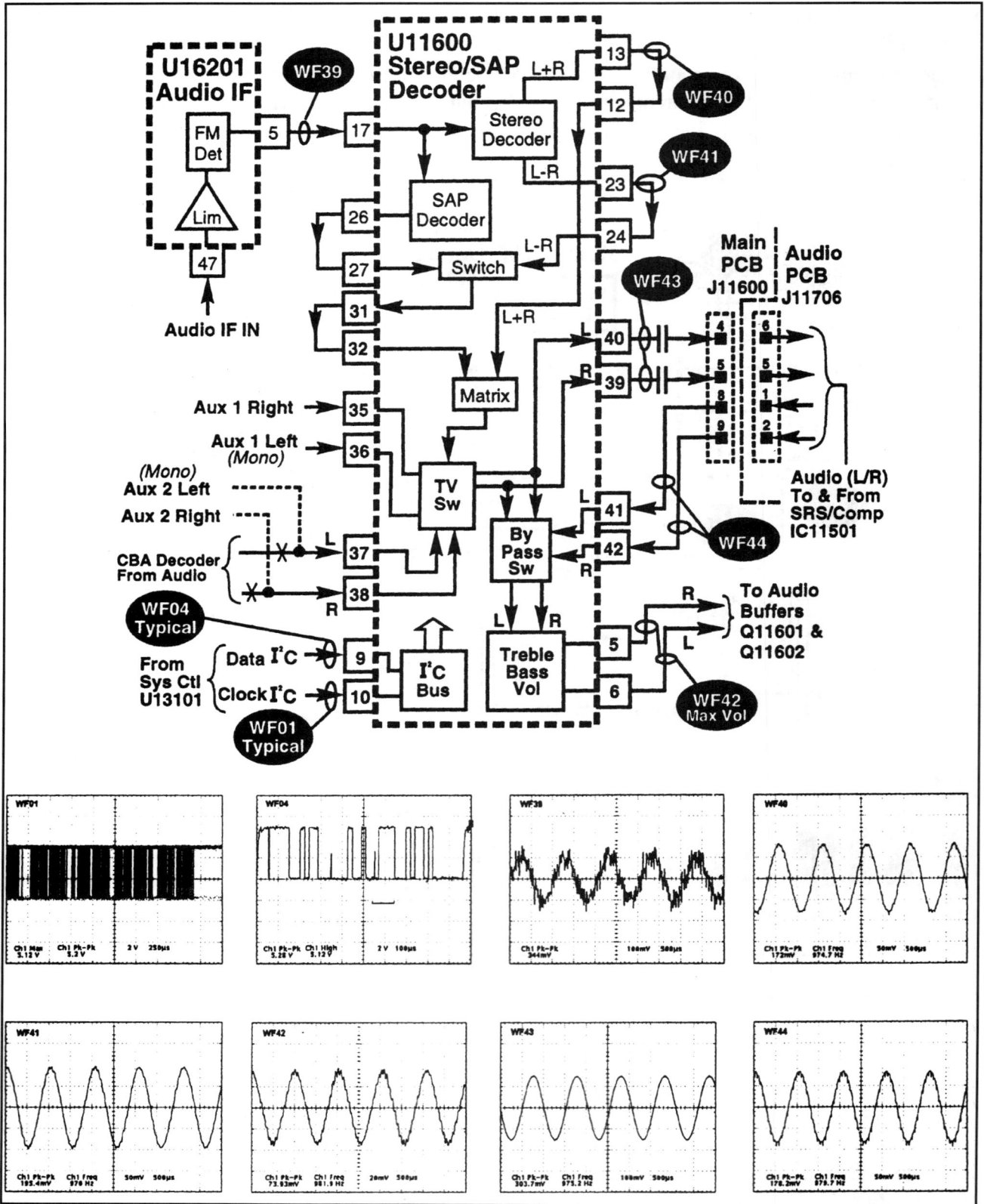

Figure 5-41. U11600 block diagram.

I will not describe in detail the inner workings of these ICs. You can easily trace the route the audio signal follows by studying *Figures 5-40* and *5-41*. To make the circuit complete, I am adding a block diagram of the audio output circuit in *Figure 5-42*. The addition of voltages and waveforms should make troubleshooting straightforward. If you like things a bit more concrete, take a look at *Figure 5-43*, a troubleshooting tree taken from the factory service literature.

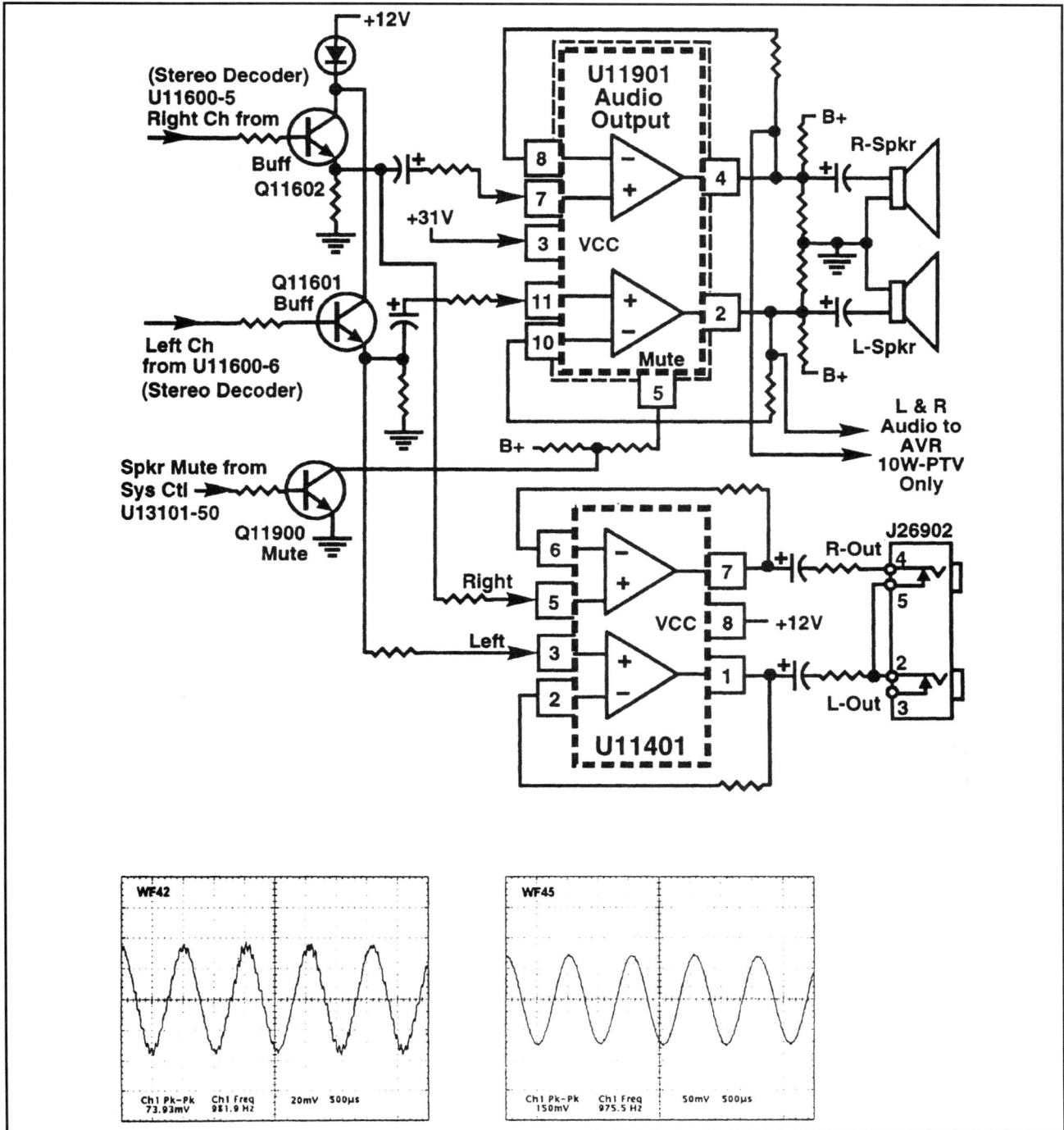

Figure 5-42. Audio output.

Audio Troubleshooting

Tune set to receive a mono audio signal. Adjust Sound Processor for Mono, Speakers on, Fixed Level Mode off. Scope Stereo IC - U11600 pin 17

Audio present — No → Trace back through SIF

Yes ↓

Scope Stereo IC - U11600 pin 39 or 40(L,R)

↓

Audio present — No → Check Vcc and all DC bias voltages on U11600. Possible wrong data sent or defective IC11600

Yes ↓

Set Volume to max. Scope Stereo IC - U11600 pin 5 & 6.

↓

Audio present — No → Check components on U11600 pins 1-4,7,8. Check components and bias on Q11601 & Q11602. Possible wrong data sent or defective IC11600

Yes

Audio present at U11901 pins 7 & 11 — No → Check parts back through Q11601 & Q11602

Yes ↓

Audio present at U11901 pins 2 & 4 — No → Troubleshoot U11900 bias and components. Pin 3 = 25-30 volts, pin 5 should be 6 volts lower than pin 3, Q11900 - b >4.5Vdc. Possible IC damage

Yes ↓

Check speakers and components in path to speakers

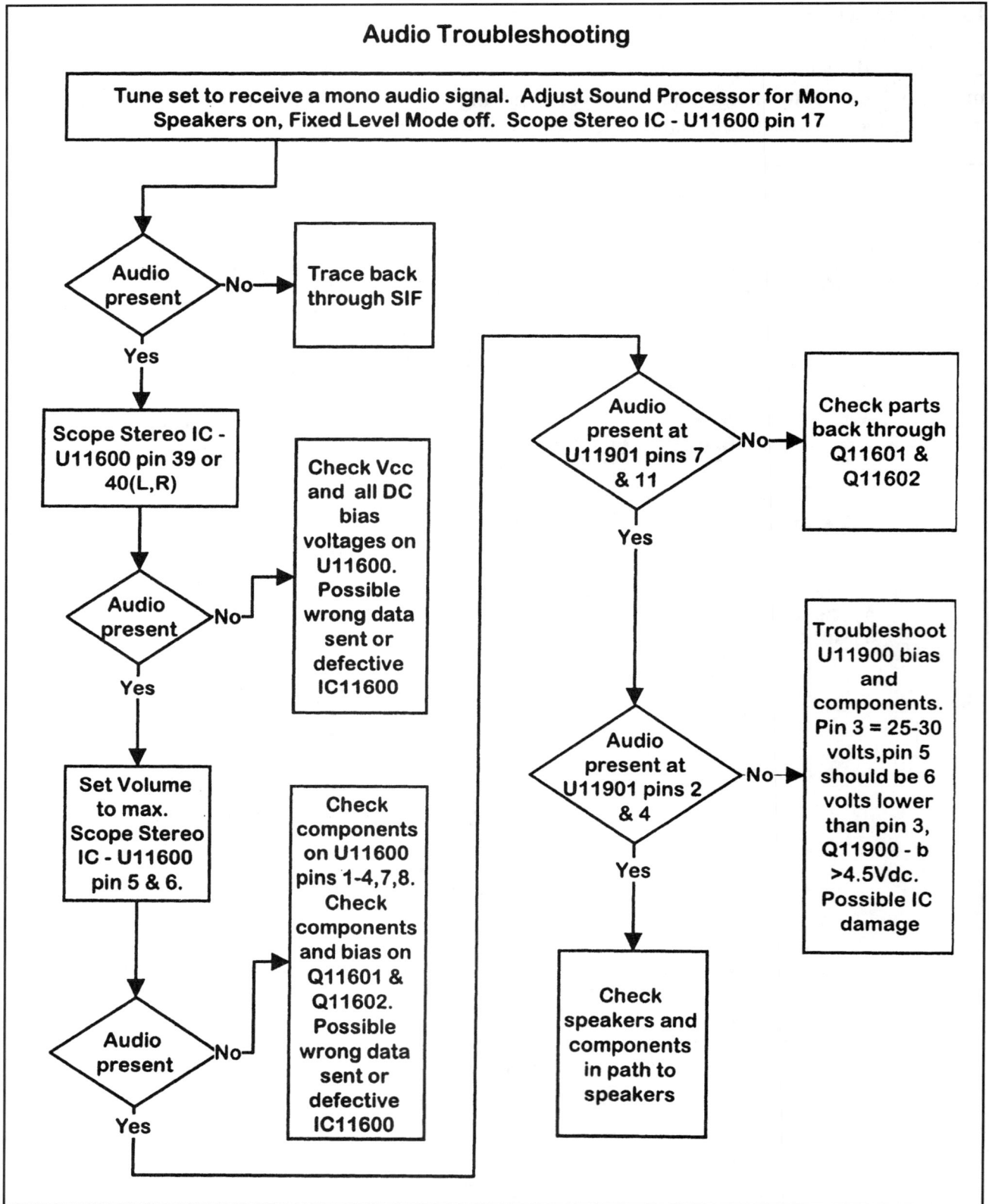

Figure 5-43. Audio troubleshooting flowchart.

Automatic Volume Reduction (AVR)

Automatic volume reduction may require a comment or two. *Figure 5-44*, a very basic diagram, will serve as the reference for this brief description. Note that the AVR circuit is found only in the PTV sets.

The circuit serves to protect the speakers or other output devices during possible overcurrent condition by lowering the signal to the audio output devices. The microprocessor polls the system over the AVR control line at a 10 Hz rate to check the status of the output level. As long as the control line stays low, the microprocessor will do nothing. If the line goes high, the volume register will decrement (i.e., go down) at a 10 Hz rate until the AVR sense line goes low. The volume register will stop counting down, and the microprocessor will activate a 5-second counter. If the line goes high again, the timer turns off and the volume countdown begins again. If the timer counts out and the sense line stays low, the microprocessor will permit the volume register to increment (i.e., count up) to the original setting.

The AVR circuit activates when the voltage on the bases of Q11902 and Q11903 reach about 4.9 volts, permitting the transistors to turn on and place a low on the base of Q11901, turning it on. When it turns on, Q11901 places a high on the AVR sense line to the microprocessor.

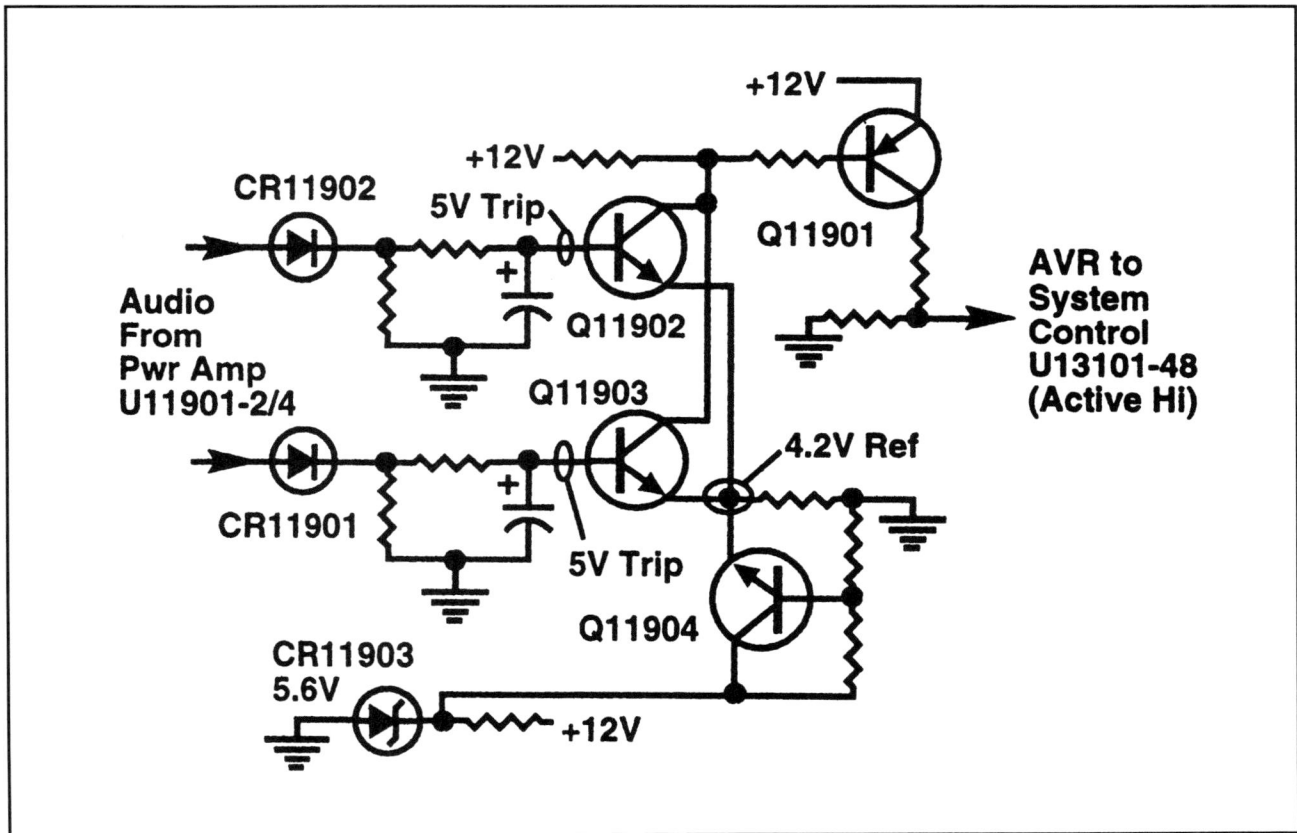

Figure 5-44. Automatic volume reduction.

SRS/Compressor Circuits:
(Sound Retrieval System/Compression)

The circuit (*Figure 5-45*) is contained on a single inline (SIP) circuit board that plugs into a connector on the main chassis of the PTV models. It is contained on a module connected to the main chassis via a cable in the CTC197. The module has the following interconnections: left and right inputs, left and right outputs, +12 volts, ground, and clock and data lines. It gives the L-R channel a moderate boost to enhance weak stereo sources and widen the image. It has a filter to keep the midrange from being overly boosted. Both paths have filters to enhance directional clues by compensating for the human ear's directional frequency response. *Figure 5-46* shows the waveforms.

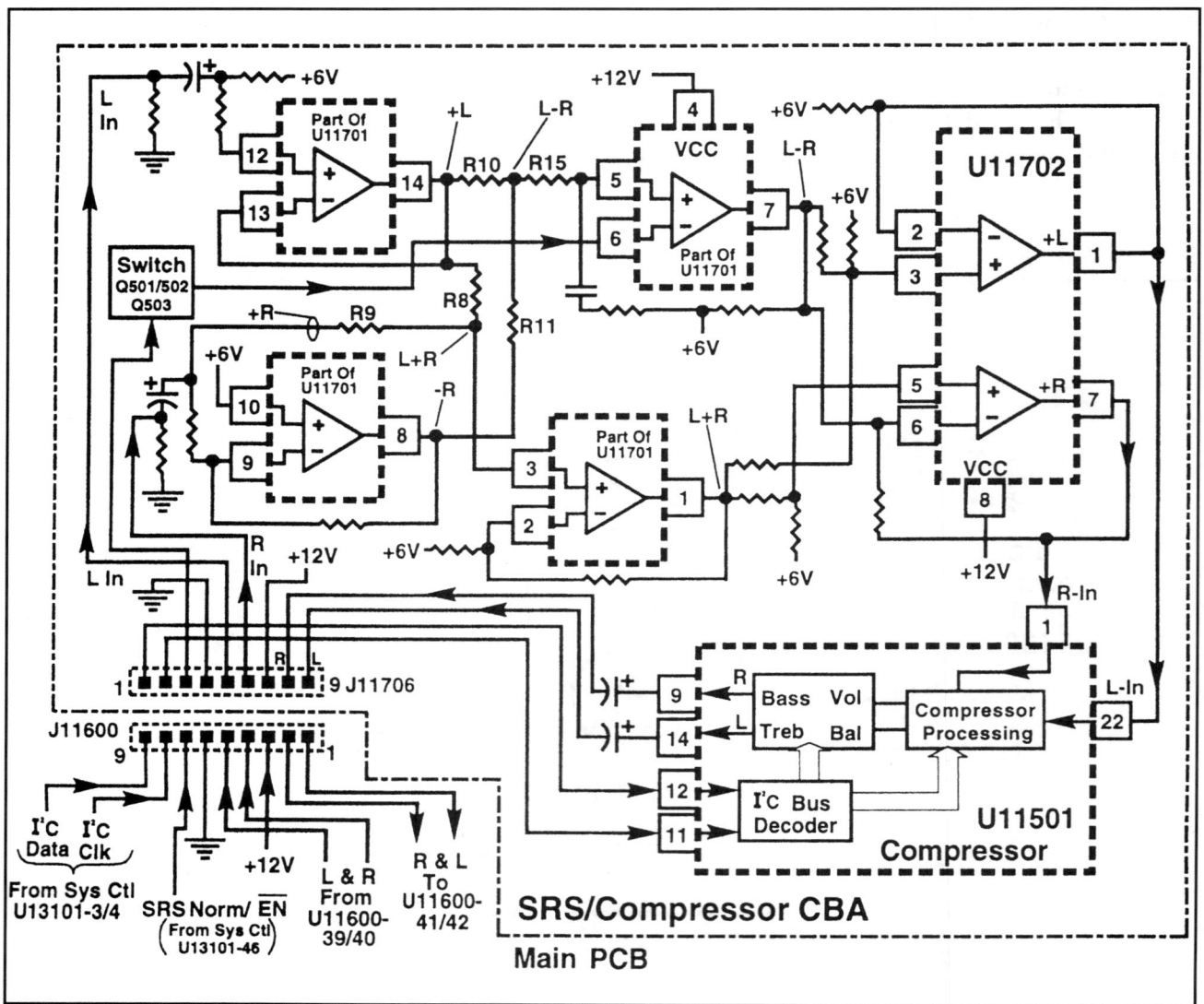

Figure 5-45. Sound retrieval system/compression.

SRS CONT LINE	SRS MODE
LOW	OFF
MID	SRS NORMAL
HIGH	SRS ENHANCED

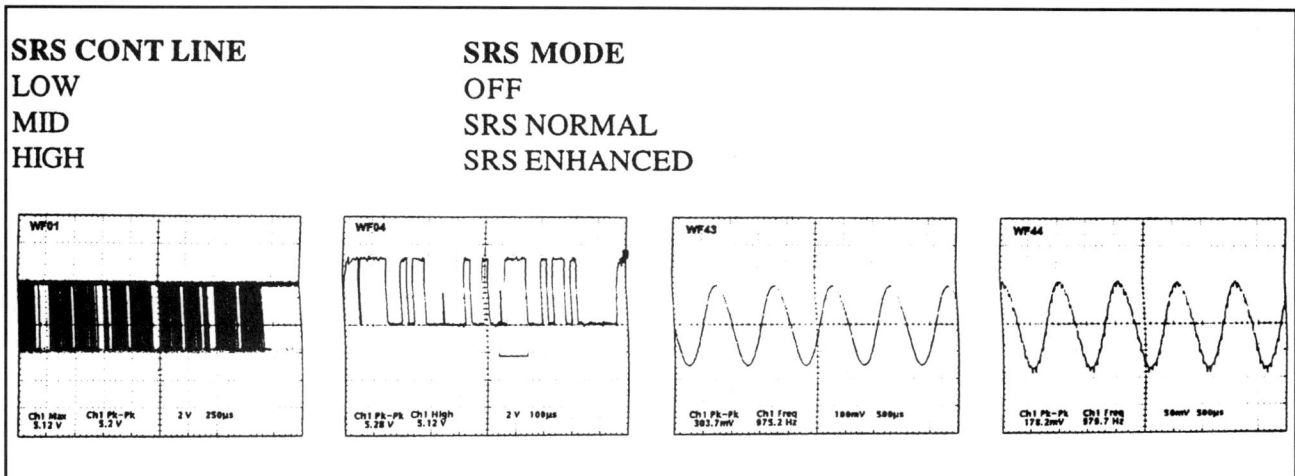

WF01		WF04		WF43		WF44	
Ch1 Max 5.12 V	Ch1 Pk-Pk 5.2V 2 V 250µs	Ch1 Pk-Pk 5.28 V	Ch1 High 5.12 V 2 V 100µs	Ch1 Pk-Pk 303.7mV	Ch1 Freq 975.2 Hz 100mV 500µs	Ch1 Pk-Pk 178.2mV	Ch1 Freq 979.7 Hz 50mV 500µs

Figure 5-46. SRS/Compressor waveforms.

Repair History

The CTC197 is so new that it has almost no repair history. The information I am about to give you comes solely from my reading and conversations; that is, it comes from research instead of a combination of research and hands-on experience.

Thomson has had some problems with the microprocessor software versions <u>before</u> 6.05. These problems include audio fading out during the last 60 seconds of the first hour if the sleep timer has been set for more than an hour, and the set repeating the previous remote command when the audio button on the remote is pressed. The cure is to replace the microprocessor with part number 237643.

There is another software-related problem. If video flashes on composite video inputs, the FPIP circuit may falsely detect the presence of S-Video and switch from composite to S-Video. This problem relates to software versions earlier than 7.29. Replace the microprocessor with the same part number.

You can check the software version by calling up the service menu. The version will be listed like this: P: 00   (software version)   V: 00.

There are several other manufacturing "bugs" that should have already been corrected. The most recent one (February 1998) concerns herringbone-like interference, which drifts through the picture on VHF channels. Check R17106 in the tuner circuit. If its value is 10 ohms, change it to a 39 ohm resistor (part number 192103) and realign channels 35, 41, 45, 48 and 50 using the Chipper Check. This will apply to televisions manufactured before December 1997.

You can tell when RCA televisions were manufactured by reading the serial number. The serial number on the back of the TV will have nine numbers. The first number will refer to the year of manufacture, and the next two numbers refer to the week of manufacture. For example, a serial number that begins 750XXXXXX will have been manufactured in the fiftieth week of 1997.

CHAPTER 6

CTC130 Through CTC157

I devoted Chapters 1 through 5 to the relatively new Thomson televisions because these are the sets we servicers will be asked to fix for the next several years. However, there are lots—and I mean lots—of older RCA/GE televisions still in consumer hands. Because those sets have given good service, their owners still ask to have them repaired, especially if the cabinet is in good shape. For example, hardly a week passes that I am not asked to service half-a-dozen or more. And just recently we sold four used CTC130 consoles to people who were delighted to get them.

I have, therefore, decided to conclude my book with a look at some of the older, popular chassis beginning with the CTC130. There are older RCAs out there that have been working for 15 to 20 years, but I have decided to use the CTC 130 as the cutoff chassis.

I will use a different methodology in this chapter as I discuss the older TVs. My approach will be, "If you have such-and-such problem, this might be your solution." And I will point you to relevant reading material which time and space will not permit me to reproduce. The service information I am about to give you comes from problems I have solved, from my reading, and from conversations with other servicer personnel. Long ago, I formulated the habit of storing information on 4 x 6" index cards. I have about four cards devoted to the CTC130 chassis alone. If I solve a difficult problem, if I hear about a unique problem, or if I read about one, I will make a note on the appropriate card and credit the source if I can. The advent of computers seems to make keeping track of information easier, but I still prefer my index cards. I have never, repeat <u>never</u>, had a file box crash or refuse to "open" when I approached it! Maybe at fifty-five I'm a bit old fashioned, but I've had enough problems with computers at work and at home to be a bit defensive about what information I want to entrust to them.

Before we get to the chassis, let's take a look at available literature.

RCA's "Unitized Chassis Technical Reference Library" is the technical training literature for the televisions we are about to discuss. The "library" comes in a two-volume set and covers Thomson's products from the CTC117 through the CTC169. It is very helpful—in fact often indispensable—but it will not take the place of schematics. I don't recall how much the library cost, but I have the impression it is fairly expensive. However, if you service RCA products, you should think seriously about adding it to your service aids. Given the publication and sale of Thomson's new "field service guides," the servicer now has another, more convenient, and probably less expensive option.

There are two sources for service literature (schematics)—Thomson's factory service literature and Sams PHOTOFACTS®. If you have ever used factory service literature, you know those folks do things differently. For example, the microprocessor and its external circuitry will not be shown complete on one page. Rather, it will be broken up in functional blocks and shown with the circuits it controls. So, if you are chasing a system control problem, you may have to go from the video IF section to the power supply section to the tuner section, etc. Such an approach can be confusing if you are not accustomed to working with it. In most instances, a Sams works fine, but there are those times when the factory literature will give you that additional something you need. For example, RCA has always included an excellent waveform guide in its literature. I service most brands of televisions, so I get a PHOTOFACT shipment every month. I therefore have available both sets of literature. Speaking from experience, I testify to the fact that Sams can be a veritable time-saver from two perspectives. It can save you time by helping to locate parts with which you are not familiar, and, since it shows the circuit in a different layout, Sams may let you see something you might otherwise have to look long and hard to find.

I will not cover any of these chassis in depth, and there will be a host of problems I won't even mention. Time and space limitations require me to do a great deal of "editing." However, I have tried to give you sufficient resources to help with most of these problems. I mentioned in the introduction a software program called FixFindr™ that has a mountain of material about each of these chassis. FixFindr uses the problem/solution format and, should you decide to purchase it, will quickly pay for itself.

The CTC 130 Chassis

I recently got a mailer from Thomson advertising their new "field service guides"—literature to which I have already referred. The field service guide for the CTC130 also covers the CTC133 and CTC140 chassis and is numbered FSG11. If you see a lot of these televisions and you don't have RCA service literature, you might consider investing in it. The price is $40.00 plus shipping and handling, but it might be worth it because the field service guide has all the information you will need to fix these sets—schematics, waveform guides, parts lists, and factory service bulletins.

Homer Davidson has an interesting and helpful article in the March 1997 issue of *Electronic Servicing and Technology*, "Servicing RCA's CTC130 Startup And Shutdown Symptoms." This back issue would be worth the price just to get Mr. Davidson's article.

MTT Problems

The MTT module can cause lots of different problems: no turn on (or off), snowy picture or no picture at all, distorted audio or audio that suddenly becomes very loud. To complicate matters, these symptoms may be intermittent. The MTT modules are quite expensive. A recent flyer advertised the MTT002A for $92.50 rebuilt. I have on occasion done business with a company in Texas that advertised their repairs for $49.95. A new one, of course, costs much more.

As an illustration, let's talk about the CTC130 that would not turn off. The MTT module is responsible for turning the chassis on and off. I could have had it rebuilt or even purchased a new one. Knowing what I know, I popped the covers off (about a 5-minute job) and checked the on/off transistor, which was leaky. I installed a new one, checked the MTT for proper operation, reassembled the TV, and wrote a bill. The extra profit went into my bank account, and the customer was delighted to get her television fixed at a price she could afford.

The good news is you can fix most of them yourself. I briefly discussed MTT repair in an article in the June 1997 issue of *Electronic Servicing And Technology*, "Servicing Television Tuner Problems." I tried to point out we techs/business owners can fix a significant number of tuners ourselves, saving our customers money and making a little extra for ourselves. *Figure 6-1* is taken from that article, and illustrates areas which are prone to problems. Did you know a host of MTT problems are caused by ringing cracks around the stakes that serve as wire wraps? Take the back off one that is giving you problems and look at the point where the stakes are soldered to the PCB. You will more than likely see tiny circular cracks around several, if not all, of the stakes (*Figure 6-1A*). Resoldering the defective connections will often solve the problem. *Figure 6-1B* shows you where to look to confirm the voltages that must be present for the TV to work.

While you have the covers off the module, check capacitor C677 and the two capacitors located close to it. It is a good idea to replace all three because these little creatures cause such problems as a dead set, no remote control function, dark raster, very low audio, and the appearance of scanning channels while the channel indicator remains where it was originally set.

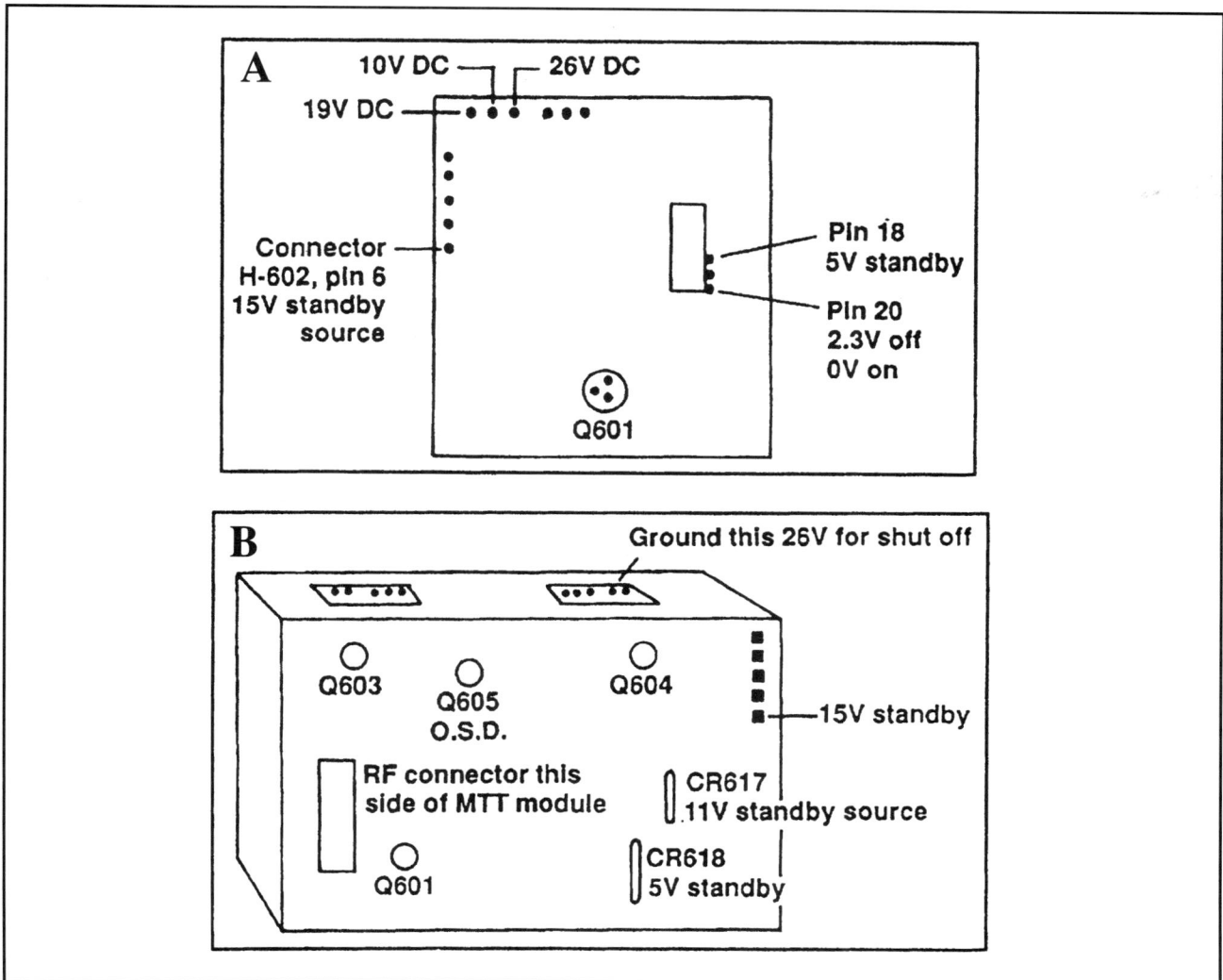

Figure 6-1. MTT repair.

Power Supply Problems

(1) Intermittent startup and when it started, the TV would play for several minutes and then shut off. R16 in the VIPUR emitter circuit had increased in value, which caused the TV to go into overcurrent shutdown.

(2) The set would pulse on and off two to three times before it came on. If the chassis were warm, the TV would come on the first time. Check capacitors C7 and C11 in the VIPUR power supply. They will usually show an elevated ESR reading. But if they check good, replace them anyway! NOTE: C11 can cause a no-start condition in which the TV just pulses. This is a 15 μF, 50-volt capacitor.

(3) Dead set. Check CR7 in the power supply. I have seen it shorted on several occasions. CR10 can also cause this problem.

Deflection Problems

(1) Raster would fill out when cold. If you turned the TV off and back on, there would be a thin vertical line across the screen indicating no vertical deflection. Problem was C502, the capacitor that helps generate the vertical ramp.

(2) Intermittent dead set with a tick-tick sound. Problem was loss of drive to the base of the horizontal output transistor due to cold solder around the terminals of the horizontal driver transformer.

(3) Dead set with blown fuse or open 15-watt surge resistor located to the right of the fuse. The problem will more than likely be a shorted horizontal output transistor. Do yourself a favor and replace the defective HOT with an OEM part. Take a look in the Appendix at the instruction sheet that is packaged with the new output transistor. It will direct you to make a few checks before you fire up the TV after replacing a defective HOT.

Picture Problems

(1) No luminance. The raster showed the usual symptoms, color elements present but washed out, retrace lines, etc. Q703 was defective. An ECG159 is a good replacement.

(2) Dark raster and filaments not glowing. R5013 was open.

(3) No raster and good audio. Turning up G2 voltage produced a slick raster with retrace lines. Defective Q403.

(4) TV would play okay for a little while and then the picture would go snowy and volume would go to full level. C672 in the MTT module, 220 µF/25 volt. This is one of the capacitors I suggested you check when you service an MTT module.

CTC140 Chassis

Homer Davidson has another excellent article in *Electronic Servicing and Technology* (July 1993) you might want to get and file, "Ten RCA CTC140 Chassis Symptoms." He discusses the following problems: dead set, the set that abruptly went dead, raw B+ but no sweep, green power light will not turn off, chassis shutdown, firing lines in the picture, heavy retrace lines, snowy picture, fuzzy picture, and standby hangup (no remote control function).

Picture Problems

(1) Grainy, ghosty picture with low contrast. Check the 12-volt source. If it is low, suspect CR2100.

(2) Intermittent picture or no picture or slick raster, or smeary picture or a picture that flickers, or a combination of these symptoms. I have seen this problem several times and each time traced it to the area in and around U2300, the dual IF IC. You can often make the problem appear or disappear by tapping on the IC. Check the bottom of the PCB for poor solder connections. "Shotgunning" the solder connections usually solves the problem, paying particular attention to the IC and the coils.

The Power Supply

(1) Dead set. LED light will come on when the TV is turned on and will not go off when the TV is turned off. Check R4101 for an increase in value.

(2) Dead set with B+ at Q4400 (the HOT) but no +16 volts on the deflection IC (U4300). Remove PW4700 diode SIP board and resolder CR4703, the diode that helps to produce the +16 volts. NOTE: This board can also cause no B+ for the HOT. The solution is the same. Remove it, check all diodes, resolder all connections, and reconnect the SIP board. I distinctly remember the time I removed the board, resoldered the connections, and installed the board. Guess what? I had to take it out a second time because one of the diodes was shorted!

(3) LEDs won't come on. Check the +12-volt line. If it is low, suspect CR4101.

(4) Dead set but green LED will come on and stay on and will not go off. Check CR4701, CR4104 and CR4101 for leaks or shorts. Any one of these diodes can cause this set of symptoms.

Deflection

(1) Failure of HOT. Check R4322 for an increase in value, and check the pins of the horizontal driver transformer for poor solder connections.

Audio

(1) Distorted audio which will come and go (intermittent). Tapping on the PCB around the dual IF IC will cause the audio to clear up and/or become distorted. The problem will almost always be poor solder. I usually "shotgun" the solder connections, especially around the coils and the IC, just as I did to clear up the video problem. I have had one instance when I had to replace the quad detector coil.

(2) If the set pulses on and off at turn on, don't forget that a shorted audio output IC may be the culprit. I worked on one for almost an hour before I discovered a shorted audio output IC! You see, I forgot the basics: Check resistances on the secondary of the power supply. A low resistance on one of the voltage source lines led me to the problem.

CTC 145/146

The Field Service Guide for the CTC145/146 chassis is FSG12. It also covers the CTC156/157 and CTC166/167, which makes it a bargain at $40.00. You will also find good information in the "unitized chassis library" I mentioned earlier. Since there are many models using this chassis, one Sams won't cover them all, but I find it helpful to designate a particular Sams as the "CTC145 Key" and keep most of my notes on it. That one Sams will assist you in making more repairs than you think!

The CTC145/146 Tuner

I have had good experience repairing the tuner for this chassis, especially if the symptom is no picture and audio or very snowy picture and scratchy audio. The tuner is not easy to remove from the chassis because of the grounding tabs. Once you get the tuner out of the chassis, you face the rather difficult task of getting into it. But the prospect of completing a profitable job makes the effort worthwhile.

After you get the tuner out of the chassis, remove the cover for the foil side of the PCB (the side facing the flyback). This will not be particularly easy and will require a solder sucker and solder wick. If the problem is lightning related, and it probably will be, you will find, about midway from top to bottom and about an inch or so from the input side of the tuner, one or more burned components. You will have to look closely to see the damage. Most of the time a lightning strike will leave a scorched mark on the tuner cover which you can use as a guide to find the trouble area. A "zero ohm" SMD resistor or small piece of hook-up wire works just fine. In some instances I have used as a fix a 10 to 12 pF capacitor from the RF input to a point past the damage.

After you complete the repair and get the set up and working, be very sure to do a safety leakage test. You don't want to return a set that does not provide isolation between the tuner input and hot chassis.

While we are on the subject, there are some tuner-related symptoms you might be interested in:

(1) The picture drifts in and out, as if the AFT refuses to lock. I have had to repair faulty connections at the FLO/K line on three occasions.

(2) The set will not change channels and appears to be stuck on one low-band VHF channel. Inspect C601 off pin 9 of U600 (in the shielded area).

(3) Autoprogramming stops near the end of the low-band VHF range and high-band VHF appears to be inoperative. Look for R126 (10 ohm at 1/4 watt) under the power supply heatsink. It may have increased in value or opened. I picked up this tip from a friend of mine who said he got bald-headed fixing the set!

Picture Problems

(1) No video or on-screen display. Check Q2706. It will more-than-likely be the culprit. An ECG transistor works like a champ here.

(2) Color problems which range from no color at all to color that comes and goes. A likely candidate is C2810 at pin 5 of U1001.

(3) Dark picture. High voltage okay. R2715 in the beam-limiter circuit may have opened or increased in value. The resistor is located off pin 11 of the flyback. I won't call the failure of R2715 a common thing, but I will see one or two TVs a year that a new resistor fixes.

(4) No OSD. If Q4501 fails, it will cause the no-OSD symptom. I have had to replace it on several occasions. Again, an ECG component will be fine.

(5) OSD is okay but the picture is very dim. Check the voltage at pin 14 of U1001. If it is low, suspect R2728.

(6) The picture is either shaded toward green or is green and may have retrace lines. The 5-volt standby source may be high. Check the base of Q5004 for 5 volts. If the voltage is high, look for a leaky CR3601.

Deflection Problems

(1) No horizontal drive. Check for poor solder around the terminals of the horizontal driver transformer.

(2) Vertical foldover. The raster may also be stretched at the top. C4505 may have opened.

(3) No vertical deflection. Check U4501, R4513 and R4515. If you still don't have vertical deflection, check C4501 because there have been reports of it shorting. CR4501 can also cause no vertical deflection

Power Supply

(1) Dead set with standby voltages running high. A possible cause is an open CR3101.

(2) The TV will not start, but you can hear the relay click. The TV may be starting up and shutting down. Try to start it at reduced voltage. If you can start it at a low line voltage, suspect Q4102, the regulator SCR.

(3) The TV makes a clicking sound but there is no evidence of video or audio. Check the 12-volt run supply. If it is missing or low, suspect CR4118.

Shut-Down Problems

(1) The TV will turn on and shut down. It may play for a while and may not. Check CR4118, which may be leaky.

(2) The set immediately shuts down at turn-on. Obviously, a lot—and I mean "a lot"—of things can cause this problem. Based on my experience, I place CR3104 near the top of the list.

(3) The dreaded "intermittent" shutdown. Again, any number of faulty components can lead to intermittent shutdown. For example, is the TV losing horizontal drive? If it is, you must troubleshoot the horizontal drive circuits. Since several circuits can cause the set to intermittently shut down, you often have to monitor certain circuits to find the problem. For example, in two instances I suspected a problem with the 12-volt run supply which caused me to hook a meter to it and let the TV play until it shut down. In both instances, CR4118 had failed under a load. I replaced it to fix the sets.

CTC149

I have had limited experience with the CTC149, and it began in a cruel way. I got one in that kept on blowing HOTs. Finding no apparent reason for this expensive problem, I called tech line for help and was given a procedure for checking drive to the HOT without blowing it. I still couldn't find anything wrong until I disconnected and isolated the CRT from the chassis. You guessed it, a shorted picture tube! My records indicate that except for this one isolated instance, repairs on the CTC149 have been routine.

CTC156/157

For some reason, the CTC156/157 was not popular in my area of the country; therefore, I have had limited experience with them. Most of the ones I have seen have had routine problems, like no vertical deflection because the vertical output IC was defective, or lightning-damaged power supplies. My records do indicate that I have serviced the following rather unusual problems.

Picture Problems

(1) A blank raster, no video and no audio. On one occasion, I found CR4118 open.

(2) No color. If the voltage at pin 5 of U1001 is low, check C2810. There have been reports that C2817 can cause the no-color problem. Check pin 3 of U1001. If it stays below 3 volts, then check this component. I owe these fixes to a friend of mine who has had more experience with the CTC157 than I have.

Power Supply Problems

(1) Pulses on and off and makes a squealing sound. I found CR4101 to be defective.

(2) The TV will not come on. C3125 at pin 1 of U3100 was defective. Q3102 can also cause this problem.

(3) Dead set with no 12 volts on the emitter of Q4162. The power supply is somewhat complicated and may require a bit of poking around to fix. The easiest way effect a fix is to take voltage measurements. In this particular instance I found that the emitter of Q4162 was missing its 12 volts, a condition caused by a defective CR4162. Don't trust your test equipment. It may check good but still be defective! Incidentally, this power supply is about the same as the one RCA uses in the CTC166/167 chassis. What I said in Chapter 1 also applies here.

Tuner-Related Problems

I have serviced two "no autoprogram" problems. In one instance the TV would receive just channel 2 as I recall. The voltage at pin 10 of U3600 was low, no tuning voltage, because C3605 was leaky. In the second instance, the TV would not autoprogram but would tune the channels if I entered the channel with the remote control. The voltage at pin 11 of U3300 was low because C3301 was leaky.

Appendix

RCA Training Tapes (VHS)

(The following list is taken from Service Data And Technical Training Publications Catalog. It can be ordered from TCE Publications 10003 Bunsen Way, Louisville, KY 40299.)

CTC195 Digital Convergence Alignment Video

CTC175/176/177 Familiarization Video

Projection Television Servicing Video

Unitized Chassis Education Manual

There are three:

1. T-UCSE-1 covers CTC117/118/120, CTC123/125/126 and updated information on the CTC131/132.

2. T-UTRL-5 volume I and volume II. These can be ordered separately or as one package. The manuals cover most chassis from the CTC140 through CTC185, including the TX81 and projection TVs.

3. T-ATC-1 covers the CTC170, PTK171, and CTC172.

Technical Training Manuals

You can order a technical training manual for just about any television RCA has manufactured in recent years. Two of the chassis I have discussed have companion troubleshooting manuals, namely the CTC 177/187 and the CTC 169. RCA also makes available certain speciality manuals, for example an in-home service guide for the CTC168/169 and a 31-inch direct TV technical training manual.

Field Service Guides

Field service guides are available for RCA products from the CTC854 through CTC169. These guides incorporate complete schematics, board and chassis views, service adjustments, parts lists, service bulletins, and more.

TV Prior-Year Service Data

You can order television literature on paper or microfiche. You can order it by year and get everything RCA manufactured during a particular year, or you can get "television service data single copies."

Service Bulletins

RCA will sell you their service bulletins. You can order the entire package (1993-1997), or you can order by the year. Incidentally, you can also order by the product—TV, VCR, camcorder, laser disk, audio, etc.

FixFinder™

The Fixfinder program is available on 3.5 floppy or as a part of the new CD-ROM format.

PartsFinder II™

PartsFinder is designed to help you find replacement parts "at the stroke of a key."

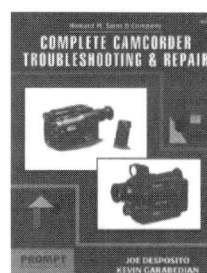

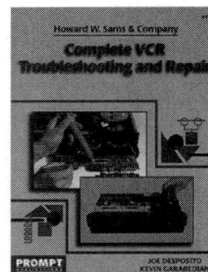

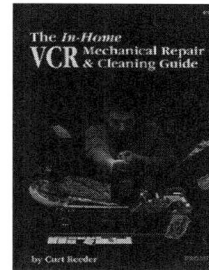

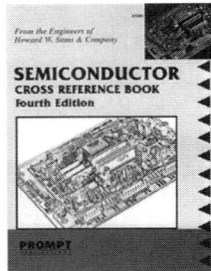

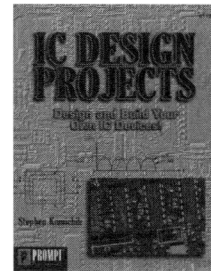

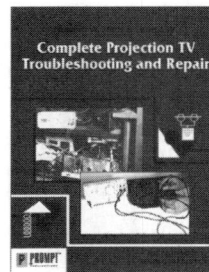

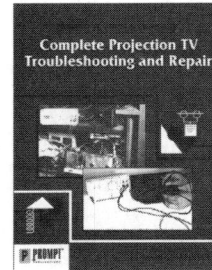